HANDBOOK OF
MAMMALS OF
MADAGASCAR

HANDBOOK OF
MAMMALS OF
MADAGASCAR

NICK GARBUTT

Princeton University Press
Princeton and Oxford

To my Dad, Mike (31st January 1940 – 3rd November 2020)
Years ago, I completely ignored his suggestion to get a proper
job. I miss him dearly, especially on the golf course, the
racecourse and during any major sporting event.

This edition published in the United States, Canada, and the Philippines in 2023 by
Princeton University Press
41 William Street, Princeton, New Jersey 08540
press.princeton.edu

First published in the United Kingdom in 2022 by Bloomsbury Publishing Plc

Library of Congress Control Number 2022943236
ISBN 978-0-691-23991-0

2 4 6 8 10 9 7 5 3 1

Design by Susan McIntyre
Maps by Julian Baker Illustrations, 2022
Typeset in 9.25 on 11pt Gill Sans MT Pro

Printed and bound in India by Replika Press Pvt. Ltd.

FSC
www.fsc.org

MIX
Paper from
responsible sources
FSC® C016779

Front cover: *Female Diademed Sifaka with young (3–4 months), Andasibe-Mantadia National Park.*
Back cover: *Female Fosa, Kirindy Forest.*
p.2–3: *A group of Ring-tailed Lemurs scamper across the rocks in Isalo National Park.*

Contents

Eastern Sucker-footed Bat.

Famous Avenue of Baobabs (Allé des Baobabs) north of Morondava.

Aye-aye in forest canopy, Daraina.

Adult Indri, the largest extant lemur, Andasibe-Mantadia National Park.

Introduction

It is more than 30 years since I first visited Madagascar in May 1991. The country had just three national parks at the time – Montagne d'Ambre, Isalo and the then recently created Ranomafana. I camped in the forest of Ranomafana for four nights (back then this was allowed at *Belle Vue*). It was late afternoon when I arrived and it didn't take long before I saw wild lemurs for the first time: a group of Red-fronted Brown Lemurs *Eulemur rufifrons* came into the trees above where I'd pitched my tent to gorge on fruiting Strawberry Guava *Psidium cattleianum*. A little later, darkness fell and I followed Brown Mouse Lemurs *Microcebus rufus* by torchlight as they bounced around lower branches looking for insects and succulent fruits. Then a Spotted Fanaloka *Fossa fossana* emerged from the undergrowth with dainty digitigrade steps and sniffed around the periphery of the clearing searching for morsels. I was enthralled, captivated, spellbound. Just how magical was this place I had come too?

More than three decades down the road and I am still under Madagascar's spell. There is simply nowhere else quite like it. In the intervening period so much has changed of course. Today the island boasts no fewer than 25 national parks plus more than 20 other major protected areas, the majority of which are possible for tourists to visit. This is testimony to the increased level of commitment from the Malagasy government and international community to conserving the island's unique biological heritage in the face of ever-escalating threats.

Furthermore, the past decades have seen a corresponding surge in the volume of research taking place within the island's forests, wetlands and other natural habitats. Even by the early 1990s there were still only a relatively small number of species that had been studied in any detail and for which anything more than scant information was available.

Most early research concentrated on the higher profile (and relatively accessible) lemurs that are unquestionably the island's flagship animals. Some of these studies began as long ago as the 1960s and 70s (e.g., Ring-tailed Lemurs *Lemur catta* in Berenty and Bezá-Mahafaly) and are ongoing, with other long-term projects beginning in the 1980s and 90s at Kirindy, Ranomafana and other sites around the island. Subsequently, gaps have been filled at an ever-increasing rate and there has now been field research investigating members of every lemur genus, and many of these have become long-term, multi-generational studies.

While lemurs undoubtedly remain the island's best-studied animals, research into Madagascar's other mammalian groups – tenrecs, bats, carnivorans and rodents – has also gathered considerable momentum. Initially, a programme of detailed biological inventories covering a breadth of wide-ranging sites in all major habitats across the island established basic species information and particularly advanced our knowledge of the distributions and habitat preferences of the smaller mammals.

More recently (over the past two decades) molecular DNA studies have advanced our understanding of the evolutionary history and relationships within and between mammal families. We now have a much clearer picture as to the origins of their ancestors (each of the four major non-volant mammal groups being derived from a single common ancestor), as well as insights into how they may have arrived on the island and over what timescale (see Biogeography), and subsequently how these original colonists then diversified into the array of species seen in the past and today. These studies have, therefore, also confirmed that similarities in body form and morphology seen in Malagasy species relative to continental forms are the result of convergent evolution.

The augmentation of knowledge has led to the publication of some landmark works in the scientific literature. One such notable volume is *The Natural History of Madagascar*, S. M. Goodman & J. P. Benstead (eds) (2003, Chicago University Press), which has been followed nearly two decades later by the even more comprehensive and impressive *New Natural History of Madagascar*, S. M. Goodman (ed) (Princeton University Press, due for publication in 2022). This monumental multi-volume work is a collection of essays and reviews covering a huge range of subjects and aspects relating to the island's biodiversity, from soils and geology to botany, invertebrate zoology, vertebrate zoology, conservation and land management. The sections covering the mammals in particular are comprehensive and hugely informative.

Mother and infant Golden-crowned Sifaka, Bekaraoka Forest, Daraina.

However, even works of this magnitude serve to highlight the gaps in our knowledge that still exist. The numbers of new species described in the past two decades, of nocturnal lemurs and small mammals in particular, is astounding (and not without controversy – see The Species Conundrum). It seems certain this trend will continue and levels of recorded species richness will increase. Further, as field techniques become more refined, it is likely that our understanding of many of the basic natural history parameters (e.g., diet, social and reproductive behaviour) of the still poorly understood species will improve.

In this third incarnation of *Mammals of Madagascar*, I have tried to incorporate all of the latest information and make each species account as up-to-date as possible. Comparison with the previous version (A. & C. Black, 2007) will highlight how far knowledge has moved on. Yet, what is presented here remains far from complete and in some cases will become dated relatively quickly. Nonetheless, this synopsis should prove a useful and informative guide for those wishing to discover more about arguably the most fascinating and unusual mammalian assemblage on Earth.

How to Use the Handbook

This handbook aims to provide a concise and up-to-date review of our knowledge of Madagascar's mammals and a practical guide for visitors to the island. Information has been drawn from a wide variety of disparate sources, many of which are primarily accessible only to the academic and conservation communities (see e-pdf for the extensive citations list).

Despite the vast amount of research that has taken place since the previous edition in 2007, still minimal basic information is known about a substantial number of Madagascar's smaller, less 'glamorous' mammal species. Hence, depth and detail in the species profiles varies considerably. Information given here is intended as a first step for those wishing to learn about the species concerned, and also as a conduit to access more detailed information through the citations.

Introductory Chapters

To set the island and its natural history in context, it is crucial to understand its biogeography: how Madagascar was created and why it is the world's oldest island. The implications and consequences of long isolation are far reaching and have been important factors in shaping the evolutionary developments of Madagascar's fauna and flora in general, and its mammal communities in particular.

This is followed by a review of the major biomes and habitat types. The traditional perspective has been that, in its pristine state, the island was almost entirely cloaked in forest, and subsequent to human arrival it has suffered ever-increasing levels of deforestation and habitat degradation. However, new research, focusing on historic patterns of climate change is challenging this narrative. Areas of grasslands, that today cover vast swathes across the Central Highlands and many lower elevation areas, may have had their origins in the climate becoming warmer and drier around 5,000 years ago: these zones later expanded under human influence as forests were felled. In the past there may have been connecting forest conduits across the Central Highlands, linking eastern and western regions, which permitted evolving mammals to expand their ranges and spread. As the Central Highlands became denuded, these corridors would have been severed and populations isolated.

Species Accounts

Species accounts for each of the mammalian orders that naturally occur on Madagascar comprise the majority of the book. Accounts are subdivided into the following categories: Measurements, Description & Identification, Habitat & Distribution, Behaviour, and Where to See.

Vernacular Names

Many species have more than one, sometimes several, vernacular names. To promote consistency, the majority of vernacular names used here follow Goodman, S. M. (ed.) (in press) *The New Natural History of Madagascar*, Princeton University Press. Numerous species are named after the region from which they were originally described, or after an individual person or family. In order to assist placing geographic distribution, priority here is given to the locality alternative, for example in the case of *Cheirogaleus andysabini*, Montagne d'Ambre Dwarf Lemur is used in preference to Sabini's Dwarf Lemur.

Scientific Names

Written in *italics* and generally from Latin or ancient Greek, the binomial scientific name comprises the genus and species names. The species name often relates to some aspect of appearance, where the species was originally described from, the region it inhabits, or refers to, and is in recognition of, an individual person (often an eminent scientist) or family. For example, with the Red-bellied Lemur *Eulemur rubriventer*, *ruber* means red in Latin, and *venter* means belly, whereas meridian derives from *meridies* in Latin, meaning south, hence *Hapalemur meridionalis*, the Southern Grey Bamboo Lemur. Many species are named after an individual or even a family: if the species name ends in '*i*', it is masculine, so is named after a man e.g., *Propithecus coquereli* (Coquerel's Sifaka); if the name ends in '*ae*' it is feminine so is named after a woman, e.g., *Microcebus berthae* (Madame Berthe's Mouse Lemur); and if it ends in '*orum*' it refers to a family e.g., *Avahi mooreorum* (Masoala Woolly Lemur) is named after the Moore family. Where a species name ends in 'sis' it refers to a region or location; e.g., *Lepilemur ankaranensis* (Ankarana Sportive Lemur) is obviously named after the Ankarana Massif, the principal region the species inhabits.

Scientific names are given with vernacular names in the title of each species' account and elsewhere, within the main body of text, only on first mention in the book. Thereafter, only the primary vernacular name is given (there are occasional necessary exceptions).

Measurements

Basic body measurements that aid field recognition are given. Where applicable, differences between the sexes or significant seasonal variations are outlined. Where information is limited, figures may relate to a very small number of specimens and are indicated by an asterisk (*).

Description & Identification

Descriptions concentrate on external features that aid field identification, including pelage characteristics and coloration, and notable morphological traits. Differences between the sexes or between adults and immatures are highlighted. Variations between populations, some of which may correlate to subspecies, are outlined where appropriate. A concise summary of the main features used in identifying a species, for instance, approximate size and body posture, is followed by comparisons with taxa that may potentially cause confusion or misidentification.

Habitat & Distribution

Madagascar's non-flying mammals are, for the most part, essentially native forest dwellers. There are a number of distinct forest types on the island and the preferred habitat type of each species is given: in some instances, species with extensive distributions have wider tolerances and may be found in more than one forest type, including also in some secondary and human-altered habitats and cultivated areas. Distributions outline in as much detail as is feasible a species' known range and also suggest elevational preferences and limits where known. Where possible, population density figures are given for preferred and alternative habitats and locations. In addition, the major threats to a species' continued survival, and conservation implications are given.

Maps

Maps illustrate a species' range and attempt to outline the approximate *actual* range by relating distributional information to approximate remaining native forest cover. This approach has its pitfalls and will be inaccurate in places. Even within shaded areas, forests are not continuous and neither are the species' populations. Further, there may also be isolated fragments of forest lying outside the shaded areas in which small remnant populations survive. However, as perhaps only 15–20% of original forest is intact, this approach should provide a far more realistic impression of present distributions. A number of species are known from only a single or handful of localities. In these cases, the localities are pinpointed

Adult male Fosa, Kirindy Forest.

localities. In these cases, the localities are pinpointed and the probable range that extends between them is suggested. Again, this is to be treated only as an approximate guide.

Map Key	Confirmed range	Probable range

Behaviour
Includes information on daily and seasonal activity patterns, social interaction, group composition and structure, range size, foraging preferences, diet, breeding habits, development and predation.

Where to See
The best localities to see a species, including any seasonal variation, are listed. Alternative localities in different parts of a species' range are also suggested to offer variety.

Threats to Madagascar's Mammals

The continued survival of the island's biodiversity in general, and mammal communities in particular, faces a variety of ever-increasing threats. Ranging from habitat loss and hunting to the threats posed by invasive species, these are outlined in this section.

Conservation and Protected Areas

Madagascar's conservation concerns and priorities are outlined and some of the solutions currently being pursued are discussed. A brief outline of the island's national parks and reserves network is given.

Important Mammal Watching Sites

These are selected on the basis of quality of potential viewing and accessibility, to provide a complete cross-section of habitat types and, therefore, cover an extensive array of mammal species. Key species at each location are given. A species highlighted in **bold type** indicates that observations can be particularly rewarding and corresponds to the recommended sites suggested in the '*Where to See*' section of the species accounts.

The Extinct Mammal Fauna

A history and brief outline of the extinct mammal forms is given, together with classification of all known species.

Arguably the most beautiful lemur: an adult Diademed Sifaka, Andasibe-Mantadia National Park.

Biogeography

How Did Mammals Arrive in Madagascar?

Madagascar is home to one of the most unusual mammalian assemblages on Earth. How this came to be is an intriguing question that has baffled naturalists for centuries and has yet to be fully resolved. To simply state that Madagascar is an island, and that its biota is different because it has evolved in isolation, is a dramatic over-simplification that reveals little of the many complexities that have been involved in the historic development and resulting composition of the island's mammalian fauna in particular, and broader faunal and floral communities in general.

One of the biggest barriers to providing answers is a gaping hole in Madagascar's fossil record: there are virtually no identifiable terrestrial mammalian fossils for the entire Cenozoic era, i.e., the last 66 million years, although there are marine mammals from this era, e.g., dolphins and sirenians (Samonds & Fordyce 2019, Samonds et al. 2019). There is a considerable catalogue of subfossils that date back to c.80,000 years before present (see Extinct Mammal Fauna), then a void covering the entire Tertiary period, until fossils appear in rocks from the Late Cretaceous (Samonds et al. 2012). The boundary of the Late Cretaceous and Tertiary, 66 million years ago (mya) (formally known as the Cretaceous-Paleogene K-Pg boundary), denotes the mass extinction event caused by a catastrophic asteroid impact (Chicxulub Crater, the Yucatán Peninsula, Mexico). The small number of Malagasy terrestrial mammal fossils from pre-Cenozoic rocks (i.e., before the K-Pg boundary), e.g., gondwanatherians like *Adalatherium hui* (Krause et al. 2020), does not include any placental mammals, and therefore, no potential ancestors for extant mammal taxa (Krause 2010). Hence, the ancestors of all the mammals we see today in Madagascar must have arrived after the mass extinction, 66 mya.

Madagascar was once part of the giant southern supercontinent, Gondwana, comprising two halves, East Gondwana containing Seychelles, the Indian subcontinent, Antarctica, Australia and Madagascar, and West Gondwana comprising what is now the African mainland, Arabia and South America. Madagascar with the Indian subcontinent attached was nestled in a position against what is now south-east Africa and Antarctica. East and West Gondwana began separating around 160–165 mya (so this is

when Madagascar separated from Africa). Then c.120 mya, Madagascar and the Indian subcontinent separated from other parts of East Gondwana with sections moving away in different directions (Ali & Aitchison 2008). From an initial more southerly position, Madagascar (and the Indian subcontinent) drifted north, reaching roughly its current position around c.90 mya. Shortly afterwards (c.88 mya), the Indian subcontinent split from Madagascar and moved north and east (approximately 1,500km), eventually colliding with Eurasia around 45 mya (forming the Himalayas). Madagascar has therefore been isolated in its approximate current position relative to mainland Africa for c.88 million years (at the time both Africa and Madagascar were c.16–17 degrees further south relative to the equator), making it the world's oldest island (Samonds et al. 2012, 2013).

Whilst Gondwana was in the process of fragmenting an extensive vertebrate fauna existed, composed largely of dinosaurs, including the ancestors of birds, other reptiles, early amphibians, and early non-placental mammals (e.g., gondwanatherians). The majority of these, including all early mammals on Madagascar, failed to survive the mass extinction event 66 mya (Krause et al. 1997), thus effectively leaving Madagascar as a 'blank canvas' full of niches waiting to be occupied.

*An Elephant Bird (*Aepyornis* sp.) egg is equivalent to approximately 160 chicken eggs.*

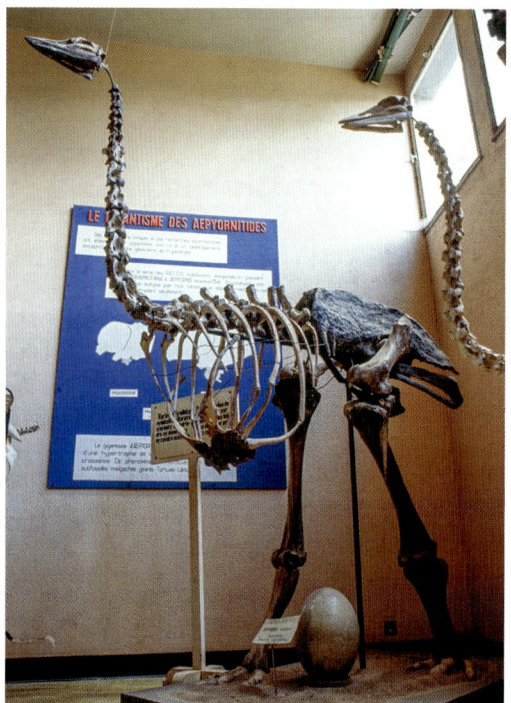

*The giant Elephant Bird (*Aepyornis *sp.) became extinct within the past 1000 years probably due to overhunting of adults and eggs by humans.*

Relic forms did survive, for instance the ancestors of flightless birds (ratites), which on Madagascar evolved into the elephant birds (Aepyornithidae), some of which became the heaviest birds ever to have lived – *Aepyornis maximus*, and *Vorombe titan* weighed *c.*600–730kg and stood over 3m tall.

At various times three main hypotheses have been proposed to explain the composition of the Malagasy mammal fauna. 1: at some point in the past, a chain of islands was revealed in the Mozambique Channel after a reduction in sea level and these acted like 'stepping stones' facilitating colonisation (Tattersall 1982, 2006b). 2: on one or more occasions, a temporary landbridge has existed between Africa, Antarctica (when it occupied a tropical position) and Madagascar, allowing free passage in both directions (McCall 1997). 3: rafting on mats of drifting vegetation (Simpson 1940, Krause 2010). The geological evidence for temporary islands or landbridges is scant (Krause *et al.* 1997, Ali & Krause 2011, Ali & Vences 2019) and neither hypothesis satisfactorily explains why so few terrestrial mammal groups successfully colonised Madagascar. If a landbridge had offered easy transit,

surely many more familiar African forms would be present on Madagascar?

Therefore, the pervading narrative has been that the ancestors of Madagascar's non-volant mammals (and the majority of other vertebrates) must have reached the island over water by rafting or swimming, with these ancestors originating on the nearest continental landmass, Africa (Krause 2010). It is widely accepted that many animals reach isolated islands in this way on clumps and mats of floating vegetation, washed out to sea from large rivers on continental mainlands (Ali & Vences 2019, Seiffert *et al.* 2020). This might seem a highly improbable means for any land-based animal to reach distant shores, but in recent times floating islands of natural debris (up to 100m across) harbouring 'marooned' animals have been reported at sea up to 240km from their place of origin (Samonds *et al.* 2012), so such phenomena do occur with some frequency. Add into the equation the immensity of geological time and apparently rare events become highly plausible – any event that is not impossible becomes probable if sufficient time elapses (Simpson 1940).

Certainly, crossing the Mozambique Channel by rafting (sweepstake dispersal) is a prime candidate for explaining the arrival of Madagascar's first mammals. However, this alone is insufficient to answer all the questions. For instance, if the forebears of Madagascar's mammal communities could make the crossing, why have other groups that proliferate on Africa not done so subsequently? We know Madagascar has been roughly in its current position relative to the African mainland for *c.*88 million years, so theoretically reaching it on floating vegetation should have been equally easy at any subsequent point in time.

Consequently, we need to consider possible reasons that may have made the conditions more favourable to achieve successful crossings at certain times, together with potential reasons that made the pioneer mammals better mariners (in terms of their ability to survive a lengthy accidental sea crossing) than mainland African species that came along later in evolutionary history.

Addressing the first issue, analysis of ocean currents provides some compelling evidence. Today and in recent times, prevailing winds and currents in the Mozambique Channel have not been in directions conducive to rafting from Africa to Madagascar. However, during the Eocene and Oligocene (*c.*56– 22 mya), Madagascar resided slightly further south, and weather patterns and ocean currents were very different. Every *c.*100 years or so, for several-week periods, there were strong currents moving north-west to south-east across the Mozambique Channel that would have carried large floating mats of vegetation to Madagascar in three to four weeks (Ali & Huber 2010). Then, as Madagascar continued its gradual northerly drift, currents shifted again, such that from the early Miocene (*c.*21 mya) to the present day, successful rafting events became far less likely.

There are just four extant native non-volant mammal groups in Madagascar: tenrecs, lemurs, carnivorans and Nesomyine rodents. Molecular DNA evidence strongly suggests each of these is derived from a single ancestor (by necessity a pregnant female or male/female pair) of African origin (Jansa & Carleton 2003, Olson & Goodman 2003, Yoder 2003, Yoder in press, Yoder & Flynn 2003, Poux *et al*. 2005). Furthermore, molecular evidence points to four independent colonisation events: ancestral primates, 50–60 mya (Yoder *et al*. 1996, Yoder 2003, Herrera & Dávalos 2016), ancestral mongoose-like carnivorans, 18–24 mya (Yoder *et al*. 2003, 2021, in press), ancestral tenrecs, 30–55 mya (Poux *et al*. 2005, Everson *et al*. 2016) and ancestral Nesomyine rodents, 20–24 mya (Poux *et al*. 2005), all of which fall within the Eocene to Oligocene timescale when ocean currents favoured dispersal from Africa to Madagascar.

Why then did only four mammal groups successfully cross the Mozambique Channel when conditions were conducive, a period lasting perhaps 25+ million years? It is clear that small body size would be advantageous in being able to stay afloat on a makeshift raft. And although large species have existed and continue to exist on Madagascar, evidence suggests these evolved at a later date (for instance some extinct lemurs), and that original colonising stock consisted only of small species. But of course there are many smaller mammal groups on Africa, which are absent from Madagascar – why? Furthermore, small body size usually correlates to a higher metabolism and the need to eat regularly. Clearly this trait would place small mammals at a disadvantage in the sea-crossing stakes. One final piece of evidence may provide the clue.

Many of Madagascar's smaller mammals have the capacity to lower their metabolism and become dormant during periods of cold and low food availability (Schmid & Stephenson 2003, Dausmann *et al*. in press). In conjunction, they can also build up fat reserves during times of plenty to sustain themselves through lean periods. This ability has been demonstrated both in mouse and dwarf lemurs (family Cheirogaleidae), which are probably very similar to Madagascar's early primate colonists, and in tenrecs (family Tenrecidae) (Kappeler 2000, Ali & Vences 2019). Also, some of the island's endemic carnivorans (family Eupleridae) are able to lay down fat reserves in their tails.

Here then may be the answer. Madagascar was colonised from Africa by small mammal species rafting on natural debris, and the only species to survive the journey were those with the capability to lower their metabolism, become dormant and/or live without food for sufficiently long periods on reserves built up previously. As it turned out, there were only four such successful events resulting in the arrival of the ancestors of lemurs, mongoose-like carnivorans, tenrecs and Nesomyine rodents.

There are potential anomalies that require explanation. The theory that all lemurs (including all extinct forms) are derived from a single colonising ancestor is well established and is supported by a considerable body of molecular evidence (Yoder *et al*. 1996, Yoder 2003, Poux *et al*. 2005). However, recent fossil evidence from mainland Africa casts some doubt. *Propotto* from the early Miocene of Kenya is seen as being related to Aye-ayes (*Daubentonia* spp.) (Gunnell *et al*. 2018): if confirmed, this necessitates that Madagascar has been colonised

Germinating seed of Barringtonia asiatica *washed up on the beach, Masoala Peninsula. Early mammal colonists probably reached Madagascar by rafting on larger clumps of vegetation.*

twice by lemurs, once producing the Aye-aye lineage and a second leading to all other lemurs, and that these colonisations were far more recent. The established view is that ancestral Malagasy lemurs crossed to Madagascar and diverged from their African ancestors *c.*50–60 mya (Yoder *et al.* 1996, Yoder 2003, Herrera & Dávalos 2016), whereas the last common ancestor of *Propotto* and *Daubentonia* is placed on the African mainland *c.*28 mya and maximum age of divergence of the other Malagasy lemurs from the *Propotto / Daubentonia* group is estimated at *c.*41 mya (Gunnell *et al.* 2018).

Further, three species of hippopotamus (*Hippopotamus lemerlei, H. laloumena* and *H. madagascariensis*) and two 'aardvark-like' mammals (*Plesiorycteropus madagascariensis* and *P. germainepetterae*) are part of Madagascar's extinct mammal fauna that do not obviously belong to any of the four ancestral mammal groups. The hippopotamuses were probably relatively recent arrivals (late Pleistocene) and were all small compared to the contemporary African Hippopotamus *Hippopotamus amphibius*, but too large to occupy rafts, therefore it its assumed their ancestors accidentally floated / swam to Madagascar (Ali & Vences 2019). The 'aardvark-like' mammals are more problematic, as their origins are unclear. They are placed in their own order Bibymalagasia (MacPhee 1994), although a recent study suggests they may be closely related to tenrecs and should be placed in the order Tenrecoidea (Buckley 2013); consequently, their origins are aligned with the ancestors of other tenrecs. However, further supporting evidence is required to corroborate this view (Goodman & Soarimalala in press).

The idea of 'sweepstake dispersal' to explain Madagascar's mammal fauna has gained widespread traction, but support is not universal. Recently, the landbridge concept has resurfaced, with the suggestion based on new geological data that three separate landbridges connecting East Africa with western Madagascar existed during the Cenozoic, each lasting *c.*5 million years. The first of these, 60–66 mya, allowed the ancestors of lemurs and tenrecs to colonise, the second 30–36 mya facilitated the movement of rodents and carnivorans, and the third, 5–12 mya, permitted hippopotamuses to reach Madagascar (Masters *et al.* 2020). Furthermore, proponents assert that factors such as small mammal ancestors entering torpor to survive sea crossings and hippopotamuses swimming across the Mozambique Channel are in no way supported by evidence. The only aspect that is crystal clear is that argument and counter-argument seem set to continue and this debate will rumble on. As new techniques for peering into the past become available some uncertainties may be resolved and new discoveries within Madagascar's Cenozoic fossil record will potentially fill the significant gaps in our knowledge of the origin and evolution of the island's mammals (Samonds *et al.* in press).

One group of extant Malagasy mammals has thus far not been mentioned, the bats (order Chiroptera). Bats are the only mammals capable of true flight, a factor that has obviously had a significant bearing on their distribution. For bats to colonise islands, distance from originating continental landmass (as with non-volant taxa) and prevailing wind direction are significant. Madagascar is located in the trade wind zone (which extends from 30°N to 30°S) and has likely been within this zone since the middle Eocene (the last *c.*45 million years). Hence prevailing winds have been from the east, potentially aiding the transfer of bats (and other volant animals) from other Indian Ocean islands, the Indian Subcontinent, South-East Asia and Australasia. Furthermore, during the austral summer the south-west portion of the Indian Ocean is regularly affected by major tropical storms (cyclones), which are capable of transporting flying animals hundreds of kilometres from their places of origin (Samonds *et al.* 2012).

Madagascar has 45 or so extant bat species (38 of which are endemic) that probably represent between 28 and 30 independent colonisation events (Goodman *et al.* in press), of both African (e.g., *Eidolon, Mops*) and Asian (e.g., *Pteropus, Rousettus, Nycteris*) origin (Goodman 2011, Samonds *et al.* 2012). Some lineages may have their origins within the Paleogene (before *c.*23 mya); however, the majority are thought to have colonised the island far more recently, with many likely during the last 5 mya (Rakotoarivelo *et al.* 2015, Foley *et al.* 2017, Samonds *et al.* in press). Species like Midas Free-tailed Bat *Mops midas* have probably moved regularly between Africa and Madagascar (Ratrimomanarivo *et al.* 2007).

Regions and Habitats of Madagascar

Madagascar has a considerable diversity of habitats and flora, with associated high degrees of endemism. Covering an area of about 587,000km², it is the world's fourth largest island (after Greenland, 2,175,600km², New Guinea, 785,750km² and Borneo, 748,170km²), yet its environmental diversity rivals that of a continent. The island has numerous physical characteristics that divide it into natural 'sectors'; these include mountain ranges and large rivers, which promote increased variation. There are rainforests in the east, deciduous forests in the north and west, dense xerophytic forests in the south, high mountain forests in the island's interior, various isolated limestone outcrops in the west harbouring more humid forests, and important lakes, marshes and wetlands in both eastern and western areas. Not surprisingly such variety has fostered the evolution of a faunal assemblage so unusual as to rival that of any comparable area on earth.

Habitat variety is the result of dramatic climatic variability, which in turn is a consequence of Madagascar's position and topography (Jury 2003). The island lies virtually entirely within the tropics and has a corresponding climate. From north to south it spans some 1,650km, between 12°S and 25°S, and there is notable variation in basic sea level temperatures: in the far north the annual mean is 27°C (81°F), while at the southern extremity the mean drops to 21.5°C (71°F).

The island consists of a backbone of Precambrian crystalline rocks running north to south. This forms the Central Highlands. To the east there is an abrupt escarpment, while to the west the slope is relatively gentle towards the Mozambique Channel. The highest mountains are in the north and rise to 2,876m (Maromokotro, the island's highest peak in the Tsaratanana Massif), and there are other scattered massifs and peaks, some of which reach over 2,000m. The eastern coastal plain between the escarpment and the Indian Ocean is generally narrow, but the coastal plain to the west is much broader.

Major Rivers

Map Key

1. Loky (Lokia) River	20. Mangoky River
2. Manambato River	21. Fiherenana River
3. Bemarivo River	22. Mangoky River
4. Antainamalana River	23. Matsiatra River
5. Anove River	24. Mananantanana River
6. Maningory River	25. Zomandao River
7. Onibe River	26. Morondava River
8. Mangoro River	27. Tsiribihina River
9. Onive River	28. Manambolo River
10. Nosivolo River	29. Manambaho River
11. Namorona River	30. Sambao River
12. Faraony River	31. Mahavavy River
13. Manampatrana River	32. Betsiboka River
14. Mananara River	33. Ikopa River
15. Mandrare River	34. Mahajamba River
16. Menarandra River	35. Sofia River
17. Linta River	36. Sandrakota River
18. Onilahy River	37. Sambirano River
19. Imaloto River	38. Mahavavy du Nord River

Prevailing weather comes from the Indian Ocean brought by the trade winds, which are forced to rise over the escarpment. Consequently, the majority of rain falls on the eastern side of the island, with the far west lying in a permanent rain shadow, so conditions gradually become drier and hotter towards the west coast.

The north–south mountain chain and east/west slope differential creates significant water drainage asymmetry. The steep eastern slopes prompt relatively short fast-flowing streams to the Indian Ocean, whereas c.70% of the islands water flows from the Central Highlands through large drainage systems to the west, which develop into long, wide meandering rivers like the Betsiboka, Mangoky, Tsiribihina and Onilahy that often terminate in broad deltas before reaching the Mozambique Channel.

These two different types of river system have important implications for the dispersal of mammals.

In eastern regions, short narrow rivers present less of a barrier to dispersal, especially when forests were continuous, as species with broad elevational tolerances would have been able to migrate around headwater regions e.g., Brown Lemur *Eulemur fulvus* (Goodman & Ganzhorn 2004). In contrast the broad meandering lower reaches of many western rivers impose significant obstacles to dispersal, resulting in some species distributions being constrained by rivers, e.g., sifakas (*Propithecus* spp.) and sportive lemurs (*Lepilemur* spp.).

During the austral summer, western regions are subject to a monsoon regime that originates in the north and whose influence dwindles towards the south. Hence, there is a double rainfall gradient over the island: declining from east to west on the one hand, and north to south on the other. As a result, the north-east is the wettest part of the island and the south-west is the driest (Jury 2003).

The combined effect is a profusion of climatic conditions, which consequently result in the variety of vegetation and habitat types. Attempts to classify the island's vegetation are principally based on region and elevation (Du Puy & Moat 1996, 2003, Gautier & Goodman 2003). Difficulties arise in classifying many eastern rainforest regions where steep slopes result in important changes in elevation over short distances. Here, the underlying substrate may vary, as does exposure to wind and sun. Thus, there are numerous localised microclimates that contradict general rules. For instance, in some sheltered valleys at higher elevations in Marojejy, forest characteristic of the 'mid-elevation type' grows above 1,400m, whereas on exposed ridgetops in Mantadia, the forest resembles 'high-elevation montane' although the elevation is only around 1,100m to 1,200m.

Until recently, a popular narrative has been that, in its pristine state, Madagascar was

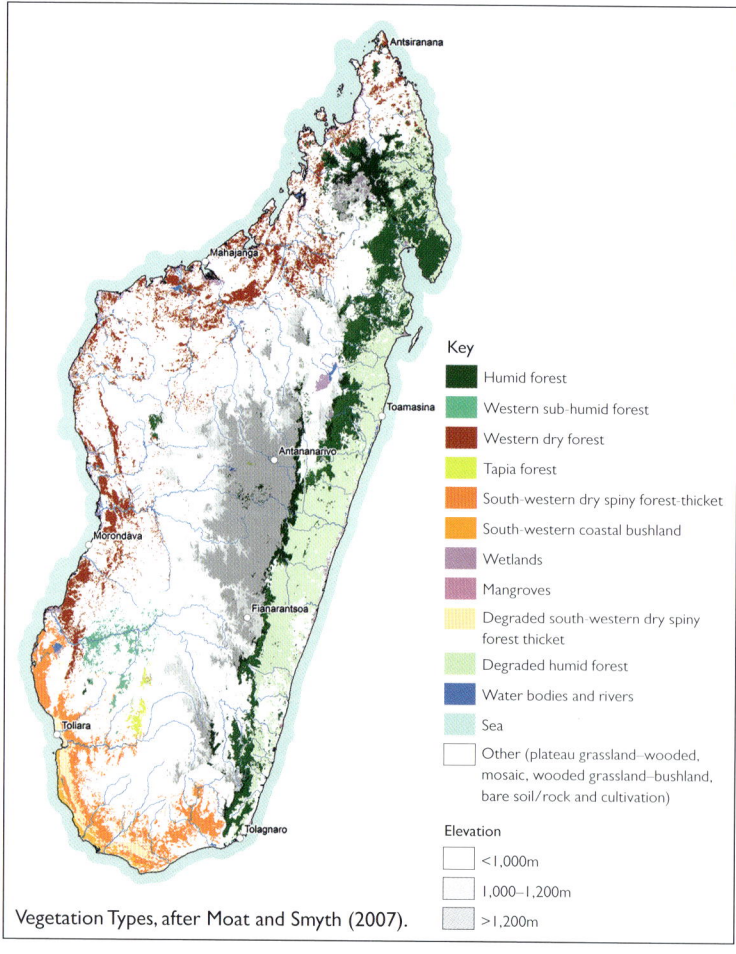

Key

- Humid forest
- Western sub-humid forest
- Western dry forest
- Tapia forest
- South-western dry spiny forest-thicket
- South-western coastal bushland
- Wetlands
- Mangroves
- Degraded south-western dry spiny forest thicket
- Degraded humid forest
- Water bodies and rivers
- Sea
- Other (plateau grassland–wooded, mosaic, wooded grassland–bushland, bare soil/rock and cultivation)

Elevation

- <1,000m
- 1,000–1,200m
- >1,200m

Vegetation Types, after Moat and Smyth (2007).

perhaps at least 80% covered in various native forest types, while remaining areas comprised marshes and lakes with very limited areas of grassland primarily in montane areas. As such, it was believed, the extensive grasslands seen today (particularly in the Central Highlands) must be largely anthropogenic in origin and have proliferated as a consequence of native forests being removed by humans after initial colonisation *c*.2,000–3,000 years ago (Gade 1996, Patterson *et al*. 1997). More than 90% of Madagascar's animal species live exclusively in forest and woodland habitats (Goodman *et al*. 2003a), and very few endemic species utilise grasslands, which certainly alludes to grasslands being a more recent phenomenon on the island.

However, current thinking suggests grasslands in the Central Highlands and northern and western regions might at least in part be natural formations (Bond *et al*. 2008, Vorontsova *et al*. 2016). Further, these formations have expanded and contracted over the last *c*.10,000 years due to climate change and only relatively recently has human impact been a factor in the expansion of grassland habitat (Quéméré *et al*. 2012, Salmona *et al*. 2017). While a number of lines of evidence indicate large areas of open grassland (comparable to East African savannas) are unlikely to be natural formations, there are suggestions that a more wooded savanna formation (similar to southern African miombo woodland) might have been widespread in the Central Highlands and would have been occupied by now extinct giant terrestrial lemurs (e.g. *Mesopropithecus* spp.), elephant birds (*Aepyornis* spp.), Malagasy 'aardvarks' (*Plesiorycteropus* spp.)

and Malagasy giant tortoises (*Aldabrachelys* spp.) (Goodman & Jungers 2014).

Madagascar today is divided into two biogeographic regions: Eastern and Western. The Eastern Region is further subdivided into four floristic domains: Eastern, Central, High Mountain and Sambirano. The Western Region is divided into two floristic domains: Western and Southern (White 1983). For practical purposes and convenience these domains translate to the broad habitat divisions outlined below.

Lowland Rainforest and Littoral Forest (Eastern Domain)

Sea level to *c*.800m, extending along east coast from north of Sambava to Tolagnaro region in south. Corresponds to areas of highest rainfall – mean 2,000–3,000mm per year, and in north-east up to *c*.6,500mm per annum, e.g., Masoala Peninsula (Kremen 2003). No clearly defined dry season, although December–March consistently wetter than other times.

Enormous species richness and diversity coupled with very high levels of endemism. Largest trees have huge buttress roots and canopy averages 30–35m, with no emergent trees. Understorey dominated by small trees, with little herbaceous growth below. Epiphytic plants such as orchids are abundant. Due to the vast scale of deforestation, this is one of most threatened vegetation types on the island: almost completely disappeared in southern and central areas, only large areas remaining are in north-east and to lesser extent one area in extreme south (Tsitongambarika).

Lowland rainforest understory dominated by Mammea *sp., Masoala Peninsula.*

Mammal diversity very high, all major taxonomic groups are represented. Examples include: Masoala National Park, Makira Natural Park, Manambo Special Reserve and Tsitongambarika Protected Area.

Mid-Elevation Montane Rainforest (Central Domain)

Occurs between *c.*800m and *c.*1,300m, with localised patches at higher elevations (occasionally to *c.*1,800m). Occurs west of lowland rainforest belt, extending parallel to east coast from south of Vohemar in north to vicinity of Tolagnaro in south, with an isolated region in the far north centred on Montagne d'Ambre Massif (summit 1,475m). Annual rainfall exceeds 1,500mm with no discernible dry season, wettest periods correspond to austral summer (December–March).

Diverse and species-rich, with levels of endemism equivalent to lowland rainforest. Canopy averages 20–25m, occasionally reaching 30m in sheltered valley bottoms. Understorey and herbaceous layers better developed than in lowland forests. Epiphytes like tree-ferns abound, and many trees are festooned with mosses and lichens. The majority of mid-elevation forest has been felled for slash-and-burn agriculture, hardwood extraction, charcoal production and firewood. What does remain constitutes the most extensive natural areas of native forest left on the island.

Mammal diversity very high, with all major taxonomic groups represented. Both Tenrecidae and Nesomyinae reach their highest levels of species richness. Madagascar's most familiar rainforest locations correspond to this habitat, including; Andasibe-Mantadia National Park, Marojejy National Park, Ranomafana National Park and Anjozorobe-Angavo Forest.

High-Elevation Montane Forest (High Mountain Domain)

Restricted to elevations between *c.*1,300m and *c.*1,800m. Occurs in discrete areas primarily west of mid-elevation montane rainforest zone; northerly limit is south of Sambava (Marojejy and Anjanaharibe-Sud Massifs) and southern extreme is Andohahela Massif (north of Tolagnaro). Annual precipitation high, much occurring as dense mist and low cloud, so often referred to as 'cloud forest'.

Species diversity lower than montane or lowland rainforests, but levels of endemism remain high. Canopy height 10–15m at lower elevations, declining gradually with altitude. Canopy layers and understorey often difficult to differentiate: forest appears as a dense single stratum. Leaves on most trees smaller than in lower-altitude rainforests and have leathery cuticles to counter desiccation. Trees encrusted with mosses and lichens. Mosses form a dense carpet on the forest floor. Extensive deforestation in most locations for slash-and-burn (*tavy*) agriculture, hardwood extraction, charcoal production and firewood. Very few areas remain.

Mammal diversity moderately high, with most major taxonomic groups represented, but species diversity reduced compared to rainforest at lower elevations, e.g., bat diversity declines dramatically above *c.*1,500m. Examples are higher elevations

Mid-elevation montane rainforest, Andasibe-Mantadia National Park.

of Ranomafana National Park (Vohiparara), areas above Camp Simpona (Camp 3) in Marojejy National Park, higher regions of Anjozorobe-Angavo Forest Corridor and Tsinjoarivo Forest.

Sclerophyllous Forest and High Mountain Thicket (High Mountain Domain)

Found above c.1,800m with highest areas (generally above 2,000m) above treeline. Restricted to five distinct locations: Tsaratanana (Maromokotro, 2,876m, Madagascar's highest peak) in north-west, Marojejy in north-east, Ankaratra in central east, Andringitra in central south-east and Andohahela in far south-east. Between c.1,800m and c.2,000m stunted trees and bushes 2–5m high, at higher elevations vegetation consists of dwarf open-bush thicket less than 2m in height dominated by Ericaceae and Asteraceae formations. Ferns comprise a significant element (Gautier & Goodman 2003). Rainfall high year-round. Daily and seasonal temperature variations; during austral winter nighttime temperatures can fall below freezing, e.g., in Pic d'Imarivolanitra (2,658m), Andringitra National Park (Rasolonandrasana & Grenfell 2003). Threatened by deforestation and anthropogenic fires.

Mammal diversity low, only groups regularly present being Tenrecidae and Nesomyinae, although some lemurs have been recorded, e.g., Ring-tailed Lemurs (up to c.2,600m) in Andringitra Massif (Goodman & Langrand 1996) and Red-bellied Lemurs *Eulemur rubriventer* in Marojejy (Garbutt 2007). Only examples are: Tsaratanana (summit 2,876m), Marojejy (2,133m),

High-elevation montane rainforest, Vohiparara, Ranomafana National Park.

Anjanaharibe-Sud (2,064m), Ankaratra (2,643m), Andringitra (2,658m) and Andohahela (1,956m). Marojejy and Andringitra are the only examples that can be visited relatively easily.

Seasonal Humid Forest (Sambirano)

An enclave of semi-humid/seasonal moist forest restricted to north-west. Lies between approximately Nosy Faly Peninsula in north and Andranomalaza River in south, incorporating west side of Tsaratanana Massif and Manongarivo Massif, and represents an extension to north-eastern rainforest zone. Annual rainfall exceeds 2,000mm, with at lower elevations a drier season in July–September.

High mountain thicket. The summit of Marojejy (2133m).

A diminishing waterhole in dry deciduous forest, Kirindy Forest.

Represents a transition between floral communities of east and west, sharing many features with both regions. High species diversity and levels of endemism. Canopy reaches 30m, at lower elevations some trees emerge to *c.*35m. Understorey and shrub layers substantial, with epiphytes, vines and creepers abundant (Gautier & Goodman 2003).

Mammal diversity moderate, most major taxonomic groups represented, with some notable exceptions, e.g., *Propithecus* spp. Examples include Manongarivo Special Reserve, Ampasindava Peninsula and Nosy Faly Peninsula (Ambato Massif).

Dry Deciduous Forest (Western Domain)

Extends from sea level to 800m, from area of Antsiranana in north to vicinity of Morombe in south-west, excluding Montagne d'Ambre Massif and the Sambirano domain (which are distinct, see above). Annual rainfall ranges from 2,000mm in north to just 500mm in south: majority falls December–March, with marked dry season lasting seven months (May–November) when many trees shed their leaves in canopy layer (Du Puy & Moat 1996).

Species diversity very high, but lower than in eastern region. Levels of endemism are higher, often around 90%. Canopy averages 12–15m but in some areas reaches 20–25m. Understorey and shrub layers well developed, herbaceous and epiphytic growth rare. In some areas forests dominated by baobabs (*Adansonia* spp.). Excellent examples of Grandidier's Baobab *A. grandidieri* occur north and south of Morondava.

Most dry deciduous forests have already been cleared and the few remaining areas are under severe threat from slash-and-burn agriculture, timber extraction and uncontrolled pasture fires (Dufils 2003). Along with Eastern Lowland Rainforest/Littoral Forest, this is the most threatened native forest type in Madagascar. Mammal diversity high, most major taxonomic groups represented, but species diversity reduced in comparison to eastern rainforest areas. Best examples at Ankarafantsika National Park, Anjajavy Forest, Kirindy Forest and Kirindy-Mitea National Park, together with variations around limestone karst formations such as Tsingy de Namoroka National Park, Ankarana Special Reserve and Tsingy de Bemaraha National Park.

Subarid Thorn Scrub or Spiny Forest (Southern Domain)

Subarid thorn scrub is often referred to as spiny forest. Extends south along coastal plain from Morombe on west coast; includes whole of southern area to a point *c.*40km west of Tolagnaro (western flank of the Anosyennes Mountains). Restricted to elevations below 400m. In Morombe region annual rainfall

*Grandidier's Baobab (*Adansonia grandidieri*) north of Morondava.*

*Baobabs (*Adansonia *sp.) growing on a remote rocky shoreline on the north west coast.*

reaches around 500mm, but declines further south, with driest areas receiving only 300mm per year. Dry season marked and long, extending up to ten months (March–December). This is highly variable and 18-month dry periods have been recorded (Fenn 2003).

Species diversity high, levels of endemism extremely so. Height and density of vegetation vary with climatic and soil conditions: in some areas trees reach 8–12m, but in others only *c.*2m. All plant species adapted for harsh desiccating conditions, with abundant thorns and small waxy leaves. Region is dominated and characterised by endemic family Didiereaceae and members of the Euphorbiaceae.

Along major watercourses, spiny forest replaced by narrow bands of riparian habitat known as gallery forest, showing vague resemblance to western deciduous forests. Canopy of similar height (up to 15m), with extensive understorey and herbaceous layer. Tamarind trees *Tamarindus indica* often dominate. Forest clearance for agriculture (manioc and maize), firewood and charcoal are major threats. In most southerly regions significant areas cleared for sisal plantations. In marginal areas browsing by goats and cattle is also detrimental.

Mammal diversity moderate. Several major taxonomic groups are notable absentees, e.g., *Eulemur* spp. Small-mammal diversity also reduced. Examples are Tsimanampetsotse National Park,

Andohahela National Park (Parcel 2), Amoron'i Onilahy Protected Area, south of Toliara and forests near Ifaty, north of Toliara, e.g., Reniala Forest (part of Ranobe-PK32 Protected Area).

At marginal localities, where western and southern formations meet, there are areas of 'transition forest' exhibiting characteristics of both areas. These include forests north of Morombe and south of Mangoky River, and Zombitse-Vohibasia National Park further inland. In the south-east there is also a transition zone between southern and eastern formations on western slopes of Anosyennes Mountains (Fenn 2003).

Spiny forest dominated by the family Didiereaceae.

Mammal diversity very low: no specialists or permanent residents. On occasion lemurs use mangroves for foraging, sleeping and moving between terrestrial forest patches. Records of c.23 lemur species (from all five families) utilising mangroves, including observations of Coquerel's Sifaka *Propithecus coquereli* 1km from dry land and Northern Giant Mouse Lemur *Mirza zaza* up to 3km from dry land (Gardner 2016).

Major mangrove areas include, Mahajamba Bay and Ambaro-Ambanja Bays and river estuaries of Tsinjomorona, Betsiboka, Mahavavy, Tsiribihina and Mangoky.

*Bottle Baobab (*Adansonia za*), Tsimanampetsotse National Park.*

Mangroves

Mangroves are forest ecosystems restricted to intertidal areas. In Madagascar, they occupy a relatively small area (c.210,000ha), with 98% occurring in estuaries and lagoons on the west coast (Roger & Andrianasolo 2003). Due to extreme and dynamic conditions, and unpalatable vegetation, they are marginal habitats for most terrestrial mammals; e.g., Proboscis Monkey *Nasalis larvatus*, endemic to Borneo, is the only mangrove specialist primate.

Grasslands and Palm Savannas

Today, c.80% of Madagascar is dominated by a mosaic of agricultural land, degraded vegetation and grasslands. In lowlands of east these are likely all anthropogenic in origin and have proliferated as forests were cleared over the last c.2,000 years (the majority probably disappearing in the last 100 years). In many areas these savannas are interspersed by stands of Traveller's Tree *Ravenala madagascariensis* (Strelitziaceae).

Grasslands and agricultural land dominate many areas at higher elevations in the east where forest has been removed and continue across the vast majority of the Central Highlands. Highland grassland is also found on exposed mountain ridges. In lower elevation western regions coarse grasses dominate, in some areas broken by stands of fire-resistant palms such as *Bismarckia nobilis*, *Hyphaene coriacea* and *Borassus madagascariensis*. Grasslands are further impoverished by deliberate anthropogenic

Mangroves on the central west coast.

fires started several times per year, to encourage new shoots for cattle grazing. Overgrazing also commonplace and locally severe, resulting in large-scale erosion, often visible as huge gullies, known in Malagasy as *lavaka*.

While grasslands in Central Highlands and western regions have certainly expanded significantly as a result of human impact and deforestation, there is growing evidence that their floral communities are natural (Bond *et al.* 2008, Vorontsova *et al.* 2016, Tiley *et al.* 2020), pre-date human colonisation and have undergone periods of expansion and contraction due to climate change over the last *c.*10,000 years (Quéméré *et al.* 2012, Salmona *et al.* 2017).

Mammal diversity and endemism very low. Largely devoid of native terrestrial mammal species. In eastern regions, in some marginal areas, tenrecs (e.g., rice tenrecs *Oryzorictes* spp.), rodents and some indigenous bats occur where trees are present. Eastern Sucker-footed Bat *Myzopoda aurita* can be locally common in areas with large stands of Traveller's Trees and *Typhonodorum* sp. (Araceae): this may be one of the few endemic mammals to have benefited from habitat modification by man (Goodman 2019). Commensal mammals, e.g., Black Rat *Rattus rattus*, House Mouse *Mus musculus* and Asian Musk Shrew *Suncus murinus*, may be common.

In some western areas Western Big-footed Mouse *Macrotarsomys bastardi* occurs locally in some grasslands. Similarly, some bats, e.g., Western Sucker-footed Bat *Myzopoda schliemanni* can utilise habitats where palms, e.g., *Bismarckia nobilis* and *Borassus madagascariensis*, are present. As with eastern regions, the commensals Black Rat, House Mouse and Asian Musk Shrew may be common.

Much of the Central Highlands is now dominated by rice cultivation and barren grasslands.

*After deforestation, vast swathes of the eastern lowlands are now dominated by Traveller's Trees (*Ravenala madagascariensis*).*

In many western and central southern regions grasslands are dominated by fire-resistant palms from the genera Bismarckia, Hyphaene *and* Borassus.

Highly vocal Red Ruffed Lemur, Masoala National Park.

The Mammals of Madagascar

Overview

The mammal fauna of Madagascar is perhaps as remarkable for the major taxonomic groupings that are absent as it is for the collection of unusual species that are present. Given the island's proximity to mainland Africa, it would be easy to assume Madagascar contains simply an impoverished version of the mammalian diversity present on its continental neighbour. This could not be further from the truth. None of the large herbivores that dominate the plains of Africa occur; carnivorans, such as wild cats and wild dogs are conspicuous by their absence, as are monkeys and apes, and a host of other smaller forms like lagomorphs (rabbits and hares).

The contemporary native Malagasy mammal fauna comprises only five major groups – tenrecs (order Afrosoricida: family Tenrecidae), bats (order Chiroptera), lemurs (order Primates: infraorders Lemuriformes and Chiromyiformes), carnivorans (order Carnivora: family Eupleridae) and rats and mice (order Rodentia: family Nesomyidae, subfamily Nesomyinae) – and is exceptional for three major reasons. Firstly, every native non-volant species (a total of c.175 species) is endemic, i.e., they occur naturally nowhere else. Even when volant species, i.e., bats (c.85% of which are endemic) are included, 213 of 220 indigenous mammal species (or c. 97%) are endemic. No other island or place on Earth boasts such a combination of mammalian species richness and endemism. Secondly, these mammals have evolved an extraordinary diversity of body forms and lifestyles often displaying significant convergence with continental forms, but also at times evolving utterly unique features (Goodman et al. 2003). And thirdly, and perhaps most remarkable of all, the diversity expressed by the non-volant orders is the consequence of just four separate colonisations, indicating how extraordinarily rare such events have been in geological history (Goodman et al. in press).

The ancestors of all Madagascar's terrestrial mammals almost certainly reached the island by rafting on natural debris floating across the Mozambique Channel. Although it is difficult to pinpoint when the first colonists arrived, molecular analyses of extant and extinct forms indicate these occurred during the Paleogene and Early Neogene, i.e., all no earlier than c.60mya (see Biogeography).

The lottery and rigours of such a crossing severely restricted mammalian invasions and only a handful of forms proved capable of arriving. From mitochondrial DNA analyses we now know each of the four major non-volant groups (tenrecs, lemurs carnivorans and rodents) is each derived from a single common ancestor and by necessity, a unique colonisation event, by a pregnant female or male/female pair (Jansa & Carleton 2003b, in press, Olson & Goodman 2003, Yoder 2003, Yoder in press, Yoder & Flynn 2003, Poux et al. 2005, Everson et al. 2016, Veron et al. in press).

Once on Madagascar, these early colonists largely had the island to themselves (as all dinosaurs, except those later evolving into birds, died out at the K-Pg boundary mass extinction) and, with a multitude of vacant niches and little competition, were able to diversify spectacularly, eventually evolving into species exploiting niches occupied elsewhere by other groups of mammals and even birds. For instance, Ring-tailed Lemur lives in a manner similar to some baboons; Giant Jumping Rat *Hypogeomys antimena* shows remarkable similarities in body form and lifestyle to rabbits and hares; Greater Long-tailed Shrew Tenrec *Microgale principula* looks and behaves like a small arboreal mouse; Fosa *Cryptoprocta ferox* in aspects of appearance and behaviour resembles a cat, whilst the bizarre Aye-aye *Daubentonia madagascariensis* extracts insect larvae from tree bark much as woodpeckers would do (a group of birds that is absent from the island).

The remaining major group, the bats (order Chiroptera) are the only truly volant mammals and are represented by 45 or so species on Madagascar (two previously listed species are no longer included in the island's bat fauna). Their capacity for flight enables them to colonise more distant shores with greater frequency and ease than any other terrestrial mammals. Nonetheless, this ability is still strongly influenced by aspects of their morphology and stochastic factors like weather (particularly wind direction). Recent phylogenetic studies have advanced our understanding of Madagascar's bats considerably and it is now believed they probably represent between 28 and 30 independent colonisations (Goodman et al. in press), with some originating in Asia and the majority from Africa.

There are some peculiarities in the extinct mammal fauna that warrant discussion. Subfossil evidence shows that in the recent past (Holocene) three species of dwarf hippopotamus (Hippopotamus madagascariensis, H. lemerlei and H. laloumena) and two 'aardvark-like' mammals (Plesiorycteropus madagascariensis and P. germainepetterae) were present on Madagascar (Goodman & Jungers 2014). The hippopotamuses, all three species of which were smaller than today's African Hippopotamus Hippopotamus amphibius, were probably relatively recent arrivals (Late Pleistocene): being semi-aquatic their ancestor probably inadvertently floated/swam to Madagascar under freak circumstances, perhaps having been washed out from a river estuary after a storm (Ali & Vences 2019). It is suspected they became extinct around 1,000 years ago as a consequence of human persecution, although some evidence suggests they might have survived as recently as 200 years ago (Godfrey 1986, Burney & Ramilisonina 1999).

The origins of the 'aardvark-like' mammals are unclear. They are placed in their own order Bibymalagasia (MacPhee 1994); however, recent investigation points to them being related to tenrecs and hence incorporated into the order Tenrecoidea (Buckley 2013). Consequently, their origins could be aligned with the original ancestor of other tenrecs. Further supporting evidence is necessary to corroborate this theory (Goodman & Soarimalala in press).

In its pristine state, prior to human arrival (perhaps 2,000–3,000 years ago), the majority of Madagascar was probably covered in forest of one type or another (there is debate and now changing perspective as to whether grasslands are natural or human-induced formations – see Regions and Habitats of Madagascar), consequently the majority of vertebrate species, especially mammals, evolved in forests and are largely forest dependent. The extent and types of forests across the island have undoubtedly altered over the millennia, as climate has changed (across glacial and interglacial periods), but the pace of these changes would have been relatively slow and species would have been able to adapt gradually. Some species would have benefited and their populations expanded, others would have suffered and their populations contracted. For instance, we know from subfossil evidence that the Greater Bamboo Lemur Prolemur simus was once widespread across western, north-western, central and eastern regions (Godfrey & Vuillaume-Randriamananatena 1986), so presumably these areas were once covered by forests containing large-culmed bamboos on which the species depends. Similarly, subfossil remains point to the range of Giant Jumping Rat, which today is incredibly restricted, extending into the Central Highlands and the far south and south-west (Goodman & Jungers 2014).

After the arrival of humans, the pace of change increased, as forests were cleared and larger mammals hunted. Over the last 200 years, and especially the past 70 years, forest clearance has reached unprecedented levels, resulting in an extraordinary pace of change. Forest-dependent mammals have been unable to adapt and their ranges have contracted and populations plummeted, in many instances to now dangerously low levels. For example, among the island's flagship mammals, the lemurs, some 72% of species (80 of 108), are now listed in the two most severe categories of threat (Endangered and Critically Endangered) by the IUCN.

Ironically, as populations dwindle alarmingly, our knowledge of the islands' mammals is burgeoning. The past quarter of a century in particular has seen huge advancements across a panoply of diciplines for all major taxonomic groups. Yet there are still many significant gaps and it is to be hoped the array of conservation initiatives across the island are successful, so that the endemic mammals on Madagascar will continue to survive and flourish well into the future, to further advance our knowledge of this fantastical assemblage.

Taxonomy

With the advent of molecular genetic techniques to elucidate evolutionary relationships and the application of the Phylogenetic Species Concept, the past 20 years has seen a remarkable increase in the number of mammal species described from the island (see The Species Conundrum). This approach is not without controversy and polarises opinion among evolutionary biologists and taxonomists. Consequently, the number of species recognised in each of the major mammal groupings may differ depending on the authority consulted. Given that much of this research is ongoing and molecular techniques are constantly being refined, it seems reasonable to presume that further revisions will be proposed in the future. Therefore, the taxonomy followed here is a synthesis of previous and recent publications.

Within the Tenrecidae 31 species are recognised, divided between three subfamilies: Tenrecinae, Geogalinae and Oryzorictinae, following Everson *et al.* (2016) and outlined in Goodman & Soarimalala (in press).

Classification of the bats follows Goodman & Soarimalala (in press): as such 45 species are currently recognised, with two species previously listed from the island – Madagascar Slit-faced Bat *Nycteris madagascariensis* and Reunion House Bat *Scotophilus* cf. *borbonicus* – no longer regarded as valid.

Herein 108 species of lemurs (Lemuriformes and Chiromyiformes) are recognised, their classification following Wilson & Mittermeier (2013), with additional species described by Hotaling *et al.* (2016) and Schüßler *et al.* (2020).

The arrangement of the carnivorans is based on Veron *et al.* (2017), who recognised three subfamilies, namely Cryptoproctinae, Euplerinae and Galidiinae, within the endemic family Eupleridae. However, eight species are outlined including Grandidier's Vontsira *Galidictis grandidieri*, which Veron *et al.* (2017), regard as a subspecies of Broad-striped Vontsira *G. fasciata*.

Following Jansa & Carleton (in press), the native rodents are placed in the endemic subfamily Nesomyinae, within the African family Nesomyidae (pouched rats, climbing mice and fat mice).

The Species Conundrum

What is a species? It's a simple question and intuitively we might assume one with a straightforward answer. We can all look at a Tiger (*Panthera tigris*), a Ring-tailed Lemur, a European Robin (*Erithacus rubecula*) or a Pedunculate Oak (*Quercus robur*) and recognise them for what they are. A Tiger is a very distinctive big cat with stripes; a Ring-tailed Lemur is a partially terrestrial primate with an instantly recognisable tail. How could they be mistaken for anything else? Tigers are clearly different from Lions, but they obviously share similarities too. Both have a suite of features or traits that together add up to being 'a big cat'. Equally we can recognise that smaller cats like Ocelots, European Wild Cats and domestic tabbies, all share similar traits that make them 'cats' and if size is factored out, Tigers and Lions are also very similar 'variations on the same theme'. We instinctively recognise all cats as being cats, which alludes to a shared evolutionary history, and as such they are grouped together in a taxonomic family, the Felidae.

The science of grouping organisms into hierarchical and related categories is called taxonomy. Defining higher-level categories like the mammals (Class Mammalia, from Latin *mamma* for breast) is relatively straightforward. Mammals all share four characteristics: they are unique in having mammary glands that in females produce milk, they possess a neocortex (a particular region of the brain), have fur or hair, and also posses three middle ear bones (ossicles). Slightly lower-tier groupings, e.g., families, like the cats, require longer lists of more specific characteristics to pinpoint them: members of the family Felidae all share 16 defining characteristics. The lowest or fundamental unit of taxonomy is species, which could be thought of as the basic units into which groups of individuals are 'packaged' (Tattersall 2007). A species is denoted in a formal system of binomial nomenclature that was originally conceived by Carl Linnaeus (first edition *Systema Naturæ*, 1735). The name of any species, which is customarily written in italics, is composed of the genus name, which is capitalised, followed by its species name, which is not capitalised. For example, the Tiger is *Panthera tigris*, while the Ring-tailed Lemur is *Lemur catta*.

When Linnaeus devised his system, the over-arching view (even among scientists) was that all living things were created (by God), unchanging and immutable. This was more than 120 years before Darwin published his theory of evolution in *On the Origin of Species* in 1859, so species *could* be placed in 'hard-edged' boxes. Of course, we now know that species change over time and evolve from one form to another. However, evolution usually takes place over vast periods, so changes within human timeframes are often imperceptible and species appear unchanging from one generation to the next.

All animals have traditionally been classified and organised into hierarchical groups; phylum, class, order, family, genus, species, based on their anatomical, morphological and behavioural characteristics. This approach helped identify the evolutionary lineages that define the majority of mammalian orders and families that are recognised today, and has resulted in the description of thousands of living and extinct species (there are around 6,500 described species of mammal living today).

When we look at a species, effectively what we are seeing is the end of a line of individuals that extends backwards into evolutionary history, in a continuous unbroken chain, one generation at a time. The individuals in each generation are conduits through which genetic material (genes) pass from one generation to the next. Samples of individuals three or five or even 20 generations back, would show little or no discernible genetic differences, but a sample of individuals taken say 200,000 generations ago (equating to *c*.5 million years for humans) would potentially be very different genetically (and possibly physically) and may, therefore, be classified as belonging to a different species.

So how is a species defined? Defining a species is like trying to place something with vague fuzzy diffuse edges into a hard-edged box. And yet it is crucial that we do so in some workable sense, as 'species' are seen as the fundamental units in taxonomy (and the majority of other disciplines within the biological sciences), and for distinguishing different animal and plant types. There have been at least 25 attempts to come up with a workable 'watertight' definition (Frankham *et al.* 2012). All have their faults and 'break down' at some point when attempting to apply them across all types of living organisms.

Traditionally, vertebrate species have been identified using the Biological Species Concept (BSC), in which species are considered to be groups of interbreeding natural populations that are reproductively isolated from other groups. Put simply, males and females of a species can mate and produce viable (fertile) offspring, but are unable to produce viable offspring if mated with a different (but similar) species. Until the late 20th century, the vast majority of vertebrate species were classified and described by a combination of their reproductive isolation from other species, and their shared anatomical, morphological and behavioural characteristics.

The development of molecular genetic techniques in the 1980s proved revolutionary and offered new and novel approaches for species identification. These techniques enable estimation of the genetic variation within populations of a species. They also permit assessment of how closely related (genetically) two species are, or, put another way, how different they are, and when populations diverged to the extent that one species began evolving into another.

Molecular genetic techniques resulted in a shift in the way many biologists identify species, with a change in focus to the genetic and evolutionary relatedness of organisms (called phylogenetics). In this approach, a species is defined as an irreducible group (cluster) of organisms, which have a shared evolutionary history, and are descended from a common ancestor: such a cluster is genetically distinct from other clusters.

Molecular genetics have significantly changed our understanding of evolutionary histories and the relationships of Madagascar's mammals. For example, they confirm the Malagasy carnivorans (family Eupleridae) evolved from a single common ancestor that arrived on the island between 18 and 24 mya (Yoder *et al.* 2003, Poux *et al.* 2005), and indicate that the Aye-aye, whose ancestry had previously been the subject of much debate and speculation, is part of the same evolutionary lineage as other lemurs, and that all can be traced back to a single colonising ancestor (Yoder *et al.* 1996, Yoder 1997, 2003). Indeed, this approach has also confirmed a common within-Madagascar ancestry for the tenrecs and the Malagasy rodents too (Poux *et al.* 2005, Everson *et al.* 2016, Jansa & Carleton in press).

While molecular techniques have certainly helped answer important evolutionary questions, they have also generated controversies, particularly regarding the description of new species. This has been especially true in Madagascar. Since the turn of the 21st century, more than 70 new mammalian species have been described from the island, including no fewer than 45 lemur species. These are largely the result of phylogenetic analysis, in which small genetic differences between populations have been used as the basis for describing a new species.

Two lemur groups in particular have contributed to this spectacular increase in the number of species. In the mid 1990s only two species of mouse lemur (*Microcebus* spp.) were recognised, the Brown Mouse Lemur *M. rufus* in eastern humid forests and Grey Mouse Lemur *M. murinus* from dry western and southern forests. Today there are 25 *Microcebus* species. Similarly, the sportive lemurs (genus *Lepilemur*) comprised seven species in the 1990s (Harcourt & Thornback 1990), but today the genus numbers 26 species. In addition, the numbers of described dwarf lemurs (*Cheirogaleus* spp.) and woolly lemurs (*Avahi* spp.) have both risen from two to nine species since the 1990s. In the vast majority of cases these apparently new species do not represent previously unknown populations that are clearly and discernibly different, but rather are subpopulations that have been shown to be different at the genetic level and consequently are given distinct species status. For example, Ganzhorn's Mouse Lemur *Microcebus ganzhorni* and Bemanasy Mouse Lemur *M. manitatra* are subpopulations previously identified within the widely distributed Grey Mouse Lemur (Hotaling *et al.* 2016). If individuals of each of the 25 mouse lemur species and each of the 26 sportive lemur species were to sit side by side on a branch it would be virtually impossible to tell them all apart.

It is perhaps not coincidental that the proliferation of species identification is among nocturnal rather than diurnal groups. Why should this be? One possible explanation is that diurnal species show diversity in visually discernible external traits as a mechanism for mate recognition. For example, the sifakas (genus *Propithecus*) have distinct colour patterns that differ across species with separate (allopatric) geographic distributions. The same is not true for nocturnal species, as visually discernible traits are of little value in the dark. If visual traits are the sole means of describing species, diversity will remain hidden. In contrast, genetic techniques allow diversity to become apparent for groups that essentially look the same. For example, *Microcebus*, *Cheirogaleus*, *Lepilemur* and *Avahi* are regarded as being morphologically cryptic genera, with species very similar / near identical in appearance, but each evolutionarily distinct (Yoder in press).

The adoption of the phylogenetic approach to species identification, the Phylogenetic Species Concept (PSC), has not been universally accepted and its validity is questioned (Chikhi & Sgarlata in press). Critics point out that in many instances, only small numbers of individuals were sampled to ascertain differences and given that every species contains a degree of variation and diversity, small measured differences may simply be a reflection of variation within a species. Furthermore, they point out that the genetic differences between individuals from newly described species are in some cases smaller than differences between individuals within previously defined species. Consequently, it has been argued that genetic differences are insufficient to warrant new species descriptions and genetic

data alone should not be used to delineate species boundaries. Some consider that the resulting 'taxonomic inflation' is invalid and about half of the currently identified species of lemurs should be merged with previously described species (Tattersall 2007, 2013, Markolf *et al.* 2011). Taking such advice on board, a more cautious rationale is beginning to be adopted, with some biologists advocating a more 'objective' integrated approach that combines genetic data with morphological, ecological, and geographic data (Zimmermann & Radespiel 2014, Chikhi & Sgarlata in press). Consequently, the number of recognised species, especially within nocturnal lemur groups, is likely to be re-evaluated in the near future.

Not only have new species been described in the past 20 years, but also many previously recognised subspecies have been elevated to full species, which has further boosted the total. A subspecies is generally recognised as a geographically localised variation within a more broadly distributed species. For instance, in the 1990s the Diademed Sifaka *Propithecus diadema* was divided into four subspecies, the nominate form *P. d. diadema*, and three subspecies, Milne-Edwards's Sifaka *P. d. edwardsi*, Silky Sifaka *P. d. candidus* and Perrier's Sifaka *P. d. perrieri*. Similarly, four subspecies were recognised within Verreaux's Sifaka *P. verreauxi*, the nominate form *P. v. verreauxi*, together with *P. v. coquereli*, *P. v. deckeni* and *P. v. coronatus*. All of these subspecies are now regarded as separate species (Mittermeier *et al.* 2006). The same is true of the brown lemurs that were once considered to be six subspecies of *Eulemur fulvus*. Consequently, today very few subspecies are recognised, as virtually all have been upgraded to full species.

There are other considerations where defining species is concerned. Conservationists generally recognise the species as the basic unit for conservation: by conserving the maximum number of species, the maximum amount of genetic diversity is by default also conserved. Conservation requires funding, which may be easier to raise if a 'new' species is seen as in peril in a remote forest. Would it be perceived with the same level of importance if it were just an outlying population of a more common and broadly distributed species? Given the critical need to preserve every worthwhile patch of remaining native Malagasy forest, this approach may perhaps be justifiable.

However, genetic analysis is not a one-way street. In some instances, scrutiny of genetic data has not led

The taxonomic status of Grandidier's Vontsira (Galidictis grandidieri) is contentious. Some authorities e.g. Veron et al. 2017, regard it as a subspecies of Broad-striped Vontsira, Galidictis fasciata grandidieri.

to an increase in the number of recognised species, but the opposite. In the past it has been suggested that the largest of the lemurs, the Indri *Indri indri* should be divided into at least two subspecies on the basis of coat coloration differences, but genetic analysis has shown there is no justification for such separation (Brenneman *et al.* 2016). Also, a more detailed look at genetic and demographic data suggests Mittermeier's Mouse Lemur *Microcebus mittermeieri* and Goodman's Mouse Lemur *M. lehilahytsara* represent divergent populations of the same species (Poelstra *et al.* 2020; Schüßler *et al.* 2020).

The taxonomy of the Malagasy carnivorans has long been debated. Although it is now clear they comprise a group originating from a single colonising ancestor (Yoder *et al.* 2003), the number of species within the group remains contentious. Genetic analysis suggests the recognition of Durrell's Vontsira *Salanoia durrelli* is invalid and it is simply a variant of Brown-tailed Vontsira *S. concolor* (Veron *et al.* 2017). Similarly, genetic data do not support

the recognition of two species of Falanouc (genus *Eupleres*) and they are now seen as the same species (Veron & Goodman 2018). Moreover, there appear to be low levels of genetic diversity between Broad-striped Vontsira *Galidictis fasciata* and Grandidier's Vontsira *G. grandidieri*, with the suggestion that these taxa are better defined as subspecies (Veron *et al.* 2017), although in this book they are still included as full species.

Ultimately, when dealing with a concept that cannot be firmly defined, there can be no absolute correct and incorrect, no right and wrong. There is simply the grey area of informed opinion that sits in between. The debate between those who prefer to define species by more holistic means and amalgamate populations in a broader sense (the lumpers), and those who prefer to express even the smallest perceivable variation with the recognition of a new species (the splitters) will undoubtedly continue, and the number of recognised species will wax and wane.

Classification of Extant Native Mammals of Madagascar

IUCN Red List criteria
NE = Not Evaluated
DD = Data Deficient
LC = Least Concern

NT = Near Threatened
VU = Vulnerable
EN = Endangered
CR = Critically Endangered

Order Afrosoricida
  Family Tenrecidae (Tenrecs and Shrew Tenrecs)
  Subfamily Tenrecinae (Spiny Tenrecs)
    Tailless Tenrec *Tenrec ecaudatus* (Schreber, 1778) LC
    Greater Hedgehog Tenrec *Setifer setosus* Martin, 1838 LC
    Lesser Hedgehog Tenrec *Echinops telfairi* (Schreber, 1778) LC
    Lowland Streaked Tenrec *Hemicentetes semispinosus* (G. Cuvier, 1798) LC
    Highland Streaked Tenrec *Hemicentetes nigriceps* Günther, 1875 LC
  Subfamily Geogalinae (Large-eared Tenrec)
    Large-eared Tenrec *Geogale aurita* Milne-Edwards & A. Grandidier, 1872 LC
  Subfamily Oryzorictinae (Furred Tenrecs)
    Mole-like Mole Tenrec *Oryzorictes hova* (A. Grandidier, 1870) LC
    Four-toed Mole Tenrec *Oryzorictes tetradactylus* Milne-Edwards & A. Grandidier, 1882 DD
    Dobson's Shrew Tenrec *Nesogale dobsoni* (Thomas, 1884) LC
    Talazac's Shrew Tenrec *Nesogale talazaci* (Major, 1896) LC
    Web-footed Tenrec *Microgale mergulus* (Major, 1896) VU
    Mountain Shrew Tenrec *Microgale monticola* Goodman & Jenkins, 1998 VU
    Dryad Shrew Tenrec *Microgale dryas* Jenkins, 1992 VU
    Short-tailed Shrew Tenrec *Microgale brevicaudata* G. Grandidier, 1899 LC
    Pygmy Shrew Tenrec *Microgale parvula* G. Grandidier, 1934 LC
    Lesser Long-tailed Shrew Tenrec *Microgale longicaudata* Thomas, 1882 LC
    Drouhard's Shrew Tenrec *Microgale drouhardi* G. Grandidier, 1934 LC
    Taiva Shrew Tenrec *Microgale taiva* Major, 1896 LC
    Pale Shrew Tenrec *Microgale fotsifotsy* Jenkins, Raxworthy & Nussbaum, 1997 LC
    Greater Long-tailed Shrew Tenrec *Microgale principula* Thomas, 1926 LC
    Cowan's Shrew Tenrec *Microgale cowani* Thomas, 1882 LC
    Naked-nosed Shrew Tenrec *Microgale gymnorhyncha* Jenkins, Raxworthy & Nussbaum, 1996 LC
    Shrew-toothed Shrew Tenrec *Microgale soricoides* Jenkins, 1993 LC
    Major's Shrew Tenrec *Microgale majori* Thomas, 1918 LC
    Gracile Shrew Tenrec *Microgale gracilis* (Major, 1896) LC
    Thomas's Shrew Tenrec *Microgale thomasi* Major, 1896 LC
    Grandidier's Shrew Tenrec *Microgale grandidieri* Olson, *et al.*, 2009 LC
    Least Shrew Tenrec *Microgale pusilla* Major, 1896 LC
    Dark Shrew Tenrec *Microgale jobihely* Goodman *et al.*, 2006 EN
    Nasolo's Shrew Tenrec *Microgale nasoloi* Jenkins & Goodman, 1999 VU
    Jenkins's Shrew Tenrec *Microgale jenkinsae* Goodman & Soarimalala, 2004 EN

Order Chiroptera
Suborder Pteropodiformes (Yinpterochiroptera)
  Family Pteropodidae (Old World Fruit Bats)
    Madagascar Flying Fox *Pteropus rufus* É. Geoffroy Saint-Hilaire, 1803 VU
    Madagascar Straw-coloured Fruit Bat *Eidolon dupreanum* (Pollen in Schlegal & Pollen, 1867) VU
    Madagascar Rousette *Rousettus madagascariensis* G. Grandidier, 1929 VU
  Family Hipposideridae (Old World Leaf-nosed Bats)
    Commerson's Leaf-nosed Bat *Macronycteris commersoni* (É. Geoffroy Saint-Hilaire, 1813) NT
    Madagascar Cryptic Leaf-nosed Bat *Macronycteris cryptovalorona* Goodman *et al.*, 2016 NE
  Family Rhinonycteridae (Trident Bats)
    Rufous Trident Bat *Triaenops menamena* Goodman & Ranivo, 2009 LC

Grandidier's Trident Bat *Paratriaenops auritus* (G. Grandidier, 1912) VU
Trouessart's Trident Bat *Paratriaenops furculus* (Trouessart, 1906) LC
Suborder Vespertilioniformes (Yangochiroptera)
Family Emballonuridae (Sheath-tailed Bats)
Western Sheath-tailed Bat *Paremballonura tiavato* (Goodman *et al.*, 2006) LC
Peters's Sheath-tailed Bat *Paremballonura atrata* (Peters, 1874) LC
Madagascar Sheath-tailed Bat *Coleura kibomalandy* Goodman *et al.*, 2012 DD
Mauritian Tomb Bat *Taphozous mauritianus* É. Geoffroy Saint-Hilaire, 1818 LC
Family Nycteridae (Slit-faced Bats)
Madagascar Slit-faced Bat *Nycteris madagascariensis** G. Grandidier, 1937 DD
Family Myzopodidae (Sucker-footed Bats)
Eastern Sucker-footed Bat *Myzopoda aurita* Milne-Edwards & A. Grandidier, 1878 LC
Western Sucker-footed Bat *Myzopoda schleimanni* Goodman *et al.*, 2007 LC
Family Molossidae (Free-tailed Bats)
Peters's Little Mastiff Bat *Mormopterus jugularis* (Peters in P. L. Sclater, 1865) LC
Malagasy Eastern Free-tailed Bat *Chaerephon atsinanana* Goodman *et al.*, 2010 LC
Malagasy Western Free-tailed Bat *Chaerephon jobimena* Goodman & Cardiff, 2004 LC
Grandidier's Lesser Free-tailed Bat *Chaerephon leucogaster* (A. Grandidier, 1869) LC
Malagasy Large White-bellied Free-tailed Bat *Mops leucostigma* (G.M. Allen, 1918) LC
Midas Free-tailed Bat *Mops midas* (Sundevall, 1843) LC
Malagasy Large-eared Free-tailed Bat *Otomops madagascariensis* Dorst, 1953 LC
Malagasy Large Free-tailed Bat *Tadarida fulminans* (Thomas, 1903) LC
Family Vespertilionidae (Vesper Bats)
Malagasy Mouse-eared Bat *Myotis goudoti* (A. Smith, 1834) LC
Dark Madagascar Pipistrelle *Neoromicia bemainty* (Goodman *et al.*, 2015) LC
Isalo Serotine *Laephotis malagasyensis* (Peterson, Eger & Mitchell, 1995) VU
Madagascar Serotine *Laephotis matroka* (Thomas & Schwann, 1905) LC
Roberts's Serotine *Laephotis robertsi* (Goodman *et al.*, 2012) DD
Dusky Pipistrelle *Pipistrellus hesperidus* (Temminck, 1940) LC
Racey's Pipistrelle *Pipistrellus raceyi* Bates *et al.*, 2006 DD
Rüppell's Pipistrelle *Vansonia rueppellii* (Fischer 1829) LC
Malagasy Western House Bat *Scotophilus tandrefana* Goodman *et al.*, 2005 DD
Marovaza House Bat *Scotophilus marovaza* Goodman *et al.*, 2006 LC
Malagasy Large House Bat *Scotophilus robustus* Milne-Edwards, 1881 LC
Reunion House Bat *Scotophilus* cf. *borbonicus*** (É. Geoffroy, 1803) DD
Family Miniopteridae (Long-fingered Bats)
Aellen's Long-fingered Bat *Miniopterus aelleni* Goodman *et al.*, 2009 LC
Griveaud's Long-fingered Bat *Miniopterus griveaudi* Harrison, 1959 DD
Glen's Long-fingered Bat *Miniopterus gleni* Peterson, Eger & Mitchell, 1995 LC
Short-tragus Long-fingered Bat *Miniopterus brachytragos* Goodman *et al.*, 2009 LC
Malagasy Northern Long-fingered Bat *Miniopterus ambohitrensis* Goodman *et al.*, 2015 LC
Major's Long-fingered Bat *Miniopterus majori* Thomas, 1906 LC
Eger's Long-fingered Bat *Miniopterus egeri* Goodman *et al.*, 2011 LC
Manavi Long-fingered Bat *Miniopterus manavi* Thomas, 1906 LC
Small Sister Long-fingered Bat *Miniopterus sororculus* Goodman *et al.*, 2007 LC
Peterson's Long-fingered Bat *Miniopterus petersoni* Goodman *et al.*, 2008 DD
Mahafaly Long-fingered Bat *Miniopterus mahafaliensis* Goodman *et al.*, 2009 LC
Griffiths's Long-fingered Bat *Miniopterus griffithsi* Goodman *et al.*, 2010 DD

Order Primates
Infraorder Lemuriformes
Family Cheirogaleidae (Mouse, Giant Mouse, Dwarf and Fork-marked Lemurs)
Grey Mouse Lemur *Microcebus murinus* (J. F. Miller, 1777) LC
Grey-brown Mouse Lemur *Microcebus griseorufus* Kollman, 1910 LC
Madame Berthe's Mouse Lemur *Microcebus berthae* Rasoloarison *et al.*, 2000 CR
Peters's Mouse Lemur *Microcebus myoxinus* Peters, 1852 VU
Golden-brown Mouse Lemur *Microcebus ravelobensis* Zimmermann *et al.*, 1997 VU
Bongolava Mouse Lemur *Microcebus bongolavensis* Olivieri *et al.*, 2007 EN
Ambarijeby Mouse Lemur *Microcebus danfossi* Olivieri *et al.*, 2007 VU

* Occurrence of species and family on Madagascar in question (Goodman *et al.* in press)
** Occurrence on Madagascar cannot be substantiated (Goodman *et al.* in press)

Antafondro Mouse Lemur *Microcebus margotmarshae* Louis *et al.*, 2008 EN
Sambirano Mouse Lemur *Microcebus sambiranensis* Rasoloarison *et al.*, 2000 EN
Nosy Be Mouse Lemur *Microcebus mamiratra* (Andriantompohavana *et al.*, 2006) EN
Montagne d'Ambre Mouse Lemur *Microcebus arnholdi* Louis *et al.* 2008 VU
Tavaratra Mouse Lemur *Microcebus tavaratra* Rasoloarison *et al.*, 2000 VU
Mittermeier's Mouse Lemur *Microcebus mittermeieri* Louis *et al.*, 2006 EN
Anjiahely Mouse Lemur *Microcebus macarthurii* Radespiel *et al.*, 2008 EN
Goodman's Mouse Lemur *Microcebus lehilahytsara* Roos & Kappeler, 2005 VU
Jonah's Mouse Lemur *Microcebus jonahi* Schüßler *et al.*, 2020 NE
Simmons's Mouse Lemur *Microcebus simmonsi* Louis *et al.*, 2006 EN
Boraha Mouse Lemur *Microcebus boraha* Hotaling *et al.*, 2016 DD
Gerp's Mouse Lemur *Microcebus gerpi* Radespiel *et al.*, 2012 CR
Marohita Mouse Lemur *Microcebus marohita* Rasoloarison *et al.*, 2013 CR
Brown Mouse Lemur *Microcebus rufus* (Lesson, 1840) VU
Jolly's Mouse Lemur *Microcebus jollyae* Louis *et al.*, 2006 EN
Anosy Mouse Lemur *Microcebus tanosi* Rasoloarison *et al.*, 2013 EN
Ganzhorn's Mouse Lemur *Microcebus ganzhorni* Hotaling *et al.*, 2016 EN
Bemanasy Mouse Lemur *Microcebus manitatra* Hotaling *et al.*, 2016 CR
Coquerel's Giant Mouse Lemur *Mirza coquereli* (A. Grandidier, 1867) EN
Northern Giant Mouse Lemur *Mirza zaza* Kappeler & Roos, in Kappeler *et al.*, 2005 VU
Hairy-eared Dwarf Lemur *Allocebus trichotis* (Günther, 1875) EN
Fat-tailed Dwarf Lemur *Cheirogaleus medius* É. Geoffroy Saint-Hilaire, 1812 VU
Montagne d'Ambre Dwarf Lemur *Cheirogaleus andysabini* Lei *et al.*, 2015 EN
Ankarana Dwarf Lemur *Cheirogaleus shethi* Frasier *et al.*, 2016 EN
Crossley's Dwarf Lemur *Cheirogaleus crossleyi* A. Grandidier, 1870 EN
Greater Dwarf Lemur *Cheirogaleus major* É. Geoffroy Saint-Hilaire, 1812 EN
Sibree's Dwarf Lemur *Cheirogaleus sibreei* (Forsyth Major, 1896) EN
Haute Matsiatra Dwarf Lemur *Cheirogaleus grovesi* McLain *et al.*, 2017 VU
Lavasoa Dwarf Lemur *Cheirogaleus lavasoensis* Thiele *et al.*, 2013 EN
Thomas's Dwarf Lemur *Cheirogaleus thomasi* (Forsyth Major, 1894) CR
Masoala Fork-marked Lemur *Phaner furcifer* (de Blainville, 1839) EN
Sambirano Fork-marked Lemur *Phaner parienti* Groves & Tattersall, 1991 EN
Pale Fork-marked Lemur *Phaner pallescens* Groves & Tattersall, 1991 EN
Northern Fork-marked Lemur *Phaner electromontis* Groves & Tattersall, 1991 EN

Family Lepilemuridae (Sportive Lemurs)
Weasel Sportive Lemur *Lepilemur mustelinus* I. Geoffroy Saint-Hilaire, 1851 VU
Betsileo Sportive Lemur *Lepilemur betsileo* Louis *et al.*, 2006 EN
Small-toothed Sportive Lemur *Lepilemur microdon* (Forsyth Major, 1894) EN
Manombo Sportive Lemur *Lepilemur jamesorum* Louis *et al.*, 2006 CR
Kalambatritra Sportive Lemur *Lepilemur wrightae* Louis *et al.*, 2006 EN
Andohahela Sportive Lemur *Lepilemur fleuretae* Louis *et al.*, 2006 EN
Mananara-Nord Sportive Lemur *Lepilemur hollandorum* Ramaromilanto *et al.*, 2009 CR
Seal's Sportive Lemur *Lepilemur seali* Louis *et al.*, 2006 VU
Masoala Sportive Lemur *Lepilemur scottorum* Lei *et al.*, 2008 EN
Daraina Sportive Lemur *Lepilemur milanoii* Louis *et al.*, 2006 EN
Ankarana Sportive Lemur *Lepilemur ankaranensis* Rumpler & Albignac, 1975 EN
Sahafary Sportive Lemur *Lepilemur septentrionalis* Rumpler & Albignac, 1975 CR
Grey-backed Sportive Lemur *Lepilemur dorsalis* J.E. Gray, 1871) EN
Nosy Be Sportive Lemur *Lepilemur tymerlachsoni* Louis *et al.*,2006 CR
Mittermeier's Sportive Lemur *Lepilemur mittermeieri* Rabarivola *et al.*, 2006 CR
Sahamalaza Sportive Lemur *Lepilemur sahamalaza* Andriaholinirina *et al.*, 2017 CR
Anjiamangirana Sportive Lemur *Lepilemur grewcockorum* Louis *et al.*, 2006 CR
Ambodimahabibo Sportive Lemur *Lepilemur otto* Craul *et al.*, 2007 EN
Milne-Edwards's Sportive Lemur *Lepilemur edwardsi* (Forsyth Major, 1894) EN
Antafia Sportive Lemur *Lepilemur aeeclis* Andriaholinirina *et al.*, 2017 EN
Tsiombikibo Sportive Lemur *Lepilemur ahmansoni* Louis *et al.* 2006 CR
Bemaraha Sportive Lemur *Lepilemur randrianasoloi* Andriaholinirina *et al.*, 2017 EN
Red-tailed Sportive Lemur *Lepilemur ruficaudatus* A. Grandidier, 1867 CR
Zombitse Sportive Lemur *Lepilemur hubbardorum* Louis *et al.*, 2006 EN
Petter's Sportive Lemur *Lepilemur petteri* Louis *et al.*, 2006 EN
White-footed Sportive Lemur *Lepilemur leucopus* (Forsyth Major, 1894) EN

Family Lemuridae ('True' Lemurs)
Eastern Grey Bamboo Lemur *Hapalemur griseus* (Link, 1795) VU
Northern Grey Bamboo Lemur *Hapalemur occidentalis* Rumpler, 1975 VU
Southern Grey Bamboo Lemur *Hapalemur meridionalis* Warter, et al., 1987 VU
Lac Alaotra Reed Lemur *Hapalemur alaotrensis* Rumpler, 1975 CR
Golden Bamboo Lemur *Hapalemur aureus* Meier, et al. 1987 CR
Greater Bamboo Lemur *Prolemur simus* (J.E. Gray, 1871) CR
Ring-tailed Lemur *Lemur catta* Linnaeus, 1758 EN
Brown Lemur *Eulemur fulvus* (É. Geoffroy Saint-Hilaire, 1796) VU
Sanford's Brown Lemur *Eulemur sanfordi* (Archbold, 1932) EN
White-fronted Brown Lemur *Eulemur albifrons* (É. Geoffroy Saint-Hilaire, 1796) VU
Red-fronted Brown Lemur *Eulemur rufifrons* (Bennett, 1833) VU
Rufous Brown Lemur *Eulemur rufus* (Audebert, 1799) VU
White-collared Brown Lemur *Eulemur cinereiceps* (A. Grandidier & Milne-Edwards, 1890) CR
Red-collared Brown Lemur *Eulemur collaris* (É. Geoffroy Saint-Hilaire, 1812) EN
Mongoose Lemur *Eulemur mongoz* (Linnaeus, 1766) CR
Crowned Lemur *Eulemur coronatus* (J.E. Gray, 1842) EN
Red-bellied Lemur *Eulemur rubriventer* (I. Geoffroy Saint-Hilaire, 1850) VU
Black Lemur *Eulemur macaco* (Linnaeus, 1766) EN
Blue-eyed Black Lemur *Eulemur flavifrons* (J.E. Gray, 1867) CR
Black-and-white Ruffed Lemur *Varecia variegata* (Kerr, 1792) CR
Red Ruffed Lemur *Varecia rubra* (É. Geoffroy Saint-Hilaire, 1812) CR
Family Indriidae (Woolly Lemurs, Sifakas and Indri)
Eastern Woolly Lemur *Avahi laniger* (J.F. Gmelin, 1788) VU
Masoala Woolly Lemur *Avahi mooreorum* Lei et al., 2008 EN
Peyrieras's Woolly Lemur *Avahi peyrierasi* Zaramody et al., 2006 VU
Betsileo Woolly Lemur *Avahi betsileo* Andriantompohavana et al., 2007 EN
Manombo Woolly Lemur *Avahi ramanantsoavanai* Zaramody et al., 2006 VU
Southern Woolly Lemur *Avahi meridionalis* Zaramody et al., 2006 EN
Western Woolly Lemur *Avahi occidentalis* (Lorenz von Liburnau, 1898) VU
Bemaraha Woolly Lemur *Avahi cleesei* Thalmann & Geissmann, 2005 CR
Sambirano Woolly Lemur *Avahi unicolor* Thalmann & Geissmann, 2000 CR
Diademed Sifaka *Propithecus diadema* Bennett, 1832 CR
Milne-Edwards's Sifaka *Propithecus edwardsi* A. Grandidier, 1871 EN
Silky Sifaka *Propithecus candidus* A. Grandidier, 1871 CR
Perrier's Sifaka *Propithecus perrieri* Lavauden, 1931 CR
Verreaux's Sifaka *Propithecus verreauxi* A. Grandidier, 1867 CR
Coquerel's Sifaka *Propithecus coquereli* (A. Grandidier, 1867) CR
Decken's Sifaka *Propithecus deckeni* Peters, 1870 CR
Crowned Sifaka *Propithecus coronatus* Milne-Edwards, 1871 CR
Golden-crowned Sifaka *Propithecus tattersalli* Simons, 1988 CR
Indri *Indri indri* (J.F. Gmelin, 1788)

Infraorder Chiromyiformes
Family Daubentoniidae (Aye-aye)
Aye-aye *Daubentonia madagascariensis* (J.F. Gmelin, 1788) EN

Order Carnivora
Family Eupleridae (Malagasy Carnivorans)
Subfamily Euplerinae
Fosa *Cryptoprocta ferox* Bennett, 1833 VU
Spotted Fanaloka *Fossa fossana* (Müller, 1776) VU
Falanouc *Eupleres goudotii* Doyère, 1835 VU
Subfamily Galidinae
Ring-tailed Vontsira *Galidia elegans* I. Geoffroy Saint-Hilaire, 1837 LC
Brown-tailed Vontsira *Salanoia concolor* (I. Geoffroy Saint-Hilaire, 1837) VU
Broad-striped Vontsira *Galidictis fasciata* (J.F. Gmelin, 1788) VU
Grandidier's Vontsira *Galidictis grandidieri* Wozencraft, 1986 EN
Narrow-striped Boky *Mungotictis decemlineata* (A. Grandidier, 1867) EN

Order Rodentia
Family Nesomyidae (Pouched Rats, Climbing Mice and Fat Mice)
Subfamily Nesomyinae (Malagasy Rodents)

White-tailed Tree Rat *Brachytarsomys albicauda* Günther, 1875 LC
Hairy-tailed Tree Rat *Brachytarsomys villosus* Petter, 1962 VU
Sleek-furred Ground Rat *Gymnuromys roberti* Forsyth Major, 1896 LC
Tsingy Tufted-tailed Rat *Eliurus antsingy* Carleton et al., 2001 DD
Carleton's Tufted-tailed Rat *Eliurus carletoni* Goodman et al., 2009 LC
Daniel's Tufted-tailed Rat *Eliurus danieli* Carleton & Goodman, 2007 LC
Ellerman's Tufted-tailed Rat *Eliurus ellermani* Carleton, 1994 DD
Grandidier's Tufted-tailed Rat *Eliurus grandidieri* Carleton & Goodman, 1998 LC
Major's Tufted-tailed Rat *Eliurus majori* Thomas, 1895 LC
Lesser Tufted-tailed Rat *Eliurus minor* Forsyth Major, 1896 LC
Western Tufted-tailed Rat *Eliurus myoxinus* Milne-Edwards, 1885 LC
White-tailed Tufted-tailed Rat *Eliurus penicillatus* Thomas, 1908 EN
Petter's Tufted-tailed Rat *Eliurus petteri* Carleton, 1994 EN
Tanala Tufted-tailed Rat *Eliurus tanala* Forsyth Major, 1896 LC
Rock-loving Tufted-tailed Rat *Eliurus tsingimbato* Jansa et al., 2019 NE
Webb's Tufted-tailed Rat *Eliurus webbi* Ellerman, 1949 LC
Eastern White-tailed Mountain Mouse *Voalavo antsahabensis* Goodman et al., 2005 EN
Northern White-tailed Mountain Mouse *Voalavo gymnocaudus* Carleton & Goodman, 1998 LC
Western Big-footed Mouse *Macrotarsomys bastardi* Milne-Edwards & G. Grandier, 1898 LC
Long-tailed Big-footed Mouse *Macrotarsomys ingens* Petter, 1959 EN
Petter's Big-footed Mouse *Macrotarsomys petteri* Goodman & Soarimalala, 2005 DD
Malagasy Mountain-dwelling Mouse *Monticolomys koopmani* Carleton & Goodman, 1996 LC
Giant Jumping Rat *Hypogeomys antimena* A. Grandidier, 1869 EN
Gregarious Short-tailed Rat *Brachyuromys ramirohitra* Forsyth Major, 1896 LC
Betsileo Short-tailed Rat *Brachyuromys betsileoensis* (Bartlett, 1880) LC
Eastern Red Forest Rat *Nesomys rufus* Peters, 1870 LC
Lowland Red Forest Rat *Nesomys audeberti* (Jentink, 1879) LC
Western Red Forest Rat *Nesomys lambertoni* G. Grandidier, 1928 EN

Non-Native Feral or Free-ranging Species

Order Rodentia
Subfamily Murinae (Old World Rats and Mice)

Black Rat *Rattus rattus* Linnaeus, 1758
Brown Rat (Sewer Rat) *Rattus norvegicus* Berkenhout, 1769
House Mouse *Mus musculus* Linnaeus, 1758

Order Carnivora – 3 species
Family Viverridae (Civets, Genets and Oyans)
Subfamily Viverrinae (Terrestrial Civets)

Small Indian Civet *Viverricula indica* (É. Geoffroy Saint-Hilaire, 1803)

Family Felidae (Cats)
Subfamily Felinae (Small Cats)

Feral Cat (Wild Cat) *Felis silvestris* Schreber, 1777

Family Canidae (Dogs)
Subfamily Caninae (True Dogs)

Domestic/Feral Dog *Canis familiaris* Linneaus, 1758

Order Eulipotyphla
Family Soricidae (Shrews)

Asian Musk Shrew *Suncus murinus* (Linnaeus, 1766)
Etruscan Shrew *Suncus etruscus* (Savi, 1822)

Order Artiodactyla
Family Suidae (Old World Pigs)

Bush Pig *Potamochoerus larvatus* (F. Cuvier, 1822)

Tenrecs and Shrew Tenrecs

Order Afrosoricida

The order Afrosoricida (a Latin-Greek compound name meaning 'looking like African shrews') includes three families of insectivorous small mammals: Potamogalidae (the otter-shrews restricted to equatorial Africa), Chrysochloridae (the golden moles confined to southern Africa), and Tenrecidae (the tenrecs and shrew tenrecs, which are naturally restricted to Madagascar). In previous taxonomic arrangements, these families were included in the orders Insectivora and later Lipotyphla, neither of which is now recognised.

Tenrecs Family Tenrecidae

Tenrecs and shrew tenrecs comprise the family Tenrecidae, which is endemic to Madagascar. Currently 31 species are recognised, subdivided between three subfamilies: Tenrecinae, Geogalinae and Oryzorictinae (Goodman & Soarimalala in press). Previous taxonomy also aligned African otter-shrews (Potamogalinae) within this grouping (Olson & Goodman 2003), but recent molecular investigations have concluded that a close evolutionary relationship is incorrect, and the African genera (Potamogale and Micropotamogale) are now placed in their own family Potamogalidae (Everson *et al.* 2016).

Tenrecs evolved from a single common ancestor that arrived on Madagascar between 30 and 55 mya (Poux *et al.* 2005, Everson *et al.* 2016). Subsequently, around the end of the Oligocene (23 mya), a massive adaptive radiation took place, such that forms diversified to fill numerous small-mammal niches occupied elsewhere today by several non-related families. Tenrec eco-morphological adaptations include parallels with semi-fossorial hedgehogs (genera *Tenrec*, *Setifer* and *Hemicentetes*), fossorial moles (genus *Oryzorictes*), semi-aquatic desmans and otter-shrews (Web-footed Tenrec *Microgale mergulus*), true shrews (genus *Microgale*), scansorial species (genus *Echinops*) and even small arboreal mice (Lesser Long-tailed Shrew Tenrec *Microgale longicaudata*, Greater Long-tailed Shrew Tenrec *M. principula* and Major's Shrew Tenrec *M. majori*) (Soarimalala & Goodman 2003). Speculation surrounds the probable causal factors driving such a spectacular radiation. The timing appears to coincide with the arrival of the first rodent and carnivoran colonists on Madagascar and may be linked to numerous other factors like climate change, increased predation (by carnivorans) and growing competition from the recently arrived rodents (Jenkins 2018).

Equally remarkable is their diversity within habitats. Humid forests of the east and highlands support the greatest diversity, with six genera and 26 species, although some of these species also occur in western dry regions. Western regions support 14 species, whilst the arid south-west is the least diverse region with eight species (Jenkins 2018). At some humid eastern sites six genera occur in the same area of forest, including up to 11 species of shrew tenrecs (*Nesogale* spp. and *Microgale* spp.) (Goodman *et al.* 2000b, Jenkins 2018). Tenrec diversity also varies with altitude, different species having different elevational preferences: mid-elevation (1,200–1,500m) rainforests support the greatest diversity. Four species appear to be highland specialists, preferring habitats above 1,200m – Highland Streaked Tenrec *Hemicentetes nigriceps*, Four-toed Mole Tenrec *Oryzorictes tetradactylus*, Least Shrew Tenrec *Microgale pusilla* and Dark Shrew Tenrec *Microgale jobihely* (Jenkins 2018).

Greater Hedgehog Tenrec, Andasibe-Mantadia National Park.

Tenrecs are herterothermic and have some of the lowest (generally less than 35°C) and most variable body temperatures in mammals, their body temperature patterns often being reptile-like (Dausmann *et al.* in press). Their basal metabolic rates are amongst the lowest seen in eutherian mammals (Stephenson & Racey 1995). They generally maintain a constant body temperature only when pregnant or lactacting; at other times body temperature is labile (Dausmann *et al.* in press). In addition, several species are known to hibernate in response to unfavourable dry and cold seasons (Levesque *et al.* 2014, Dausmann *et al.* 2020). For example, Lesser Hedgehog Tenrec has one of the lowest recorded body temperatures (around 31°C) and, when not reproductively active, enters daily torpor irrespective of season or ambient temperature. In the warm season (austral summer), this species rewarms passively from torpor tracking daily fluctuations in ambient temperature, thus decreasing its energy expenditure. It also synchronises activity phases and torpor bouts among individuals (Dausmann *et al.* 2020).

Scent marking is particularly important for tenrecs: the cloacal region is rubbed against the substrate (perineal dragging) (Jenkins 2018). This not only occurs when individuals are exploring their environment but also during encounters between individuals and courtship. Tenrecs also appear to favour latrine sites, depositing urine and faeces at specific locations. In some species (genera *Tenrec*, *Setifer* and *Hemicentetes*) these are often close to burrow entrances.

As a group, tenrecs are quite vocal and emit a variety of grunts, squeaks, hisses, squeals, chirps and tongue clicks. Tongue clicks in three species that have been studied more closely (Lowland Streaked Tenrec *Hemicentetes semispinosus*, Tailless Tenrec *Tenrec ecaudatus* and Dobson's Shrew Tenrec *Nesogale dobsoni*) are believed to relate to echolocation (Jenkins 2018).

Some of the widely distributed tenrec species are perhaps the most numerous native mammals on the island; nonetheless on the whole the group faces significant threats, such as habitat loss through swidden agriculture, hunting and incidental capture in fishing traps (Stephenson *et al.* 2019). Longer term, climate change will also alter tenrec habitats and is likely to adversely affect populations. Six species with restricted ranges are already regarded as threatened: four are Vulnerable (Dryad Shrew Tenrec *Microgale dryas*, Web-footed Tenrec *M. mergulus*, Mountain Shrew Tenrec *M. monticola* and Nasolo's Shrew Tenrec *M. nasoloi*) and two are Endangered (Jenkins's Shrew Tenrec *M. jenkinsae* and Dark Shrew Tenrec).

Tenrecs are regarded as 'game' species under Malagasy law. Between May and October, they are widely hunted across all regions of the island (Reuter *et al.* 2016c). Unsurprisingly, the larger species, Tailless Tenrec *Tenrec ecaudatus*, Greater Hedgehog Tenrec *Setifer setosus* and Lesser Hedgehog Tenrec *Echinops telfairi* are particularly targeted. They are generally dug from their burrows while aestivating, caught in traps or hunted with dogs (Garcia & Goodman 2003, Golden 2009). They are hunted for subsistence, and are particularly important for rural and lower income communities (Reuter *et al.* 2016b), but are also widely traded and sold to restaurants and urban markets in considerable quantities (Jenkins *et al.* 2011, Reuter & Sewall 2016). In some areas, smaller tenrec species are not eaten as they are protected by local taboos (*fady*), which can relate to them appearing more 'mouse-like', as rodents are not widely consumed (Reuter & Sewall 2016).

Larger tenrec species, like Tailless Tenrec are widely hunted, especially in more remote rural areas where they constitute a major source of protein.

Spiny Tenrecs Subfamily Tenrecinae

Four genera comprise the 'spiny tenrecs' (subfamily Tenrecinae), *Tenrec*, *Setifer*, *Echinops* and *Hemicentetes*. All are more or less hedgehog-like in appearance, having variably spiny pelages, but have considerably shorter (or absent) tails than true hedgehogs (Erinaceidae). In the wild, they range in size from c.40g to 1,200g (captive individuals are often heavier). They are widely distributed across the various forest types of the island and are primarily terrestrial, although one species, Lesser Hedgehog Tenrec, does exhibit semi-arboreal (scansorial) characteristics. Except streaked tenrecs (*Hemicentetes* spp.), they are mainly solitary although they do sometimes hibernate in small groups.

The evolution of spines (modified hair) is an obvious anti-predator adaptation: the tendency towards 'becoming spiny' probably evolved as tenrecs moved from forests into more open environments and became larger in size, and were therefore potentially more vulnerable to predators (Stankowich & Stensrud 2019).

Adult Highland and Lowland Streaked Tenrecs and juvenile Tailless Tenrecs possess a stridulating organ: an area of robust hollow quills on the back that are rubbed together to produce sound (sometimes audible, sometimes high-frequency ultrasound). Beneath this area is a disc of muscle that enables fast vibration of the spines. These sounds are detectable by conspecifics up to 10m away. With streaked tenrecs (*Hemicentetes* spp.) that live in family units or groups, this may help maintain contact between foraging animals (Jenkins 2018). It serves a similar purpose in juvenile Tailless Tenrecs that live in large litter groups, but is lost when they become adult and solitary.

TAILLESS TENREC

Tenrec ecaudatus (Schreber, 1778)

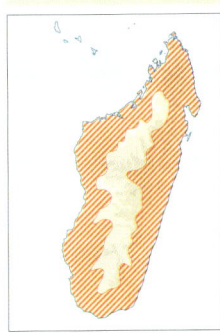

Other name: Common Tenrec

MEASUREMENTS
Head-body: 180–350mm. Tail: 10–15mm. Weight: 600–1,200g, in captivity up to 1,500g; males larger than females. (Nicoll 2003, 2009, Jenkins 2018)

DESCRIPTION & IDENTIFICATION Adults rotund with shortish limbs, hindlimbs longer than forelimbs. No visible external tail. Pelage sparse: a mixture of coarse hairs, softer bristles and bristly spines, the latter more apparent on crown and nape, which forms a mane and crest (which is erectile). The spines are two-toned: mostly pale with a dark brown to blackish band that imparts an overall grizzled appearance. Basic colour creamy-brown to grey-brown, with reddish-brown areas. Males appear paler. Face and venter may be light tan.

Snout long and pointed with conspicuous long dark vibrissae. Gape wide (span up to 10cm), supporting well-developed musculature and canines long (up to 15mm); a snapping defensive bite inflicts deep wounds (Nicoll 2003, 2009, Goodman *et al.* in press a). Overall appearance reminiscent of tailless versions of Caribbean solenodons (Solenodontidae) or North American Opossum *Didelphis virginiana*.

Infants and juveniles much darker, sometimes almost blackish-brown, with paler cream spines arranged in five longitudinal rows, giving overall streaked appearance (Nicoll 1982). Coloration and pattern very similar to that of adult streaked tenrecs and is likely cryptic, but may also serve as a warning to potential predators (Jenkins 2018). In juveniles alone, a broad band of spines midway down the back forms the stridulating organ. The spines, streaks and stridulating organ are lost at the first moult (*c*.35–70 days), replaced by the adult pelage (Nicoll 2009).

HABITAT & DISTRIBUTION Perhaps the most widely occurring native non-volant mammal (Goodman *et al.* in press). Occurs in all native forest regions, including arid south-west, from sea level to montane

Adult Tailless Tenrec, Andasibe-Mantadia National Park.

regions (c.1,700m); absent from high mountain zones (Nicoll 2003, Goodman et al. in press). Also a range of modified habitats, agricultural fields, secondary woodland, grassland and urban environments, providing there is sufficient cover (Ganzhorn et al. 1990, Nicoll 2009, Jenkins 2018). Prefers environs with brush/undergrowth and free-standing water. Introduced to Comoros, Seychelles, Réunion and Mascarene Islands (during 19th century), as food for plantation workers (Nicoll 2003).

An important human food resource and hunted extensively across the island, not only for immediate consumption but also traded to restaurants in major urban areas (Nicoll 2003, 2009, Reuter & Sewall 2016). In some regions, Tailless Tenrec provides a significant nutritional contribution to subsistence-level families (Golden et al. 2014), with the result that populations in some areas have declined alarmingly or even disappeared, e.g., Masoala Peninsula (Goodman et al. in press a). IUCN Least Concern.

BEHAVIOUR Adults mainly nocturnal, activity peaking at 18.00–21.00 and 01.00–05.00 h, with different sex and age classes sometimes partially diurnal (Nicoll 1982, 1983). Spends day in natural cavity under tree roots, excavated burrow or fallen hollow log. Individuals shift sleep sites over a given period, and a single site may be utilised by different individuals separately (Nicoll 2009). In dry forests (Kirindy) average female home range (three individuals) c.3ha, with much overlap (Randriamahady 2012). Considerable seasonal variation in home range, correlated to food availability and reproductive state; range can become up to three times larger (Goodman et al. in press a).

Adapted for some digging and burrowing: forages in topsoil and leaf litter, vigorously scraping with forelimbs and 'rooting' with long, powerful snout to break up soil fragments. Omnivorous, feeding on insects, especially beetle larvae and termites (Ade 1996, Soarimalala & Goodman 2003), also other invertebrates, fruits and even small vertebrates (Nicoll 1982). Prey detected by combination of smell, sound and touch, the last via long, sensitive vibrissae swept side to side (Nicoll 2003). Fruit particularly important towards end of wet season (February– April) when they fatten up for hibernation. At onset of hibernation fat may constitute 50% of body mass (Nicoll 2009). Appears to utilise latrine sites: at a specific place they dig a small hole, defecate and urinate, wipe their rear on the ground, then kick back soil to cover their deposits (Jenkins 2018).

Has better eyesight than other tenrecs but relies on well-developed tactile and olfactory senses. When alarmed erects a mane on upper back (longer, coarser hair than the rest of body) and holds mouth wide open to expose teeth; simultaneously may emit a low hissing sound in conjunction with squeaks and squeals.

Body temperature very variable, fluctuating with ambient temperature, and enters bouts of daily torpor. Hibernates in austral winter (May–October), due to unfavourable climate and decreased food availability (Nicoll 1985), in excavated deep (1–2m) narrow burrow, plugged with soil (Nicoll 2003), either singly or possibly in groups. Breathing rate drops to 30 breaths per minute, body temperature tracks soil temperature, and can fall to 11°C (cf. in summer months when active body temperature c.35°C). When body temperature drops below 20°C during hibernation, individuals may suspend breathing for up to 45 minutes (Treat et al. 2019). Sleeps continuously with no periodic arousal (exceptional among hibernating animals) (Lovegrove et al. 2014b); entirely reliant on fat reserves during this period (Nicoll 1982, 1986, Treat et al. 2019). In austral spring males emerge from torpor on average one month before females. Hibernation lasts 5–6 months, but in extreme environments of south-west may extend to nine months (Lovegrove et al. 2014a, Goodman et al. in press a).

Solitary for most of year: male/female associations are brief (1–3 days) and occur during mating season in austral spring (late September– early November). Receptive females reduce nightly ranges (from c.1ha to <0.1ha), minimising movement and remaining close to sleep site. Thought to concentrate her scent, increasing likelihood of location by male (Nicoll 1982). Several males may converge and fight for access to female. Larger males tend to dominate. Births occur in wet season (December–February), peaking in January in east and slightly later in west, when invertebrate food base is at peak (Nicoll 2001, 2009). Gestation period 56–64 days (Eisenberg & Gould 1970).

Reproduction is remarkable and prolific. Maximum litter size is 32 (maximum embryo count 34); in captivity 31 young have been reared (Eisenberg & Gould 1984). Wild litters smaller and vary with age of mother and habitat: larger litters associated with more seasonally variable habitats; in the dry western area 25 is average, in humid eastern regions 20 is more normal (Nicoll 1983, 2003, Nicoll & Racey 1985, Goodman et al. in press a).

Female Tailless Tenrec, with young. Note broadly similar appearance of young to streaked tenrecs (Hemicentetes sp.).

Mothers feed young from 17 pairs of nipples (the most of any mammal) (Goodman *et al.* 2003). Young are altricial: spines become visible at *c*.7 days, eyes open between nine and 14 days. Emerge from nest at 18–20 days and forage in a group with the mother, remaining together for up to 55 days. From around 35 days juveniles begin foraging in smaller more diffuse groups, sometimes without mother, until they become independent (Goodman *et al.* in press a).

Demand for milk is high: females draw on all their fat reserves during lactation (Nicoll 2003). Mothers and young regularly extend foraging beyond darkness into daylight hours, when striped coloration of young increases camouflage and reduces predation. Adult females with young darker than males for same reason. Young stridulate standing on hindlegs, with crest and dorsal spines erect. These spines rub together to produce pulse of sound that functions as communication within family units (Jenkins 2018).

Moult to adult coloration begins between 35 and 70 days and there is a shift to nocturnal activity. Infant mortality is high: rarely do more than ten offspring survive. Fully grown by three to four months (Nicoll 2003).

Predators include Fosa, especially in wet season when Tailless Tenrec is more active (Hawkins 2003), large raptors e.g., Madagascar Harrier-Hawk *Polyboroides radiatus* and Madagascar Buzzard *Buteo brachypterus*, and large boas (*Sanzinia* and *Acrantophis* sp.) (Goodman & Langrand 1996b, Karpanty & Goodman 1999).

WHERE TO SEE Potentially visible in most reserves in west or east during warmer, wetter months (November–April), particularly Andasibe-Mantadia, Montagne d'Ambre, Kirindy and Ankarafantsika. Also encountered at night in and around towns and villages.

GREATER HEDGEHOG TENREC
Setifer setosus   (Schreber, 1778)

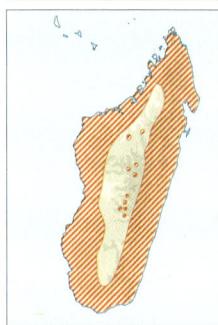

MEASUREMENTS
Head-body: 105–240mm.
Tail: 9–17mm. Weight:
110–350g, mean 280g.
(Eisenberg & Gould 1970,
Goodman *et al.* 2003,
Jenkins 2018)

DESCRIPTION & IDENTIFICATION Body rotund with short legs. Overall grey-brown in colour. Completely covered in stiff non-detachable spines, except face and underparts, which are covered in paler sparse hair. Spines creamy-white, dark grey-brown towards tip, with a white point. Darker melanistic forms reported. Snout moderately long and pointed with many vibrissae (Eisenberg & Gould 1970). Appearance is of a small long-bodied hedgehog; striking similarity to true hedgehogs (Erinaceidae) an example of convergent evolution.

HABITAT & DISTRIBUTION Found throughout various native forests types in east, west, north and south, up to 2,250m; range overlaps with Lesser Hedgehog Tenrec in south and south-west (Goodman *et al.* 2013, Stephenson *et al.* 2016k). Not associated with wetland and marsh areas. Also recorded in degraded and fragmented forests, agricultural and cultivated zones, and urban areas, including in Central Highlands (Jenkins 2018). IUCN Least Concern.

BEHAVIOUR Nocturnal and omnivorous. Forages by snuffling on ground in leaf litter or similar loose substrate, but also has limited climbing capabilities. Prey primarily detected by scent; diet includes insects, grubs, worms, other invertebrates and fruit (Soarimalala & Goodman 2003). Also scavenges around refuse in urban areas (Eisenberg & Gould 1970).

Excavates short tunnels underneath logs or tree roots ending in leaf-lined nest chamber where it sleeps curled with head tucked under body. Also climbs to nest in low-level tree holes (Levesque *et al.* 2012). There is a latrine site close to burrow entrance. Active throughout year, but individuals in more seasonal areas (dry regions of north, west and south and higher elevation locations) do reduce levels of activity and become torpid for short periods during the austral dry season (Jenkins 2018).

Generally solitary: male home ranges (*c.*14ha) are larger than those of females (*c.*7ha), and are non-exclusive with considerable overlap with several individuals of the opposite sex, perhaps suggesting a promiscuous mating system (Levesque *et al.* 2012). Males and females may form temporary associations during breeding season; they communicate via a

Greater Hedgehog Tenrec, Ranomafana National Park.

series of grunts, squeaks and chirps. Mating occurs late September to mid October (Eisenberg & Gould 1970). Nest-building activity increases the week prior to birth. Nest usually a closed tree hole 2m or more off ground. Female uses this for 20–25 days, towards the end of which young are already foraging with her. After this, female changes nest sites daily (Levesque *et al.* 2012). Litters 1–4 (average three) (Eisenberg 1975). Gestation varies with ambient conditions, from 51 to 69 days; higher temperatures result in a shorter gestation (Eisenberg 1975). Young closely tended for first two weeks; eyes begin to open at nine days and are fully open at 14 days, then begin to take solid food and accompany mother outside burrow on short foraging excursions. Females may still be lactating 30–40 days after giving birth, by which time they are pregnant again. In dry forest areas, females can be at some stage of reproduction for the entire warm, wet period (October–April) (Levesque *et al.* 2012).

Similar to true hedgehogs, anti-predator response is to curl into a tight ball, with spines outermost and no vulnerable parts exposed. Such behaviour successfully deters smaller carnivorans but is often ultimately unsuccessful against Fosa (Hawkins 2003) and large snakes, e.g., boas (*Sanzinia* and *Acrantophis* spp.).

WHERE TO SEE May be encountered on night walks in many of the popular reserves and forest areas, e.g., Montagne d'Ambre, Masoala, Andasibe-Mantadia, Ranomafana, Ankarafantsika, Daraina, Nosy Mangabe, Ankarana and Kirindy Forest.

LESSER HEDGEHOG TENREC
Echinops telfairi   (Martin, 1838)

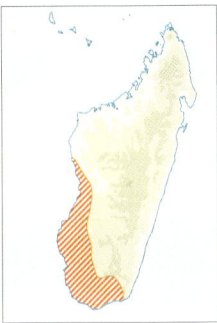

MEASUREMENTS
Head-body: 104–165mm. Tail: 10–13mm. Weight: 50–230g, mean 140g. (Eisenberg & Gould 1970, Goodman *et al.* 2003, Jenkins 2018)

DESCRIPTION & IDENTIFICATION Body rotund (more so than Greater Hedgehog Tenrec), almost entirely covered with dense sharp spines (absent on

Lesser Hedgehog Tenrec, Tsimanampetsotse National Park.

face and belly). Coloration variable, from very pale grey to deep slate-grey. Face and underparts sparsely furred and paler, sometimes almost white. Ears relatively large (compared to Greater Hedgehog Tenrec) and rounded. Legs short, forefeet, hindfeet and digits relatively long and claws short but deeply curved. These, along with other anatomical modifications, are adaptations to climbing habits (Jenkins 2018). Similar to Greater Hedgehog Tenrec, but smaller, with less pointed snout.

HABITAT & DISTRIBUTION Confined to dry forests of west and xerophytic spiny forest and gallery forest in south and south-west, from sea level to 1,350m (Goodman *et al.* 2013, Stephenson *et al.* 2016a). Additionally recorded from western dry portion of Andohahela region in south-east (Goodman *et al.* 1999) and from grasslands, disturbed and non-native habitats far from intact forest (Stephenson *et al.* 2016a). Range overlaps that of Greater Hedgehog Tenrec. IUCN Least Concern.

BEHAVIOUR Nocturnal and arboreal, but terrestrial at times (Jenkins 2018). Surprisingly agile and capable of climbing thin branches when short tail is used as a brace). All four feet have long digits and curved claws for gripping bark and branches. When climbing hindfeet are more important, ankles extremely flexible and can be reversed (Jenkins 2018). Forages alone in dense shrubbery, both on ground and branches. Feeds on insects, larvae and fruits (Eisenberg & Gould 1970). Uses sharp claws to excavate insects from cracks in bark and similar crevices.

Lesser Hedgehog Tenrecs are surprisingly agile climbers, Ifaty Forest.

During very dry season, when food is scarce, may become torpid for 3–5 months (May–September) (Godfrey & Oliver 1978, Nicoll & Thompson 1987). Hibernates, either singly or with two or three others, in hollow log or tree hole, several metres above ground: nest lined with grasses and leaves that are carried in the mouth and arranged to form a neat cup. Sleeps in a tightly curled ball. During the wet summer activity period Lesser Hedgehog Tenrec also enters a torpor on a daily basis for 12–18 hours: torpid period begins before midnight, then body temperature and metabolic rate gradually decline to a minimum around noon, followed by slow increase to a maximum when activity begins after dark. During austral summer ambient temperatures fluctuate between daytime highs (c.36°C) and nighttime lows (5°C), and correspondingly body temperature of Lesser Hedgehog Tenrec fluctuates over c.1–3°C (Lovegrove & Génin 2008, Dausmann *et al.* 2020).

Generally solitary. Males can be very aggressive towards each other, but males and females are more tolerant when together. Mating begins in October, shortly after emergence from torpor (Poppitt *et al.* 1994). Females build the natal nest a week before birth and continue to add to nest even after young born (Eisenberg & Gould 1970). Gestation period c.60–68 days, litter size one to six (Jenkins 2018). Young weigh less than 10g at birth and are naked. Open their eyes between seven and nine days and at ten days will follow mother to nest entrance. Female retrieves offspring that fall from nest. Take solid food from 14 days, weaned at c.3 weeks and begin foraging with mother at this time, continuing until c.4–5 weeks old (Jenkins 2018). Sexual maturity reached after first period of hibernation (Godfrey & Oliver 1978).

Like true hedgehogs (Erinaceidae), initial anti-predator response is to erect spines on crown and thrust head forward, sometimes mouth gaping, hissing and bucking. They then roll into a tight impenetrable ball, protecting the vulnerable parts (Eisenberg & Gould 1984).

WHERE TO SEE Berenty Reserve, Anjhampolo Spiny Forest, Parcels 2–3 of Andohahela National Park and spiny forests around Mangily and Ifaty (Ranobe-PK 32 Protected Area), north of Toliara, Bezà-Mahafaly Special Reserve and Tsimanampetsotse National Park south of Toliara are the best places to look for Lesser Hedgehog Tenrec.

LOWLAND STREAKED TENREC
Hemicentetes semispinosus   (G. Cuvier, 1798)

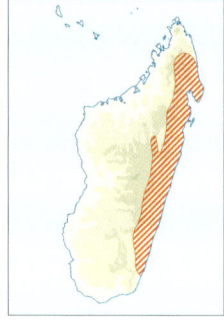

MEASUREMENTS
Head-body: 130–190mm.
Tail: none. Weight:
90–220g, mean 130g.
(Stephenson 1991, 2003b,
Goodman *et al.* 2003)
Head-body: 104–176mm.
Tail: none. Weight:
76–108g. (Jenkins 2018)

Adult Lowland Streaked Tenrec, Masoala National Park.

DESCRIPTION & IDENTIFICATION A medium-small, fairly slender tenrec with a pronounced pointed snout. Upperparts, nape and crown completely covered by spines/quills interspersed with sparse long guard hairs. Overall blackish-brown, with longitudinal creamy-yellow to golden-yellow stripes and golden crown. Quills barbed and detachable (except stridulating organ), prominent and longer on crown (nuchal crest). A narrowing yellowish band runs from crown over forehead and nose to near the tip of snout. Snout, long, bare and moderately mobile. Underparts are chestnut-brown to buff-brown, with softer hair and few spines. Forelimbs strong and well developed, obvious adaptations to digging. Broadly similar in appearance to juvenile Tailless Tenrec, but larger and spinier.

HABITAT & DISTRIBUTION Restricted to lowland, mid-altitude and occasionally higher-elevation rainforests of eastern region, from sea level to 1,550m (Goodman *et al*. 2000a,c) and even c.2,050m (Jenkins 2018). Also, outside native forests in degraded and agricultural areas, including anthropogenic grasslands, villages and urban zones (Eisenberg & Gould 1970, Stephenson *et al*. 2016d). IUCN Least Concern.

BEHAVIOUR Resemblance to immature Tailless Tenrecs not coincidental, as Lowland Streaked Tenrec is partially active during daylight hours, when its coloration and pattern provide a degree of crypsis (Stephenson 2003b, 2007). Semi-fossorial: forages among leaf litter, seeking earthworms, insect larvae, especially beetle larvae (Coleoptera) and other soft-bodied invertebrates. This more specialised diet correlates with a substantially reduced dentition. Peak foraging may be during daylight or evening and early morning (Jenkins 2018).

Potentially active year-round, but often less so during winter months (May–October) when able to reduce body temperature to within 2°C of ambient, to conserve energy yet remain active (Stephenson & Racey 1994). Also, may become torpid if prevailing conditions (temperature and food supply) are unsuitable. If temperatures remain higher and conditions are favourable able to breed year-round but more often young born during wet season (Stephenson 2003b, 2007, Stephenson and Racey 1994). Therefore, there is considerable variation in behaviour across this species' range.

Sleeps in excavated burrow, often near stream or waterbody, up to 150cm in length. This descends to a depth of 15cm and contains a nest chamber.

Female Lowland Streaked Tenrec with young, Andasibe-Mantadia National Park.

Entrance plugged with leaves, with a latrine site close by (Eisenberg & Gould 1970).

Gestation period 55-63 days (Stephenson 2003b, 2007); litter size varies between two and eleven in wild (average six) (sometimes larger in captivity). Young weigh c.8g at birth (Stephenson et al. 1994a) and develop quickly, reaching sexual maturity at 35–40 days. Able to reproduce in same season as their birth. Family groups may produce several litters per season and number >20 individuals from three related generations (Eisenberg & Gould 1970). These multi-generational family groups form perhaps the most complex social systems of any Afrotherian (Stephenson 2003b, 2007).

Forages in subgroups or singly. When together, communicate by vibrating quills of stridulating organ to produce low-frequency ultrasonic sounds (similar to dry grass being rubbed together) (Gould 1965, Wever & Herman 1968): important in keeping mother and offspring in contact with one another in undergrowth.

When threatened, erects crest of spines on crown (nuchal crest), bucks head violently and stamps its forefeet, attempting to embed detachable barbed quills into snout of would-be predator. May also emit a buzzing vocalisation. Predators include Fosa, Spotted Fanaloka, Ring-tailed Vontsira *Galidia elegans* and large snakes (Goodman et al. 1999b, Hawkins 2003).

WHERE TO SEE Can be seen year-round, but more readily during warmer wetter months (November–April). Potentially encountered in any eastern rainforest reserve or area, including grounds and gardens of tourist lodges, e.g., Andasibe-Mantadia, Anjozorobe-Angavo Forest Corridor, Ranomafana, Masoala, and Marojejy.

DESCRIPTION & IDENTIFICATION Similar in size, shape and coloration to Lowland Streaked Tenrec, but less spiny with more developed underfur and guard hairs giving 'woolly' or 'fluffy' appearance (Stephenson 2003b, 2007), presumably because of colder temperatures at higher elevations. Base colour near-black; streaks more whitish-cream than yellowish. Nuchal crest and nape well developed with long cream-coloured spines, but less 'robust' than Lowland Streaked Tenrec. Crown and forehead black with no median stripe (cf. Lowland Streaked Tenrec). Underparts creamy-white soft, bristly hair. Snout tip (rhinarium) bare and moderately mobile. Forelimbs well developed and adapted for digging.

HABITAT & DISTRIBUTION Restricted to montane and sclerophyllous forests, montane heath and adjacent habitats at higher elevations (1,200–2,350m) on southern half of eastern escarpment and central plateau. Also recorded around villages and small towns far from forest (Stephenson et al. 2016c). Generally, Lowland and Highland Streaked Tenrecs do not occur sympatrically. Where ranges overlap, e.g., Andringitra and Ivohibe Massifs, they occur at different elevations (Highland Streaked Tenrec above Lowland Streaked Tenrec) (Stephenson 2003b, 2007). However, there are records of both species living side by side (allopatry) in Tsinjoarivo Forest (Mahatsinjo) at 1,550m (Goodman et al. 2000b,c). IUCN Least Concern.

BEHAVIOUR Semi-fossorial and strictly nocturnal, with peak activity three or four hours after darkness. Marked daily rhythm: day spent in burrow, body temperature rises during late afternoon prior to peak activity early evening (Stephenson 2003b). Diet primarily consists of earthworms, beetle larvae

HIGHLAND STREAKED TENREC
Hemicentetes nigriceps   (Günther, 1875)

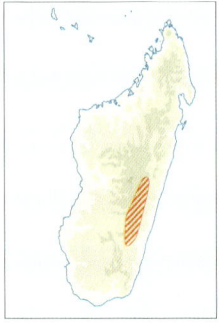

MEASUREMENTS
Head-body: 120–160mm. Tail: none. Weight: 70–160g, mean 100g. (Stephenson 1991, 2003b, 2007, Goodman et al. 2003) Head-body: 124–174mm. Tail: none. Weight: 39–129g. (Jenkins 2018)

Adult Highland Streaked Tenrec, Andringitra National Park.

Female Highland Streaked Tenrec with young, Andringitra National Park.

(Coleoptera) and other soft-bodied invertebrates, correlating to their reduced dentition.

Hibernation more marked than in Lowland Streaked Tenrec: enters profound torpor between May and October (Stephenson & Racey 1994). Torpor stimulated by reductions in ambient temperature, day length and food supply, but also influenced by endogenous rhythms. Breeds October–March. Gestation 55–58 days, litter size in wild one to five (Stephenson 2003b, Goodman *et al*. 2003, Jenkins 2018). Females become sexually mature after 35 to 40 days.

Erects nuchal crest when threatened and bucks head violently to deter approaching potential predators. These include Fosa, Ring-tailed Vontsira and large snakes (Goodman *et al*. 1997, Stephenson *et al*. 2016c). Other aspects of behaviour are presumed similar to Lowland Streaked Tenrec.

WHERE TO SEE Common in Andringitra Massif, south-west of Fianarantsoa. Night walks around Antanifotsy and Camp Belambo can be productive between November and April.

Large-eared Tenrec Subfamily Geogalinae

Currently a subfamily containing a single species, Large-eared Tenrec *Geogale aurita*. The evolutionary history of *Geogale* and consequently its taxonomy is contentious. Previously, some authorities have placed it within the Oryzorictinae; however, the most recent molecular research provides strong evidence to support subfamily designation (Everson *et al*. 2016). This research also suggests cryptic diversity within the genus and consequently further species may await description.

In common with other tenrecs, the Large-eared Tenrec exhibits a number of characteristics probably typical of the earliest eutherian (placental) mammals, including relatively low metabolic rate and body temperature, retention of a common uro-genital opening (cloaca), abdominal testes, and reliance on auditory and olfactory sensing (Stephenson 2002, 2003a).

LARGE-EARED TENREC
Geogale aurita
(Milne-Edwards & A. Grandidier, 1872)

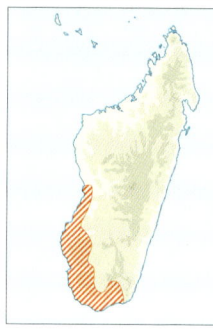

MEASUREMENTS
Head-body: 51–76mm.
Tail: 28–41mm. Weight:
5–9g, mean *c.*7g.
(Stephenson 1991, 1993,
Goodman *et al.* 1999, 2003,
Jenkins 2018)

DESCRIPTION & IDENTIFICATION Very small with large prominent rounded ears, a triangular head and pointed snout. Pelage soft, short but not dense. Upperparts pale grey to pale reddish-brown, underparts buffy-white to creamy-white, sometimes with yellowish-orange flecks on flanks. Tail 50% of head-body length, scaly and covered in fine hairs (Stephenson 1993). One of the smallest tenrecs: general appearance intermediate between that of a shrew and a field mouse.

HABITAT & DISTRIBUTION Found mainly in dry deciduous forests of west and spiny forests of south and south-west, together with gallery forests in this region, to elevations around 870m. Also recorded in anthropogenic grassland far from native forest (Stephenson *et al.* 2016b). Northern limit appears to be Tsiribihina River (Soarimalala 2011). Further

investigations in dry regions north of the latter may extend range. In extreme south-east occurs in spiny forests (Parcel 2) of Andohahela National Park (Goodman *et al.* 1999). There is a possible record of Large-eared Tenrec from humid forest (Jenkins 2018). IUCN Least Concern.

BEHAVIOUR Strictly nocturnal and able to enter daily torpor at all times of year, but more frequently and for longer periods during dry season (austral winter). Often sleeps inside fallen logs and crevices in rotting wood, usually alone, but occasionally sexes found in close proximity (Stephenson 1991, 2003a).

During foraging large ears are extended: sound is important in detecting prey. A variety of invertebrates eaten, with a marked preference for termites (Stephenson *et al.* 1994a, 2002). This species appears largely dependent on termite mounds, especially in grassland areas away from forests.

Large-eared Tenrec is one of the most extreme heterothermic tenrecs, its body temperature closely paralleling ambient temperature throughout the year. During pregnancy and lactation, female body temperatures increase (Stephenson & Racey 1993a). Compared to other similar-sized small mammals, including other tenrecs, the resting metabolic rate of Large-eared Tenrec is *c.*50% of the expected value (Stephenson & Racey 1993a,b). This may be linked to the termite-based diet (Stephenson & Racey 1992, 1995).

Mating occurs late September and March. Gestation ranges from 54 to 69 days, in some cases females become temporarily torpid during pregnancy and embryonic development is arrested

Large-eared Tenrec, Kirindy Forest.

(Stephenson 1993, Stephenson et al. 1994a). Births occur November–February. Litter size one to five (average three or four). Young highly altricial, weigh <1g and are born naked with eyes closed. Eyes open between 21 and 33 days. Weaning and independence after c.35 days (Stephenson 1993b).

Large-eared Tenrec is unique within Tenrecidae; females can conceive a second litter while suckling their first (post-partum oestrous) (Stephenson 1993). They can produce up to ten offspring per year (average 6–7). This strategy allows reproductive output to be maximised when favourable breeding conditions prevail (Stephenson 1993).

Predated by Barn Owl *Tyto alba* and Madagascar Long-eared Owl *Asio madagascariensis* (Goodman et al. 1993a,b) and probably also by snakes and small carnivorans (Stephenson 2003a).

WHERE TO SEE Most likely to be seen at Bezà-Mahafaly Reserve (where it appears to be particularly abundant), Kirindy Forest and Zombitse National Park.

Large-eared Tenrec, Bezà-Mahafaly Special Reserve.

Furred Tenrecs Subfamily Oryzorictinae

A subfamily comprising three diversified genera, *Oryzorictes*, *Nesogale* and *Microgale*, the latter having radiated spectacularly into 21 species. All are small to medium-sized, ranging from Pygmy Shrew Tenrec *Microgale parvula* (c.2–5g) to Web-footed Tenrec (c.60–110g).

A number of species provide good examples of convergent evolution as they demonstrate similar morphological and ecological traits to those found in unrelated groups elsewhere: *Oryzorictes* resemble true moles (Talpidae) and golden moles (Chrysochloridae), several *Microgale* species are reminiscent of true shrews (Soricidae) and Web-footed Tenrec exploits a similar niche to desmans (Talpidae) and otter-shrews (Potamogalidae).

MOLE TENRECS OR RICE TENRECS

GENUS *Oryzorictes*

Oryzorictes is a specialised and distinctive genus currently comprising two species, commonly referred to as mole tenrecs, because of their obvious resemblance to true moles (Talpidae), or 'rice tenrecs' due to their association with rice paddies. *Oryzorictes* is derived from Greek, meaning 'rice burrower'.

Four-toed Mole Tenrec appears confined to high-elevation zones above c.2,000m, whereas Mole-like Mole Tenrec is widespread throughout humid eastern regions from sea level to c.2,000m. Recent molecular work strongly suggests this is an over-simplification and three cryptic lineages exist within *O. hova*. With further investigation, new species may be described based on largely distinct north-eastern, central-eastern and south-eastern ranges (Everson et al. 2018, Olson & Soarimalala 2018, Everson et al. in press).

Characteristically rotund, mole tenrecs are easily distinguished from other furred tenrecs by their stockiness, stout and robust limbs, broad naked snout, small ears and tiny eyes. The pelage is very soft, dense and velvety, but coloration is highly variable, even within species (Goodman 2003c, Everson *et al.* in press). They are fossorial; their extremely strong forelimbs, broad spade-like forefeet and reduced eyes and ears are all clear adaptations for burrowing. Unlike true moles (Talpidae), mole tenrecs have longish tails, equivalent to *c.*50% head-body length. The two species are easily distinguished from one another by the number of digits on the forelimbs: Mole-like Mole Tenrec has five, Four-toed Mole Tenrec has four.

They are principally associated with eastern humid forest and marshy areas, but also occur in deforested areas converted to rice paddies (Goodman 2003c). They forage underground or in leaf litter and humus. Their diet consists primarily of worms (Annelida), insects, other soil-dwelling invertebrates, and perhaps some vegetable matter (Jenkins 2018). They appear to be mainly nocturnal but may be active at all times of day in their extensive burrow systems.

Known predators include Madagascar Red Owl *Tyto soumagnei* (Goodman & Thorstrom 1998) and some snakes, for example the fossorial colubrid *Pseudoxyrhopus* (Goodman *et al.* 1999).

Because of their lifestyle they are rarely seen. Most known specimens have been caught in forest floor pitfall traps at night, often after heavy rain when they appear more active (Goodman 2003c).

MOLE-LIKE MOLE TENREC
Oryzorictes hova　　　　　(A. Grandidier, 1870)

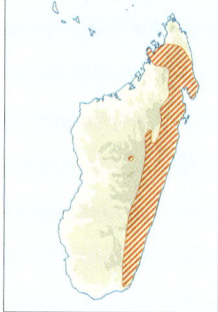

Other names: Mole-like Rice Tenrec, Hova Mole Tenrec

MEASUREMENTS
Head-body: 85–124mm. Tail: 38–62mm. Weight: 28–59g, mean *c.*35g. (Goodman 2003c, Goodman *et al.* 2003, Jenkins 2018, Everson *et al.* in press)

DESCRIPTION & IDENTIFICATION The fur is soft, dense and velvety, and highly variable in colour. Upperparts dark grey-brown and blackish-brown to light tan-brown, underparts dark brown to buffy-white. On Nosy Mangabe albinos have been recorded (Goodman 2003). Tail naked, pale in colour and relatively short (40–50% head/body length). Forefeet have five digits. *O. talpoides* is currently regarded as a junior synonym of *O. hova*, but ongoing investigations may result in its re-elevation to species (Everson *et al.* 2018).

Mole-like Mole Tenrec.

HABITAT & DISTRIBUTION A variety of environs from lowland rainforest to sclerophyllous forest approaching the treeline, from sea level to c.2,000m. Often found in wetter areas like valley bottoms, natural marshes and similar areas altered to rice paddies (Goodman 2003, Randriamoria 2017).

Range extends across most of eastern humid forest region, from Andohahela in south, to Masoala, Makira, Anjanaharibe-Sud, Marojejy in north, including island of Nosy Mangabe, and Sambirano region (Manongarivo) (Stephenson 1995a, Goodman & Soarimalala 2002, Goodman 2003c). Also recorded in forest fragments in Central Highlands (Ambohitantely) (Goodman & Rakotondravony 2000). IUCN Least Concern.

BEHAVIOUR Observed using muzzle to probe beneath leaf litter and humus when foraging (Eisenberg & Gould 1970). In captivity observed dragging earthworms below soil surface before consumption (Stephenson 1994b). Widely recognised by villagers tilling rice paddies, when disturbed from burrows. Burrow networks thought to damage crops by uprooting seedlings. Based on embryo counts the maximum litter size is four (Goodman 2003c).

Four-toed Mole Tenrec.

FOUR-TOED MOLE TENREC
Oryzorictes tetradactylus
(Milne-Edwards & A. Grandidier, 1882)

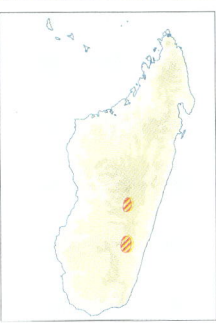

Other name: Four-toed Rice Tenrec

MEASUREMENTS
Head-body: 106–122mm. Tail: 42–57mm. Weight: 29.5–36g. (Goodman 2003c, Jenkins 2018, Everson *et al*. in press)

DESCRIPTION & IDENTIFICATION Pelage long, soft and dense. Mainly dark brown, with mottled areas of lighter brown on upperparts. Four digits on forelimbs and tail is short relative to head-body length (Goodman 2003c).

HABITAT & DISTRIBUTION Known from a few locations in high mountain areas at eastern edge of central highlands. Most are outside forest and above treeline, e.g., ericoid thicket at 2,050m on Plateau d'Andohariana in Andringitra National Park, and marshy areas at 2,450m near foot of Pic d'Imarivolanitra (formerly Pic Boby), Madagascar's second highest peak (2,658m) (Goodman *et al*. 1996c, Goodman & Rasolonandrasana 2001). IUCN Data Deficient.

BEHAVIOUR No specific information.

SHREW TENRECS
GENERA *Nesogale* and *Microgale*

As the vernacular suggests, shrew tenrecs show close visual similarity to true shrews (Soricidae). Currently 23 species are recognised, divided between two genera, *Nesogale* and *Microgale*. Their taxonomy has been refined several times (MacPhee 1987, Stephenson 1995b, Olson *et al*. 2004) and in recent years several new species have been described. Further, the inventory likely is incomplete: ongoing mitochondrial DNA analysis suggests many existing species may harbour currently unnamed cryptic diversity (Everson *et al*. 2020b, Jenkins *et al*. in press).

Recent molecular analysis has further clarified relationships in the group. Two species, *dobsoni* and *talazaci*, previously included in *Microgale* form a distinct evolutionary lineage (clade) and are now placed in the resurrected genus *Nesogale* (Everson *et al*. 2016). *Microgale* is the most speciose tenrec genus, with 21 extant species (and one extinct species, *M. macpheei* – see The Extinct Mammal Fauna), and shows remarkable morphological variation, ranging in size from Pygmy Shrew Tenrec *Microgale parvula* (*c*.2–5g) to the carnivorous Talzac's Shrew Tenrec *Nesogale talazaci* (*c*.35g) and the relatively large Web-footed Tenrec (*c*.60–110g) (Jenkins *et al*. in press). Further, some species have become partially arboreal: Lesser Long-tailed Shrew Tenrec, Greater Long-tailed Shrew Tenrec and Major's Shrew Tenrec have prehensile tails to help grip twigs and branches (Goodman *et al*. 2003).

Web-footed Tenrec is the largest species and based on its morphological adaptations to an aquatic lifestyle was previously placed in the monotypic genus *Limnogale* (Benstead & Olson 2003). Perhaps surprisingly, molecular analysis does not support this distinction, instead strongly suggesting that it is actually an aquatic *Microgale* (Everson *et al*. 2016).

Shrew tenrecs are perhaps the most problematic Malagasy mammals to specifically identify. Size, relative tail length and pelage colour are the only external characters of use, and considerable overlap exists between species. Often, accurate identification is possible only after detailed morphological examination that is well beyond the scope of all, bar experienced field scientists.

GENERAL DESCRIPTION & IDENTIFICATION Pelage soft and velvety. Head and upperparts buffy to olive-brown, dark russet-brown or slate grey-black. Ears prominent and project well above the fur. Generally, tail length is longer than head-body length, except for Short-tailed Shrew Tenrec *Microgale brevicaudata* which has a relatively short tail, and Lesser Long-tailed Shrew Tenrec and Greater Long-tailed Shrew Tenrec which have exceptionally long tails. Forelimbs not well developed, and all limbs have five digits.

HABITAT & DISTRIBUTION Prefer primary forests and generally associated with dense vegetation; some species have been found in disturbed, non-forested and agricultural areas and marshes (Jenkins 2003). Species richness greatest within eastern humid forests, where the majority of species occur exclusively (two *Nesogale*, 16 *Microgale*). Two species, Short-tailed Shrew Tenrec and Major's Shrew Tenrec have been recorded in both humid eastern regions and drier western ones. Dobson's Shrew Tenrec has been recorded in dry forests of extreme north (Montagne des Français) as well as in most eastern rainforest regions, and three further species, Nasolo's Shrew Tenrec, Grandidier's Shrew Tenrec *M. grandidieri* and Jenkins's Shrew Tenrec appear exclusive to dry regions of the west and south-west.

Five species – Pygmy Shrew Tenrec, Lesser Long-tailed Shrew Tenrec, Pale Shrew Tenrec *M. fotsifotsy*, Dobson's Shrew Tenrec and Talzac's Shrew Tenrec – occur over the entire length of the eastern rainforest belt, from Montagne d'Ambre in north to Andohahela in south (Raxworthy & Nussbaum 1994, Jenkins *et al*. 1997, Goodman *et al*. 1999), whilst numerous species' ranges cover the majority of rainforest regions from Andapa Basin (Marojejy and Anjanaharibe-Sud) in the north to the Anosyenne Mountains (Andohahela) in the south (Jenkins 2003, Jenkins *et al*. in press).

Shrew tenrecs occur in a variety of habitats – lowland and mid-altitude rainforest, high-altitude mossy sclerophyllous forest and montane ericoid scrub – over a very wide elevational range, from sea level to above *c*.2,500m (Jenkins 2003, 2018). Some species are found across a broad elevational range e.g., Talzac's Shrew Tenrec, Pygmy Shrew Tenrec and Cowan's Shrew Tenrec *M. cowani* all occur from lowland forests to high mountain habitats (Jenkins 2003, Jenkins *et al*. in press). Others are more specialised, e.g., Mountain Shrew Tenrec prefers montane habitats between 1,550m and 1,950m (Goodman & Jenkins 1998, 2000), although it has recently been recorded as low as 500m (Jenkins *et al*. in press). Throughout the eastern rainforests, middle elevations between 1,200m and 1,500m support the greatest species diversity (Stephenson 1995a, Jenkins 2003) and this may correspond to maximum invertebrate diversity at these altitudes (Goodman *et al*. 1996a,b).

In humid forest areas, as many as 11 shrew tenrecs (two *Nesogale*, nine *Microgale*) may occur sympatrically and these generally include Cowan's Shrew Tenrec, Naked-nosed Shrew Tenrec *M. gymnorhyncha*, Lesser Long-tailed Shrew Tenrec, Pygmy Shrew Tenrec, Pale Shrew Tenrec and Shrew-toothed Shrew tenrec *M. soricoides* (Jenkins 2003, 2018). Within these communities, there is some separation along elevational gradients and clear morphological differences account for further ecological niche separation (e.g., the food preferences of larger species are different to smaller species).

Nonetheless, accounting for niche separation for species very similar in size and morphology is challenging. How such fine ecological niche separation is achieved is not yet fully understood. Where communities are very diverse, the medium-sized species, like Cowan's Shrew Tenrec, Drouhard's Shrew Tenrec M. drouhardi and Taiva Shrew Tenrec M. taiva, are most common. Larger species like Dobson's Shrew Tenrec and Talzac's Shrew Tenrec appear relatively rare (Jenkins 2003).

BEHAVIOUR Thought to be solitary and mainly nocturnal or crepuscular. Most shrew tenrecs are primarily terrestrial, foraging among leaf litter, root tangles and fallen branches, but they may also exhibit limited scansorial capabilities and climb lianas, low shrubs and forest understorey (Jenkins 2003). Three species – Lesser Long-tailed Shrew Tenrec, Greater Long-tailed Shrew Tenrec and Major's Shrew Tenrec – have morphological adaptations (long prehensile tail, elongated hindfeet and digits) to climbing and are certainly scansorial and possibly partially arboreal (Jenkins 2018).

In rainforest regions births coincide with the onset of the rainy season (mainly late October to early February), so that juveniles can take advantage of the seasonal increase in invertebrates (Jenkins & Goodman 1999, Jenkins et al. in press). Based on embryo counts in wild-caught specimens, litter sizes range from one to four (Jenkins 2018).

Shrew tenrecs rely on their secretive habits and ability to conceal themselves to avoid predation. They are predated by several of the islands' native carnivorans including Fosa, Spotted Fanaloka, Ring-tailed Vontsira and Narrow-striped Boky Mungotictis decemlineata. In high mountain zones like Andringitra, Nesogale and Microgale species comprise a significant proportion of Fosa diet (Goodman et al. 1997b). Owls, e.g., Barn Owl and Madagascar Red Owl, and snakes are also known to prey on shrew tenrecs (Goodman et al. 1997a, Goodman & Thorstrom 1998). Also, there is a record of Scaly Ground Roller Geobiastes squamiger eating a large shrew tenrec (possibly Talzac's Shrew Tenrec) (Rakotoarisoa & Be 2004).

DIET The order of magnitude size differences across shrew tenrec species is reflected in a varied diet. Their diet is primarily invertebrate-based, with crickets and grasshoppers (Orthoptera), beetles (Coleoptera) and the larvae of both being predominant (Soarimalala & Goodman 2003).

Larger species like Talzac's Shrew Tenrec, Dobson's Shrew Tenrec and, to a lesser extent, Drouhard's Shrew Tenrec are carnivorous at times, eating small vertebrates such as frogs. There is also evidence (from captures in pitfall traps) that both Nesogale species and larger Microgale eat smaller Microgale species. How frequently such cannibalistic behaviour occurs in the wild is unclear (Jenkins 2018). Web-footed Tenrec has the most specialised diet, dominated by aquatic insect larvae, other aquatic invertebrates and some small aquatic vertebrates.

WHERE TO SEE Because of their small size and secretive lifestyle, shrew tenrecs are rarely observed in a 'casual' way. Chance encounters on night walks tend to be very fleeting.

DOBSON'S SHREW TENREC
Nesogale dobsoni   (Thomas, 1884)

MEASUREMENTS
Head-body: 79–130mm. Tail: 89–128mm. Weight: 20.5–47.5g (significant seasonal variation). (Jenkins et al. 1996, in press, Goodman & Jenkins 1998, 2000, Goodman et al. 1999, 2003, 2013)

DESCRIPTION & IDENTIFICATION Large, tail roughly equal or slightly longer than head-body length. Pelage quite long, upperparts brown to grey-brown, underparts grey with buff wash. Ears prominent. Tail has grey-brown upperside, paler buff below (Goodman & Jenkins 2000). Fat reserves stored in tail: becomes particularly thickset towards end of rainy season, gradually thinning as reserves diminish (Jenkins 2018).

HABITAT & DISTRIBUTION Widespread. Recorded in eastern rainforests and east-Central Highlands including forest edge and areas (Jenkins 2003, 2018), from Montagne des Français in far north to Andohahela in south (Goodman et al. 2013, Stephenson et al. 2016f). Wide elevational tolerance from sea level to 2,500m (above treeline) (Stephenson et al. 2016f). IUCN Least Concern.

Where sympatric with close relative Talzac's Shrew Tenrec, e.g., at Anjanaharibe-Sud and Marojejy

Dobson's Shrew Tenrec, Ankazumivady.

in north-east, Dobson's Shrew Tenrec prefers slightly lower elevations, but at times both species have overlapping home ranges (Goodman & Jenkins 1998, 2000). Tolerant of disturbance: found in non-forested areas and agricultural zones in central highlands.

BEHAVIOUR Forages on ground among leaf litter, detecting prey by sound. Some evidence for a basic echolocation system to help foraging in dark undergrowth (Eisenberg & Gould 1970). Also trapped on branches over 1m above ground, suggesting some scansorial activity (Jenkins 2003). Feeds on insects and larvae (especially Orthoptera); also carnivorous, eating frogs and worms. Eats other smaller shrew tenrecs (*Microgale* spp.) when in pitfall traps.

Solitary and shows some territorial behaviour: individuals normally spread out, evenly distributed and highly antagonistic towards each other. Males

Talazac's Shrew Tenrec, Andasibe-Mantadia National Park.

and females may establish stable relationship during breeding season, in austral spring/summer. Gestation 58–64 days, litter size 1–6 (Stephenson *et al.* 1994a, Goodman *et al.* 2003).

Only shrew tenrec known to accumulate fat reserves in body and tail prior to austral winter: body weight may increase to >80g (Eisenberg & Gould 1970). Does not enter full torpor in winter, but does become inactive, constructing nest and sleeping, eating less, and lowering body temperature (Stephenson *et al.* 1994b); an adaptation to higher elevations with marked seasonal climatic variation.

TALAZAC'S SHREW TENREC
Nesogale talazaci   (Major, 1896)

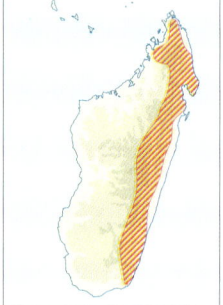

MEASUREMENTS
Head-body: 105–189mm. Tail: 103–162mm. Weight: 24.5–56g, mean 36g. (Goodman *et al.* 2003, Jenkins 2003, 2018, Jenkins *et al.* in press)

DESCRIPTION & IDENTIFICATION Large, robust shrew tenrec with tail slightly longer than head-body length. Head large, ears relatively small. Pelage soft and dense. Upperparts mid-brown to dark brown, underparts grey with reddish-buff wash (Goodman & Jenkins 1998, 2000). Tail relatively long but does not thicken with fat reserves. Similar to Dobson's Shrew Tenrec, but averages larger, although overlap does occur.

HABITAT & DISTRIBUTION One of the most widely distributed shrew tenrecs. Recorded throughout eastern rainforest region from Vondrozo in south to Montagne d'Ambre in north, including Masoala and humid forests of the Sambirano (Goodman *et al.* 1996b, Goodman & Jenkins 1998, Soarimalala 2018). Prefers primary forests, but tolerates some disturbance. Has broad altitudinal range: sea level to *c.*2,000m (Goodman *et al.* 2003). In areas of sympatry with Dobson's Shrew Tenrec, appears to favour higher elevations. IUCN Least Concern.

BEHAVIOUR Carnivorous, with small vertebrates (like frogs) part of diet, as well as insects (Stephenson et al. 1994a,b). Forages on ground and in lower undergrowth (Stephenson 1993a), and may dig burrow systems. Shows no seasonal inactivity and does not lay down fat reserves, a reflection of its more stable rainforest habitat (Stephenson & Racey 1993b). Males and females may occur together year-round and breed in austral spring/summer. Gestation 58–63 days; litter size one to three (average two) (Stephenson et al. 1994a).

WEB-FOOTED TENREC
Microgale mergulus   (Major, 1896)

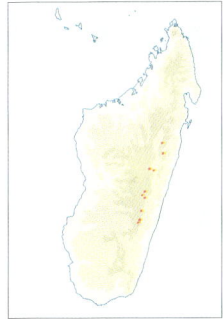

Other name: Aquatic Tenrec

MEASUREMENTS
Head-body: 116–170mm.
Tail: 128–161mm.
Weight: 60–110g, mean 80g. (Stephenson 1994c, Benstead & Olson 2003, Goodman et al. 2003)

DESCRIPTION & IDENTIFICATION Previously regarded as sole member of genus *Limnogale*: recent molecular analysis has established close evolutionary relationship with shrew tenrecs, hence inclusion within *Microgale* (Everson et al. 2016). Its morphology, ecology and behaviour relate to semi-aquatic lifestyle, which has drawn parallels with otter-shrews (Potamogalinae) of West and Central Africa, and desmans (Talpidae) of Europe. These traits reflect convergent evolution (Jenkins et al. in press).

The largest Oryzorictinae species. Pelage dense and soft. Head and upperparts greyish-brown with subtle reddish and blackish guard hairs. Underparts paler yellowish-grey. Eyes and ears small and all but hidden beneath dense fur. Head broad and flattened, muzzle blunt, appearing slightly 'swollen', with long, conspicuous vibrissae. Forefeet fringed and toes webbed to base of claws. Tail stout at base becoming laterally compressed (flattened) towards tip and covered in dense hairs; forms a distinct 'keel' (Benstead & Olson 2003). All features, including relatively large body size, are adaptations to semi-aquatic mode of life: the only Malagasy mammal exploiting this niche (Stephenson 1994c). Molecular analysis surprisingly suggests closest evolutionary relationship with Pygmy Shrew Tenrec (Everson et al. 2018).

Web-footed Tenrec.

HABITAT & DISTRIBUTION Associated with clear faster-flowing streams within humid forests of central-eastern and eastern highland areas. Range limits poorly understood: confirmed from just ten sites, from Sihanaka Forest in north (south of Lac Alaotra) to region of upper Iantara River, to east of Andringitra Massif in south, at altitudes between 450m and 2,000m (Benstead et al. 2001). Total area of occupancy estimated at c.2,000km² (Stephenson et al. 2016e). Most of these rivers flow east to Indian Ocean; however, one site at eastern edge of Central Highlands (headwaters of Mania River) is on west side of island divide (Benstead & Olson 2003). In Ranomafana area, found in streams no longer bordered by native forest (Benstead et al. 2001). IUCN Vulnerable.

BEHAVIOUR Nocturnal and carnivorous. Day spent in burrows, excavated in bank close to water's edge. Burrows c.10cm in diameter and lined with grass (Malzy 1965). Emerges shortly after sunset, but activity period appears to be variable. Occasionally, remains active all night, but at other times may return to burrow for prolonged rest periods lasting up to four hours. Always returns to burrows 1–1.5 hours before sunrise (Benstead et al. 2001).

During activity, appears to remain within same stream channel and spends majority of time in water. Nightly distance travelled within stream varies widely from day to day and between individuals, but can be up to 1,550m. Web-footed Tenrecs inhabiting narrow streams appear to travel longer distances than those in broader streams. Total home range, based on area of streambed utilised, c.7,000m² (Benstead et al. 2001).

Unusual among tenrecs in using specific latrine sites; repeatedly defecates on favoured prominent boulders in streams or around fallen trees. Piles of faeces can become visible and are an obvious sign of activity. Unknown if latrines demarcate territorial boundaries or are communal (Benstead & Olson 2003).

In Ranomafana area, feeds heavily on aquatic insect larvae, particularly mayflies (Ephemeroptera), caddisflies (Trichoptera), dragonflies and damselflies (Odonata) and to a lesser extent crayfish and frog tadpoles (Benstead et al. 2001). In other areas, frogs and freshwater shrimps are reportedly important dietary components (Malzy 1965, Eisenberg & Gould 1970). When swimming, uses hindfeet for propulsion and laterally compressed tail as a rudder. Sweeping motions of head help cover large areas and prey

is located using sensitive vibrissae (Benstead et al. 2001). Prey caught in mouth and brought to water's edge, held in forefeet and consumed (Malzy 1965). Foraging bouts are successive short-duration dives to streambed, lasting no longer than 15 seconds (Benstead & Olson 2001).

Young are born in a nest between December and January, when relative water temperature of rivers is highest (Malzy 1965); an average litter size of two to three is suggested (Eisenberg & Gould 1970, Jenkins et al. in press).

Does not appear to become dormant: studies around Ranomafana National Park found active animals in October, November, May, July and August (Stephenson 1994c, Benstead et al. 2001).

WHERE TO SEE Apparently rare and secretive, and very difficult to see. Ranomafana National Park offers a possibility, although considerable effort and luck are required. Has been seen in the stream by village of Ambatolahy, 1km from park entrance. Most conspicuous feature of Web-footed Tenrec presence is prominent latrine sites. When walking along watercourses in eastern montane rainforest always keep an eye out for faeces on boulders along streams.

MOUNTAIN SHREW TENREC
Microgale monticola   Goodman & Jenkins, 1998

Other name: Montane Shrew Tenrec

MEASUREMENTS
Head-body: 72–92mm.
Tail: 98–117mm. Weight:
12–17.5g. (Goodman &
Jenkins 1998, 2000)

DESCRIPTION & IDENTIFICATION Medium-sized, with tail longer than head-body length (c.120%). Upperparts dark brown, slightly grizzled with dark guard hairs. Underparts dark brown with grey-buff underfur. Tail dark brown above, slightly paler below. Feet dark brown: forefeet slightly broadened and hindfeet elongate with lengthened digits.

HABITAT & DISTRIBUTION Appears to prefer mountain habitats: higher-altitude rain and sclerophyllous montane forests. Currently known from mountainous regions near Andapa; Anjanaharibe-Sud at 1,550–1,950m, Marojejy at 1,625–1,875m (Goodman & Jenkins 1998, 2000). Also recorded at lower elevations (500–720m) further south in Makira (Goodman *et al.* 2013). Appears to be one of commonest *Microgale* at these localities (Goodman & Jenkins 2000). Has close resemblance to Thomas's Shrew Tenrec *M. thomasi*, but does not occur sympatrically: Thomas's Shrew Tenrec known from higher elevations in south-east humid forests. Mountain Shrew Tenrec is ecological equivalent in north (Jenkins 2003). IUCN Vulnerable.

BEHAVIOUR Appears strictly terrestrial. Wild-caught individuals contained 1–2 embryos. One female appeared pregnant and lactating simultaneously, something not recorded in *Microgale* previously (Jenkins 2018).

Mountain Shrew Tenrec, Marojejy National Park.

DRYAD SHREW TENREC
Microgale dryas   Jenkins, 1992

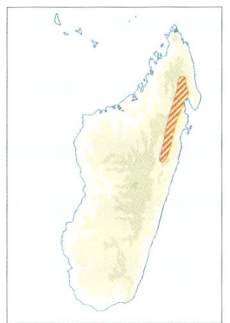

Other name: Tree Shrew Tenrec

MEASUREMENTS
Head-body: 105–114mm. Tail length: 68–71mm. Weight: *c.*30–49g. (Jenkins 1992, 2018)

DESCRIPTION & IDENTIFICATION Upperparts grizzled, dark reddish-brown or grey-brown with long dark guard hairs. Merges to grey underparts. Tail uniform grey and *c.*60% head-body length. Proboscis long with naked tip (Jenkins 1992, 2018). Distinguished from other *Microgale* by unusual dorsal pelage.

HABITAT & DISTRIBUTION Known from only few sites in centre-east and north-east, including Anjanaharibe-Sud, Makira, Marotandrano, Ambatovaky and An'ala, at elevations between 540m and 1,260m (Soarimalala & Goodman 2011, Goodman *et al.* 2013). IUCN Vulnerable.

BEHAVIOUR Terrestrial and probably fossorial.

SHORT-TAILED SHREW TENREC
Microgale brevicaudata   G. Grandidier, 1899

MEASUREMENTS
Head-body length: 68–88mm. Tail length: 30–45mm. Weight: 9–13g. (Goodman & Soarimalala 2004, Jenkins 2018)

DESCRIPTION & IDENTIFICATION Small to medium-sized with relatively short tail (less than *c.*60% head-body length). Pelage short and coarse. Upperparts grizzled, brown with buffy-brown speckles, underparts pale grey-brown (Goodman & Jenkins 2000). Very short tail distinctive.

HABITAT & DISTRIBUTION Unusual as found in both humid eastern and dry western biomes. Recorded in varied habitats from lowland rainforests to dry deciduous forests, even degraded forests and agricultural areas, from near sea level to 1,150m (Jenkins 2003). Range extends from around Masoala and Marojejy, north to Montagne d'Ambre, then through coastal areas of north-west and west as far south as Bemaraha Plateau (Goodman & Soarimalala 2004, Goodman *et al.* 2011b, Soarimalala 2018). IUCN Least Concern.

Short-tailed Shrew Tenrec, Tsingy de Bemaraha National Park.

BEHAVIOUR Terrestrial and probably semi-fossorial.

PYGMY SHREW TENREC
Microgale parvula   G. Grandidier, 1934

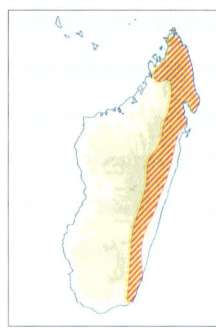

MEASUREMENTS
Head-body: 45–65mm. Tail: 46–66mm. Weight: 2–5g. (Jenkins 2003, 2018)

DESCRIPTION & IDENTIFICATION Very small and distinctively dark. Tail slightly less than or equal to head-body length. Head very small, delicate and elongate. Dorsal pelage dark brown, almost black in places, ventrally dark grey-brown. Tail and feet uniform dark grey-brown (Goodman & Jenkins 1998, 2000, Goodman et al. 1999b). Molecular analysis surprisingly suggests closest evolutionary relationship with Web-footed Tenrec (Everson et al. 2018).

Can only be confused with Least Shrew Tenrec *M. pusilla*: Pygmy Shrew Tenrec is dark grey-brown above and only slightly paler below, whilst Least Shrew Tenrec is reddish buffy-brown dorsally with abrupt transition to grey-brown venter.

HABITAT & DISTRIBUTION In all intact native eastern forests from Montagne d'Ambre in north to Andohahela in the south, at elevations from 30m to >2,000m (Raxworthy & Nussbaum 1994, Jenkins et al. 1996, 1997, Goodman & Jenkins 1998, 2000, Goodman et al. 1999b, Soarimalala 2018). Also recorded in some drier forests and in degraded habitats and forest fragments (Goodman et al. 1996c, 2013). IUCN Least Concern.

BEHAVIOUR Mainly terrestrial. Based on embryo count, litter size up to four (Goodman et al. 2003).

LESSER LONG-TAILED SHREW TENREC
Microgale longicaudata   Thomas, 1882

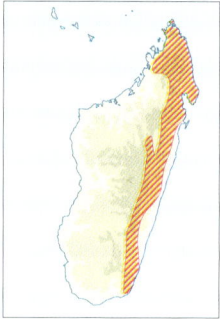

MEASUREMENTS
Head-body: 59–80mm. Tail: 136–158mm. Weight: 5.5–11g. (Jenkins et al. 1996, Goodman & Jenkins 1998, 2000, Goodman et al. 1999, Jenkins 2018)

DESCRIPTION & IDENTIFICATION Extreme body proportions make this the most easily recognised *Microgale* (along with Greater Long-tailed Shrew Tenrec). Body size small, but tail extremely long (more than 2× head-body length). Dorsal pelage dark brown with reddish-brown wash, underparts grey with reddish-buff wash. Tail grey-brown above, reddish-buff below. Tail tip naked and prehensile, indicating an arboreal lifestyle. Digits, forefeet and hindfeet proportionately long (Jenkins 2003).

Further molecular investigation may reveal 'cryptic' species within *M. longicaudata*, including '*M. prolixacaudata*' of northernmost localities (Anjanaharibe-Sud, Marojejy, Manongarivo and Montagne d'Ambre) (Olson et al. 2004, Soarimalala & Goodman 2011, Nicoll & Ratsifandrihamanana 2014).

HABITAT & DISTRIBUTION Lowland, mid- and high-altitude rainforest from Montagne d'Ambre in the north to Andohahela in the south (Raxworthy & Nussbaum 1994, Jenkins et al. 1996, Goodman & Jenkins 1998, 2000, Goodman et al. 1999b). Appears equally common on ridgetops, slopes or valley

Lesser Long-tailed Shrew Tenrec.

bottoms, perhaps reflecting its arboreal lifestyle. Has an extreme altitudinal tolerance, from *c*.500m to *c*.2,500m (Goodman *et al.* 2013). Appears commoner at higher elevations. IUCN Least Concern.

BEHAVIOUR Several morphological features (prehensile tail, elongate limbs and digits) indicate this species is scansorial and semi-arboreal. Nocturnal: a captive individual slept in a nest of leaves and became active around dusk (Jenkins 2018). It may forage on the ground, but also climbs and forages in vegetation and is able to jump amongst branches (Jenkins 2003). Diet includes variety of insects (Orthoptera, Coleoptera and Hymenoptera) and other small invertebrates.

DROUHARD'S SHREW TENREC
Microgale drouhardi   G. Grandidier, 1934

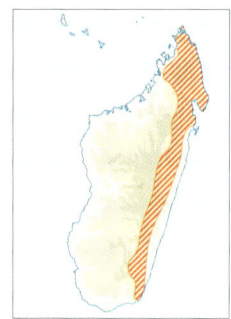

Other name: Striped Shrew Tenrec

MEASUREMENTS
Head-body: 62.5–83mm.
Tail: 53–83mm. Weight:
c. 8–14g. (Jenkins *et al.*
1997)

DESCRIPTION & IDENTIFICATION Medium-sized with distinct dark brown mid-dorsal stripe. Tail slightly shorter than head-body length. Variable colour among populations; upperparts dark grey-brown to light rufous-brown with some yellowish speckles, underparts silver-grey with buff or reddish-buff wash. Transition moderately distinct, extending from head to base of tail. Tail bicoloured: dark grey dorsally and buff ventrally (Jenkins *et al.* 1997).

Considerable size variation between populations; those from central-eastern regions (Ambatovaky and Zahamena) appreciably smaller than elsewhere. Distinctive dorsal stripe helps differentiate from other *Microgale* species.

HABITAT & DISTRIBUTION Lowland and mid-altitude rainforests of north, north-west (including Manongarivo and Tsarantanana) and east, from Montagne d'Ambre in north to Andohahela in south-east. A recent record from Masoala Peninsula extends previously known range (Soarimalala 2018). Elevational range from *c*.30m (Masoala) to above 2,300m (Tsaratanana). Many specimens collected in valley bottoms (Jenkins *et al.* 1997). In Montagne d'Ambre, where several species of shrew tenrec occur, Drouhard's Shrew Tenrec is the most common, comprising up to 70% of the *Microgale* community (Jenkins 2003). IUCN Least Concern.

BEHAVIOUR Terrestrial and fossorial. Conceals itself under leaves and vegetation on forest floor. Active and can move rapidly beneath this layer. Diet includes variety of insects (Orthoptera, Coleoptera, Dermaptera and Hymenoptera) and other small invertebrates.

Drouhard's Shrew Tenrec, Manongarivo Reserve.

TAIVA SHREW TENREC
Microgale taiva   Major, 1896

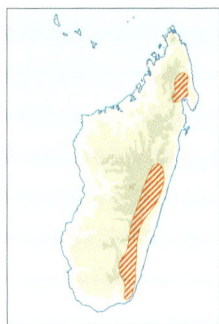

MEASUREMENTS
Head-body: 61–89mm.
Tail: 66–95mm. Weight:
10.5–16, mean 12.5g.
(Jenkins *et al.* 1996,
Goodman *et al.* 2003,
Jenkins 2018)

DESCRIPTION & IDENTIFICATION Medium-sized, with moderately long tail, equal to or slightly longer than head-body length. Dorsal pelage dark russet-brown with buffy-brown speckling, ventrally grey-brown. Tail dark grey above, slightly paler grey below (Jenkins *et al.* 1996).

HABITAT & DISTRIBUTION Prefers rainforest habitats from lowlands (*c.*500m) to high mountain zones (*c.*2,500m). Known from two separate eastern populations: in north-east (Makira, Masoala, Anjanaharibe-Sud, Marojejy) and centre-east (Mantadia) south to Andohahela. Apparently absent from centre-east to north-east (Goodman *et al.* 2013). Further investigation may reveal 'cryptic' species within *M. taiva* (Stephenson *et al.* 2016j). IUCN Least Concern.

BEHAVIOUR Diet includes variety of insects (Orthoptera, Coleoptera and Hymenoptera) and other small invertebrates, including spiders and worms. Known to consume smaller shrew tenrecs in pitfall traps, but cannibalism is unlikely to be natural behaviour (Jenkins 2018).

PALE SHREW TENREC
Microgale fotsifotsy   Jenkins *et al.*, 1997

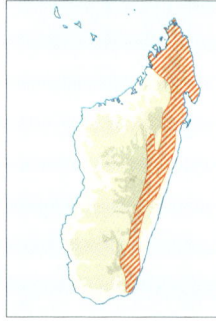

Other name: Pale-footed Shrew Tenrec

MEASUREMENTS
Head-body: 59–85mm.
Tail length: 69–94mm.
Weight: 7–15g. (Jenkins *et al.* 1997, Goodman *et al.* 1999b, Goodman & Jenkins 2000, Goodman & Soarimalala 2004)

DESCRIPTION & IDENTIFICATION Small to medium-sized, with tail approximately equal to head-body length. Pelage soft and silky; upperparts grizzled yellowish grey-brown with dark guard hairs, underparts pale grey with buff or reddish wash. Ears prominent and pale. Tail bicoloured, grey-brown above, light grey-buff below, ending in distinctive thin pencil-line of fine white hairs. Fore- and hindfeet light brown with very pale digits: *fotsifotsy* is

Malagasy for pale or whitish (Jenkins *et al.* 1997). Light body coloration, relatively large ears, conspicuous pale feet and tail tip help differentiate from all other *Microgale* species. Molecular analysis suggests closest evolutionary relationship with Nasolo's Shrew Tenrec (Everson *et al.* 2018). Initial genetic studies suggest Pale Shrew Tenrec may warrant division into two species, one broadly distributed in eastern rainforests and a second restricted to far north (Everson *et al.* 2020a).

HABITAT & DISTRIBUTION Primarily in humid forest, but also transitional and dry forest. Prefers lowland and mid-altitude rainforest (400–1,500m), but also occurs in high-altitude forests up to 2,500m (Goodman & Rasolonandrasana 2001, Goodman *et al.* 2003, Stephenson *et al.* 2016g). Range extends from far north (Montagne d'Ambre) to extreme south (Andohahela), including humid Sambirano region in north-west (Raxworthy & Nussbaum 1994, Jenkins *et al.* 1997, Goodman *et al.* 1999b). IUCN Least Concern.

BEHAVIOUR Probably mainly terrestrial, but has some scansorial adaptations (Jenkins 2003). Diet includes crickets and grasshoppers (Orthoptera), beetles (Coleoptera), and presumably other insects.

GREATER LONG-TAILED SHREW TENREC *Microgale principula* Thomas, 1926

MEASUREMENTS
Head-body: 69–89mm.
Tail: 144–171mm. Weight:
8.5–14g. (Goodman
& Jenkins 1998, 2000,
Goodman *et al.* 1999b,
Jenkins 2018)

DESCRIPTION & IDENTIFICATION Very distinctive: medium-sized with tail *c.*2× head-body length. Distal portion of tail naked, with broad scales and prehensile. Upperparts reddish-brown, underparts grey with buff wash. Digits elongate. Clearly adapted to semi-arboreal lifestyle.

HABITAT & DISTRIBUTION Prefers lowland and mid-altitude rainforest to *c.*1,900m. Recorded length of eastern rainforest belt, from Marojejy and Masoala in north to Andohahela in south (Goodman & Jenkins 1998, 2000, Goodman *et al.*

Greater Long-tailed Shrew Tenrec, Ranomafana National Park.

1999, 2013, Soarimalala 2018). At several localities, e.g., Marojejy, Anjanaharibe-Sud and Andohahela, occurs sympatrically with Lesser Long-tailed Shrew Tenrec. While habitat overlap does occur at middle elevations, Greater Long-tailed Shrew Tenrec prefers lower altitudes, whilst Lesser Long-tailed Shrew Tenrec prefers upper elevations (Goodman & Jenkins 1998, 2000, Goodman et al. 1999b). IUCN Least Concern.

BEHAVIOUR Several morphological features, including long prehensile tail and elongate digits, indicate a scansorial to semi-arboreal way of life. An agile climber and forages in vegetation above ground level. Tail used as a counter-balance and it coils around twigs. Also able to jump short distances between branches (Jenkins 2003).

COWAN'S SHREW TENREC
Microgale cowani   Thomas, 1882

MEASUREMENTS
Head-body: 66–95mm. Tail: 54–87mm. Weight: 10.5–17g. (Jenkins et al. 1996, Goodman & Jenkins 1998, 2000, Goodman et al. 1999b, Jenkins 2018)

DESCRIPTION & IDENTIFICATION Medium-sized, with moderately short tail, shorter than head-body length. Dorsal pelage dark brown to rufous-brown with some paler russet flecking that is reduced

Cowan's Shrew Tenrec, Anjanaharibe-Sud.

on rump. Upperparts appear speckled brown. Underparts grey with reddish-brown wash. Tail bicoloured: dark brown dorsally, reddish-buff ventrally (Jenkins et al. 1996).

HABITAT & DISTRIBUTION Prefers mid-altitude rainforest and high-altitude mossy sclerophyllous forest; also found above treeline in ericoid scrub. Recorded in non-forested, grassland and agricultural areas (Jenkins 2003). Range extends length of eastern rainforest, including eastern escarpment of central highlands, from Andohahela in south to Marojejy in north (Jenkins et al. 1996, Goodman et al. 2000, Soarimalala et al. 2010). Recorded from 530m to over c.2,500m, appears most common between 1,200m and 1,800m. In several locations, e.g., Marojejy, Anjanaharibe-Sud and Andohahela, where several shrew tenrec species occur sympatrically, Cowan's Shrew Tenrec often most common, comprising 25–50% of *Microgale* communities (Jenkins 2003). IUCN Least Concern.

BEHAVIOUR Forages in leaf litter, moving in cryptic fashion. Forelimbs and claws allude to some digging (Jenkins 2018). Females may become sexually mature while still dentally immature. Probably gives birth between late October and December. Litter size two to three (Goodman et al. 2003, Jenkins 2018). Diet includes crickets and grasshoppers (Orthoptera), beetles (Coleoptera) and worms.

NAKED-NOSED SHREW TENREC
Microgale gymnorhyncha   Jenkin et al., 1996

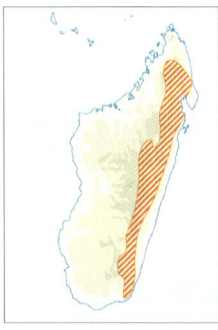

MEASUREMENTS
Head-body: 75–101mm. Tail: 59–75mm. Weight: 13.5–26g, mean 16g. (Jenkins et al. 1996, Goodman & Jenkins 1998, 2000, Goodman et al. 1999b, 2003)

DESCRIPTION & IDENTIFICATION Moderately large with tail shorter than head-body length. Dorsal pelage dense, soft and velvety, dark grey-brown, with dark brown guard hairs. Ventral pelage dark grey. Head long and gracile, and muzzle very long, forming naked proboscis which extends

Naked-nosed Shrew Tenrec, Andringitra National Park.

well beyond mouth. Eyes and ears very small and virtually concealed in pelage. Tail dark grey above, paler below. Forefeet broad with claws (Jenkins *et al*. 1996). Differs from all congeners, except Gracile Shrew Tenrec *M. gracilis*, in having elongate head, long muzzle, small concealed ears, broad forefeet with large claws. Naked-nosed Shrew Tenrec slightly smaller, yet stockier than Gracile Shrew Tenrec, and has shorter tail.

HABITAT & DISTRIBUTION Broad distribution on east side of Central Highlands and along eastern escarpment, from Anjanaharibe-Sud and Marojejy in north to Andohahela in far south (Jenkins *et al*. 1996, Goodman & Jenkins 2000). Recorded between *c*.600m and *c*.2,500m: prefers mid-altitude rainforest and higher-elevation mossy cloud forest at 1,000–2,000m (Goodman & Rasolonandrasana 2001, Goodman *et al*. 2013). Widespread but relatively rare; comprises no more than 5–6% of the *Microgale* community at any location (Jenkins 2003). IUCN Least Concern.

BEHAVIOUR Shares several adaptive features with Gracile Shrew Tenrec: elongate proboscis-like muzzle, small eyes, reduced ears and broadened forefeet with large claws, all indicating a semi-fossorial lifestyle and a diet consisting of soil invertebrates (Jenkins 2003, 2018).

SHREW-TOOTHED SHREW TENREC *Microgale soricoides*   Jenkins, 1993

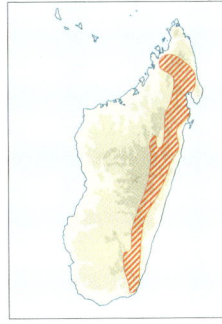

MEASUREMENTS
Head-body: 66–103mm. Tail: 81–112mm. Weight: 14–22g. (Jenkins 1993, 2018, Jenkins *et al*. 1996, Goodman & Jenkins 1998, 2000, Goodman *et al*. 1999)

DESCRIPTION & IDENTIFICATION Medium-large with slightly shorter tail than head-body length. Upperparts light buff-brown, slightly grizzled, and paler grey-brown underparts with reddish-buff wash. Tail grey-brown above, paler buffy-brown below; tip may be white (Jenkins 1993, Goodman & Jenkins 2000). Dentition unlike other *Microgale*: name derives from distinctive first upper and lower incisors, which closely resemble some shrews (e.g., *Suncus* spp. and *Crocidura* spp.) (Jenkins 2018).

HABITAT & DISTRIBUTION Widespread and common: from Manongarivo and Tsaratanana in north-west and Anjanaharibe-Sud and Marojejy in north-east, to Andohahela in south, including eastern edge of central plateau (Goodman & Jenkins 2000,

Goodman *et al.* 2013). At elevations from *c.*675m to over *c.*2,500m, but seems to prefer higher altitudes (Goodman *et al.* 2003, 2013). IUCN Least Concern.

BEHAVIOUR Probably semi-scansorial: collected on branches and in tangles of vines *c.*2m above ground (Jenkins 2003). Massive incisors indicate predatory tendencies: in pitfall traps, Shrew-toothed Shrew Tenrec has partially eaten other *Microgale* species (Goodman & Jenkins 2000).

MAJOR'S SHREW TENREC
Microgale majori   Thomas, 1918

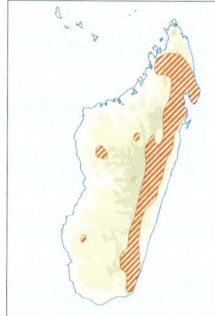

Other name: Major's Long-tailed Shrew Tenrec

MEASUREMENTS
Head-body: 52–66mm.
Tail: 102–134mm. Weight: 5–8g. (Jenkins 2018)

DESCRIPTION & IDENTIFICATION Originally described in 1918, later regarded as synonymous with Lesser Long-tailed Shrew Tenrec (Morrison-Scott 1948, MacPhee 1987). Recent research confirms it as a valid species (Olson *et al.* 2004). In appearance, a smaller version of Lesser Long-tailed Shrew Tenrec. A small *Microgale* with very long prehensile tail (*c.*170% head-body length). Upperparts dark brown, with reddish-brown tones, underparts dark grey with buff wash. Tail bicoloured: grey-brown above, reddish-buff below (Jenkins 2018).

HABITAT & DISTRIBUTION Widespread: in humid forests from Manongarivo in north-west and entire length of eastern rainforest belt, at elevations between *c.*780m to *c.*2,500m, also in isolated locations in western dry forests (Ambohijanahary Special Reserve, Analavelona Forest) (Olson *et al.* 2004, Yoder *et al.* 2005, Stephenson *et al.* 2016h). Also recorded in degraded habitats of centre-east (Randriamoria *et al.* 2015, Randriamoria 2017). Further studies required to establish taxonomic status and ranges of the three prehensile-tailed shrew tenrec species, Major's Shrew Tenrec, Lesser Long-tailed Shrew Tenrec and Greater Long-tailed Shrew Tenrec. IUCN Least Concern.

Major's Shrew Tenrec, Andringitra National Park.

BEHAVIOUR Based on morphology, probably scansorial or semi-arboreal. Other behaviour unknown.

GRACILE SHREW TENREC
Microgale gracilis   (Major, 1896)

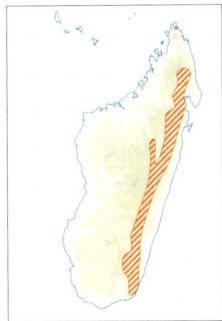

MEASUREMENTS
Head-body: 85–105mm. Tail: 75–93mm. Weight: 19.5–33g. (Jenkins *et al.* 1996, Goodman & Jenkins 2000, Goodman *et al.* 1999, Jenkins 2018)

DESCRIPTION & IDENTIFICATION A largish *Microgale*, with tail shorter than head-body length. Head elongate and gracile, and muzzle/proboscis very long. Eyes and ears very small and partially obscured by pelage, which is soft and velvety. Upperparts dark brown with buff speckles, underparts dark grey with buff wash. Tail bicoloured: dark brown dorsally and light brown ventrally. Feet broad with stout claws (Jenkins *et al.* 1996, Goodman *et al.* 1999, Goodman & Jenkins 2000).

HABITAT & DISTRIBUTION Throughout eastern rainforests at elevations of c.900m to c.2,100m: from Marojejy in north (Goodman & Jenkins 2000) to Andohahela in south (Goodman *et al.* 1999b). Appears to prefer higher-elevation montane forest, but also found in degraded areas (Jenkins *et al.* 1996, Soarimalala & Goodman 2011). IUCN Least Concern.

BEHAVIOUR Several morphological features, e.g., elongate proboscis, reduced eyes and ears, and broadened feet with large claws, suggest this species' habits are semi-fossorial. Probably feeds on soil invertebrates (Jenkins 2018).

THOMAS'S SHREW TENREC
Microgale thomasi   Major, 1896

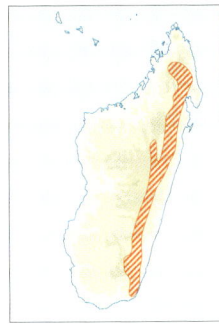

MEASUREMENTS
Head-body length: 75–112mm. Tail length: 59–80mm. Weight: 19.5–26g, mean 22g. (Goodman *et al.* 1999, 2003, Jenkins 2018)

DESCRIPTION & IDENTIFICATION Moderately large with tail slightly shorter than head-body length. Upperparts speckled dark rufous-brown, and underparts noticeably paler with reddish wash. Ears prominent. Tail bicoloured and covered in dense hair (Goodman *et al.* 1999).

HABITAT & DISTRIBUTION Eastern rainforest and montane forest regions, from Tsaratanana and Marojejy in north to Andohahela in south (Goodman *et al.* 1999b, 2013). Mainly between 800m and 2,000m, but also at lower and higher elevations (Jenkins 2018). Also, degraded habitats in central-east (Randriamoria *et al.* 2015). IUCN Least Concern.

BEHAVIOUR Terrestrial; forages on forest floor. In pitfall traps has shown cannibalistic behaviour, eating smaller *Microgale* species. Litter size of two suspected (Goodman *et al.* 2003).

Thomas's Shrew Tenrec, Ranomafana National Park.

GRANDIDIER'S SHREW TENREC

Microgale grandidieri   Olson et al., 2009

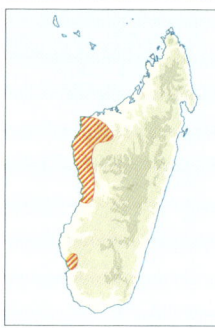

MEASUREMENTS
Head-body: 60–80mm.
Tail: 33–43mm. Weight:
8–10g. (Olson *et al.* 2009)

DESCRIPTION & IDENTIFICATION Very short tail (<50% head-body length) distinctive: only *M. brevicaudata* (from which this species was split) has similarly short tail. Dorsal pelage soft and finely textured: brindled (agouti), dark brown mixed with lighter brown. Ventral fur very fine, medium grey, with clear demarcating line. Ears relatively short. Muzzle relatively short. Vibrissae black (Olson *et al.* 2009).

HABITAT & DISTRIBUTION Occurs in dry deciduous forest, arid spiny forest and more humid gallery forest at elevations between c.50m and c.430m. Currently known from small number of localities in west and south-west, from Namoroka area in north to

Onilahy River in south, including forest surrounding limestone karst (*tsingy*) formations (Namoroka and Bemaraha) and Kirindy-Menabe region (Olson *et al.* 2009, Soarimalala 2011, Goodman *et al.* 2013, Nicoll & Ratsifandrihamanana 2014). Has also been found in degraded forest areas. Recorded sympatrically with similar Short-tailed Shrew Tenrec (Beanko Forest) (Goodman *et al.* 2011b). Further surveys at intermediate localities may add to species' range. IUCN Least Concern.

BEHAVIOUR Evidence suggests that it becomes seasonally torpid: Grandidier's Shrew Tenrec only found active in warm, wet season. Presumed to aestivate during cooler dry season when food availability much reduced (Olson *et al.* 2009).

LEAST SHREW TENREC

Microgale pusilla   Major, 1896

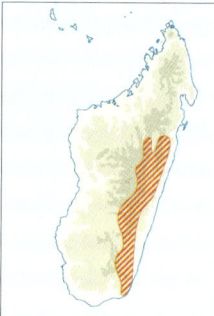

Other name: Lesser Shrew Tenrec

MEASUREMENTS
Head-body: 47–63mm.
Tail: 62–85mm. Weight:
3–5g. (Goodman & Soarimalala 2004, Jenkins 2018)

DESCRIPTION & IDENTIFICATION Recognised by very small size and relatively long tail (130–160% head-body length). Upperparts reddish buffy-brown and grizzled, underparts grey-brown; transition abrupt and noticeable. Tail dark grey-brown with slightly paler underside. Similar in size to Pygmy Shrew Tenrec, but paler.

HABITAT & DISTRIBUTION Found in variety of habitats: humid, montane and littoral forests, damp grassland and sedge marshes. Able to survive in degraded habitats and agricultural zones (Jenkins 2003). Common in suitable habitat (Stephenson *et al.* 2016i). Range extends from central-eastern regions to southern tip of island, mainly at c.530–1,700m, but recently also recorded at higher elevations (Jenkins 2018). IUCN Least Concern.

BEHAVIOUR Forages on ground and in lower branches of undergrowth.

Least Shrew Tenrec, Tsinjoarivo.

DARK SHREW TENREC
Microgale jobihely   Goodman *et al.* 2006

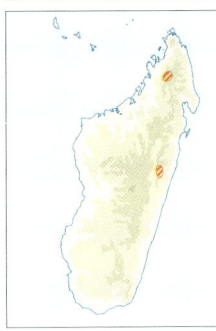

Other name: Northern Shrew Tenrec

MEASUREMENTS
Head-body: 53–80mm. Tail: 44–57mm. Weight: 7–10g. (Goodman *et al.* 2006d)

DESCRIPTION & IDENTIFICATION Pelage soft and dense. Upperparts a mixture of black and reddish-brown giving agouti (brindled) appearance. Underparts finer and softer, paler tan to grey-brown, with gradation between dorsal and ventral coloration. Tail relatively short (60–90% head-body length) and brown, but slightly darker above (Goodman *et al.* 2006d).

HABITAT & DISTRIBUTION Dense humid montane forest at elevations between 1,000m and c.1700m. Seems able to tolerate some anthropogenic disturbance. Restricted range: currently known from just three locations: south-west slope of Tsaranatana Massif (c.1,420–1,680m) in north-west, Ambatovy in centre-east, and Ambohitantely in Central Highlands (Goodman *et al.* 2006d, Soarimalala *et al.* 2010, Jenkins *et al.* in press). IUCN Endangered.

BEHAVIOUR No specific information.

NASOLO'S SHREW TENREC
Microgale nasoloi   Jenkins & Goodman, 1999

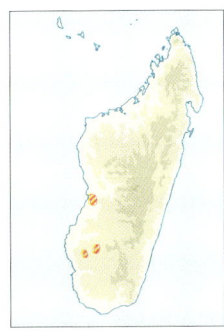

MEASUREMENTS
Head-body: 70–81mm. Tail: 50–62mm. Weight: 5.9–14g. (Jenkins & Goodman 1999, Jenkins 2018).

DESCRIPTION & IDENTIFICATION Pelage soft, upperparts slate-grey, slightly darker grey below. Overall grey coloration distinguishes it from other *Microgale*. Ears moderately large and prominent. Tail grey and relatively short (60–80% head-body length) (Jenkins & Goodman 1999, Jenkins 2018).

HABITAT & DISTRIBUTION Dry deciduous and slightly transitional humid forests in west and south-west, between 80m and 1,050m. Known from handful of locations: Vohibasia and Analavelona Massif in south-west (where species was discovered) (Jenkins & Goodman 1999), and Menabe-Kirindy region in central west (Soarimalala & Goodman 2008). IUCN Vulnerable.

BEHAVIOUR Probably terrestrial and scansorial. Trapped female contained three embryos. Other behaviour unknown.

JENKINS'S SHREW TENREC
Microgale jenkinsae   Goodman & Soarimalala, 2004

MEASUREMENTS*
Head-body: 59–62mm. Tail: 79–81mm. Weight: 4.9–5.3g. (Goodman & Soarimalala 2004) *based on two subadult specimens.

DESCRIPTION & IDENTIFICATION Very small with tail 130–140% head-body length. Fur dense and soft. Upperparts tannish-brown with darker flecks (agouti). Underparts paler grizzled slate-grey (Goodman & Soarimalala 2004). Ears relatively long (up to 18mm) and prominent. Tail dark brown, paler on underside.

HABITAT & DISTRIBUTION Known only from transitional dry deciduous forests and dense spiny bush of Mikea region, between Morombe and Manombo River in south-west, at elevations up to 80m (Goodman & Soarimalala 2004). Surveys at other similar sites in region failed to identify additional localities. IUCN Endangered.

BEHAVIOUR Presumed to be primarily terrestrial.

Bats

Order Chiroptera

Bats are among the most diverse and geographically dispersed mammals. They form the largest mammalian aggregations and may also be the most abundant mammals. There are more than 1,400 species: among mammals, only the Order Rodentia (the rodents) contains a greater number of species.

Bats have traditionally been divided into two major subgroups (suborders): the Megachiroptera or 'megabats' and Microchiroptera or 'microbats', with a range of characteristics defining each group, including enhanced visual acuity in megabats and complex laryngeal echolocation systems (ultrasound vocalisation produced in the larynx) in microbats (Teeling et al. 2000). Previously, the megabats contained a single family Pteropodidae (the fruit bats) and the 'micrcobats' were represented by 18 families (Simmons 2005). Recent molecular investigation has revealed rather different evolutionary relationships resulting in these divisions being reassessed. Two suborders are now recognised: Pteropodiformes (also called Yinpterochirpotera) containing seven families, and Vespertilioniformes (also called Yangochiroptera) represented by 14 families (Wilson & Mittermeier 2019).

Bats are the only mammals capable of powered flight (other 'flying' mammals, e.g., flying squirrels, can only glide). Obviously, this has been a major factor in their wide distribution (O'Brien 2011). Their ability to migrate has resulted in reduced levels of endemism in the bats of Madagascar compared to other mammalian orders on the island: currently 45 or so species are known, of which 39 are endemic (cf. 100% endemism for non-volant mammals), the remainder being shared with Africa and are Afrotropical in origin (Racey et al. 2009). Nonetheless, bat endemism (c.85%) is extremely high compared to chiropteran assemblages on similar-sized islands elsewhere (e.g., New Guinea 16% and Borneo 6.5%) and is testament to Madagascar's relative isolation in space and time (Racey et al. 2009, Cardiff & Jenkins 2016). At higher taxonomic levels only one of nine families is endemic, Myzopodidae.

Unlike the non-volant indigenous mammal groups in Madagascar (lemurs, carnivorans, tenrecs and Nesomyine rodents), which are each derived from a single colonising ancestor (Poux et al. 2005), the bat assemblage on the island (45 or so extant species, as two previously listed species are no longer regarded as valid) probably represent 28–30 independent colonisations events, with the majority of founders originating in Africa and a smaller number in Asia (Goodman et al. in press b).

Bats occur in all native forest areas on the island and some species can survive in degraded and anthropogenic areas, including the Central Highlands well away from remaining native forests. This may be partially explained by their frequent use of edge habitats for hunting (Cardiff & Jenkins 2016). However, their presence and activity decline dramatically in high mountain areas above 1,500m.

Suborder Pteropodiformes (Yinpterochiroptera)

The suborder Pteropodiformes (also referred to as Yinpterochiroptera) contains seven families, including three represented in Madagascar: Pteropodidae (Old World fruit bats), Hipposideridae (Old World leaf-nosed bats) and Rhinonycteridae (trident bats). The Pteropodidae lack complex echolocation, whereas the Rhinonycteridae and Hipposideridae possess sophisticated ultrasound echolocation systems and were previously placed within the superseded suborder Microchiroptera.

Old World Fruit Bats Family Pteropodidae

This diverse family of primarily fruit and nectar-feeding bats contains 191 species; their origins lie in Australasia and date back >30 million years (Giannini 2019). Pteropodidae includes the largest bat species and the majority are larger than other types of bats, but with considerable size variation in the family: weights range from c.15g to 1,500g. In most species the tail is short or rudimentary.

In general, they lack ultrasonic echolocation and have large eyes (with a reflective tapeta) and excellent vision, together with a well-developed sense of smell. Three species, arranged across three genera, occur in Madagascar: Madagascar Flying Fox Pteropus rufus, Madagascar Straw-coloured Fruit Bat Eidolon dupreanum and Madagascar Rousette Rousettus madagascariensis. All are important pollination and seed-dispersal agents (Andrianaivoarivelo et al. in press): in deforested areas where trees like large baobabs (Adansonia sp.) are 'marooned', fruit bats may be their only pollinators (MacKinnon et al. 2003). The flexibility of the three species and their willingness to feed on introduced, as well as native, fruits and plants, has certainly contributed to their ability to survive in degraded landscapes.

Each of the three genera of fruit bats in Madagascar represents independent colonisation events of the island, Madagascar Straw-coloured Fruit Bat from Africa, with the two other species, Madagascar Flying Fox and Madagascar Rousette, being of Asian origin (Goodman et al. in press b).

All fruits bats in Madagascar are very susceptible to disturbance and colonies should not be approached too closely. Also, when in the vicinity of a roost (especially in caves), it is advisable to not touch or allow contact with bat urine and faeces.

Flying Foxes

GENUS *Pteropus*

The genus *Pteropus* contains more than 60 species: they are often called flying foxes (Almeida et al. 2014, Giannini 2019). They include the largest bats: some attain weights of 1,500g and wingspans up to 1.7m. Their origins lie in South-East Asia, Australasia, and islands in the western Pacific. They have spread to the western Indian Ocean islands, but have not reached the African mainland.

Islands in the western Indian Ocean are remarkable for the radiation of *Pteropus*. Several species occur (or did until recently) on islands in the region: Madagascar, Madagascar Flying Fox *P. rufus*; Mauritius and Réunion, Greater Mascarene Flying Fox *P. niger*; Rodrigues, Rodrigues Flying Fox *P. rodricensis*; Aldabra, Aldabra Flying Fox *P. aldabrensis*; Seychelles, Seychelles Flying Fox *P. seychellensis seychellensis*; Comoros and Mafia Island (off Tanzania), Seychelles Flying Fox *P. seychellensis comorensis*; Anjouan and Moheli islands (Comoros Archipelago), Livingstone's Flying Fox *P. livingstonii*; and Pemba Island (off Tanzania), Pemba Flying Fox *P. voeltzkowi* (Cheke & Dahl 1981, Giannini 2019).

Flying foxes roost in trees, often in large colonies and may use the same sites year after year, resulting in defoliation of the trees. During the day, colonies are generally noisy as individuals jostle for prime spaces. They are easily disturbed and will take to the air if threatened. At dusk they leave the roost in search of fruit trees in which to feed. Strong flyers, they can cover large distances, sometimes flying for more than two hours in one direction. They are almost entirely frugivorous, feeding on fruit pulp and juices, although some also feed on flowers, nectar and occasionally leaves. They eat, rest and digest their food at the feeding site before returning to the roost site by dawn.

Madagascar Flying Foxes, near Anjajavy.

MADAGASCAR FLYING FOX

Pteropus rufus   (É. Geoffroy Saint-Hilaire, 1803)

Other names: Malagasy Flying Fox, Madagascar Fruit Bat

MEASUREMENTS
Head-body: 230–310mm. Forearm: 155–175mm. Tail: none. Wingspan: 1,000–1,250mm. Weight: males *c.*650–750g, females: *c.*500–550g. (MacKinnon *et al.* 2003, Giannini 2019)

DESCRIPTION & IDENTIFICATION Largest bat in Madagascar (and one of largest in the world). Males slightly larger than females. Basic body colour russet brown; upper chest, shoulders and head lighter, varying between rufous-brown and yellowish golden-brown; around eyes and muzzle darker. Pronounced muzzle gives face a 'fox-like' appearance. Eyes large, ears long and pointed, prominent and widely separated. Wings slate-grey to black. There is a claw on the second digit in addition to the thumb.

Pelage coloration, larger size and absence of tail immediately differentiates Madagascar Flying Fox from other fruit bat species (*Eidolon* and *Rousettus*) on the island. (Bennett & Russ 2001).

HABITAT & DISTRIBUTION In all regions where substantial tracts of native forest remain, including eastern rainforest, western dry forest and southern spiny forest, from sea level to *c.*1,200m (Giannini 2019); may be more prevalent in humid forest. Many major roosts are within 100km of the sea, near the coast or on offshore islands; the most remote coastal areas between Morombe on the west coast and Antsiranana in the far north, probably hold the largest number of roosts (MacKinnon *et al.* 2003). *P. rufus* is dependent on trees for roosting, so widespread deforestation has a severe negative impact on populations (Hutcheon 2003). Can survive in human-altered landscapes, but probably now absent from Central Highlands, where human populations are highest (Racey 2016).

Hunted for food throughout the island (often with shotguns): particularly vulnerable at roost sites and when feeding in trees adjacent to villages (Rakotonandrasana & Goodman 2007, Racey *et al.* 2009). Probably consumed far more in urban

Madagascar Flying Foxes are regularly trapped for food.

Madagascar Flying Foxes often feed on cultivated fruits like mangoes, as well as papayas, lychees and guavas.

than rural areas, which alludes to a bushmeat trade (Jenkins & Racey 2008, Jenkins *et al*. 2011, Olesky *et al*. 2015a). Many traditional roosts now abandoned: desertion occasionally caused by natural events like cyclones, but more often due to deforestation, human disturbance and especially over-hunting (Jenkins *et al*. 2007a, Andrianaivoarivelo *et al*. in press).

Overall population in Madagascar declining: currently (as of 2019) estimated at *c*.200,000, which represents a *c*.35% decline in last 20 years (Andrianaivoarivelo *et al*. in press). IUCN Vulnerable.

BEHAVIOUR By day, rests in colonies (called camps), up to several thousand strong (range 10–10,000, average 400–500) (MacKinnon *et al*. 2003, Andrianaivoarivelo *et al*. in press), in tree canopy in primary and secondary forests, forest fragments (usually remote fragments), mangroves and occasionally plantations. Number of bats at a roost varies seasonally (Jenkins *et al*. 2007a). Largest roost recorded in Madagascar was in Ambondrombe (Menabe Region), with up to 10,000 in 2009.

Prefers to hang on outermost branches (probably linked to predator avoidance) (MacKinnon *et al*. 2003, Jenkins *et al*. 2007b, Long & Racey 2007, Rakotoarivelo & Randrianandriananina 2007). Roosts often sited close to water bodies, which probably helps with orientation and navigation during nocturnal bouts of foraging (Jenkins *et al*. 2007b, Oleksy 2014).

Activity at roost sites is constant: colonies very noisy and generally heard before being seen. Chatter intensifies if disturbed. Potential predators, like Fosa, may cause whole colony to take flight (MacKinnon *et al*. 2003). Large raptors circling above roosts cause distress. Likely predation by Madagascar Harrier-Hawk *Polyboroides radiatus* has been recorded (Goodman & Pidgeon 1992). Bats often fly to different localities within the roost; when landing, there are noisy and agonistic exchanges between incoming and incumbent individuals.

During early daylight hours, often hangs with wings outstretched to absorb heat. Later, when temperatures are higher, will lick inner and outer wing membranes and hang with them outstretched, to reduce body temperature by evaporation. May also 'fan' wings to keep cool: this is common in mothers with suckling or roosting young (Wells 1995).

At dusk flies to foraging sites up to 34km away (Long 2001, Olesky *et al*. 2015b). Average flight speed *c*.32km/h, with speeds of >60km/h recorded

(Long 2001, Oleksy 2014). Several 'scouts' may fly ahead of main group to locate suitable fruiting trees. In Berenty, in south-east, home ranges approaching 60,000ha have been recorded (Oleksy 2014). Bats on offshore islands fly to neighbouring islands or back to mainland to feed. Has excellent vision and does not navigate by echolocation.

Fruit juices dominate diet (60–65%): obtained by squeezing pieces of fruit pulp in mouth, then swallowing the juice and rejecting remaining pulp fibres. Very soft fruit pulps may be swallowed, along with some seeds. Wide variety of native fruits are eaten (>35 species recorded); *Ficus* spp. (Moraceae) are particularly relished and seeds germinate better having passed through gut of Madagascar Flying Fox (Hutcheon 2003, Racey *et al* 2009, Oleksy 2014). Crops like papayas, lychees and guava also eaten (Raharimihaja *et al*. 2016). When fruit scarce, leaves are eaten (up to 18%); these contain proteins that are deficient in a fruit-based diet. Flowers and pollen also an important dietary component (17–35%; >25 species recorded, including baobabs *Adansonia* spp., and kapoks *Ceiba* spp.) (Baum 1995, 1996, Raheriaisena 2005, Andriafidison *et al*. 2006b). In southern Madagascar (Berenty region) pollen from commercial sisal crops is a dominant food source (Long & Racey 2007). Flying foxes are important agents of both seed dispersal and pollination in forests and fragmented ecosystems (Long 2001, Bollen & van Elsacker 2002, Hutcheon 2003, Bollen *et al*. 2004).

Usually inverts body to defecate and urinate, to avoid soiling itself. Especially vocal when sexually active (April–May), also males urinate onto chest and neck, and rub it into the fur with their feet and forearms. Males pursue females by crawling along branches with thumbs and feet. Young born in October–November; singletons the norm, twins very occasional (MacKinnon *et al*. 2003, Andrianaivoarivelo *et al*. in press).

WHERE TO SEE Most accessible colony is at Berenty Reserve, west of Tolagnaro, where 800–1,800 bats roost in gallery forest (numbers vary seasonally). Also, colonies on Nosy Ravina south of Nosy Mangabe; Nosy Tanikely off Nosy Be; and in mangroves near Anjajavy. **Colonies should never be approached too closely (less than 100m), as this causes distress and can be particularly detrimental when females are pregnant or lactating.**

STRAW-COLOURED FRUIT BATS

GENUS *Eidolon*

A genus now placed in its own subfamily Eidolinae (Almeida *et al.* 2016) and containing two species; African Straw-coloured Fruit Bat *E. helvum*, Africa's most widely distributed fruit bat, found on the Arabian Peninsula and throughout sub-Saharan Africa; and Madagascar Straw-coloured Fruit Bat *E. dupreanum,* restricted to Madagascar (Giannini 2019). Molecular studies have revealed Madagascar Straw-coloured Fruit Bat is derived from African Straw-coloured Fruit Bat, with divergence probably occurring after initial colonisation of Madagascar, estimated at mid to late Miocene (*c.*7–14 mya) (Shi *et al.* 2014). Both species are gregarious and roost by day in colonies in large trees, buildings and caves. On mainland Africa some seasonal colonies attain huge proportions, containing >10 million bats (Kasanka National Park, Zambia).

Straw-coloured Fruit Bats have a more pointed head than other fruit bats and very long narrow wings adapted for flying long distances during seasonal migrations. As with all other Pteropodidae fruit bats, the wings are also used for climbing in branches at roost sites.

MADAGASCAR STRAW-COLOURED FRUIT BAT

Eidolon dupreanum

(Pollen in Schlegel & Pollen, 1867)

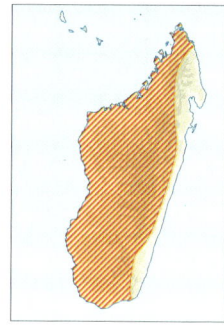

Other name: Malagasy Straw-coloured Fruit Bat

MEASUREMENTS
Head-body: 190–215mm. Forearm: 115–135mm. Tail: 14–22mm. Wingspan: 750–950mm. Weight: *c.*250–400g. Males *c.*10% larger than females. (Bennett & Russ 2001, MacKinnon *et al.* 2003, Giannini 2019)

DESCRIPTION & IDENTIFICATION Pelage short, covering head, dorsum and venter. Neck fur longer and woollier. Face in front of and below eyes is bare. Overall colour of head and body medium grey-brown, upperparts may be dull yellow, underparts often tawny-olive. Reddish-cinnamon patches form collar around neck; more conspicuous in males, but otherwise sexes alike. Wings long, pointed, narrow and dark blackish-brown. At rest wings folded back,

Straw-coloured fruit bats (genus Eidolon*) are strong flyers.*

Colony of Madagascar Straw-coloured Fruit Bats, Ankarana Special Reserve.

tips folded in. Second digit clawed (Giannini 2019). Madagascar Straw-coloured Fruit Bat larger than African Straw-coloured Fruit Bat: the second largest bat in Madagascar. Significantly smaller size, dull coloration, long narrow wings and small tail help distinguish from Madagascar Flying Fox.

HABITAT & DISTRIBUTION Widespread, but patchily distributed: in all regions across island but seemingly less numerous in eastern rainforest areas. Not recorded above 1,200m (MacKinnon *et al*. 2003). Able to survive in highly degraded areas: numerous roosts in Central Highlands where surrounding areas have been deforested, but tend to be much smaller (<100 bats) than roosts in forested areas (Ratrimomanarivo 2007, Andriafidison *et al*. 2008). Likely that distribution correlates to areas with rocky outcrops and crevices that are preferred roosting locations (Goodman *et al*. 2005). Widely hunted for bushmeat, especially in Central Highlands: a common technique is to smoke them out of their roosts or trap them at foraging sites (Jenkins & Racey 2008, Olesky *et al*. 2015a). IUCN Vulnerable.

BEHAVIOUR Nocturnal. Day spent at roost sites: can be canopy of dense stands of trees, but more commonly in crags, cliffs, rock fissures and especially caves with high, wide entrances (MacKinnon *et al*. 2003, Goodman *et al*. 2005). Sites deep in large caves offer stable conditions that buffer against extremes in temperatures and humidity (Cardiff 2006). Colonies typically 10–500 individuals (mean 200), although three roosts up to 1,400 known from Ankarana (MacKinnon *et al*. 2003, Andrianaivoarivelo *et al*. in press). Roosting bats are restless and noisy. and regularly move from place to place. Leave at dusk to search for fruiting trees. Roost size varies seasonally, particularly in semi-desert south-west: smallest during dry months when food is limited, which suggests some inter-island migration (MacKinnon *et al*. 2003, Andrianaivoarivelo *et al*. in press).

Ranges widely in search for food. Large roosts rarely close to one another, often >60km apart, suggesting bats forage up to 30km from roost site. Diet primarily fruit pulp and juice (native and non-native species), also nectar, blossoms and young shoots (Picot 2005, Picot *et al*. 2007). Known to feed

Corpse of Madagascar Straw-coloured Fruit Bat beneath a cave roost, Ankarana Special Reserve.

at flowers of baobabs (*Adansonia* spp.) and kapok trees (*Ceiba* spp.) (Andriafidison *et al.* 2006b). Tip of long tongue is 'brush-like' to extract nectar and pollen from deep flowers. Pulp of *Borassus* palm fruit is important and soft wood is occasionally chewed for moisture. Diet varied; faecal samples reveal pollen from 40 plant species and fruits of more than 30 species including several non-natives (Picot *et al.* 2007). Important seed disperser, with germination perhaps aided by passage through gut (Picot *et al.* 2007) and a key pollinator, including species like baobabs (*Adansonia* spp.) (Du Puy 1996, MacKinnon *et al.* 2003, Andriafidison *et al.* 2006b).

Births occur in December and January; single infant weighs *c*.50g. Large numbers of females may gather to form 'nursery' colonies. Thought to live more than 20 years.

WHERE TO SEE Roosts in *Raphia* palms in the Parc Botanique et Zoologique de Tsimbazaza in Antananarivo. Another well-known roost is at Lac Tritriva, near Antsirabe. The largest roosts in forested areas are in Ankarana Reserve (MacKinnon *et al.* 2003, Andriafidison *et al.* 2008), the most visible being in *Grotte des Chauve-souris*. **Approaching roost sites too closely causes major disturbance. Always remain distant. When near a roost do not touch or come into contact with bat urine or faeces.**

ROUSETTE FRUIT BATS
GENUS *Rousettus*

A genus containing eight species, widely distributed from the Indian Subcontinent to South-East Asia as far as Borneo, Sulawesi, New Guinea and Solomon Islands; also, Arabian Peninsula, and across Africa, including the Comoros (Bergmans 1994, Giannini 2019). Populations on Madagascar formerly considered subspecies of African mainland populations, but genetic studies confirm Madagascar form is distinct (Goodman *et al.* 2010d). Based on recent genetic studies, the colonisation origins of *Rousettus* in Madagascar are likely Asian (Stribna *et al.* 2019).

Rousette bats produce very high-pitched sounds by clicking their tongues (with an audible 'buzz'); this forms the basis of a simple but precise lingual echolocation system and is used only for orientation in dark caves where they roost (Holland *et al.* 2004, Yovel *et al.* 2010, 2011, Schoeman & Goodman 2012). Vision and olfaction remain the primary senses for food detection.

MADAGASCAR ROUSETTE
Rousettus madagascariensis
(G. Grandidier, 1929)

Other name: Malagasy Rousette

MEASUREMENTS
Head-body: 115–142mm. Forearm: 65–80mm. Tail: 14–25mm. Wingspan: 425–520mm. Weight: *c*.30–85g. (Bennett & Russ 2001, MacKinnon *et al.* 2003, Giannini 2019)

DESCRIPTION Pelage longish and dense, less so on neck, throat and shoulders. Upperparts greyish-brown with some reddish-brown hues, underparts paler grey-brown. Muzzle rather pointed; ears relatively short. Wings proportionately broad. Sexually dimorphic: males larger (Andrianaivoarivelo 2012, Goodman *et al.* 2017). A small-sized delicate fruit bat; the smallest fruit bat on Madagascar and smallest *Rousettus* in African region. In flight could be mistaken for a large 'insectivorous' or 'microbat', but eyeshine and direct darting flight distinguish it from the back-and-forth flight of 'microbats'.

HABITAT & DISTRIBUTION Endemic to Madagascar. Throughout eastern rainforests, including some offshore islands, and dry forests of west, north-west and far north in primary and degraded areas, forest edge and cultivated areas. Absent from Central Highlands, dry south and south-west, although recorded in semi-arid transition zone (Parcel 3) in Andohahela National Park (MacKinnon *et al.* 2003, Goodman *et al.* 2005, Randrianandrianina *et al.* 2006, Rakotonandrasana & Goodman 2007). Majority of records from lower elevations; not recorded above 1,150m. While probably not dependent on native forest (Goodman *et al.* 2005a), Madagascar Rousette is threatened by habitat destruction, which causes abandonment of roosts (Andrianaivoarivelo *et al.* 2011b). Also suffers considerable bushmeat hunting pressure throughout Madagascar (Goodman *et al.* 2008, Cardiff *et al.* 2009, Golden 2009, Andrianaivoarivelo *et al.* 2011a). IUCN Vulnerable.

Madagascar Rousette, Ankarana Special Reserve.

Colony of Madagascar Rousettes, Ankarana Special Reserve.

BEHAVIOUR Shows marked preference for roosting in deep caves well beyond twilight zone, but also in large trees and tree holes (Cardiff 2006, Andrianaivoarivelo et al. in press). Roosting requirements appear very specific and localised distribution probably correlates to proximity to caves (Andriafidison et al. 2006, Andrianaivoarivelo et al. 2019, Rakotoarivelo & Randrianandrananina 2007, Rakotonandrasana & Goodman 2007). Larger colonies range from a few hundred to >1,000 bats.

Initial studies reveal three distinct vocalisations: two associated with social contexts, and echo-clicks produced when leaving roosts and flying towards cave entrance that are sensory and potentially are incipient echolocation used to navigate in darkness (Schoeman & Goodman 2012).

Feeds on fruit juices, soft fruit pulp, flower parts, nectar and leaves. Fruit juice and pulp ingested, but fibre rejected. Fruits of endemic, rather than commercial species, are preferred (Razafindrakoto 2006, Andrianaivoarivelo et al. 2007, 2011b, 2012). Able to grasp fruit in mouth and fly off before consuming it. At times feeds more intensively on flowers than fruits, e.g., *Ceiba pentandra* flowers in Kirindy-Mitea (Andriafidison et al. 2006b). Flies considerable distances each night to find sufficient food. Roosts sites are often several km from food sources and round trips approaching 30km recorded (Giannini 2019). Madagascar Rousette is an important agent of seed dispersal, especially of endemic *Ficus* spp. (Moraeceae) (Andrianaivoarivelo et al. 2012a,b).

Pregnancy and births occur in wetter summer months (December–March) when food availability higher. Singletons are norm, occasionally twins. Weaning at *c.*8 weeks, sexually mature at one year (Giannini 2019). Predators of *Rousettus madagascariensis* include Barn Owl *Tyto alba* (Goodman & Griffiths 2006).

WHERE TO SEE Caves in Ankarana in north are the best places to see this species, especially *Grotte des Chauve-souris*. Other roosts include caves in Bemaraha National Park and Anjohibe near Mahajanga. Tourist visits to *Rousettus* colonies, especially ones deep in caves, do cause significant disturbance (Cardiff et al. 2012). **Visits should be discouraged. Always remain at least 15m from roosting bats and do not shine torches directly.**

Old World Leaf-nosed Bats Family Hipposideridae

A family occurring throughout the Old World tropics, from Africa, through South-East Asia and into Australasia; seven genera and 88 species. Easily recognised by a well-defined and elaborate nose leaf (hence sometimes called 'roundleaf' bats) and large pointed ears. All emit echolocation calls via the nostrils, where the nose leaf acts to focus and modify echolocation signals. Carnivorous and insectivorous, catching prey on the wing or hawking from a perch. All strictly nocturnal. Closely related to trident bats (Rhinonycteridae) and horseshoe bats (Rhinolophidae) (Monadjem 2019a).

Two extant species are known from Madagascar, including one recently described cryptic species (Goodman et al. 2016). Previously placed in genus *Hipposideros*, but both extant species in Madagascar now placed in genus *Macronycteris* (Foley et al. 2017). One extinct species is also known, Anjohibe Leaf-nosed Bat *M. besaoka* (Samonds 2007, Goodman et al. 2016b).

VOCALISATIONS

The echolocation calls of Hipposideridae are high duty-cycle, narrow-band, and long constant-frequency (Russ et al. 2003). The calls comprise up to three harmonics with most energy in the second harmonic (Kofoky et al. 2009). Commerson's Leaf-nosed Bat *Macronycteris commersoni* emits calls with peak frequency at *c.*66.6 kHz and long duration of *c.*20 ms, although there may be variation between the sexes and geographically (Ramasindrazana et al. 2015). The peak frequency of larger males is at 68.6 kHz, with smaller females *c.*72.9 kHz.

COMMERSON'S LEAF-NOSED BAT
Macronycteris commersoni
(É. Geoffroy Saint-Hilaire, 1813)

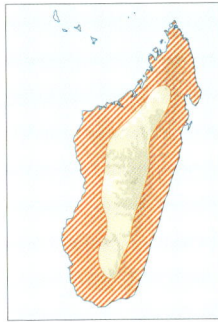

Other names: Commerson's Roundleaf Bat

MEASUREMENTS
Head-body: 104–110mm.
Forearm: 80–103mm.
Tail: 31–45mm. Wingspan: 475–560mm. Weight: 40–100g. Males larger than females. (Peterson *et al.* 1995, Ramasindrazana *et al.* 2015)

DESCRIPTION Pelage dense and short; upperparts and head pale grey-brown to reddish-brown, some individuals orange-brown (Goodman *et al.* 2014b), underparts paler, with very pale patches on flanks and underarms. Dark band on shoulders. Ears large, long and falcate with rounded tips. Nose leaf large, elaborate and distinctive. Commerson's Leaf-nosed Bat is largest insectivorous, i.e., 'non-fruit bat', in Madagascar. Large size, large ears and prominent leaf-nose distinctive. Slightly larger than but difficult to differentiate from Madagascar Cryptic Leaf-nosed Bat *M. cryptovalorona* (Foley *et al.* 2017).

DISTRIBUTION Endemic to Madagascar and widely distributed around entire island. Found in most native forest types, including rain, dry, littoral, gallery and degraded forests, and zones bordering cultivated areas (Goodman *et al.* 2005a, Raharinantenaina *et al.* 2008, Goodman & Ramasindrazana 2013). Mostly occurs below 850m, but recorded to 1,350m (Goodman 1996, Eger & Mitchell 2003, Rakotoarivelo *et al.* 2015). Bats from northern regions larger than from south (Ranivo & Goodman 2007). Threats include habitat loss and hunting, being particularly vulnerable at roosts (Jenkins & Racey 2008, Cardiff *et al.* 2009). Suffers considerable hunting pressure in most areas, with tens of thousands taken annually (Goodman 2006, Goodman *et al.* 2008c, 2011). IUCN Near Threatened.

BEHAVIOUR Most roost sites in caves or rock crevices (especially in west and south) and to lesser extent large hollow trees (Raharinantenaina *et al.* 2008). In eastern rainforest will roost singly on tree branches below dense canopy (Goodman *et al.* 2014, Razafimanahaka *et al.* 2016). Will occasionally utilise

Commerson's Leaf-nosed Bat, Marojejy National Park.

open roofs of rural buildings. Colonies numbering several hundred known from caves in Ankarana (McHale 1987); most colonies contain fewer than 20 individuals. Each bat roosts slightly apart – at wingtip distance – from its neighbour, rather than in a tight cluster. During breeding season, females roost separately from males (maternity roosts), with males and non-breeding females together in separate caves (Monadjem 2019a). Some populations may migrate within Madagascar, as some roost sites appear to be seasonally vacated (Ranivo & Goodman 2007, Ramasindrazana *et al.* 2015).

Recent research offers an alternative explanation for species' apparent 'disappearance'. In highly seasonal areas of west and south-west, during extreme dry periods Commerson's Leaf-nosed Bat becomes torpid and remains roosting in caves for long periods (c.2 months) (Reher *et al.* 2018), surviving on fat reserves accumulated during wet season (Reher *et al.* 2019). Indeed, this physiological response is deployed routinely throughout year: Commerson's Leaf-nosed Bat enters shortest, regular torpor bouts of any mammal studied to date

Commerson's Leaf-nosed Bat hunting, Tsimanampetsotse National Park.

(Dausmann *et al.* in press). When resting in warm conditions (24–34°C), they regularly shut down metabolism repeatedly for mini bouts of torpor lasting just a few minutes (Reher & Dausmann 2021).

A study in littoral forest (Tampolo) suggests home ranges are larger for females (*c.*42ha) than males (*c.*32ha), with core areas (85% of records) being *c.*32ha and *c.*15ha respectively. Male home ranges show little overlap, while some female home ranges overlap considerably (Razafimanahaka *et al.* 2016).

Generally, hunts in lower levels of forest, often hawking along trails and similar open areas for large insects. Diet dominated by beetles (Coleoptera) and, to a lesser extent, bugs (Hemiptera) but a variety of other large insects also taken and perhaps even small frogs (Razakarivony *et al.* 2005, Kofoky *et al.* 2007, Rakotoarivelo 2008). In dry areas hunts close to water (Rakotoarivelo *et al.* 2007, Bader *et al* 2015). Uses tree trunks or horizontal branch perches from which they sally out, returning to consume the prey (when most often seen). When suspended, they move their head side to side searching for prey (typical of large predatory bats) (Russ & Bennett 1999). In rainforest regions also suspected of predating small vocalising frogs along narrow streams (Eger & Mitchell 2003). During rainy season, when

food plentiful, lays down significant fat reserves (Goodman 2006, Rakotoarivelo *et al.* 2007).

Predated by a variety of birds and mammals, including Madagascar Long-eared Owl and Grandidier's Vontsira *Galidictis grandidieri* (Goodman *et al.* 1993a, Andriatsimietry *et al.* 2009).

WHERE TO SEE May be encountered at night hawking/perch-feeding along forest trails and around campsites at many locations: Ankarana, Kirindy, Ampijoroa, Masoala and Nosy Mangabe are more reliable sites.

MADAGASCAR CRYPTIC LEAF-NOSED BAT *Macronycteris cryptovalorona*
(Goodman *et al.*, 2016)

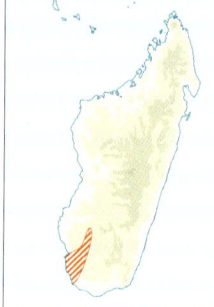

Other names: Madagascar Cryptic Roundleaf Bat

MEASUREMENTS
Total length: 89–92mm.
Forearm: 80–81mm.
Tail: 30–34mm. Wingspan: n/a. Weight: 26–42.5g.
(Goodman *et al.* 2016)

DESCRIPTION Pelage short and dense. Upperparts dark brown to reddish-brown, underparts paler. Whitish armpits. Ears long and narrow. Nose leaf large and elaborate. Smaller than Commerson's Leaf-nosed Bat and difficult to differentiate (Goodman *et al.* 2016).

DISTRIBUTION Endemic to Madagascar. In dry deciduous forest and spiny forest. Known from only a few localities in south-west: occurs sympatrically with Commerson's Leaf-nosed Bat at some (Goodman *et al.* 2016, Reher *et al.* 2018). IUCN Not evaluated.

BEHAVIOUR Little information available, but probably similar to *M. commersoni*.

WHERE TO SEE No known sites. Any *Macronycteris* sp. seen in potential localities like Isalo and Lac Tsimanampetsotse National Parks may correspond to this species.

Trident Bats Family Rhinonycteridae

The trident bats are a recently recognised family of insectivorous bats (formerly placed in Hipposideridae) (Foley *et al.* 2015, Armstrong *et al.* 2016). Rhinonycteridae contains nine species arranged across four genera and is distributed throughout the Middle East and tropical Africa. The complex nose leaf structure is unique to, and characteristic of, the group. Posterior part of nose leaf in all species bar one, including all Malagasy species, has three tall prominent projections that form a trident-like structure (Benda 2019). Trident bats use sophisticated echolocation calls produced in the larynx and emitted through the nose. All are strictly nocturnal.

Three species, split between two genera, occur on Madagascar, with a fourth species, Goodman's Trident Bat *Triaenops goodmani* now extinct (Samonds 2007). A relatively poorly known group: all are aerial insectivores. They broadly resemble leaf-nosed bats (*Macronycteris* sp.), but differ by smaller size and distinctive trident nose leaf. They prefer drier habitats; not recorded in humid eastern and Central Highland regions. Although small, all species are strong flyers.

Species in Madagascar have likely resulted from two separate colonisation events from Africa, the first leading to Rufous Trident Bat *Triaenops menamena* (and *T. goodmani*) and the second to the two species comprising the genus *Paratriaenops* (Russell *et al.* 2007, 2008a).

VOCALISATIONS

Rhinonycteridae emit high duty-cycle, narrow-band, and relatively short constant-frequency echolocation calls terminating in brief frequency-modulated (FM) elements (Russ *et al.* 2003). Calls have a maximum of three harmonics, with the majority of energy in second harmonic (Kofoky *et al.* 2009). Larger male Rufous Trident Bats *Triaenops menamena* emit lower peak frequency calls than smaller females, where as in Grandidier's Trident Bat *Paratriaenops auritus* and Trouessart's Trident Bat *P. furculus* smaller males emit higher peak frequency calls than larger females (Ramasindrazana *et al.* 2013).

RUFOUS TRIDENT BAT

Triaenops menamena   Goodman & Ranivo, 2009

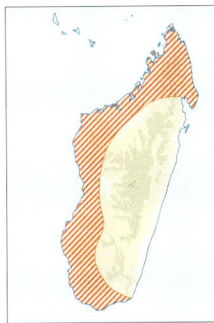

MEASUREMENTS
Head-body: 56–66mm.
Forearm: 46–56mm.
Tail: 28–35mm. Weight:
6–16g. (Benda 2019)

DESCRIPTION Upperparts and underparts similar. Variable in colour, from brownish to rufous brownish-grey, sometimes brighter rufous-orange. Generally darker than congeners in Madagascar. Has complex nose leaf, round or pentagonal in shape. Ears, short, wide and pointed, and roughly triangular (Goodman 2011, Ramasindrazana *et al.*

2013). Previously designated *T. rufus* (Goodman & Ranivo 2009). Easily identifiable by its echolocation call (Kofoky *et al.* 2007).

DISTRIBUTION Endemic. Known primarily from dry western and south-western forest regions, including karst (*tsingy*) formations, at lower elevations (below 550m), with fewer records in northern, southern and central regions, including the humid north-east and south-east (Goodman & Ranivo 2009). IUCN Least Concern.

BEHAVIOUR Strictly nocturnal and insectivorous. Hawks for insects in flight, including moths, beetles, bugs, cockroaches and flies, mainly within forest interior, less often at forest edge (Razakarivony *et al.* 2005, Rakotoarivelo *et al.* 2007). Becomes active at sunset, exiting roost after dusk in large numbers. Starts to return to roost in large numbers after 21.00. Roosts mainly in caves, especially in karst (*tsingy*) regions but also in hollow trees

Rufous Trident Bat, Tsimanampetsotse National Park.

including baobabs (*Adansonia* spp.). Roosts may be very large; largest recorded >40,000 individuals (Olsson *et al* 2006, Kofoky *et al*. 2007). During cold, dry season regularly enters short-term torpor during the day, before becoming active to forage at night (Reher *et al*. 2019). Predation by large boas recorded (Cardiff & Jenkins 2016).

GRANDIDIER'S TRIDENT BAT
Paratriaenops auritus   (G. Grandidier, 1912)

MEASUREMENTS
Head-body: 47–64mm.
Forearm: 43–51mm.
Tail: 19–28mm. Weight:
5–8g. (Benda 2019)

DESCRIPTION Pelage brownish-orange to reddish-golden. Upperparts slightly darker than underparts. Ears large and tail short. Ears and nose leaf pale greyish. Large nose leaf has three tall projections (Ramasindrazana *et al*. 2013). Previously included in genus *Triaenops*, now regarded as phylogenetically distant, hence alignment in *Paratriaenops* (Benda & Vallo 2009).

DISTRIBUTION Endemic. Restricted to dry forests in extreme north and north-west, from sea level to 600m. Southern limit Andrafiamena Mountains (Goodman *et al*. 2005, Goodman & Ramasindrazana 2013). Total range less than 20,000km² (Monadjem *et al*. 2017a). Locally abundant in parts of range, but scarce in other areas, probably related to location of roost caves (Goodman *et al*. 2005, Robinson *et al*. 2006). Appears dependent on forest and suitable caves for roosting (Goodman *et al*. 2005). IUCN Vulnerable.

BEHAVIOUR Strictly nocturnal. Little information available. Shows marked preference to roost in deep, narrow caves, beyond twilight zone (Cardiff 2006).

Roosts may contain over 1,000 and exceptionally up to 2,000 animals (Monadjem *et al.* 2017a).

TROUESSART'S TRIDENT BAT
Paratriaenops furculus (Trouessart, 1906)

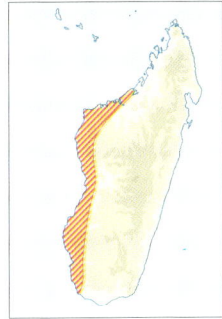

MEASUREMENTS
Head-body: 50–60mm.
Forearm: 42–49mm.
Tail: 19–27mm. Weight:
3–9g. (Benda 2019)

DESCRIPTION Relatively small. Drab grey-brown upperparts (some have russet tones), light grey underparts. Large nose leaf has three prominent projections. Ears moderately large, broad at their bases, slightly falcate and pointed at tips (Benda 2019). Previously placed in genus *Triaenops*, now regarded as phylogenetically distant and aligned in *Paratriaenops* (Benda & Vallo 2009).

DISTRIBUTION Possibly endemic. In Madagascar, restricted to narrow band in west, from Anjohibe in north to south-west. Associated with coastal deciduous, spiny and gallery forests, and degraded areas. Prefers karst (*tsingy*) areas. Does not extend inland to any extent above elevations of *c.*140m (Goodman *et al.* 2005, Ranivo & Goodman 2006, Kofoky *et al.* 2009). Also possibly recorded on Cosmoledo and Aldabra Atolls in outer Seychelles (Hutson 2004), which casts doubt on validity of endemism. IUCN Least Concern.

BEHAVIOUR Obligate cave-roosting species, often in large colonies of up to *c.*10,000 individuals (Olsson *et al.* 2006). Sometimes emerges before total darkness, with large-scale roost exits after dark (Olsson *et al.* 2006). Hunts by hawking, mainly in forest interior, rarely at edge. Diet mainly moths (Lepidoptera) and beetles (Coleoptera), with some bugs and cockroaches (Razakarivony *et al.* 2005). Appears to require relatively intact forest for foraging (Kofoky *et al.* 2007). Periodically enters torpor bouts lasting up to seven days (Reher *et al.* 2019). A study in Saint Augustin region recorded mean home ranges of *c.*48ha (range *c.*24.5–108ha) (Manjaoazy 2018). Remains of this species have been identified in Bat Hawk *Macheiramphus alcinus* pellets (Goodman *et al.* 2016a, Razakaratrimo *et al.* 2019).

Trouessart's Trident Bat.

Suborder Vespertilioniformes (Yangochiroptera)

The suborder Vespertilioniformes (also known as Yangochiroptera) contains 14 families, six (possibly five – see below) of which occur in Madagascar and one, Myzopodidae, is endemic. All species possess complex echolocation used for orientation and to locate, identify, track and capture prey. Most calls are ultrasonic and beyond the range of adult human hearing. Sounds are produced by contracting the muscles in the larynx and, for most of the Vespertilioniformes, are emitted via the mouth. Call features, such as frequency, bandwidth, duration and pulse interval, are all related to ecological niche and can also be used to distinguish between some species.

Diet varies widely among species and includes invertebrates, fruit, nectar, blood and small vertebrates. Species in Madagascar feed primarily on insects and other invertebrates. They are almost exclusively nocturnal.

Sheath-tailed Bats Family Emballonuridae

Around 54 species in 14 genera, widespread in tropical and subtropical regions of the Old and New Worlds. Emballonuridae is represented by four species divided between three genera in Madagascar, three of which are endemic (Bonaccorso 2019). It is thought each of the genera, *Paremballonura*, *Coleura* and *Taphozous*, represent separate colonisation events from Africa to Madagascar (Goodman *et al.* in press). All are small to medium-sized bats characterised by a shortish tail, which extends freely beyond the interfemoral membrane, but is retracted into a sheath in the membrane during flight.

VOCALISATIONS
Emballonuridae produce low duty-cycle, short constant-frequency or quasi-constant echolocation calls, terminating with frequency-modulated (FM) elements (Russ *et al.* 2003). Calls consist of up to five harmonics, with majority of energy in the second (fundamental) harmonic (Kofoky *et al.* 2009). Peak echolocation frequency of Western Sheath-tailed Bat *Paremballonura tiavato* is 54.2 kHz and Peters's Sheath-tailed Bat *P. atrata* is 52.9kHz, showing some overlap (Kofoky *et al.* 2009). When distressed Peters's Sheath-tailed Bat emits calls with broad FM components and short inter-pulse duration (Russ *et al.* 2004). Madagascar Sheath-tailed Bat *Coleura kibomalandy* has a peak echolocation frequency of 33.8 kHz, a band width of 2.9 kHz, and duration of 4.3 ms (Goodman *et al.* 2012a).

WESTERN SHEATH-TAILED BAT
Paremballonura tiavato   (Goodman *et al.*, 2006)

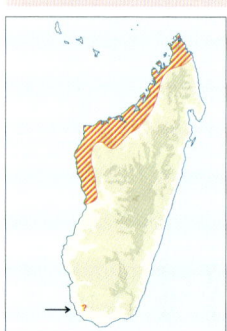

Other name: Rock-dwelling Sheath-tailed Bat

MEASUREMENTS
Head-body: 40–46mm.
Forearm: 35–41mm.
Tail: 15–18mm. Weight: 2.7–3.8g. (Goodman *et al.* 2006a)

DESCRIPTION Pelage long, slightly shaggy and silky. Upperparts medium greyish-brown, slightly paler on lower body. Underparts paler buff-brown. Noticeably paler than Peters's Sheath-tailed Bat (Goodman *et al.* 2006). Ears long (11–15mm) and slightly pointed. Previously in genus *Emballonura*, now assigned to new genus *Paremballonura* (Goodman *et al.* 2006a, 2012a).

DISTRIBUTION Endemic to Madagascar. Associated with intact native forest and believed to be forest dependent (Goodman *et al.* 2005a, 2006a). Recorded from sea level to 350m, from drier regions of north and north-west (Daraina and Ankarana) south to at least Tsingy de Bemaraha National Park, including islands of Nosy Be and Nosy Komba (Goodman et al. 2005a, 2006a). Also isolated record from Toliara in south-west may correspond to this species (Monadjem et al. 2017b). IUCN Least Concern.

BEHAVIOUR One of first species to emerge at dusk and begins foraging in forest understorey and above streams before darkness falls (Robinson et al. 2006). Slow, deliberate, delicate flyer. Diet includes moths and other insects (Razakarivony et al. 2005). Roosts in cave entrances and rock overhangs (not necessarily in complete darkness) in small numbers, usually fewer than 20 individuals (Cardiff 2006, Goodman et al. 2006, Kofoky et al. 2007). A single offspring is born in February. Specific name 'tiavato' comes from Malagasy meaning 'likes rocks'.

PETERS'S SHEATH-TAILED BAT
Paremballonura atrata   (Peters, 1874)

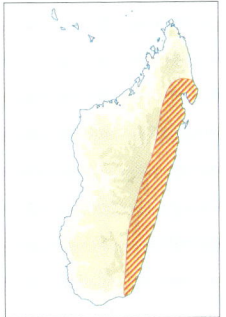

MEASUREMENTS
Head-body: 45–49mm.
Forearm: 37–41mm.
Tail: 18–20mm. Weight: 3.5–5g. (Bennett & Russ 2001, Eger & Mitchell 2003)

DESCRIPTION Small, uniformly dark slate-grey to black, generally slightly paler below. Nasal appendages poorly developed, snout quite pointed. Ears prominent, broad and rounded, with a distinct 'notch' near tip. Eyes relatively large. At rest, wingtip folded back onto wing (Bennett & Russ 2001). Originally in genus *Emballonura*, now assigned to *Paremballonura* (Goodman et al. 2012a).

DISTRIBUTION Endemic to Madagascar. Closely associated with humid native forest. Recorded from sea level to c.900m throughout eastern rainforest belt, and parts of east-Central Highlands, from Masoala Peninsula and Makira in north, to Tolagnaro in south (Goodman et al. 2006, Randrianandrianina et al. 2006). IUCN Least Concern.

BEHAVIOUR Roosts in caves or rock crevices in humid forest areas, invariably within twilight zone

(Goodman et al. 2006, Randrianandrianina et al. 2006, Jenkins et al. 2007b). Also roosts below exposed tree roots. In roosts often associates with Rufous Trident Bat and long-fingered bats (*Miniopterus* spp.). Roosts in forested areas contain 10–30 bats. Also known to roost in old and new buildings (López-Baucells et al. 2017).

Recorded foraging at 2–8m above forest edge and over rivers (Russ & Bennett 1999, Bennett & Russ 2001). Diet consists of flies (Diptera), ants (Hymenoptera), beetles (Coleoptera), bugs (Hemiptera) and other insects (Rasoanoro et al. 2015). A single offspring is born November (Monadjem et al. 2017c).

MADAGASCAR SHEATH-TAILED BAT
Coleura kibomalandy   (Goodman, et al. 2012)

MEASUREMENTS
Head-body: c.63–64mm.
Forearm: 48–52mm.
Tail: 11–18mm. Wingspan: 220mm. Weight: 8.4–12.5g (Goodman et al. 2012)

DESCRIPTION Pelage long and shaggy. Upperparts dark brown, underparts greyish-cream to pure white. Muzzle pointed, naked and black. Ears large and rounded, eyes large. Wings pale brown and translucent.

Madagascar Sheath-tailed Bat.

DISTRIBUTION Endemic. Known from only two sites, Ankarana in extreme north and Namoroka in north-west, at 20–115m. Generally associated with karst (*tsingy*) formations and may occur locally in other dry regions (Goodman *et al.* 2012a, Goodman & Ramasindrazana 2013, Goodman 2017b). Malagasy populations once regarded as African Sheath-tailed Bat *C. afra* (Goodman *et al.* 2008b) until description as endemic species (Goodman *et al.* 2012a, Goodman 2017b). IUCN Data Deficient.

BEHAVIOUR Probably crepuscular, foraging in and around forest edge. Diet moths and other insects. Roosts deep in limestone caves in colonies of up to *c.*500 individuals (Goodman *et al.* 2008b). Scant breeding data suggest females produce single offspring, perhaps twice a year (Goodman & Ramasindrazana 2013). Predation of this species recorded by Bat Hawk (Goodman *et al.* 2016a, Razakaratrimo *et al.* 2019) and Barn Owl (Goodman *et al.* 2008b).

MAURITIAN TOMB BAT
Taphozous mauritianus
É. Geoffroy Saint-Hilaire, 1818

MEASUREMENTS
Head-body: 76–88mm.
Forearm: 58–65mm.
Tail: 15–28mm. Weight: 20–36g. (Eger & Mitchell 2003, Bonaccorso 2019)

DESCRIPTION Pelage sleek with sheen. Medium-sized with mottled, grizzled grey upperparts and head, distinctive creamy-white underparts. Juveniles generally darker than adults. Head flattish and triangular, face bare below and in front of eyes. Ears broad and moderately rounded. Mature males have conspicuous gular (throat) pouch.

DISTRIBUTION Occurs throughout sub-Saharan Africa, but not in Horn of Africa, nor south of a line extending approximately from southern Angola to southern Mozambique (presence uncertain in equatorial Central Africa). Also, islands of Mauritius, Réunion, Assumption, Aldabra and Madagascar. In

Mauritian Tomb Bats roost either singly or more commonly in small clustered groups, Ankarafantsika National Park.

Madagascar, found throughout most regions and habitats below 900m, except driest areas of far north and extreme south (Eger & Mitchell 2003, Monadjem et al. 2017e, Bonaccorso 2019). IUCN Least Concern.

BEHAVIOUR Daytime roosts include shaded tree trunks, crevices in tree trunks, rock faces and both interior and exterior walls of buildings. Preferred sites provide overhead shelter and protection from predators (Bonaccorso 2019, O'Malley et al. 2020). Roosts singly, or more usually in small clusters of 2–5 bats. When disturbed moves quickly sideways in a crab-like manner or will fly to adjacent site. Hunts mainly after dark. An aerial feeder in open spaces and over water. Diet includes ants (Hymenoptera), moths, beetles and a variety of other insects (Taylor 2000, Razafimalala 2017). Generally encountered in small groups of between six and 12 individuals. Known predators include Barn Owl and Bat Hawk (Goodman et al. 1993b, 2016a, Razakaratrimo et al. 2019).

WHERE TO SEE Regularly found roosting on trunks of large trees in campsite at Ampijoroa (Ankarafantsika National Park). Also seen in rock crevices along Manambolo River gorge near Bekopaka (Tsingy de Bemaraha National Park).

Slit-faced Bats Family Nycteridae

A monogeneric family containing 15 species found throughout sub-Saharan Africa, along Nile Valley and into Arabian Peninsula. Also in South-East Asia, including Sumatra, Java and Borneo. Characterised by two unique features: a T-shaped cartilaginous process on the end vertebrae that is enclosed by the tail membrane, and folds of skin flanking a furrow along top of muzzle from nostrils to a pit in middle of forehead. Fur and nose leaf conceal this hollow, but obvious in the skull (Monadjem 2019b). Sometimes called hollow-faced bats. Ears typically large and long. All are strictly nocturnal. Highly manoeuvrable flyers and insectivorous/carnivorous. Larger taxa can take arthropods like spiders and scorpions. Roost singly or in small groups. Nycteris are easily recognisable because of distinctive face and long ears, but are very susceptible to disturbance and rarely seen. A single endemic species has been documented previously from Madagascar.

MADAGASCAR SLIT-FACED BAT
Nycteris madagascariensis   G. Grandidier, 1937

MEASUREMENTS
No details available.

DESCRIPTION No details available.

DISTRIBUTION Originally described from two specimens collected in 1912 from 'la Vallée du Rodo', in the past interpreted as the Irodo River valley, near Analamerana in the far north (Goodman 2011). Despite several subsequent surveys of the Ankarana and Analamerana areas, this species has not been recorded on the island since its original description (Goodman et al. 2005a, Robinson et al. 2006).

Recent molecular studies of the original specimen have revealed close alignment with the African mainland population of Large-eared Slit-faced Bat *Nycteris macrotis* (Demos et al. 2021). Being poor long-distance flyers, slit-faced bats rarely disperse to colonise offshore islands. This, in conjunction with the molecular results, casts considerable doubt as to whether the specimens described as *N. madagascariensis* originated in Madagascar and that the species is valid. Therefore, this species should no longer be regarded as occurring on Madagascar (Goodman et al. in press b).

Sucker-footed Bats Family Myzopodidae

A family endemic to Madagascar comprising two species. The name is derived from adhesive discs on the wrists and ankles that facilitate clinging to leaves or other smooth surfaces. These discs work by wet adhesion (not suction) and are kept moist by a network of glands (Riskin & Racey 2010, Ralisata *et al.* in press). Unlike convergent adaptations on Disc-winged Bats (family Thyropteridae) from South America, the suckers of *Myzopoda* are not on stalks.

Sucker-footed bats are small, insectivorous bats, with long, broad ears. Two unique morphological features distinguish the genus; a mushroom-shaped process (tragus) partly enclosing the ear opening and the conspicuous adhesive discs (Ralisata *et al.* in press). They do not have nose leafs. Unlike most bats that roost head down in a hanging position, sucker-footed bats roost head up when clinging to leaves: this increases adhesion efficiency (Riskin & Racey 2010). In flight they are largely noiseless, with wingbeats like a large fast-moving moth (Goodman 2019).

There is fossil evidence of Myzopodidae from Africa (Egypt and Tanzania) dating from the late Eocene and early Oligocene (up to 36 mya) (Gunnell *et al.* 2014, 2015). Why the family subsequently became extinct on the African mainland and how it originally colonised Madagascar from the mainland are unknown (Goodman *et al.* in press b).

VOCALISATIONS

The echolocation calls of Eastern Sucker-footed Bat *Myzopoda aurita* are distinct, consisting of very shallow, near-constant frequency sweeps, broken into a series of components, terminating with frequency-modulated (FM) elements (Gopfert & Wasserthal 1995, Russ *et al.* 2003, Ralisata 2018). The low duty-cycle echolocation calls have long duration (23 ms) with peak frequency at *c*.42 kHz (Russ *et al.* 2003). First harmonic dominant with fundamental harmonic occasionally seen at *c*.21 kHz. The distress call consists of a series of 'trills' (Gopfert & Wasserthal 1995). The echolocation calls of Western Sucker-footed Bat *M. schliemanni* have similarly complex structures with peak frequency at *c*.45.6 kHz (Goodman *et al.* in press b).

EASTERN SUCKER-FOOTED BAT
Myzopoda aurita

Milne-Edwards & A. Grandidier, 1878

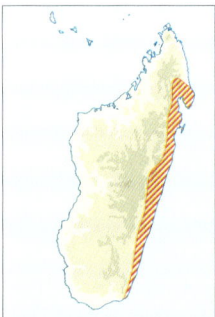

Other name: Madagascar Sucker-footed Bat

MEASUREMENTS
Total length: 106–120mm.
Forearm: 43–54mm.
Tail: 44–50mm. Weight: 8–9.5g. (Goodman *et al.* 2007, Ralisata in press)

DESCRIPTION Pelage moderately dense and long, upperparts mid-brown to golden dark-brown, underparts noticeably paler light brown. Ears large (30–35mm in length) and separate, with a mushroom-shaped tragus. Upper lip protrudes over lower. Tail long and extends beyond tail membrane. Thumb also has a small vestigial claw. Conspicuous suctions discs on wrists and ankles.

DISTRIBUTION Endemic. Restricted to humid eastern regions, from around Marojejy in north to Tolagnaro in south. Most records from lowlands and coasts; highest at 970m near Andasibe (Goodman *et al.* 2007). Lowland and littoral forests, forest edge, marshland and degraded areas, especially where there are large stands of Traveller's Tree *Ravenala madagascariensis* (Strelitziaceae) and *Typhonodorum* spp. (Araceae) (Schliemann & Goodman 2003, Goodman *et al.* 2014c). Traveller's Trees and similar species have proliferated where deforestation is acute; it is possible this is one of the few endemic mammals to have benefited from habitat modification by man (Eger & Mitchell 2003, Goodman 2019). Consequently, Eastern Sucker-footed Bat may be locally common. IUCN Least Concern.

Female Eastern Sucker-footed Bat, roosting in unfurled leaf of Traveller's Tree.

BEHAVIOUR The 'suction' pads allow *Myzopoda* to attach itself to and climb smooth vertical surfaces, such as large leaves, especially of Traveller's Tree. These organs also contain glands that secrete directly onto the surface of the suction pads and aid adhesion so entire body weight can be supported (Göpfert & Wasserthal 1995, Riskin & Racey 2010, Schliemann & Goodman 2011).

Principal roost sites dominated by Traveller's Trees. Roosts primarily in taller trees, mostly deep within unfurled leaves, with head uppermost, using stiff projecting tail as a prop (Ralista *et al.* 2015). Individual leaves may contain 1–51 individuals. Males and females roost separately: one site

contained nearly 600 males and no females. Sexes may separate according to subtle ecological and elevational differences. Bats change roosts every 1–5 days (Goodman *et al.* 2014b, Ralisata *et al.* 2010, 2015, Ralisata 2018).

Forages within areas varying from 4–148ha per night, much of time (70%) in degraded wooded/grassland habitat. Initially forages close to roost (first hour), then moves further afield, moving around 860m to 1,800m per night (Ralista *et al.* 2010). Diet primarily insects, dominated by moths (Lepidoptera), beetles (Coleoptera) and cockroaches (Blattaria), with smaller quantities of spiders also taken, which presumably means they are gleaned from vegetation, unlike insects taken on the wing (Ralista *et al.* 2009, Ramasindrazana *et al.* 2010, Ralisata *et al.* in press). Echolocation calls different to Western Sucker-footed Bat (Goodman 2019).

Breeding appears seasonal: male pseudo-scrotum swells October–November (Ralisata 2018) and juveniles captured February–March and October–November (Ralista *et al.* in press).

WESTERN SUCKER-FOOTED BAT
Myzopoda schliemanni   Goodman *et al.*, 2007

Other name: Schliemann's Sucker-footed Bat

MEASUREMENTS
Total length: 92–118mm.
Forearm: 45–51mm.
Tail: 42–54. Weight:
6.5–10.3g. (Goodman *et al.* 2007a, Ralisata in press)

DESCRIPTION Smaller and paler than Eastern Sucker-footed Bat. Pelage longish. Upperparts buff-brown, underparts paler mouse-grey. Prominent funnel-like ears with mushroom-shaped process (tragus) at base of each. Characteristic suckers on wrists and ankles. Tail extends well beyond tail membrane (Goodman *et al.* 2007a).

DISTRIBUTION Endemic. Known only from west and north-west; from Ankaboka in north to Andranomanintsy in south, including Ankarafantsika and Namoroka National Parks (Goodman *et al.* 2007a, Russell *et al.* 2008b). Not found at several survey sites further south (Monadjem 2017e). Found in dry deciduous forest and adjacent degraded areas from sea level to *c.* 200m (Goodman 2019). Very common at some marshes dominated by *Bismarckia* palms (Goodman *et al.* in press b). IUCN Least Concern.

BEHAVIOUR Roosts primarily in furled fronds of Bismarck Palms *Bismarckia nobilis*. Roost sites change frequently. Unlike Eastern Sucker-footed Bat, males and females do not appear to segregate when roosting. Individual palms contained 1–32 individuals (Goodman 2019, Ralisata *et al.* in press). In Namoraka National Park has been recorded roosting deep inside caves (Kofoky *et al.* 2006).

Forages in open and degraded areas, sometimes with marshy areas and open water (Rakotoarivelo & Randrianandrananina 2007, Goodman 2019);

mosaics of Bismarck Palms and grasslands preferred (Ralisata *et al.* in press). Foraging range varies at 15–205ha, with individuals travelling up to 3.6km from day roost (Ralisata *et al.* in press). Based on faecal samples, diet dominated by moths (Lepidoptera), cockroaches (Blattaria), beetles (Coleoptera) and ants (Hymenoptera) (Rajemison & Goodman 2007, Kofoky 2009, Ralisata *et al.* in press). Echolocation calls of Western Sucker-footed Bat similar to, but different from, those of congener (Goodman 2019).

Pregnant females captured between late October and mid December. Births recorded late November and early December (Ralisata *et al.* in press).

Predators of Western Sucker-footed Bat include Bat Hawk (Goodman *et al.* 2016a, Razakaratrimo *et al.* 2019).

Free-tailed Bats Family Molossidae

Worldwide distribution across subtropical and tropical regions: a diverse family with *c.*126 species divided across 22 genera. These bats gather in some of the largest vertebrate aggregations known; colonies of >20 million individuals roost in some caves (Taylor 2019). Robust and small to medium-sized bats with long narrow wings; strong flyers and aerial hunters. They lack a nose leaf and large proportion of their thick tail projects beyond the tail (interfemoral) membrane.

After recent revisions, and species descriptions, a family represented in Madagascar by eight

Guano from bat colonies within buildings, which may include free-tailed bats, is often collected and is a valuable fertiliser for crops such as rice.

species and five genera, *Mormopterus*, *Chaerephon*, *Mops*, *Otomops* and *Tadarida*. On the basis of recent phylogenetic studies, these are thought to represent seven separate colonisation events (Lamb *et al.* 2011, Ammerman *et al.* 2012), with separate arrivals for the monospecific genera *Mormopterus* and *Otomops* and five colonisation events by *Chaerephon*, *Mops* and *Tadarida* (Goodman *et al.* in press b).

VOCALISATIONS

Molossidae emit low duty-cycle, constant-frequency or quasi-constant echolocation calls. Most species have calls with low frequencies, with those of Malagasy Large-eared Free-tailed Bat *Otomops madagascariensis* falling within the range of human hearing (<20 kHz) (Russ *et al.* 2003, Manjoazy 2018).

The calls of four species of Molossidae in Madagascar have been recorded. The peak frequencies of Grandidier's Free-tailed Bat *Chaerephon leucogaster* at 32.0kHz and Malagasy Large White-bellied Free-tailed Bat *Mops leucostigma* at 32.8khz overlap, but their calls differ in bandwidth (31.2 and 20.7 kHz, respectively) and duration (9.0 and 5.9 ms, respectively) (Russ *et al.* 2003, Manjoazy 2018). The call of Peters's Little Mastiff Bat *Mormopterus jugularis* is variable: in open habitats the quasi-constant calls have a peak frequency of *c.*24 kHz, whereas in closed habitats calls show a relatively steep frequency-modulated (FM) sweep (Russ *et al.* 2003).

Malagasy Large-eared Free-tailed Bat produces calls with a low peak frequency (15–16.4 kHz), intermediate bandwidth (31.4 kHz) and intermediate duration (8.7 ms) (Russ *et al.* 2003, Manjoazy 2018). The species' distress calls are a series of low-frequency FM sweeps repeated at regular intervals (Russ *et al.* 2003).

PETERS'S LITTLE MASTIFF BAT

Mormopterus jugularis   (Peters in Sclater, 1865)

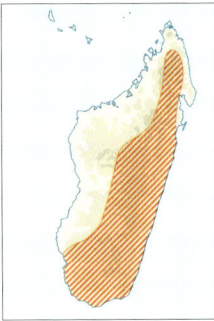

Other names: Peters's Wrinkle-lipped Bat, Peters's Goblin Bat

MEASUREMENTS
Head-body: 60–70mm.
Forearm: 30–39mm.
Tail: 20–38mm. Wingspan: 262–288mm. Weight: 8–17g. (Eger & Mitchell 2003, Taylor 2019)

DESCRIPTION Pelage soft and dense. Upperparts uniform reddish-brown to brown; underparts, including throat, much paler brown, approaching dirty white. Head flattened, muzzle blunt and slightly upturned. Ears small, triangular with rounded tips and separate at bases. Tragus small but not concealed (Taylor 2019).

DISTRIBUTION Endemic. Widespread, but patchy distribution; mainly in southern, eastern and central regions, fewer records in north. In dry and humid forest areas and anthropogenic degraded habitats, up to *c.*1,750m (Ratrimomanarivo *et al.* 2009,

Monadjem *et al.* 2017j, Taylor 2019). Often roosts in old buildings (Razafindrakoto *et al.* 2010). IUCN Least Concern.

BEHAVIOUR A crevice-roosting species adapted to synanthropic living: frequently uses old and new

Peters's Little Mastiff Bat, roosting.

Peters's Little Mastiff Bat, hawking over rice paddy.

buildings, as well as caves, rock crevices and cliff faces (Goodman *et al.* 2005, Andrianaivoarivelo *et al.* 2006, Ratrimomanarivo *et al.* 2009, López-Baucells *et al.* 2017). Usually occurs in colonies of fewer than 100 individuals, but colonies exceeding 22,000 recorded in caves at Ankarana (Taylor 2019). Often shares synanthropic roost sites with other Molossidae (Andrianaivoarivelo *et al.* 2006). Forages in open areas; diet dominated by beetles (Coleoptera) and bugs (Hemiptera), with fewer moths (Lepidoptera) and flies (Diptera) (Andrianaivoarivelo *et al.* 2006, Randrianandrianina *et al.* 2006).

Recorded predators include Barn Owl, Madagascar Long-eared Owl and Bat Hawk (Goodman *et al.* 1993a,b, 2016a, Razakaratrimo *et al.* 2019).

MALAGASY EASTERN FREE-TAILED BAT

Chaerephon atsinanana   Goodman *et al.*, 2010

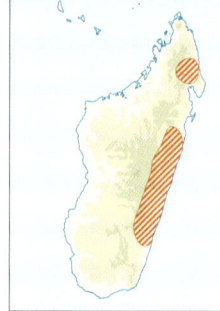

Other name: Madagascar Free-tailed Bat

MEASUREMENTS
Head-body: 61–67mm. Forearm: 37–42mm. Tail: 27–39mm. Weight: 9–17g. (Goodman *et al.* 2010a)

Malagasy Eastern Free-tailed Bat, hunting over rice paddy.

DESCRIPTION Pelage short. Upperparts blackish-brown, underparts dark brown, throat brown. Typically, a stripe of longer whitish or beige hairs on each flank. Previously included with *C. pumilus* (Goodman *et al.* 2010a, Taylor 2019).

DISTRIBUTION Endemic. Found in east, approximately from Farafangana to Toamasina and Makira, Anjanaharibe-Sud and Marojejy. Recorded to 1,100m (Goodman 2017a). IUCN Least Concern.

BEHAVIOUR Few data available. Roosts in old and new buildings including houses and churches (Razafindrakoto *et al.* 2010, López-Baucells *et al.* 2017). Diet includes beetles (Coleoptera), moths (Lepidoptera), bugs (Hemiptera) and flies (Diptera) (Andrianaivoarivelo *et al.* 2006, Rasoanoro *et al.* 2015).

MALAGASY WESTERN FREE-TAILED BAT

Chaerephon jobimena Goodman & Cardiff, 2004

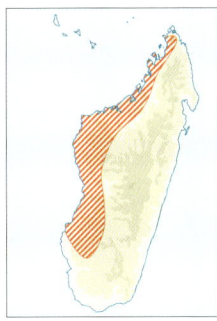

Other name: Black-and-Red Free-tailed Bat

MEASUREMENTS
Head-body: 75–79mm.
Forearm: 45–48mm.
Tail 32–51mm. Weight: 12–16g. (Goodman & Cardiff 2004)

DESCRIPTION Dorsal pelage dense and velvety. Two colour morphs; in one, upperparts rich chocolate-brown, in other they are distinctly rufous. Throat similar colour to dorsal regions, rest of underparts brownish-grey: colour transition is diffuse. Muzzle covered in hair and relatively blunt, upper lips have 5–6 wrinkles each side. Ears longer than congeners in Madagascar (Goodman & Cardiff 2004).

DISTRIBUTION Recorded at widely separated sites in dry deciduous forest up to 870m, including limestone karst (*tsingy*) in Ankarana in north, Namoroka in west, and Isalo and Zombitse National Parks in centre-south (Goodman & Cardiff 2004, Goodman *et al.* 2005a). IUCN Least Concern.

BEHAVIOUR Little information available. Roosts in caves, but also in hollow trees like baobabs

(*Adansonia* spp.) and synanthropic sites (Goodman & Cardiff 2004). Colonies up to 1,200 bats in caves at Ankarana and up to 40 in buildings near Namoroka National Park (Monadjem *et al.* 2017g).

GRANDIDIER'S LESSER FREE-TAILED BAT

Chaerephon leucogaster   (A. Grandidier, 1869)

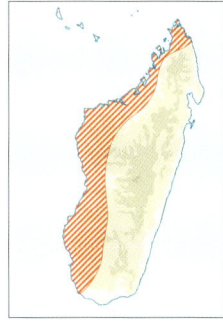

Other name: Madagascar White-bellied Free-tailed Bat

MEASUREMENTS
Head-body: *c.*53–62mm.
Forearm: 33–37mm.
Tail: 26–37mm. Wingspan: 262–280mm. Weight: 6–10g. (Eger & Mitchell 2003, Goodman & Cardiff 2004)

DESCRIPTION Pelage short. Dorsal region, throat and chest dark to mid-brown, belly and abdomen generally pale greyish-white. Ears rounded and joined by band of skin. Smallest Malagasy representative of group. Forms a species complex with Little Free-tailed Bat *C. pumilus*, Seychelles Free-tailed Bat *C. pusillus* and Malagasy Eastern Free-tailed Bat (Taylor 2019).

DISTRIBUTION Wide, scattered distribution across West, Central and East Africa. Also, Pemba, Zanzibar and Comoros islands. In Madagascar primarily recorded in drier south-west, west and north including Nosy Be and Nosy Komba. One record

Grandidier's Lesser Free-tailed Bat.

from east coast (Eger & Mitchell 2003, Goodman et al. 2005a, Taylor 2019). Found in dry open habitat, degraded forest and urban areas from sea level to c.900m.

BEHAVIOUR Little information available. Roosts mainly in old buildings (Eger & Mitchell 2003), either in single-species colonies or with Malagasy Large White-bellied Free-tailed Bat *Mops leucostigma* and Peters's Little Mastiff Bat (Goodman & Cardiff 2004). Less frequently in natural roosts under bark of dead trees. Known prey species of Barn Owl and Bat Hawk (Goodman et al. 1993b, 2016a, Razakaratrimo et al. 2019).

MALAGASY LARGE WHITE-BELLIED FREE-TAILED BAT

Mops leucostigma   (G. M. Allen, 1918)

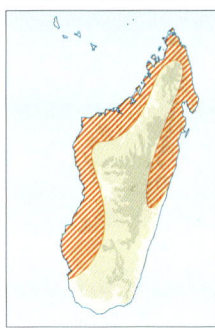

Other name: Malagasy White-bellied Free-tailed Bat

MEASUREMENTS
Head-body 70–90mm. Forearm: 40–47mm. Tail: 32–49mm. Wingspan: 315–347mm. Weight: 12–28g. Males larger than females (Eger & Mitchell 2003, Taylor 2019).

DESCRIPTION Pelage short and virtually absent between shoulders. Upperparts brown to greyish-brown, underparts paler with variable amounts of white. Upper lip has 7–8 well-defined wrinkles. Ears relatively short with skin joining them across crown.

DISTRIBUTION Endemic. Widespread; in eastern, south-western, western and northern regions up to 1,300m, but appears absent from far south and Central Highlands (Eger & Mitchell 2003). In native forest and anthropogenic areas, including urban zones. Also recorded on two islands, Mohéli and Anjouan in Comoros: origins of these populations unknown (Monadjem et al. 2017i). IUCN Least Concern.

BEHAVIOUR Roosts communally, colony size 7–25 in natural sites (caves and hollow trees), up to 350 individuals in buildings (Andriafidison et al. 2006a, López-Baucells et al. 2017, Taylor 2019). Hawks over

rivers, open areas and even beaches (Russ & Bennett 1999). Diet includes beetles (Coleoptera), moths (Lepidoptera), bugs (Hemiptera) and flies (Diptera) (Andrianaivoarivelo et al. 2006, Rabarison 2016). Recorded predators include Bat Hawk (Goodman et al. 2016a, Razakaratrimo et al. 2019).

MIDAS FREE-TAILED BAT

Mops midas   (Sundevall, 1843)

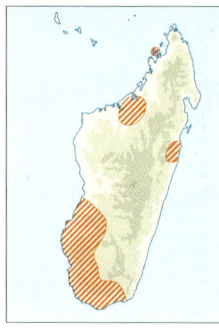

Other name: Midas Bat

MEASUREMENTS
Head-body: c.89–94mm. Forearm: 59–67mm. Tail: 37–56mm. Wingspan: 448mm. Weight: 38–69g. (Eger & Mitchell 2003, Taylor 2019)

DESCRIPTION Pelage short but sparse, especially on shoulders. Upperparts brown to greyish-brown, below paler and tawnier. Overall appearance of uniform coloration with slight 'frosting' or flecking. Reddish and orange-red colour morphs also described. Ears relatively long. Characterised by band of skin joining ears across crown. Upper lip has 5–6 distinct wrinkles. Robust skull and heavy jaw.

DISTRIBUTION Occurs at numerous but disconnected localities, from south-west Arabian Peninsula south through savannas of sub-Saharan Africa (Taylor 2019). In Madagascar widespread but discontinuous in drier areas; in northern, western and south-western regions below 150m (Ratrimomanarivo et al. 2007). IUCN Least Concern.

BEHAVIOUR Little information available. Habits largely unknown in Madagascar, but probably similar to African populations, where it prefers open woodland and savanna habitats. Small colonies found roosting in palm fronds and hollow trees, larger colonies (up to 600 in Madagascar) in old buildings (Monadjem et al. 2017d). Barn Owl is a known predator (Goodman & Griffiths 2009).

MALAGASY LARGE-EARED FREE-TAILED BAT

Otomops madagascariensis Dorst, 1953

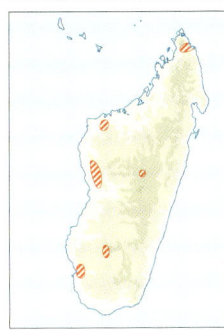

Other names: Madagascar Free-tailed Bat, Madagascar Giant Mastiff Bat

MEASUREMENTS
Head-body: 89–96mm. Forearm: 57–66mm. Tail: 35–50mm. Wingspan: 415–454mm. Weight: 17.5–29.5g. Males larger than females (Eger & Mitchell 2003, Taylor 2019).

DESCRIPTION Largish bat notable for flattened head and large ears. Pelage soft and dense. Upperparts vary from dark brown to reddish-brown, underparts slightly paler brown. There may also be a greyish area on nape and upper dorsal region. Ears 30–35mm long and directed forward at very oblique angle, projecting beyond snout. Face pinkish and pig-like. Wings long and narrow.

DISTRIBUTION Endemic. In dry deciduous and spiny bush areas from sea level to 800m. Known from disjunct localities in far north, north-west, west and south-west, and Antananarivo (Eger & Mitchell 2003, Goodman *et al.* 2005a). IUCN Least Concern.

BEHAVIOUR A fast-flying, high aerial feeder, preying primarily on moths (Lepidoptera) and beetles (Coleoptera) with smaller quantities of flies (Diptera),

bugs (Hemiptera) and mayflies (Ephemeroptera) (Andriafidison *et al.* 2007, Tsibara 2011). Can be both solitary and gregarious. Roosts in caves and hollow trees; up to 100 recorded in caves in limestone karst formations (Goodman *et al.* 2005a). In marine caves south of Toliara has been recorded roosting together with Trouessart's Trident Bat *Paratriaenops furcula*, Malagasy Mouse-eared Bat *Myotis goudoti*, Manavi Long-fingered Bat *Miniopterus manavi*, Major's Long-fingered Bat *M. majori* and Glen's Long-fingered Bat *M. gleni* (Peterson *et al.* 1995). Predated by Barn Owl and Bat Hawk (Goodman *et al.* 1993b, 2016a, Razakaratrimo *et al.* 2019).

MALAGASY LARGE FREE-TAILED BAT *Tadarida fulminans* (Thomas, 1903)

Other name: Malagasy Free-tailed Bat

MEASUREMENTS
Head-body: *c.*79–95mm. Forearm: 56–61mm. Tail: 53–66mm. Wingspan: 421–438mm. Weight: 23–49g. (Eger & Mitchell 2003, Taylor 2019)

DESCRIPTION Larger than most other African free-tailed bats. Sexes distinct: upperparts of males reddish-brown, in females dark chocolate-brown; underparts either tawny-pink (males) or white (females). Wings noticeably paler than body. Head relatively small and narrow with comparatively small ears. Lips lack obvious wrinkles (Taylor 2019).

DISTRIBUTION Found in disparate localities across East, Central and southern Africa (westermost in Democratic Republic of Congo), but originally collected near Fianarantsoa in Madagascar in 1903. Recorded from sea level to 2,000m (East African Rift) (Monadjem *et al.* 2017f). In Madagascar only a handful of records, mostly from central-southern regions, Isalo National Park and Tolagnaro region (Goodman *et al.* 2005a, Jenkins *et al.* 2007b). IUCN Least Concern.

BEHAVIOUR Few data available. In Africa roosts in small colonies of 30 or more in rock crevices and rocky outcrops. In Madagascar colonies may be *c.*20–50, but always fewer than 100 individuals (Monadjem *et al.* 2017f). A fast agile flyer (Cotterill & Fergusson 1993).

Malagasy Large-eared Free-tailed Bat.

Vesper Bats Family Vespertilionidae

The most abundant and widespread bat family, comprising 54 genera and nearly 500 species. More than one third of all bat species are considered Vespertilionidae. Also referred to as 'evening bats' or 'common bats'. Found throughout the world, except polar regions and some very remote islands. Has colonised all habitats from arid deserts to rainforests, and to elevations at limits of tree growth. Remarkably similar morphologically, characterised by small-medium stout bodies, no elaborate nasal appendages (with a few exceptions), long wings and tail completely, or almost completely, contained within the interfemoral membrane (Moratelli & Burgin 2019).

In Madagascar currently represented by 12* species in six genera: *Myotis*, *Neoromicia*, *Laephotis*, *Pipistrellus*, *Vansonia* and *Scotophilus*. The systematics and taxonomy of Afro-Malagasy vesper bats remain partially unresolved and it is likely that further cryptic species will be described in the future (Goodman *et al.* in press b). On the basis of published phylogenetic analyses these species represent eight separate colonisations of Madagascar from Africa (Goodman *et al.* in press b).

*Réunion House Bat *Scotophilus borbonicus* is not included. Despite exhaustive surveys, it has not been recorded on Réunion for *c.*140 years (Andriafidison *et al.* 2019), although it was once thought common (Cheke & Dahl 1981). On Madagascar known only from a single questionable specimen collected near Toliara in 1868 and referred to as *S.* cf. *borbonicus* (Goodman at al. 2005b). Species may be extinct and occurrence on Madagascar unsubstantiated (Goodman *et al.* in press b).

VOCALISATIONS

Members of Vespertilionidae emit low duty-cycle, frequency-modulated, quasi-constant frequency or constant-frequency echolocation calls. Their adjustable echolocation calls afford most species flexibility in their foraging behaviour, switching between open-air and clutter-edge habitat space (Fenton 1990).

Echolocation calls of six species exhibit considerable overlap: Isalo Serotine *Laephotis malagasyensis* (peak frequency 57.9 kHz, bandwidth 32.5 kHz, duration 1.6 ms), Madagascar Serotine *L. matroka* (peak frequency 57.5 kHz, bandwidth 37 kHz, duration 2.0 ms), Roberts's Serotine *L. robertsi* (peak frequency 53 kHz, bandwidth 32.1 kHz, duration 2.7 ms), Dark Madagascar Pipistrelle *Neoromicia bemainty* (peak frequency 50.2 kHz, bandwidth 47.6 kHz, duration 2.4 ms), Dusky Pipistrelle *Pipistrellus hesperidus* (peak frequency 54.0 kHz, bandwidth 48.6 kHz, duration 2.7 ms) and Racey's Pipistrelle *P. raceyi*) (peak frequency 56.1 kHz, bandwidth 78.3 kHz, duration 2.7 ms), although detailed analysis also revealed differences between species (Goodman *et al.* 2015b).

The Malagasy Mouse-eared Bat *Myotis goudoti* emits calls with a high peak frequency (64.4 kHz), very broad bandwidth (75.8 kHz) and short duration (3 ms) (Russ *et al.* 2003, Kofoky *et al.* 2009). The distress calls of this species are distinct from its echolocation calls, with a peak frequency of *c.*40 kHz and bandwidth of *c.*25 kHz (Russ *et al.* 2004).

The peak frequencies of Malagasy Western House Bat *Scotophilus tandrefana* and Marovaza House Bat *S. marovaza* show limited overlap (48.2 kHz and 45.9 kHz, respectively), with bandwidth (48.3 kHz and 26 kHz, respectively) and duration (3.0 ms and 6.4 ms, respectively) (Kofoky *et al.* 2009). Malagasy Large House Bat *S. robustus* emits echolocation calls with a 35–38 kHz peak frequency, 32.5 kHz bandwidth and 5 ms duration (Russ *et al.* 2003, Kofoky *et al.* 2009).

Malagasy Mouse-eared Bat, often emerges soon after dusk.

MALAGASY MOUSE-EARED BAT
Myotis goudoti   (A. Smith, 1834)

Other name: Malagasy Myotis

MEASUREMENTS
Total length: 90–100mm.
Forearm: 37–40mm.
Wingspan: unknown.
Weight: 5–6g. (Goodman 1998, Eger & Mitchell 2003)

Malagasy Mouse-eared Bat.

DESCRIPTION Pelage soft and dense. Upperparts sombre dark brown to ginger, underparts paler greyish-brown. Head somewhat flattened, with large rounded ears and short muzzle. Tragus reaches just beyond middle of ear (Bennett & Russ 2001).

DISTRIBUTION Endemic. Widely distributed across all regions including Central Highlands and islands of Nosy Be and Nosy Komba. Reported from sea level to 800m in lowland and mid-altitude montane forest, with one record in higher montane forest at 1,600m (Peterson *et al.* 1995, Eger & Mitchell 2003). IUCN Least Concern.

BEHAVIOUR Roosts in caves and rocky crevices (Goodman 2011), generally in small colonies, often with other species, particularly long-fingered bats (*Miniopterus* spp.). Roosts of up to 1,000 in Namoroka (Goodman *et al.* 2005a). One of the first bats to emerge at dusk. Often hunts along streams in forest. In eastern regions, diet principally beetles

(Coleoptera) and moths (Lepidoptera), along with ants (Hymenoptera), lacewings (Neuroptera) and spiders (Araneae) (Rasoanoro *et al.* 2015, Rabarison 2016), whereas at two western localities (Tsingy de Bamaraha and Namoroka), termites (Isoptera) and beetles (Coleoptera) dominated (Razakarivony *et al.* 2005). Predated by Bat Hawk (Goodman *et al.* 2016a).

DARK MADAGASCAR PIPISTRELLE
Neoromicia bemainty   (Goodman *et al.*, 2015)

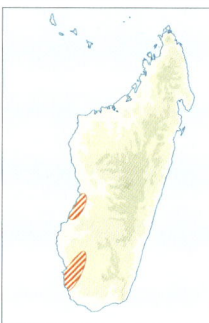

Other name: Kirindy Pipistrelle

MEASUREMENTS
Head-body: *c.*42–47mm. Forearm: 29–32mm. Tail: 33–36mm. Weight: 2.8–4.4g. (Goodman *et al.* 2015a).

DESCRIPTION Small pipistrelle. Pelage notably shaggy, upperparts dark tan to dark chocolate-brown with some grey streaking, underparts slightly paler medium tan. Bare face and ears, dark brown. Ears rounded at tips. Originally described as *Hypsugo bemainty*, but recent taxonomic revision now places it in genus *Neoromicia* (Monadjem *et al.* 2020). Species name '*be – mainty*' derived from Malagasy, meaning 'notably dark' (Goodman *et al.* 2015a).

DISTRIBUTION Endemic. Currently known only from sites in centre-west and south-west. Occurs

Dark Madagascar Pipistrelle.

in dry forest and degraded forest, and can tolerate human-altered habitats (Goodman 2017c). IUCN Least Concern.

BEHAVIOUR Little data available. In dry forest (Kirindy), termites (Isoptera), moths (Lepidoptera) and beetles (Coleoptera) dominate diet (Rakotondramanana *et al.* 2015). Presence of termites in diet likely seasonal, corresponding with wet season emergence and consumption of alates (Goodman *et al.* in press b). Predated by Bat Hawk (Goodman *et al.* 2016a, Razakaratrimo *et al.* 2019).

ISALO SEROTINE
Laephotis malagasyensis
                (Peterson, Eger & Mitchell, 1995)

MEASUREMENTS
Head-body: n/a. Forearm: 30.1–32mm. Tail: 30.4mm. Weight: n/a. (Goodman & Ranivo 2004)

DESCRIPTION Upperparts dark brown, underparts paler buff-grey to brown-grey, becoming paler towards tail (Goodman & Ranivo 2004). Originally described as endemic subspecies of Somali Serotine *Neoromicia somalicus* from Africa (Peterson *et al.* 1995). Elevated to species based on differing coloration, external measurements (it is larger) and cranial and dental features (Goodman & Ranivo 2004), a distinction questioned by some authors (Ibáñez & Juste 2019). Recent taxonomic revision moves this species from *Neoromicia* to *Laephotis* (Monadjem *et al.* 2020).

DISTRIBUTION Endemic. Known from only four sites in Sakaraha and Isalo National Park area in centre-south at 450–700m (Goodman & Ranivo 2004). IUCN Vulnerable.

BEHAVIOUR Little information available. Collected in palm savanna, specimens from Isalo caught around perimeter of massif (Canyon des Singes). Associated with dry open habitats. Diet insects.

MADAGASCAR SEROTINE

Laephotis matroka   (Thomas & Schwann, 1905)

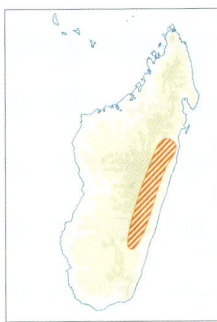

MEASUREMENTS
Head-body: *c.*46–54mm.
Forearm: 28–36mm. Tail:
29–35mm. Weight: *c.*9g.
(Ibáñez & Juste 2019).

DESCRIPTION Pelage soft and dense. Upperparts dark brown, underparts brownish-grey, paler still towards abdomen. Ears triangular with rounded tips. Originally placed in genus *Eptesicus*, then in *Neoromicia* based on bacular (penis bone) morphology (Bates *et al.* 2006). With recent taxonomic revision now in genus *Laephotis* (Monadjem *et al.* 2020). Name *matroka* refers to colour of dorsal fur and means 'dark' in Malagasy.

DISTRIBUTION Endemic. Appears restricted to eastern Central Highland regions; approximately Ambohitantely/Zahamena in north, to Andringitra/Zazafotsy in south, between 970m and 1,300m (Bates *et al.* 2006). In humid montane forest, degraded forest and anthropogenic grassland areas. IUCN Least Concern.

BEHAVIOUR Little information available. Known to roost in buildings (López-Baucells *et al.* 2017). Emerges at dusk. At Anjozorobe in eastern Central Highlands beetles (Coleoptera) and moths (Lepidoptera) dominated the diet (Rakotondramanana *et al.* 2015).

ROBERTS'S SEROTINE

Laephotis robertsi   (Goodman *et al.* 2012)

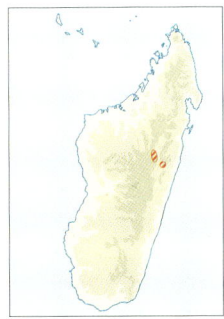

MEASUREMENTS
Head-body: *c.*53–58mm.
Forearm: 34.5–38mm.
Tail: 31–35mm. Weight:
n/a. (Goodman *et al.*
2012b)

DESCRIPTION Upperparts shades of dark brown, underparts similar but perhaps fractionally paler. Females usually larger than males (Goodman *et al.* 2012b). Distinctly larger than other vesper bats in Madagascar: only species that can be identified using external characteristics (Goodman *et al.* 2015b). Originally assigned to *Neoromicia melckorum* until described as new species, *N. robertsi* (Goodman *et al.* 2012b). After recent taxonomic revision now in genus *Laephotis* (Monadjem *et al.* 2020).

DISTRIBUTION Endemic. Currently known only from two localities in east. Anjozorobe-Angavo and Andasibe-Mantadia National Park. In forest and partially degraded habitats, at 900–1,300m. IUCN Data Deficient.

BEHAVIOUR Little information available. Diet includes beetles (Coleoptera), ants (Hymenoptera), moths (Lepidoptera) and some aquatic insects (Trichoptera) (Rakotondramanana *et al.* 2015).

DUSKY PIPISTRELLE

Pipistrellus hesperidus   (Temminck, 1940)

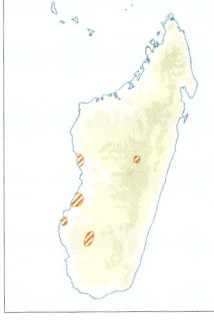

Other name: African
Pipistrelle

MEASUREMENTS
Head-body: *c.*65–68mm.
Forearm: 48–50mm.
Tail: 54–63mm. Weight:
12–15.5g. (Moratelli &
Burgin 2019)

DESCRIPTION Pelage soft and dense, coloration highly variable. Upperparts range from greyish-brown to reddish-brown or dark brown, sometimes nearly black. Underparts often paler, cream, creamy-orange, but can be darker, from reddish-orange to dark brown. Generally, underparts are paler than upperparts. Face naked and dark brown, ears have rounded tips.

DISTRIBUTION Found on east side of Africa, from Eritrea, through Rift Valley to Mozambique and South Africa. Also at scattered localities in southwest, Central and West Africa. In Madagascar recorded primarily on western side of island. Found in a variety of habitats: rainforest, montane forest, dry forest, grasslands and coastal areas from sea level to c.3000m (Ethiopian Highlands). Distribution strongly influenced by annual precipitation. Absent from driest areas (Moratelli & Burgin 2019). IUCN Least Concern.

BEHAVIOUR Little information available. Roosts in small groups (up to 12 individuals). Emerges at dusk and active throughout night. Feeds on wide variety of small insects. Bat Hawk is a known predator (Goodman *et al.* 2016a, Razakaratrimo *et al.* 2019).

RACEY'S PIPISTRELLE
Pipistrellus raceyi   Bates *et al.*, 2006

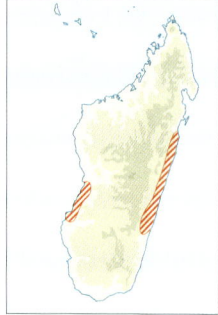

MEASUREMENTS
Head-body: 46–52mm.
Forearm: 27–33mm.
Tail: 22.9–34mm. Weight: 3.8–5.8g. (Bates *et al.* 2006)

DESCRIPTION Pelage relatively long. Upperparts pale rufous, head clearly darker. Underparts buff-brown to yellowish-brown. Face naked and pinkish, darkening towards short muzzle tip. Ears short and rounded. Eastern individuals larger than those from west.

DISTRIBUTION Endemic. Currently known from just four sites and two apparently distinct populations: two sites in lowland centre-east regions (Kianjavato and Tampolo Forests) and two in centre-west coastal regions (Mikea and Kirindy Forests), all below 80m (Bates *et al.* 2006). In humid forest and agricultural areas in east, intact and degraded dry forest in west. Further investigation likely to reveal broader distribution (Bates *et al.* 2006). Eastern and western populations may warrant separation into subspecies or species (Goodman *et al.* 2015b). IUCN Data Deficient.

BEHAVIOUR Few data available. Roosts in buildings and forest. Shares roosts with other species, e.g., Dark Madagascar Pipistrelle and Malagasy Mouse-eared Bat. In central south-east lowlands (Kianjavato) diet mainly beetles (Coleoptera) and flies (Diptera) with smaller numbers of moths (Lepidoptera) and Homoptera, whereas at a lowland western site

diet was primarily termites (Isoptera), beetles (Coleoptera) and moths (Lepidoptera), with some ants (Hymenoptera) (Rakotondramanana *et al.* 2015, Rasoanoro *et al.* 2015). Presence of termites in diet is likely seasonal, corresponding with wet season emergence and consumption of alates (Goodman *et al.* in press b). Predated by Bat Hawk (Goodman *et al.* 2016a, Razakaratrimo *et al.* 2019).

RÜPPELL'S PIPISTRELLE
Vansonia rueppellii   (Fischer, 1829)

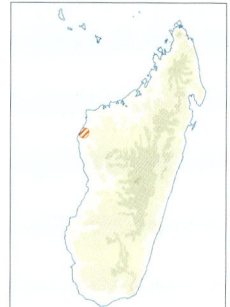

Other Name: Rüppell's Bat

MEASUREMENTS
Head-body: c.43–56mm.
Forearm: c.28-37mm.
Tail: 25-41mm. Weight: c.4–9g. (Moratelli and Burgin 2019)

DESCRIPTION Pelage soft, dense and slick. Upperparts range from grey, to greyish-brown to sepia-brown, with silvery sheen. Underparts white, throat also white, sometimes pale russet-brown. Face blackish. Ears relatively long relative to *Pipistrellus* spp. Wings generally translucent pale grey, sometimes vary from whitish to medium brown.

DISTRIBUTION Known from scattered localities through Africa and Middle East. Primarily found in woodland and savannas, also recorded in deserts and dense montane forest. In Madagascar recently recorded in west in disturbed marshland dominated by *Bismarckia* palms. IUCN Least Concern.

Malagasy Western House Bat.

BEHAVIOUR Little information available. An acrobatic flyer. Generally forages over water or marshy areas at dusk or at night. Prey predominantly Lepidoptera, Diptera, Coleoptera, Trichoptera and Hymenoptera.

MALAGASY WESTERN HOUSE BAT
Scotophilus tandrefana   Goodman *et al.*, 2005

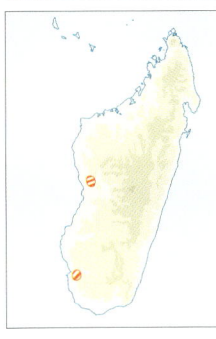

Other names: Western Yellow Bat, Malagasy Yellow Bat

MEASUREMENTS
Head-body: *c*.65–68mm.
Forearm: 44–47mm.
Tail: 43–46mm. Weight:
c.14g. (Goodman *et al.*
2005b)

DESCRIPTION Relatively small. Pelage long and soft. Upperparts rich dark chocolate-brown, underparts, chest and throat paler medium brown. Muzzle relatively short and rounded; black ears also short (Goodman *et al.* 2005b).

DISTRIBUTION Known from only two locations below 100m: Tsingy de Bemaraha National Park in centre-west, and Mahabo and Sarodrano in south-west (Goodman *et al.* 2005b). In dry forest and forest edge including grassland. All records from areas close to intact forest. Species name '*tandrefana*' is Malagasy meaning 'from the west'. IUCN Data Deficient.

BEHAVIOUR No information available.

MAROVAZA HOUSE BAT
Scotophilus marovaza   Goodman *et al.*, 2006

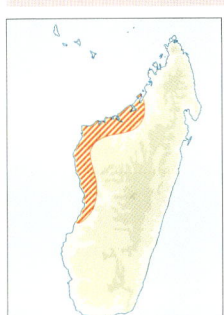

Other name: Marovaza Yellow Bat

MEASUREMENTS
Head-body: *c*.62–68mm.
Forearm: 41–45mm.
Tail: 38–45mm. Weight:
c.12.5–16.8g. (Goodman *et al.* 2006c)

DESCRIPTION Pelage short. Upperparts reddish-brown with distinctly paler brownish-red band across central back. Underparts and throat paler brownish-yellow. Muzzle broad and short, with pug-like appearance.

DISTRIBUTION Currently known at several localities in west, from Morondava area to Marovaza north of Mahajanga, but probably more widespread in lowlands of west (Goodman *et al.* 2006c, Rakotoarivelo & Randrianandriananina 2007, Kofoky *et al.* 2009). Dry forest, fragmented forest and anthropogenic savannas in proximity of forest. To 200m. IUCN Least Concern.

BEHAVIOUR Little information available. Roosts in palm fronds (*Bismarckia* spp.) and in roofs of buildings made from same material. Groups of up to five (Ratrimomanarivo & Goodman 2005, Goodman *et al.* 2006c). Diet apparently dominated by beetles (Coleoptera) with smaller quantities of bugs (Hemiptera), moths (Lepidoptera) and ants (Hymenoptera) (Ravelojaona 2017). Other aspects unknown. Recorded predation by Bat Hawk (Goodman *et al.* 2016a, Razakaratrimo *et al.* 2019).

MALAGASY LARGE HOUSE BAT
Scotophilus robustus   Milne-Edwards, 1881

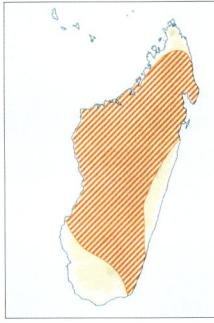

Other name: Robust Yellow Bat

MEASUREMENTS
Head-body: *c*.93–98mm.
Forearm: 62–65mm.
Tail: 55–70mm. Weight:
40.5–49g. (Eger & Mitchell 2003, Goodman *et al.* 2005b)

DESCRIPTION Largest species of Vespertilionidae in Madagascar. Heavy-bodied with robust head. Upperparts and underparts mid-brown. Skull large; jaws and teeth powerful.

DISTRIBUTION Endemic. Widespread; recorded in north-east (Marojejy and Masoala), centre-east (Tsinjoarivo and Anjozorobe), south-east, west (Bemaraha) and south (Zombitse). Associated with all native forest types up to 1,400m, except spiny forest in south-west (Pont & Armstrong

1990, Razolozaka *et al.* 1994, Bayliss & Hayes 1999, Goodman *et al.* 2005 a,b, Ratrimomanarivo & Goodman 2005). Appears rare throughout range (Eger & Mitchell 2003, Monadjem *et al.* 2017k). IUCN Least Concern.

BEHAVIOUR Few data available. Roosts around habitation and in natural sites (Ratrimomanarivo 2005). A strong flyer; leaves roosts around dusk. Diet varied, largely insect-based: in east includes moths (Lepidoptera), cockroaches (Blattodea), beetles (Coleoptera) and ants (Hymenoptera), while at a western locality beetles (Coleoptera) comprised >70% of diet (Rakotondramanana *et al.* 2015, Rasoanoro *et al.* 2015, Ravelojaona 2017). Predated by Bat Hawk (Goodman *et al.* 2016a, Razakaratrimo *et al.* 2019).

Malagasy Large House Bat.

Long-fingered Bats Family Miniopteridae

Previously, long-fingered bats were placed in Vespertilionidae; however, detailed genetic analysis has confirmed they are sufficiently distinct to be placed in their own family, Miniopteridae (Miller-Butterworth *et al.* 2007). A widely distributed monogeneric (*Miniopterus*) Old World family containing 38 species. The second phalanx (finger bone) of the longest digit (third finger) is around three times that of the first finger bone and is so elongate that the wings fold back on themselves when hanging at rest. Small to medium-sized bats, with simple muzzle (no nasal processes or nose leaf) and tail contained within membrane. All species are fast, manoeuvrable flyers that pursue insects in flight, often feeding at low levels (Ibáñez & Juste 2019). Also known as bent-winged bats, between 2007

and 2015 nine new species were described from Madagascar, and 12 species are now recognised from the island. All are very similar and expert knowledge needed to identify to species level.

Vocalisations
Long-fingered bats emit low duty-cycle, frequency-modulated echolocation calls. Peak echolocation frequency of species falls between 40.1 and 134.8 kHz, with duration ranges of 2.1–5.0 ms (Russ *et al.* 2003, Ramasindrazana *et al.* 2011, Goodman *et al.* 2015b).

   The distress calls of Major's Long-fingered Bat *Miniopterus majori* and Manavi Long-fingered Bat *M. manavi* are distinct from their echolocation calls, with peak frequencies of *c.*30 and 35 kHz, respectively, and steep frequency-modulated (FM) bandwidths of *c.*50 and 60 kHz, respectively, and very short inter-pulse durations (Russ *et al.* 2004).

Long-fingered Bats Miniopterus *spp. are very difficult to identify in the field to species level.*

AELLEN'S LONG-FINGERED BAT
Miniopterus aelleni   Goodman *et al.*, 2009

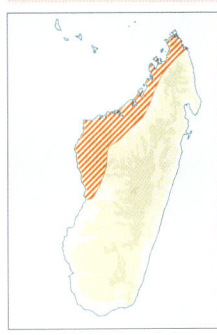

Other name: Aellen's Bent-winged Bat

MEASUREMENTS
Head-body: *c.*48–49mm.
Forearm: 35–38mm.
Tail: 39–44mm. Weight:
3.7–5.4g. (Goodman *et al.*
2009c)

DESCRIPTION Upperparts medium to dark brown, underparts slightly mottled. Head can be slightly paler.

DISTRIBUTION Endemic to Madagascar and Comoros. In Madagascar occurs in deciduous forest areas, including karstic (*tsingy*) formations and degraded anthropogenic areas from central west to northern tip of the island up to 1,350m (Goodman & Ramasindrazana 2013). Also on island of Anjouan in Comoros archipelago, where it occurs in more humid forests up to 690m (Goodman *et al.* 2010c). IUCN Least Concern.

BEHAVIOUR Little information available. Like other long-fingered bats, probably feeds on insects caught on the wing. Caves appear to be preferred roost sites (Goodman 2017d).

GRIVEAUD'S LONG-FINGERED BAT *Miniopterus griveaudi*   Harrison, 1959

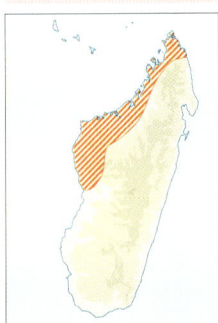

Other name: Griveaud's Bent-winged Bat

MEASUREMENTS
Head-body: *c.*50–51mm.
Forearm: 35–38mm.
Tail: 35–43mm. Weight:
4.1–7.1g. (Ibáñez & Juste 2019).

DESCRIPTION Upperparts dark brown, some individuals noticeably darker, others more reddish-brown (more prevalent on Comoros than Madagascar). Underparts greyish-buff with mottled

Several Miniopterus *sp. roost in cave systems in Ankarana Special Reserve.*

appearance. Head slightly paler than upperparts. Previously aligned with *M. minor* before being elevated to separate species (Juste *et al.* 2007).

DISTRIBUTION Endemic to Madagascar and Comoros. In dry western, north-west and northern regions of Madagascar up to *c.*400m, in native forest and cultivated areas. Also occurs in Comoro Islands (Grande Comoro and Anjouan) (Goodman *et al.* 2009a,b, 2010c). IUCN Data Deficient.

BEHAVIOUR Little information available. Probably forages in open areas and catches insects in flight. Roosts in caves in karst (*tsingy*) formations (Ibáñez & Juste 2019).

GLEN'S LONG-FINGERED BAT
Miniopterus gleni   Peterson, Eger & Mitchell, 1995

Other name: Glen's Bent-winged Bat

MEASUREMENTS
Head-body: *c.*68mm.
Forearm: 47–50mm.
Tail: 52–63mm. Weight:
10.5–17.5g. (Ibáñez & Juste 2019).

DESCRIPTION Largest long-fingered bat in Madagascar. Upperparts uniform dark chocolate-brown, underparts slightly paler chocolate-brown.

DISTRIBUTION Endemic. Recorded in all regions, including Central Highlands, but not in extreme south and south-west. In latter does not occur south of Onilahy River, where it is replaced by *M. griffithsi* (Eger & Mitchell 2003, Goodman *et al.* 2010c). Occurs in all forest types and anthropogenic areas up to *c.*1,250m. Larger populations in areas with extensive cave systems, such as the north and south-west. IUCN Least Concern.

BEHAVIOUR Little data available. Seems less gregarious than other *Miniopterus*; typical roosts of 2–8 individuals (Robinson *et al.* 2006), but colonies up to 90 in Ankarana (Monadjem *et al.* 2017h). Roosts in caves and known to share them with several other bat species (Cardiff 2006, Goodman 2006, Robinson *et al.* 2006, Kofoky *et al.* 2007).

SHORT-TRAGUS LONG-FINGERED BAT
Miniopterus brachytragos   Goodman *et al.*, 2009

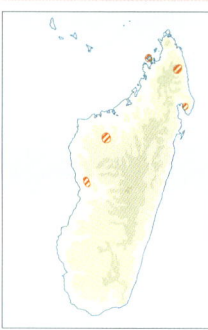

Other names: Madagascar Long-fingered Bat, Madagascar Bent-winged Bat

MEASUREMENTS
Head-body: c45–49mm.
Forearm: 35–38mm.
Tail: 38–43mm.
Weight: 2.9–6.3g.
(Goodman *et al.* 2009c)

DESCRIPTION Pelage short but not dense. Upperparts medium brown to dark brown. Underparts slightly mottled. Described as part of *M. manavi* complex, which probably contains several species, some not closely related to *M. manavi* (Christidis *et al.* 2014), one of these being Short-tragus Long-fingered Bat (Goodman *et al.* 2009c).

DISTRIBUTION To date recorded at seven sites in west, north and north-east encompassing both dry forest and humid forest locations, from near sea level to 600m. Also found in adjacent disturbed and anthropogenic areas (Ramasindrazana *et al.* 2011,

Goodman & Ramasindrazana 2013, Goodman 2017f). IUCN Least Concern.

BEHAVIOUR Little information available. Roosts in caves, often in vicinity of native forest (Goodman 2017f). Like other long-fingered bats, probably catches moths and other insects in flight (Ibáñez & Juste 2019).

MALAGASY NORTHERN LONG-FINGERED BAT
Miniopterus ambohitrensis   Goodman *et al.*, 2015

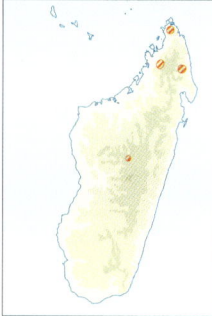

Other names: Montagne D'Ambre Long-fingered Bat, Montagne D'Ambre Bent-winged Bat

MEASUREMENTS
Head-body: *c.*53–55mm.
Forearm: 37–42mm.
Tail: 40–47mm. Weight: 5.3–7.7g. (Goodman *et al.* 2015b)

DESCRIPTION Pelage mixed medium and dark brown, chest and head sometimes tinged rufous-brown. Very similar to Aellen's Long-fingered Bat, except for rufous areas and slightly larger size. Previously included in *M. manavi* species complex (Goodman 2017e).

Miniopterus sp. in caves near Iharana, south of Ankarana Special Reserve.

DISTRIBUTION Known from four sites, three in far north (including Mt D'Ambre and Marojejy National Parks) and one in Central Highlands (Ambohitantely). All are wetter montane forest areas at 810–1,570m (Goodman *et al.* 2015b). Recorded in considerable numbers at these sites (Goodman 2017e). IUCN Least Concern.

BEHAVIOUR Little information available. Roosts in caves and rock crevices. Similar to other long-fingered bats, probably preys on moths and other insects in flight (Ibáñez & Juste 2019).

MAJOR'S LONG-FINGERED BAT
Miniopterus majori   Thomas, 1906

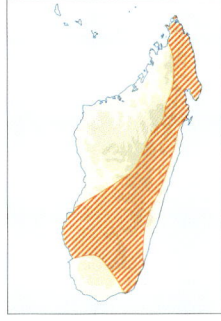

Other name: Major's Bent-winged Bat

MEASUREMENTS
Head-body: *c.*60–61mm.
Forearm: 43–47mm.
Tail: 51–60mm. Weight:
8.4–12.5g. (Ibáñez & Juste 2019)

DESCRIPTION Upperparts dark chocolate-brown, with slightly paler brown underparts. Broadly similar to Small Sister Long-fingered Bat *M. sororculus*, but slightly larger (Ibáñez & Juste 2019).

DISTRIBUTION Endemic. Wide but patchy distribution: mainly in east, also Central Highlands and south-west up to *c.*1,200m. Occurs in humid forest and spiny forest areas (Eger & Mitchell 2003). Records from Comoros archipelago probably refer to bats collected in Madagascar (Goodman & Maminirina 2007). IUCN Least Concern.

BEHAVIOUR Little information available. Gregarious, roosts in caves, often in tightly packed groups with congeners, Glen's Long-fingered Bat *M. gleni*, Manavi Long-fingered Bat *M. manavi* and Small Sister Long-fingered Bat *M. sororculus* (Peterson *et al.* 1995, Goodman *et al.* 2007b). Occasionally also recorded in tree holes. Very rapid flyer. Main prey moths (Lepidoptera), bugs (Hemiptera), small beetles (Coleoptera) and, to lesser extent, other insects (Ibáñez & Juste 2019).

EGER'S LONG-FINGERED BAT
Miniopterus egeri   Goodman *et al.*, 2011

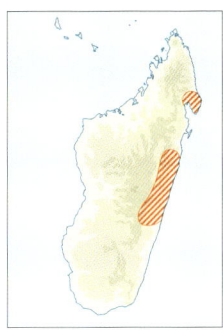

Other name: Eger's Bent-winged Bat

MEASUREMENTS
Head-body: *c.*49–51mm.
Forearm: 37–40mm.
Tail: 40–44mm. Weight:
4.2–7.6g. (Goodman *et al.* 2011a)

DESCRIPTION Pelage slightly long and dense. Mainly medium to dark brown, lighter brown in places. Differentiation from Peterson's Long-fingered Bat *M. petersoni* and Small Sister Long-fingered Bat *M. sororculus* subtle. Previously included in *M. petersoni* and, prior to that, *M. manavi* (Goodman *et al.* 2011a, Goodman 2017g).

DISTRIBUTION Found in native forest and anthropogenic habitats up to *c.*1,300m. Known from centre-east and Masoala Peninsula. IUCN Least Concern.

BEHAVIOUR Little information available. Presumed to roost in caves (Goodman 2017g). Like other *Miniopterus*, probably catches moths and other insects in flight (Ibáñez & Juste 2019).

MANAVI LONG-FINGERED BAT
Miniopterus manavi   (Thomas, 1906)

Other name: Manavi Bent-winged Bat

MEASUREMENTS
Head-body: *c.*51mm*.
Forearm: 37.6–39.2mm.
Tail: 39mm*. Weight: 6.4g*.
(Ibáñez & Juste 2019)
*single individual.

DESCRIPTION Pelage medium length. Upperparts mainly blackish-brown. Underparts similar but slightly paler. Some individuals more rufous and may represent a morph (Ibáñez and Juste 2019).

Manavi Long-fingered Bat.

DISTRIBUTION Endemic. Appears restricted to south Central Highlands. In native forest, degraded forest and cultivated areas (rice paddies) at *c.*900–1,500m. Previously regarded as most widespread *Miniopterus* in Madagascar. Recent genetic study has revealed a species complex with numerous new species now recognised (Goodman *et al.* 2009b). IUCN Least Concern.

BEHAVIOUR Little information available. Roosts in caves and large hollow trees (Andriafidison *et al.* 2006a). Diet small insects like moths (Lepidoptera), bugs (Hemiptera), beetles (Coleoptera) and ants (Hymenoptera) (Ibáñez & Juste 2019).

SMALL SISTER LONG-FINGERED BAT
Miniopterus sororculus   Goodman *et al.,* 2007

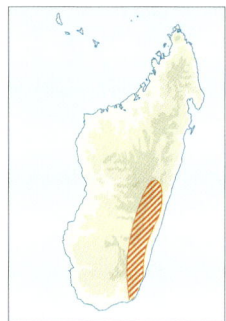

Other names: Sororcula Long-fingered Bat, Sororcula Bent-winged Bat

MEASUREMENTS
Head-body: *c.*54–57mm.
Forearm: 42–45mm. Tail:
51–58mm. Weight: 7–9.1g.
(Goodman *et al.* 2007b)

DESCRIPTION Dense, long, silky pelage. Upperparts rich chocolate-brown to medium brown, underparts slightly paler. Formerly included in Lesser Long-fingered Bat *M. fraterculus* of East Africa (Goodman *et al.* 2007b).

DISTRIBUTION Endemic. In Central and Southern Highlands and south-east. Recorded between 40m and 2,200m; most records above 900m (Goodman *et al.* 2007b, 2008a). Occurs in native forests, degraded forests, grassland and cultivated areas. IUCN Least Concern.

BEHAVIOUR Little information available. Roosts in caves and rocky crevices; at times, shares roost sites with Major's Long-fingered Bat and Manavi Long-fingered Bat (Goodman *et al.* 2007b). Presumed to feed on small insects like moths (Lepidoptera), bugs (Hemiptera) and beetles (Coleoptera).

PETERSON'S LONG-FINGERED BAT
Miniopterus petersoni   Goodman *et al.,* 2008

Other name: Peterson's Bent-winged Bat

MEASUREMENTS
Head-body: *c.*46–49mm.
Forearm: 38–43mm.
Tail: 39–50mm. Weight:
4.2–8.2g. (Goodman *et al.* 2008a)

DESCRIPTION Pelage slightly long, medium brown to dark brown. Formerly included with Small Sister Long-fingered Bat *M. sororculus* (Goodman *et al.* 2011).

DISTRIBUTION Endemic. Appears restricted to south-east. Occurs in humid and littoral forest and transitional drier areas, from sea level to *c.*550m (Goodman *et al.* 2008a). IUCN Data Deficient.

BEHAVIOUR Little information available. Presumed to be similar to other congeners in Madagascar.

MAHAFALY LONG-FINGERED BAT
Miniopterus mahafaliensis   Goodman *et al.*, 2009

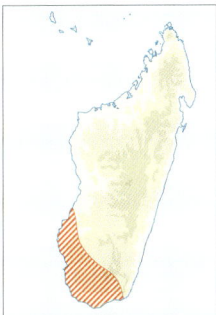

Other name: Mahafaly Bent-winged Bat

MEASUREMENTS
Head-body: *c.*48–49mm.
Forearm: 35–40mm.
Tail: 38–48mm. Weight: 3.8–7.3g. (Goodman *et al.* 2009c)

Mahafaly Long-fingered Bat, Tsimanampetsotse National Park.

DESCRIPTION Pelage long and dense. Upperparts medium brown, underparts paler brown-grey. Formerly included with Manavi Long-fingered Bat *M. manavi* (Goodman *et al.* 2009b).

DISTRIBUTION Endemic. Restricted to south and south-west, in dry deciduous forest, spiny forest and open anthropogenic areas from sea level to *c.*950m (Goodman & Ramasindrazana 2013, Goodman 2017i). IUCN Least Concern.

BEHAVIOUR Little information available. Uses caves and possibly hollow trees for roosts. Presumed to forage in open areas or close to forest edge, and feed on small insects like other long-fingered bat species (Goodman 2017i).

GRIFFITHS'S LONG-FINGERED BAT
Miniopterus griffithsi   Goodman *et al.*, 2010

Other name: Griffiths's Bent-winged Bat

MEASUREMENTS
Head-body: *c.*65–68mm.
Forearm: 48–50mm.
Tail: 54–63mm. Weight: 12–15.5g. (Goodman *et al.* 2010c)

DESCRIPTION A large long-fingered bat; in Madagascar only Glen's Long-fingered Bat is slightly larger. Upperparts around head and neck lighter brown, rest of upperparts medium brown. Underparts paler brown.

DISTRIBUTION Endemic. Currently known from three locations in far south, between 15 and 110m (Goodman & Ramasindrazana 2013). Does not occur north of Onilahy River, where it is replaced by sister species Glen's Long-fingered Bat (Goodman *et al.* 2010c). IUCN Least Concern.

BEHAVIOUR Little data available. Roosts in caves. Catches insects on the wing. Forages in open habitats and close to forest edge (Goodman 2017h).

Lemurs

ORDER PRIMATES
Suborder Strepsirrhini

Infraorders Lemuriformes and Chiromyiformes

Carl Linnaeus, who founded the modern binomial system of biological nomenclature, first coined the term '*lemur*', probably referring to the *lemures* of Roman mythology that were spectres or ghosts, exorcised during the festival of Lemuria. However, Linnaeus' initial use, in 1754, was in reference to the Slender Loris as *Lemur tardigradus*, where he explained, 'I call them lemurs, because they go around mainly by night, in a certain way similar to humans and roam with a slow pace' (Linnaeus 1754, Dunkel *et al.* 2012). The first use of '*lemur*' with respect to a Malagasy primate was in the tenth edition of the *Systema Naturæ*, the starting point for zoological nomenclature, which listed the genus *Lemur* as including the Slender Loris, Ring-tailed Lemur *Lemur catta* and Philippine Colugo *Lemur volans* (Linnaeus 1758), (hence today colugos, family Cynocephalidae, are sometimes confusingly and erroneously referred to as 'flying lemurs'). Subsequently, '*lemur*' has become widely accepted as an encompassing vernacular for all of Madagascar's primates.

Lemurs are placed in the suborder Strepsirrhini, meaning 'wet-nosed' (rhinarium or tip of snout) primates, in contrast to Haplorrhini, the dry-nosed' primates, represented by tarsiers, monkeys, apes and humans. Strepsirrhini is further divided into three infraorders, Chiromyiformes, the aye-ayes, Lemuriformes, representing all other lemurs, and Lorisiformes comprising galagos and pottos (Galagidae) from Africa and lorises (Lorisidae) from Asia.

Strepsirhinine primates (tooth-combed primates) are collectively and commonly known as '*prosimians*' (literally meaning 'before the monkeys') and are restricted to the Old World, where all species are largely confined to tropical arboreal habitats, the sole exception being the Ring-tailed Lemur *Lemur catta* that is partially terrestrial.

When exactly the earliest primates evolved is contentious. Fossil evidence points to primates emerging around the beginning of the Eocene, *c.*50–60 mya, whilst molecular evidence suggests an earlier date in the late Cretaceous, *c.*75–85 mya, thus predating the Cretaceous-Paleogene (K-Pg) boundary *c.*66 mya (Pozzi *et al.* 2014). The traditional view is that the earliest primates, which may have been similar in some ways to several present-day small nocturnal lemurs, originated soon after the K-Pg boundary, signifying the asteroid impact that led to the abrupt extinction of dinosaurs, and subsequent radiation of flowering plants (angiosperms). This unlocked a new swathe of ecological niches that placental mammals (Eutheria), including primates (as well as birds), could exploit (Rose 2006). Consequently, groups of animals that had previously remained small and unobtrusive while dinosaurs ruled the roost underwent rapid adaptive radiations soon after the mass extinction *c.*66 mya (Wible *et al.* 2007, O'Leary *et al.* 2013).

Ring-tailed Lemurs bask in early morning sun, Berenty Private Reserve.

Sometime shortly after early primates arrived on the scene, more advanced forms evolved that were better able to exploit their environment. This lineage eventually became monkeys and apes (suborder Haplorrhini), and was able to out-compete and oust the more primitive forms. Critically, prior to this, some of the early lemur-like primates inadvertently rafted from Africa 50–60 mya, and became isolated on Madagascar, their ability to survive such a journey perhaps aided by the ability to enter torpor (as do several extant nocturnal lemurs) (Yoder *et al.* 1996, Yoder 2003, in press, Kappeler 2000). Their cousins elsewhere were driven either to a solely nocturnal existence (later evolving into galagos, pottos and lorises) or to extinction by the new arrivals, but on Madagascar they found refuge (as Haplorrhine primates failed to reach Madagascar, until the arrival of humans). Once in residence, the early lemur-like primates rapidly evolved to exploit their new home to best advantage. This resulted in spectacular diversification and speciation: contemporary lemurs display an extraordinary array of morphologies, behaviours, physiologies and lifestyles, which rival those of monkeys and apes combined.

Even within some genera there has been considerable speciation, for instance 12 species of 'true' or 'typical' lemurs (genus *Eulemur*) are currently recognised. Many of these have very specific ranges, and boundaries between species are often major rivers or similar geographical barriers. Indeed, there is no location on the island where more than two *Eulemur* species naturally occur together. The boundaries between some taxa are less distinct and there are apparently narrow zones of hybridisation.

There has been major taxonomic revision of lemur groups (Groves 2001, Mayor *et al.* 2004), together with the discovery and description of a plethora of new species in the past two decades, although the validity of a number of these remains contentious (Tattersall 2007, 2013) (See 'The Species Conundrum'). Currently 108 extant species are recognised in 15 genera and five families (Cheirogaleidae, Lepilemuridae, Lemuridae, Indriidae and Daubentoniidae), all of which are endemic to Madagascar.

Because of their diversity, the evolutionary affinity of the lemurs has been the subject of considerable debate. Much of this has revolved around whether or not they are derived from a single ancestor, the implication being that if they do not form a natural evolutionary group (a clade) then there must have been two or more colonisation events between Madagascar and mainland Africa (or indeed another continental landmass).

Based largely on molecular evidence, the pervading narrative is that a single colonisation event occurred c.50–60 mya, when a single ancestor (by necessity a pregnant female or male/female pair) arrived from Africa by rafting (sweepstake dispersal) and from this, all contemporary (and extinct) lemurs subsequently evolved (Yoder et al. 1996, Yoder 2003, in press, Krause 2010, Kistler et al. 2015).

In this perspective, the position of the bizarre Aye-aye *Daubentonia madagascariensis* has been problematic. Molecular evidence points to its inclusion within the natural lemur group, from which it is assumed to have diverged very shortly after colonisation (Yoder 1997). However, some fossil evidence alludes to an alternative sequence of events. Apparent Strepsirrhine fossils from Kenya, *Propotto*, and Egypt, *Plesiopithecus*, are seen as related to the Aye-aye, with a common ancestor occurring on Africa c.28 mya (Gunnell et al. 2018). This would, therefore, require two colonisation events, the first giving rise to the Lemuriformes (all lemurs except Aye-ayes) and a second, much later, colonisation of a Chiromyiform ancestor that gave rise to Aye-ayes. Intriguing as this hypothesis is, it fundamentally contradicts and conflicts with all other molecular and fossil evidence that strongly affirms the notion of a single lemur colonisation event in the Early Paleogene (Yoder in press).

Spectacular as the present-day assemblage of lemurs is, it is only a shadow of the array that once existed on pristine Madagascar. The original lemur fauna included at least 17 additional species that are now all sadly extinct (Godfrey & Jungers 2003, Goodman & Jungers 2014). Without exception these species were larger than extant forms and some, like Giant Sloth Lemurs (*Archaeoindris* spp.), attained body weights up to 200kg. What is more, many of these species probably survived until well after humans arrived on Madagascar and their demise was likely a consequence of over-hunting and habitat destruction (Burney 1997, 1999).

Lemurs are by far the most intensively studied of Madagascar's mammals. Detailed behavioural and ecological investigation began in the 1960s, although it was not until the mid 1980s that interest in Madagascar really blossomed. Since then, there has been a huge advancement of knowledge and the discovery of numerous new species. Encouragingly, this trend shows no sign of abating; as remote forests are explored exciting discoveries are still being made.

All lemurs are threatened by habitat loss, forest fragmentation, hunting and in some instances illegal collection for the local pet trade (Golden 2009, Reuter et al. 2016 a,b, Reuter & Schaefer 2017, Eppley et al. 2020d, Golden et al. in press). Hunting is particularly acute in some areas, the extent of which has largely been overlooked in the past. Larger species like sifakas (*Propithecus* spp.) and ruffed lemurs (*Varecia* spp.) are obvious targets, but so too are many of smaller species (Garcia & Goodman 2003, Goodman & Raselimanana 2003, Golden et al. in press). In many areas, lemurs are hunted in remote localities to provide bushmeat both locally and to more distant larger villages and towns (Golden 2005). The incidence of lemur pet keeping has also proliferated disturbingly in recent times, with estimates suggesting c.30,000–50,000 individuals being kept in Malagasy households, c.50% of these in urban areas, largely away from forests, alluding to a significant trade (Reuter et al. 2019). Alarmingly, one of the drivers for this may be linked to tourism, with pet lemurs being used as revenue-generating tools.

Mouse, Giant Mouse and Dwarf Lemurs
Family Cheirogaleidae

The family Cheirogaleidae is arguably one of the most intriguing groups of primates. The family's origins date back some 25–30 million years, and among the present-day species are the world's smallest primates, the mouse lemurs (genus *Microcebus*); the only primates that are compelled to hibernate, the dwarf lemurs (genus *Cheirogaleus*); and the only primates that are known to predate members of their own zoological family, the giant mouse lemurs (genus *Mirza*).

With recent taxonomic revisions and additions of new taxa, there are currently 41 described species in this family (See 'The Species Conumdrum'). All are small to small-medium sized nocturnal lemurs, divided between five genera; *Microcebus* (mouse lemurs), *Allocebus* (hairy-eared dwarf lemur), *Mirza* (giant mouse lemurs), *Cheirogaleus* (dwarf lemurs) and *Phaner* (fork-marked lemurs).

They are characterised by medium to medium-long bodies, long tails, short legs and a horizontal body posture. All run and jump quadrupedally. They may be both solitary and gregarious when active, and sleep alone or in groups, often in tree holes or dense clusters of leaves. The family is represented throughout the various native forest types in Madagascar, including some high-elevation forests in the mountains. At least one species is likely to be present in any patch of native forest of appreciable size and in most areas two or more species live sympatrically.

All mouse lemurs (Microcebus sp.) are very agile climbers and leapers. Golden-brown Mouse Lemur, Ankarafantsika National Park.

MOUSE LEMURS

GENUS *MICROCEBUS*

Mouse lemurs comprise the genus *Microcebus* and are the most diminutive lemurs (along with genus *Allocebus*) and the smallest primates in the world, ranging from c.30 g to 85 g (with seasonal variation) (Kappeler *et al.* in press a). Males and females are not dimorphic. As their vernacular name implies, they are all vaguely 'mouse-like' and broadly similar in appearance.

Mouse lemurs first evolved c.10 million years ago (Yoder & Yang 2004): they have subsequently diversified spectacularly to exhibit a wide array of ecological adaptations that have allowed them to colonise all suitable native forest types, including secondary and partially degraded areas across the island (Radespiel 2006, Knoop *et al.* 2018), even mangroves in some west coast areas (Gardner 2016).

They are unquestionably the most abundant group of primates on Madagascar (Kappeler & Rasoloarison 2003, Kappeler *et al.* in press a), although some species appear susceptible to even subtle alterations in habitat and have suffered noticeable recent population declines (Henkel *et al.* 2020).

When active and foraging, they are primarily solitary, although a higher order social framework and complexity does exist: population 'nuclei' form, where the range of a 'dominant' male overlaps several females within a core area and 'subordinate' males are restricted to the periphery (Kappeler *et al.* in press a). Mouse lemurs favour the mid and lower forest layers with tangles of thin branches and lianas (fine branch niche), often at forest edges, and regularly descend to ground level to hunt insects in leaf litter (Radespiel *et al.* 2006, Siemers *et al.* 2007, Lahann 2008). Mouse lemurs are very agile and run rapidly in bursts along branches, and leap considerable distances, often in pursuit of prey (Kappeler *et al.* in press a). Some species apparently prefer forest edges, e.g., Golden-brown Mouse Lemur *Microcebus ravelobensis* (Burke & Lehman 2014), whereas others favour smaller open spaces within forests, such as along trails or treefall gaps, where foraging opportunities may be better (Kappeler *et al.* in press a).

They sleep during the day, usually in tree holes, tangles of dense vegetation or nests of dry leaves, either alone or in small groups. As far as is known, all species, especially those living in seasonally dry or cold habitats enter daily torpor or prolonged torpor during the austral winter (May–October) to conserve energy and water, relying on fat reserves laid down during the austral summer, and consequently lose weight (Dausmann 2014, Kobbe *et al.* 2014). Torpor episodes may be regular, irregular or daily; various patterns often occur within a population and during a single season, hence, torpor is a very flexible response to current internal and external conditions (Dausmann *et al.* in press).

Mouse lemur diet is primarily fruit, tree gums, nectar, flowers, insects and their excretions, and other small invertebrates. Compared to other members of Cheirogaleidae, they have greater dexterity and visual acuity, enabling them to catch insects on the wing. Mouse lemurs are predated by some native carnivorans, e.g., Fosa and Ring-tailed Vontsira, various snakes, raptors, owls, occasionally large vangas (Vangidae) (Goodman *et al.* 1993c) and even giant mouse lemurs (genus *Mirza*) (Schliehe-Diecks *et al.* 2010).

Mouse lemurs are highly seasonal breeders (constrained by unfavourable conditions such as low temperatures and reduced food availability), although the timing of reproduction varies between species according to the seasonality of their habitat. Typically, infant weaning coincides with maximum food availability (Evasoa *et al.* 2018). Some species breed only once per year, e.g., Madame Berthe's Mouse Lemur *M. berthae*, while others may breed up to three times, e.g., Ganzhorn's Mouse Lemur *M. ganzhorni* (Lehann *et al.* 2006). In all species, females are able to reproduce in their first year, so mate and become pregnant prior to one year of age (Evasoa *et al.* 2018).

They are also highly vocal and have a variety of calls for social contact, reproduction and raising alarm against predators; these include, trills, purrs, whistles, chatters and high-pitched squeaks (some beyond range of human hearing). Alarm calls may be cryptic, so as not to reveal the individual's whereabouts or several individuals in different locations will call in succession, making them even harder to pinpoint (Zimmermann 2010, 2016, 2018). In locations with sympatric mouse lemur species, differences in vocalisations may be a criterion by which species differentiate.

At one time this genus was thought to contain just two species with non-overlapping ranges: the Grey Mouse Lemur *M. murinus* from drier regions of the north, west and south, and Brown Mouse Lemur *M. rufus* in humid rainforest regions of the east (Harcourt & Thornback 1990).

However, the situation is now known to be far more complex. With the advent of molecular genetic analyses, it has become clear that significant mouse lemur diversity has been previously overlooked (Zimmermann & Radespiel 2014). Wide-ranging surveys including of remote forests over the past 20 years have revealed this so-called 'cryptic' diversity and resulted in the identification and description of numerous new species. Studies in southern, western and northern forests have revealed genetically distinct populations and the description of nine new species; Tavaratra Mouse Lemur *M. tavaratra*, Sambirano Mouse Lemur *M. sambiranensis*, Madame Berthe's Mouse Lemur *M. berthae*, Golden-brown Mouse Lemur *M. ravelobensis*, Nosy Be Mouse Lemur *M. mamiratra*, Bongolava Mouse Lemur *M. bongolavensis*, Ambarijeby Mouse Lemur *M. danfossi*, Antafondro Mouse Lemur *M. margotmarshae* and Montagne d'Ambre Mouse Lemur *M. arnholdi* (Zimmermann *et al.* 1997, 1998, Yoder *et al.* 2000b, Andriantompohavana *et al.* 2006, Olivieri *et al.* 2007, Louis *et al.* 2008), along with the re-evaluation to species level of two previously described taxa, Peters's Mouse Lemur *M. myoxinus* and Grey-brown Mouse Lemur *M. griseorufus* (Schmid & Kappeler 1994, Rasoloarison *et al.* 2000, Yoder *et al.* 2000b, Hapke *et al.* 2003).

The situation in the eastern rainforests is similar with 12 new species being described; Goodman's Mouse Lemur *M. lehilahytsara* (Roos & Kappeler 2005), Mittermeier's Mouse Lemur *M. mittermeieri*, Jolly's Mouse Lemur *M. jollyae*, Simmons's Mouse Lemur *M. simmonsi* (Louis *et al.* 2006), Anjiahely Mouse Lemur *M. macarthurii* (Radespiel *et al.* 2008), Gerp's Mouse Lemur *M. gerpi* (Radespiel *et al.* 2012), Marohita Mouse Lemur *M. marohita*, Anosy Mouse Lemur *M. tanosi* (Rasoloarison *et al.* 2013), Boraha Mouse Lemur *M. boraha*, Ganzhorn's Mouse Lemur *M. ganzhorni*, Bemanasy Mouse Lemur *M. manitatra* (Hotaling *et al.* 2016) and Jonah's Mouse Lemur *M. jonahi* (Schüßler *et al.* 2020). In some cases, these new species have been identified only from precise locations and consequently their distributions and range limits are currently uncertain.

There is much debate surrounding the taxonomy of this group and how *Microcebus* species should be delineated and defined (Markolf *et al.* 2011, Yoder *et al.* 2016a,b) (see 'The Species Conundrum'). The application of molecular genetic analyses is starting to unravel some of the complexities and shed light on evolutionary relationships within the genus. Mouse lemurs are thought to have initially evolved 7–10 mya (Yang & Yoder 2003) and, while most species outwardly look very similar, molecular differences reveal that populations have separated, becoming genetically isolated and distinct over time (Poelstra *et al.* 2020, Schüßler *et al.* 2020), with large rivers often presenting ecological barriers to gene flow and fostering speciation (Yoder in press). In some instances, species exhibit clear behavioural differences supporting their taxonomic status (Evasoa *et al.* 2019). Examples include co-occurring species with different vocalisations (e.g., reproductive advertisement calls and predator-avoidance calls), that breed at different times of the year, reinforcing their genetic isolation (Braune *et al.* 2008, Evasoa *et al.* 2018, Hasiniaina *et al.* 2020). Further, there is growing evidence that the spectacular speciation witnessed in mouse lemurs has occurred rapidly in the relatively recent past, perhaps in response to significant climate change (Sgarlata *et al.* 2019).

This intricate situation will only become clearer after further fieldwork and comparative evaluations of many populations. Therefore, the current taxonomy of the genus *Microcebus* should be regarded as 'work in progress', with revisions potentially to follow. There is strong evidence that ecotones (swift transitions between one major habitat type and another) and topographic barriers (e.g., large rivers, high mountains) are significant factors in delineating boundaries between different species' distributions (Rakotondranary *et al.* 2011) and that climate warming over recent millennia, resulting in the contraction of humid forest and expansion of dry forest, has had significant bearing on species' distributions (Sgarlata *et al.* 2019). Mouse lemur populations in drier western forests are known to occur at higher densities than those in humid eastern forests, which may be linked to greater habitat fragmentation, higher levels of co-occurrence with congeners and increased seasonality and resource variability, fostering species better able to cope with environmental change (Setash *et al.* 2017).

At present *Microcebus* contains 25 species. In most cases, the species are very similar in appearance and the most reliable means of tentative field identification may be geographic location. However, in some localities two species are sympatric. In dry western regions Grey Mouse Lemur is known to co-occur in five areas with one other mouse lemur species (Grey-brown Mouse Lemur, Madame Berthe's Mouse Lemur, Peters's Mouse Lemur, Golden-brown Mouse Lemur and Bongolava Mouse Lemur) (Radespiel 2016) and in precise but different northern localities Antafondro Mouse Lemur and Sambirano Mouse Lemur, Nosy Be Mouse Lemur and Montagne d'Ambre Mouse Lemur, and Montagne d'Ambre Mouse Lemur and Tavaratra Mouse Lemur have been recorded in adjacent forest fragments (Sgarlata *et al.* 2019). Similarly in the north-east, sympatry has been established for Mittermeier's Mouse Lemur and Anjiahely Mouse Lemur (Makira Forest) and Goodman's Mouse Lemur and Jonah's Mouse Lemur (Mananara-Nord, Ambavala) (Schüßler *et al.* 2020).

Like all lemurs, and native mammals, their continued survival is threatened by habitat destruction, especially forest clearance and seasonal burning around the periphery. Evidence from some fragmented western forests (Ankarafantsika) suggests that mouse lemurs have some capability to survive forest fires via particular behavioural traits, in that they gain protection and sanctuary in cavities within larger trees (Ramsay *et al.* 2020).

As small mammals, mouse lemurs are targeted prey for numerous predators, including the endemic carnivorans, e.g., Fosa, Ring-tailed Vontsira, Narrow-striped Boky and Grandidier's Vontsira, several raptors and owls, e.g., Henst's Goshawk *Accipiter henstii*, Madagascar Harrier-Hawk, Madagascar Long-eared Owl, Barn Owl and Madagascar Red Owl, a predatory endemic passerine, Hook-billed Vanga *Vanga curvirostris*, and several snakes (Goodman & Ganzhorn in press). They are even predated by another species of Cheirogaleidae, Coquerel's Giant Dwarf Lemur *Mirza coquereli* that actively stalks *Microcebus* spp. in a 'cat-like' manner in the canopy (Schliehe-Diecks *et al.* 2010).

Mouse lemurs are easily distinguished in the field from all other nocturnal lemurs by their appreciably smaller size and generally rapid movements, the one exception being Hairy-eared Dwarf Lemur *Allocebus trichotis*, which is very similar in size, shape and posture. Given a clear view, the distinctive ear-tufts of Hairy-eared Dwarf Lemur are diagnostic. Confusion with nocturnal, tree-climbing rodents, like tufted-tailed rats (*Eliurus* spp.) is possible (with poor or fleeting views), but generally only reflection from a single eye is seen with rodents, whereas the forward-facing eyes of mouse lemurs always show as two points of light. Tufted-tailed rats also have appreciably longer, more pointed snouts and proportionally longer, often largely naked tails.

Armed with a strong head torch or hand torch, it is possible to find mouse lemurs on night walks in just about any appreciably sized patch of native or

secondary forest. Their reflected eyeshine is bright orange-red and their high-pitched, chirrup-like and squeaking vocalisations are often heard. When seen, most will scurry and jump rapidly through the forest understorey, but some will, at times, rest motionless when caught in a torch beam.

GREY MOUSE LEMUR
Microcebus murinus　　　(J. F. Miller, 1777)

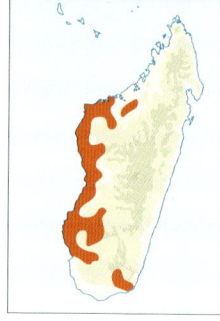

Other name: Lesser Mouse Lemur

Malagasy names: Tsidy, Koitsiky, Titlivaha, Pondiky

MEASUREMENTS
Head-body: 120–140mm. Tail: 125–150mm. Weight: mean *c.*60g, range 40–75g (varies seasonally). (Schmid & Kappeler 1994, Rasoloarison *et al.* 2000)

DESCRIPTION & IDENTIFICATION One of the larger dry-forest mouse lemurs. Upperparts and tail variable greyish-brown to brownish-grey, sometimes slightly browner towards tip. Underparts and throat dull beige to cream/off-white. Nose-bridge and between eyes creamy-white. Ears relatively long (protruding noticeably), fleshy and visible. Face rounded and eyes relatively large. Tail similar to head-body length. No sexual dimorphism, but seasonal differences in body weight: males heavier than females during breeding season, but females heavier than males outside this period.

Sympatric with several other mouse lemur species in different parts of its range: Grey-brown Mouse Lemur, Madame Berthe's Mouse Lemur, Peters's Mouse Lemur, Golden-brown Mouse Lemur and Bongolava Mouse. Distinguished from all of these by long, highly visible ears. Also, larger than Madame Berthe's Mouse Lemur and Peters's Mouse Lemur. Smaller and more active than other nocturnal species like dwarf lemurs (*Cheirogaleus* spp.) and giant mouse lemurs (*Mirza* spp).

HABITAT & DISTRIBUTION Dry deciduous, semi-humid deciduous, moist lowland, transitional, littoral and spiny forest, and some degraded forests and secondary vegetation, from sea level to *c.*800m; can even be found in degraded roadside and scrub-type vegetation and gardens on edge of villages

(Ganzhorn 1989, 1995, Radespiel 2000). Occupies the 'fine-branch' niche and prefers lower levels and understorey, where thin branches and lianas are tangled and dense (Kappeler & Rasoloarison 2003).

Occurs throughout forests of west, from Onilahy River in south, to region of Ankarafantsika and Bongolava Massif in north-west (Reuter *et al.* 2020a): the Sofia River is probably northern limit. South of Onilahy River there are populations in Bezá-Mahafaly (Rasoazanabary 2004) and an apparently isolated and disjunct population in south-east, either side of Mandrare River and in moist coast, transitional and gallery forests around Andohahela (Rasoloarison *et al.* 2000, Kappeler & Rasoloarison 2003), although with recent taxonomic additions and alterations to the genus *Microcebus* in this region, the picture may change. IUCN Least Concern.

BEHAVIOUR Nocturnal and omnivorous, largely solitary when foraging and occupies home range of c.0.3–2ha, although that of male may be twice the size of female range. Home ranges of both sexes may overlap each other (although females less so than males) and male home range always overlaps with that of at least one female (Eberle & Kappeler 2002). Male home range size increases by up to three times in breeding season (Buesching *et al.* 1998, Fietz 1999a, Radespiel 2000, Eberle & Kappeler 2002, Kappeler & Rasoloarison 2003). Related females appear to loosely 'cluster', occupying home ranges adjacent to one another, whilst males move away from natal area to occupy home ranges overlapping unrelated females (Wimmer *et al.* 2002, Radespiel *et al.* 2003).

Can be extremely abundant, reaching apparent densities of several hundred individuals/km^2 (Radespiel 2000, Eberle & Kappeler 2002). However, abundance not uniform and even in apparently similar habitat it may occur in localised concentrations or 'population nuclei' (Schmid & Kappeler 1994). In areas where Grey Mouse Lemur is sympatric with a congener, e.g., Golden-brown Mouse Lemur in Ankarafantsika, it may be dominant in encounters for space and food resources (Thorén *et al.* 2011).

Diet varies, mainly insects, especially beetles, but also moths, mantids and bugs, as well as fruits, flowers, nectar (so act as pollinators) and leaves. Also eats sap and gum of *Euphorbia* and *Terminalia* trees, and occasionally small vertebrates like tree frogs, geckos and chameleons. Seasonally important are excretions of fulgorid bug larvae *Flatida coccinea* that cluster on branches (resembling flowers) and

Grey Mouse Lemur, Kirindy Forest.

are most abundant during dry season, when other food sources are scarce (Corbin & Schmid 1995, Dammhahn & Kappeler 2008a). These insects are more common at forest edges, with female Grey Mouse Lemurs altering their ranging behaviour away from forest interiors to specifically feed on bug excretions.

By day sleeps in tree holes lined with leaf litter or purpose-built spherical nests constructed of dead leaves and moss among twigs in dense undergrowth (Kappeler 1998). May use between three and nine different tree holes, which can be shared and used for five or more consecutive days (Radespiel *et al.* 2003). Sleeping group composition changes seasonally: in breeding season (September–October) males and females may sleep together, at other times females sleep in groups with dependent offspring, while males sleep alone or in 'pairs' (Schmid & Kappeler 1998, Radespiel *et al.* 2003). During winter up to 16 individuals may sleep together when torpid.

During daytime period of inactivity may enter temporary torpor and is able to reduce metabolic rate and body temperature to ambient (as low as 7°C recorded) (Ortmann *et al.* 1997). During cooler winter months (May–August), chooses tree holes very close to ground level, where ambient temperatures are lower and more stable, permitting them to remain torpid for longer and conserving body resources (Schmid & Speakman 2000, Schmid 2001).

In dry season (May–September) females in some regions become totally inactive and remain dormant in tree holes for several weeks, and up to five months (although they are active periodically), to conserve energy and reduce predation. Onset of dormancy highly flexible and modified according to local food availability (proxy of climatic variation): if food is plentiful females delay torpor and adjust to climatic change (Vuarin *et al.* 2015). Males rarely are inactive for more than a few days and are fully active several weeks before females, allowing them to establish hierarchies for access to breeding females (Schmid & Kappeler 1998, Rasoazanabary 2001, Génin 2003).

In secondary and degraded forests populations adversely affected, as there are fewer tree holes and reduced opportunity to lower metabolic rate, enter torpor and save energy. This may lead to increased stress and mortality (Ganzhorn & Schmid 1998).

During the austral summer, both sexes lay down large reserves of fat in hindlegs and tail – up to 35% of body weight – which sustains them through the dry season and periods of aestivation (Feitz 1998, Schmid 1999).

Mating system is multi-male/multi-female, with females becoming receptive every 45–55 days between September and March (Eberle & Kappeler 2002, 2004). Females advertise oestrous by distinctive high-frequency calls and scent-marking (Buesching *et al.* 1998). During this period male testes increase significantly in size (up to eight times). Gestation *c*.60 days, twins the norm, but litters up to four recorded. Birth occurs in tree hole or leaf nest, and timed to coincide with onset of rainy season (Fietz 1999a). Two or more females may form 'breeding groups', cooperatively raising their young (Eberle & Kappeler 2006). Infants are 'parked' in nests but may be carried by mother in her mouth until six weeks. Offspring independent within two months, and females able to breed within first year. Independent males disperse away from natal area more than females, to counter likelihood of inbreeding (Huchard *et al.* 2017). Generation time

c.2.5 years (Radespiel *et al.* 2019). Reproductive lifespan 5–11 years (Kappeler & Rasoloarison 2003, Dammhahn 2021).

Predators include raptors and particularly owls, carnivorans like Ring-tailed Vontsira and Fosa, and snakes (Goodman *et al.* 1993a,b,c, Goodman 2003); occasionally predated by Coquerel's Giant Mouse Lemur (Schliehe-Diecks *et al.* 2010).

WHERE TO SEE Potentially seen in any decent patch of forest within range. Easy to see at several locations, the best being Kirindy Forest, Ankarafantsika National Park (Ampijoroa) and the gallery forests in Berenty Private Reserve (adjacent spiny forests are occupied by Grey-brown Mouse Lemur).

GREY-BROWN MOUSE LEMUR
Microcebus griseorufus Kollman, 1910

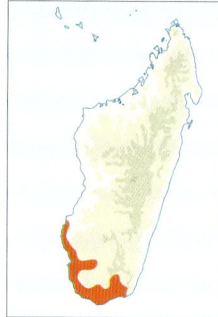

Malagasy names: Tsidy, Pondiky

MEASUREMENTS
Head-body: 115–130mm. Tail: 135–155mm. Weight: mean *c*.60g, range 45–85g (varies seasonally). (Rasoloarison *et al.* 2000, Génin 2008)

DESCRIPTION & IDENTIFICATION A large mouse lemur with distinctive coloration. Dorsal areas pale grey contrasting noticeably with cinnamon-brown mid-dorsal stripe from head or neck to tail tip. Tail cinnamon-brown above, grey below. Cinnamon-brown/rufous also on central crown extending onto face and around eyes. Pale grey/whitish line on nose, extending between eyes and onto forehead. Underparts creamy-white, with distinct change in colour between dorsal and ventral regions.

Sympatric with Grey Mouse Lemur in some locations, e.g., Bezà-Mahafaly where both species occur in spiny and gallery forests (although Grey-brown Mouse Lemur more common), and in Mikea Forest where both occur in transitional forest (between deciduous and spiny forest). In Berenty both species occur but are apparently separated by habitat preference, Grey-brown Mouse Lemur in spiny forest, Grey Mouse Lemur in gallery forest. At such defined habitat boundaries (ecotones) there is evidence Grey Mouse Lemur partially encroaches into habitat of Grey-brown Mouse Lemur and that hybridisation occurs (Gligor *et al.* 2009, Hapke *et al.* 2011).

HABITAT & DISTRIBUTION Variety of arid southern habitats; spiny forest, dry thorn scrub, gallery forest, dry deciduous forest and coastal scrub, from sea level to 250m (Steffens *et al.* 2017, Ganzhorn *et al.* 2020a). Appears more common in spiny forest (Rasoazanabary 2004). Precise limits of range unclear. Occurs south of Morombe to Tsimanampetsotse in extreme south-west, inland to Bezà-Mahafaly south of Onilahy River, and to Andohahela in south-east (Ganzhorn *et al.* 2020a). May occur further west in Zombitse, Vohibasia and Isalo National Parks, where potentially sympatric with Grey Mouse Lemur.

In south-east, a zone of hybridisation between Grey-brown Mouse Lemur and Grey Mouse Lemur occurs in two ecologically different contact zones: one in transitional forest between dry spiny bush and humid littoral forests south of the southern Anosyennes Mountains (Gligor *et al.* 2009), and second in western rain shadow of Anosyennes Mountains near Mangatsiaka, in mosaic of dry spiny bush and gallery forests (Hapke *et al.* 2011). IUCN Least Concern.

BEHAVIOUR Nocturnal and arboreal. Behaviour heavily influenced by highly seasonal and unpredictable environment (short wet season, prolonged cooler dry season). Diet varied; gum very important, especially in dry season or periods of drought, as are flowers and insects and other arthropods; in wet season eats fruits and gums in similar proportions (Rasoazanabary 2004, Génin 2008). When food availability high, feeds excessively to lay down fat reserves, subsequently may enter prolonged periods of torpor in dry cool season (Génin 2008). During dry season, adapts its response to thermoregulatory requirements, with a variety of energy-saving strategies including irregular short bouts of torpor, regular daily torpor, prolonged torpor lasting a few days or extended torpor lasting several weeks (aestivation) (Kobbe *et al.* 2011). During torpor, metabolism and body temperature drop significantly (as low as 6.5°C) and track ambient temperatures (Kobbe & Dausmann 2009).

More humid habitats with greater food availability support larger, healthier populations than drier habitats where food is more limited. Also, home range size may vary seasonally, from *c.*0.3ha in dry season to 0.5ha in wet season (Génin 2008), and is

Grey-brown Mouse Lemur, Tsimanampetsotse National Park.

adjusted according to food availability, but in dry habitats home range size has an upper limit, which consequently may limit food availability. As climate warms and habitats become drier this may adversely affect populations (Bohr *et al.* 2013).

Female home ranges overlap with those of *c.*3 males, whereas male ranges overlap with *c.*5 females. Mating occurs over an extended season (September–January). Males mate-guard females during this period (Génin 2008). Females often associate in same-sex pairs and because oestrous is not synchronous, they are able to reciprocally care for offspring. Gestation lasts *c.*52 days, with litter size up to three. Males disperse from natal group. Same-sex groups sometimes combine to form larger sleep groups: in dry season more than 12 individuals have been observed in single tree hole (Génin 2008).

In areas of sympatry with Grey Mouse Lemur, microhabitat separation occurs, with Grey-brown Mouse Lemur preferring forests with smaller trees, and Grey Mouse Lemur is found more in forest with larger trees. This is likely linked to Grey Mouse Lemur utilising tree holes as sleeping sites. Grey-brown Mouse Lemur is also more likely to remain active during the cool, dry season (Rakotondranary & Ganzhorn 2011, Rakotondranary *et al.* 2011).

Grey-brown Mouse Lemur is regularly predated by Barn Owl and Madagascar Long-eared Owl (Goodman *et al.* 1993a,b, Rasoloarison *et al.* 2000).

WHERE TO SEE Most easily seen in spiny forests of Berenty Reserve. Also readily seen, even by day in its sleeping sites, at Arboretum d'Antsokay near Toliara. Other suitable habitats within range are worth exploring, e.g., coastal spiny forests around Ifaty, north of Toliara and spiny forest in Tsimanampetsotse National Park.

MADAME BERTHE'S MOUSE LEMUR
Microcebus berthae   Rasoloarison *et al.*, 2000

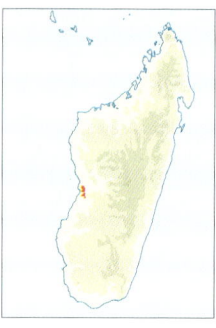

Malagasy name: Tsidy

MEASUREMENTS
Head-body length: 90–95mm. Tail: 130–140mm. Weight: mean 33g, range 24.5–38g. (Schwab 2000, Dammhahn & Kappeler 2005)

DESCRIPTION & IDENTIFICATION Smallest mouse lemur and smallest primate in the world: named in recognition of Malagasy primatologist Madame Berthe Rakotosamimanana. Pelage short and dense.

Upperparts and tail rufous to cinnamon-brown with distinct orange tinge, underparts creamy-white. Well-defined tawny dorsal stripe down back to end of tail. Head more orange and brighter than body. Dull whitish patch above nose to forehead (Rasoloarison *et al.* 2000). Tail relatively long and more densely furred than other mouse lemurs.

Can be confused only with Grey Mouse Lemur, which is noticeably larger (adults): also, Madame Berthe's Mouse Lemur is brown/orange (rather than grey), with small ears, relatively large eyes, a more pointed nose and relatively long tail.

HABITAT & DISTRIBUTION Dry deciduous forest (occasionally secondary forest) below 150m. Range now restricted to forests of Kirindy and Ambadira within Menabe-Antimena Protected Area: total area less than *c*.800km². Within this Madame Berthe's Mouse Lemur is patchily distributed, reflecting species' ecological specialisation, seasonal variation and sensitivity to forest intrusion and human disturbance. Mean density 80 individuals/km² (Schäffler & Kappeler 2014a). Extensive (and ongoing) forest reduction and fragmentation (Zinner *et al.* 2014, Hudson *et al* 2019) cause of severe recent population decline: total perhaps fewer than *c*.40,000 individuals (Schäffler & Kappeler 2014a). IUCN Critically Endangered.

Madame Berthe's Mouse Lemur, the world's smallest primate, Kirindy Forest.

Madame Berthe's Mouse Lemur, with Grey Mouse Lemur, Kirindy Forest.

BEHAVIOUR Nocturnal, arboreal and solitary. Females and males occupy separate home ranges, with the sexes' ranges overlapping. Male home ranges (c.4.9ha) approximately twice size of female's (c.2.5ha). At night, when active, covers considerable distances: males c.4.4km, females c.3.2km (double mean distance covered by sympatric Grey Mouse Lemur) (Dammhahn & Kappeler 2005, 2010). This probably relates to more specialised diet of insect excretions, which are spatially more dispersed resources. Both sexes enter daily, but not prolonged torpor, reducing metabolism and body temperature, to reduce energy expenditure and conserve resources (Schmid et al. 2000).

Prefers to sleep singly in small leaf nest in dense leaf and liana tangles 2.5–12m above ground: also sleeps in stable groups (female/female, male/male and multi-female/male) (Dammhahn & Kappeler 2005). Also uses old nests of giant mouse lemurs (*Mirza* spp.), but less frequently tree holes, probably because of increased competition from larger sympatric nocturnal lemurs (Grey Mouse Lemur and dwarf lemurs *Cheirogaleus* spp.) (Schwab & Ganzhorn 2004). Sleeping singly helps avoid predation, as groups in nest would be more conspicuous. Sleeping sites reused by both sexes, but more often by males (Dammhahn & Kappeler 2005).

Omnivorous: feeds on fruits, gums, flowers and insects, and particularly reliant on sugary excretions of fulgorid bug larvae, especially during dry season when other resources scarce (Dammhahn & Kappeler 2008b). Diet broadly similar to Grey Mouse Lemur, but more specialised, with up to c.80% insect excretions (Dammhahn & Kappeler 2008a,b). They avoid competition by spatial separation, rather than different resource use, hence their patchy distribution (Dammhahn & Kappeler 2010).

Precise details of breeding behaviour unknown: a promiscuous mating system, with strong male competition (relatively large testes suggests sperm competition), is suspected, with female offspring remaining associated with natal area (philopatry) and male young dispersing (Dammhahn & Kappeler 2005). Breeding begins in November, later than sympatric Grey Mouse Lemur, which starts in October.

Madame Berthe's Mouse Lemur is more susceptible to predation than sympatric Grey Mouse Lemur: known predators include Barn Owl, Madagascar Long-eared Owl, Fosa, Narrow-striped Boky and large snakes (Rasoloarison et al. 1995, Dammhahn & Kappeler 2008a).

WHERE TO SEE Can be difficult to locate within very limited range. Night walks in Kirindy Forest, north-east of Morondava, can be productive. Also, forests near Camp Amoureaux slightly further south are worth exploring.

PETER'S MOUSE LEMUR
Microcebus myoxinus   Peters, 1852

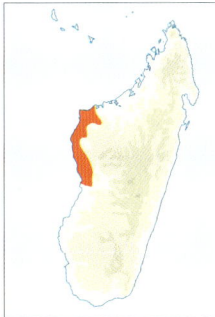

Other names: Western Rufous Mouse Lemur, Pygmy Mouse Lemur

Malagasy names: Tsidy, Malajira

MEASUREMENTS
Head-body: 120–130mm. Tail: 140–150mm. Weight: 40–55g. (Rasoloarison et al. 2000)

DESCRIPTION & IDENTIFICATION A noticeably small mouse lemur. Upperparts and tail rufous-brown with distinctive reddish-brown mid-dorsal stripe. Head more rufous. Tail relatively long, darker at tip. Underparts pale grey. Cinnamon patch

Peters's Mouse Lemur, Tsingy de Bemaraha National Park.

between eyes darkens to tawny crown. Prominent dark eyebrows. Ears relatively short (Rasoloarison *et al.* 2000). Within its range, sympatric with Grey Mouse Lemur, which is larger and mainly grey, rather than rufous-brown in colour.

HABITAT & DISTRIBUTION Dry deciduous forest and degraded forests adjacent to savanna, to 900m. Also records from coastal mangroves well away from deciduous forest (Hawkins *et al.* 1998, Baden *et al.* 2014a). Range now thought to extend apparently discontinuously from the northern Menabe region, north of Tsiribihina River, in centre-west to Soalala Peninsula and Baie de Baly area, with Betsiboka River the northern limit (Rasoloarison *et al.* 2000, Olivieri *et al.* 2007, Dammhahn *et al.* 2009). IUCN Vulnerable.

BEHAVIOUR Yet to be studied in detail. Probably similar to other mouse lemurs in western regions. Nocturnal and arboreal. Diet probably includes fruits, insects and their excretions. Breeding starts in October (Evasoa *et al.* 2018).

WHERE TO SEE Only locality with relatively easy access is Tsingy de Bemaraha National Park. Best to search forested areas near park entrance and close to village of Bekopaka on north bank of Manambolo River. Unprotected forest patches north of Tsiribihina River and south of Manambolo River are also worth searching. For the more adventurous, Tsingy de Namoroka National Park and remaining forest patches on Soalala Peninsula are worth searching.

GOLDEN-BROWN MOUSE LEMUR
Microcebus ravelobensis

Zimmermann *et al.*, 1997

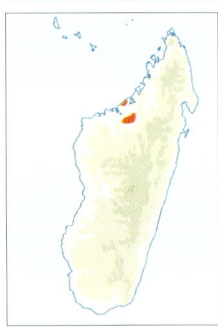

Other name: Lac Ravelobe Mouse Lemur

Malagasy name: Tsidy

MEASUREMENTS
Head-body: 120–139mm.
Tail: 144–172mm. Weight: mean 72g, range 59–110g.
(Zimmermann *et al.* 1998, Rasoloarison *et al.* 2000)

DESCRIPTION & IDENTIFICATION One of the largest western mouse lemurs, with Bongolava Mouse Lemur and Ambarijeby Mouse Lemur.

Head and upperparts rufous to golden-brown and distinctly mottled; mid-dorsal stripe indistinct. Underparts paler, yellowish-white and slightly mottled. Tail longer than body, thin, darker brown than body and darkens at tip. Head relatively small; faint white stripe extends from lower forehead to tip of muzzle. Dark brown rings around eyes. Ears long and sparsely furred. Sympatric with similarly-sized Grey Mouse Lemur: difficult to distinguish in field, but rufous, rather than grey, pelage and relatively longer tail with dark tip are features to look for.

HABITAT & DISTRIBUTION Dry deciduous forest, secondary and degraded forest up to *c.*500m. Prefers forests with lower canopy and higher density of lianas than Grey Mouse Lemur. Known from limited area centred on Ankarafantsika region, 120km south-east of Mahajanga in north-west. First collected adjacent to Lac Ravelobe at Ampijoroa, hence specific name, *ravelobensis*. Also occurs in Mariarano Forest near coast north of Mahajanga (Zimmermann *et al.* 1997, 1998, Guschanski *et al.* 2007, Olivieri *et al.* 2007, Blanco *et al.* 2020d). Occurrence very patchy, related to microhabitat preferences and forest edge effects (Rendigs *et al.* 2003, Burke & Lehman 2014, Radespiel *et al.* 2018). In favoured forests, density estimated at 930 individuals/km^2 (Rakotondravony & Radespiel 2009). Due to severe habitat loss and alteration, and probable increase in human presence, populations have declined massively, including very recently (Olivieri *et al.* 2008, Henkel *et al.* 2020). Ankarafantsika is bisected by major road (*Route Nationale* 4), which acts like 'forest edge' and impacts dispersal and movement of Golden-brown Mouse Lemur: females will not cross the road, but males do (Ramsay *et al.* 2019). IUCN Vulnerable.

BEHAVIOUR Nocturnal and arboreal. Occurs sympatrically with Grey Mouse Lemur, but with noticeable behavioural and microhabitat differences and preferences between them. Golden-brown Mouse Lemur more active than Grey Mouse Lemur: moves from branch to branch by leaping, rather than quadrupedally like Grey Mouse Lemur. Where species overlap, regionally endemic Golden-brown Mouse Lemur appears subordinate to more broadly distributed Grey Mouse Lemur (Thorén *et al.* 2011a).

Forest structure and floristic composition influences species distribution: Golden-brown Mouse Lemur occurs more in forest with dense lianas and herb cover, and is less tolerant of degraded fragmented habitats, preferring larger continuous

Diet includes gum, insect excretions, insects, arthropods, nectar, fruits and leaves. Gum appears to be a stable food source throughout year. Diet broadly similar to Grey Mouse Lemur, but overlap only partial with differences to reduce competition; insect excretions are more important to Golden-brown Mouse Lemur, fruits more prevalent in diet of Grey Mouse Lemur (Thorén *et al.* 2011b).

Males and females occupy home ranges with considerable overlap within and between sexes. The sexes sleep together in mixed groups of related animals, with sleeping sites changing but group composition remaining largely stable (Radespiel *et al.* 2003, 2009, Weidt *et al.* 2004). Often sleeps in nests constructed of dead leaves (building takes *c.*45–70 minutes): occurs more in cool dry season and may be linked to thermoregulation and when females are rearing young, inferring increased infant protection (Thorén *et al.* 2010).

Breeding begins in August, before Grey Mouse Lemur starts in September (Schmelting *et al.* 2000). Gestation not known, but probably *c.*60 days. Able to produce two litters per year and offspring can reproduce in first season following birth at *c.*9 months old (Schmelting *et al.* 2000, Randrianambinina *et al.* 2003).

WHERE TO SEE Forest around Lac Ravelobe at Ampijoroa Forestry Station, within Ankarafantsika National Park, is most accessible locality. If possible concentrate efforts around *Circuit Coquereli* (where Grey Mouse Lemur also occurs) and *Circuit du Lac* where Golden-brown Mouse Lemur predominates.

Golden-brown Mouse Lemur, Ankarafantsika National Park.

forests, whereas Grey Mouse Lemur tolerates forest edge and fragmentation, and also occurs in forest interiors with larger trees (Rendigs *et al.* 2003, Sehen *et al.* 2010, Chanu *et al.* 2012, Andriatsitohaina *et al.* 2020). This is reflected in preference for sleep sites with Golden-brown Mouse Lemur choosing nests in tangles of branches, lianas and vines, and constructed of dead leaves, whereas Grey Mouse Lemur prefers tree holes (Radespiel *et al.* 2003). Further, Golden-brown Mouse Lemur remains more active during dry season (May–December) and does not store fat in its tail like Grey Mouse Lemur (Randrianambinina *et al.* 2003).

BONGOLAVA MOUSE LEMUR
Microcebus bongolavensis   Olivieri *et al.*, 2007

Malagasy name: Tsidy

MEASUREMENTS
Head-body: 90–132mm.
Tail: 147–174mm. Weight:
mean *c.*55g, range *c.*45–65g. (Olivieri *et al.*, 2007, Mittermeier *et al.* 2010)

DESCRIPTION & IDENTIFICATION Relatively large reddish/rufous mouse lemur, similar in appearance to Golden-brown Mouse Lemur and Ambarijeby

Bongolava Mouse Lemur.

Mouse Lemur. Fur dense and short. Head uniform rufous in some individuals; others have rufous areas over eyes, with crown pale grey. Distinct white stripe between eyes. Upperparts and tail rich brownish-maroon, sometimes with faint dorsal line. Underparts creamy-white (Olivieri *et al.* 2007). Within its limited range Bongolava Mouse Lemur is sympatric with Grey Mouse Lemur, which is similarly large but mainly grey, rather than rufous-brown in colour.

HABITAT & DISTRIBUTION Dense primary western deciduous forest in uplands. Known from three forest fragments, including Bongolava and Ambodimahabibo Forests, between the Sofia and Mahajamba Rivers in north-west. These rivers may be barriers confining species' range (Olivieri *et al.* 2007, Schwitzer *et al.* 2013b). Genetic evidence strongly suggests population has suffered colossal constriction (by approximately two orders of magnitude) over the last 1,500 years, which ties in with massive human-induced deforestation and habitat fragmentation over the last 500 years (Olivieri *et al.* 2008, Vieilledent *et al.* 2018). Current known range less than 1,100km². IUCN Endangered.

BEHAVIOUR Nocturnal and arboreal. Females in oestrous August–September and pregnant October–November; males show main reproductive activity September–November (Evasoa *et al.* 2018). Diet likely includes gum, insect excretions, invertebrates, nectar and fruits. Other aspects likely similar to other mouse lemurs in western regions.

WHERE TO SEE Type locality, Ambodimahabibo and Bongolava Forests, and adjacent fragments. This is an extremely difficult area to reach.

AMBARIJEBY MOUSE LEMUR
Microcebus danfossi   Olivieri *et al.*, 2007

Other name: Danfoss's Mouse Lemur

Malagasy name: Tsidy

MEASUREMENTS
Head-body: *c.*120–136mm.
Tail: *c.*152–173mm.
Weight: mean 63g, range *c.*50–75g. (Olivieri *et al.* 2007, Mittermeier *et al.* 2010)

DESCRIPTION & IDENTIFICATION A larger-bodied reddish mouse lemur; very similar to Golden-brown Mouse Lemur and Bongolava Mouse Lemur. Dorsal fur and tail maroon with orange tinge, sometimes with indistinct mid-dorsal line. Head rufous in some individuals; others have grey crown with triangular rufous area above eyes. White stripe between eyes distinct. Underparts creamy-white (Olivieri *et al.* 2007).

Anbarijeby Mouse Lemur, Anjajavy Private Reserve.

HABITAT & DISTRIBUTION Primary, secondary and degraded deciduous forest. Known from forest fragments in Antsohihy region in north-west, between Sofia River to south and Maevarano River to north (probable limits of range) from sea level to *c.*780m. Some forest fragments are only *c.*50ha in extent (Olivieri *et al.* 2007, Randrianambinina *et al.* 2010). Within these population densities estimated at *c.*2–5 individuals/ha: higher densities correspond to more disturbed fragments, perhaps consequence of crowding or edge effects (Randrianambinina *et al.* 2010). IUCN Vulnerable.

BEHAVIOUR Nocturnal and arboreal. Females in oestrous and pregnant in September; males show main reproductive activity September–October (Evasoa *et al.* 2018). Diet likely includes gum, insect excretions, invertebrates, nectar and fruits. Other aspects probably similar to other congeners from western regions.

WHERE TO SEE Anjajavy Private Reserve on north-west coast is the most accessible location. This species can be encountered on night walks in forests with densely tangled understorey, within easy access of the hotel. Anjiamangirana Forest is also close to *Route Nationale* 6 and is easily reached (Radespiel 2021). Other forests within range, e.g., Ambarijeby, Antonibe, Marosakoa, Bekofafa and Ankaramikely, are much more difficult to reach.

ANTAFONDRO MOUSE LEMUR
Microcebus margotmarshae   Louis *et al.*, 2008

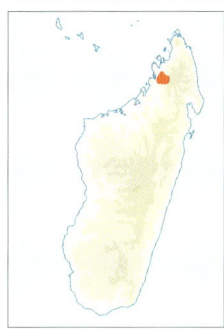

Other name: Margot Marsh's Mouse Lemur

Malagasy name: Tsidy

MEASUREMENTS
Head-body: *c.*110–120mm. Tail: *c.*140mm. Weight: *c.*40–50g. (Louis *et al.* 2008)

DESCRIPTION & IDENTIFICATION Medium-small reddish mouse lemur. Upperparts and tail reddish-orange. Head bright rufous-orange, darker fur around muzzle, distinct white patch on nose between eyes. Underparts creamy-white. Ears relatively small. Tail medium-long (Louis *et al.* 2008).

HABITAT & DISTRIBUTION Dry deciduous forest, including at high elevations. Originally described from Antafondro Classified Forest, with range

Antafondro Mouse Lemur.

thought to extend from Sambirano River in north to Andranomalaza River in south, including high-elevation forests of Tsaratanana Nature Reserve (Andriantompohavana *et al.* 2006, Louis *et al.* 2008, Radespiel 2021). Total range probably less than 1,100km². IUCN Endangered.

BEHAVIOUR Nocturnal and arboreal. Males and females show signs of reproductive activity August–September (Evasoa *et al.* 2018). Other aspects of behaviour probably similar to congeners from western regions.

WHERE TO SEE No easy locations. Any reasonable-sized forest within its range is worth exploring.

SAMBIRANO MOUSE LEMUR
Microcebus sambiranensis
Rasoloarison *et al.*, 2000

Malagasy names: Tsidy, Tsitsihy, Vokimbahy

MEASUREMENTS
Head-body: 110–120mm. Tail: 135–145mm. Weight: mean *c.*45g, range *c.*38–50g. (Rasoloarison *et al.* 2000, Andriantompohavana *et al.* 2006)

DESCRIPTION & IDENTIFICATION A medium-small mouse lemur. Dorsal pelage cinnamon-orange to rufous with indistinct amber mid-dorsal stripe beginning at shoulder, extending to tip of tail. Crown more amber-orange. Underparts dull whitish-beige with dark grey underfur. Tail densely furred, especially towards darker tip. Ears short. Distinct dark eye-rings with pale grey-beige patch running from nose tip, between eyes to forehead (Rasoloarison *et al.* 2000).

HABITAT & DISTRIBUTION Humid lowland and mid-altitude forests of Sambirano region at *c.*300–1,600m. Also seen in secondary and degraded habitats, and in some cases secondary forest with dense foliage may be preferred to old-growth forest (Hending *et al.* 2017a). Currently known from area between Andranomalaza River and Sambirano River in north-west, centred on Manongarivo Reserve and adjacent Ampasindava Peninsula, including forests

Sambirano Mouse Lemur.

of Mahilaka-Maromandia and Anabohazo (Louis *et al.* 2006a, Randriatahina *et al.* 2014). Collected on both west and east sides of Manongarivo Massif (Goodman & Soamalala 2002). Total range probably no more than *c.*1,500km². IUCN Endangered.

BEHAVIOUR Nocturnal and arboreal. Initial studies highlight some subtle behavioural differences with other *Microcebus* species. Both male and female Sambirano Mouse Lemurs generally sleep alone (sleeping with another male or female only *c.*15–20% of time), preferring leaf nests/foliage sleep sites (>90%) in densely foliated trees, instead of tree holes, which they use far less frequently. This may relate to competition with other larger nocturnal lemurs (e.g. *Lepilemur* and *Mirza*) for tree holes (Hending *et al.* 2017b).

Home ranges *c.*1–1.5ha, with ranges of several individuals overlapping. Sleep sites located more around periphery of home range and are re-used at times. At night 100–300m covered while foraging mainly in core of home range (Hending *et al.* 2017b). Diet likely includes gum, fruits, insect excretions, insects, other invertebrates and nectar. Has been observed pursuing and catching moths and cockroaches (Hending *et al.* 2018b).

In common with other mouse lemurs, Sambirano Mouse Lemur has a considerable vocal repertoire, with field studies identifying five different call types, similar in acoustic structure and function to calls of other mouse lemurs but specifically distinct (Hending *et al.* 2017a).

WHERE TO SEE Anabohazo Forest within Sahamalaza-Iles Radama National Park on the Ampasindava Peninsula is perhaps the best place to look. Alternatively, Manongarivo Special Reserve.

NOSY BE MOUSE LEMUR
Microcebus mamiratra

Andriantompohavana *et al.*, 2006

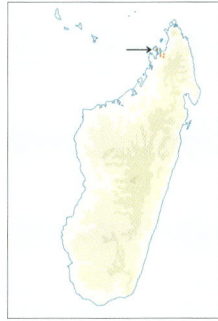

Other name: Claire's Mouse Lemur
Malagasy name: Valovi

MEASUREMENTS
Head-body: 105–139mm.
Tail length: 147–169mm.
Weight: mean *c.*60g, range *c.*42–73g.
(Andriantompohavana *et al.* 2006, Olivieri *et al.* 2007)

DESCRIPTION & IDENTIFICATION Medium-sized mouse lemur: similar to Sambirano Mouse Lemur but larger. Dorsal fur and tail light reddish-brown, brighter reddish-brown on crown, nape and upper dorsum. Tail darkens at tip. Clear whitish stripe on muzzle and between eyes. Underparts white to cream. Some have faint greyish-brown mid-dorsal stripe (Andriantompohavana *et al.* 2006). *M. lokobenisis* (Olivieri *et al.* 2007) is junior synonym.

HABITAT & DISTRIBUTION Restricted to humid Sambirano-type forests, and secondary forest, on island of Nosy Be, in particular Lokobe Reserve (Andriantompohavana *et al.* 2006, Olivieri *et al.* 2007, Hasiniaina *et al.* 2018), and similar humid forest fragments on nearby mainland, including Manehoka and riparian forests along the Mahavavy River near Ambakirano (Sgarlata *et al.* 2019). IUCN Endangered.

BEHAVIOUR Nocturnal and arboreal. Recent studies indicate that like other mouse lemurs, this is a vocal species with a complex acoustic repertoire. Calls differ within and between the sexes, and dominant

Nosy Be Mouse Lemur, Lokobe Reserve.

individuals vocalise more often than subordinates (Hasiniaina *et al.* 2018). Reproduction appears less seasonal than congeners in north-western regions: males and females show signs of breeding activity between June and November (Evasoa *et al.* 2018). Other aspects of behaviour likely similar to other mouse lemurs in western and northern regions.

WHERE TO SEE Lokobe Reserve on Nosy Be is only readily accessible locality. Search forests accessed via the beach at Ampasipohy. Not an easy species to find.

MONTAGNE D'AMBRE MOUSE LEMUR
Microcebus arnholdi

Louis *et al.* 2008

Other name: Arnhold's Mouse Lemur

Malagasy name: Tsidy

MEASUREMENTS
Head-body: 110–126mm.
Tail: 106–136mm. Weight: *c.*50g. (Louis *et al.* 2008)

DESCRIPTION & IDENTIFICATION Medium-sized mouse lemur with dorsal pelage a mix of dark brown, rufous and grey. Dark stripe along dorsal line to tail base. Tail darkens towards tip. Underparts creamy-white, with greyish underfur. Head more reddish, with darker brown muzzle and around eyes. Distinct white patch between eyes (Louis *et al.* 2008).

HABITAT & DISTRIBUTION Humid montane forest and drier seasonally humid forest. Originally described from higher-elevation forests in Montagne d'Ambre National Park and Forêt d'Ambre Reserve in far north (Louis *et al.* 2008). Also reported further south near Ambanja (Weisrock *et al.* 2010) and other humid forest fragments near Montagne d'Ambre and between Loky and Fanambana Rivers (Sgarlata *et al.* 2019). The latter locations partially overlap with distribution of Tavaratra Mouse Lemur. Ranges may segregate by habitat: Montagne d'Ambre Mouse Lemur being restricted to relict humid forests, Tavaratra Mouse Lemur in dry forests (Sgarlata *et al.* 2019). Further fieldwork required to clarify ranges and boundaries between these two species. IUCN Vulnerable.

BEHAVIOUR Nocturnal and arboreal. Other aspects of behaviour probably similar to other mouse lemurs in northern regions.

WHERE TO SEE Montagne d'Ambre National Park and adjacent Forêt d'Ambre Reserve are the best locations to see this species.

TAVARATRA MOUSE LEMUR
Microcebus tavaratra   Rasoloarison *et al.*, 2000

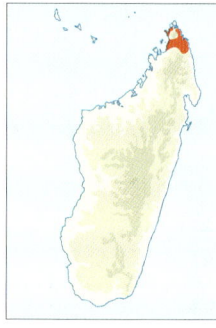

Other name: Northern Rufous Mouse Lemur

Malagasy name: Tsidy

MEASUREMENTS
Head-body: 113–139mm.
Tail: 145–167mm. Weight: mean *c.*60g, range 48–83.5g. (Rasoloarison *et al.* 2000, Yoder *et al.* 2000)

DESCRIPTION & IDENTIFICATION Medium-large mouse lemur with dark brown-cinnamon back, slightly paler flanks, rufous coloration around head and distinct mid-dorsal stripe continuous from crown to base of tail. Crown rufous. Pale whitish patch on nose. Underparts whitish-beige, underfur dark grey. Superficially similar to Golden-brown Mouse Lemur, but smaller and their ranges do not overlap.

Montagne d'Ambre Mouse Lemur.

Tavaratra Mouse Lemur, forests near Daraina.

for Tavaratra Mouse Lemur 28–473 individuals/km² (median 184 individuals/km²) (Sgarlata *et al.* 2020a). IUCN Vulnerable.

BEHAVIOUR Nocturnal and arboreal. Surveys of Loky-Manambato Protected Area suggest Tavaratra Mouse Lemur maintains high levels of genetic diversity across fragmented forest patches, suggesting substantial movement and gene flow is being maintained (Aleixo-Pais *et al.* 2019). Other aspects of behaviour likely similar to congeners in dry western and northern regions.

WHERE TO SEE Forests near Daraina, 45km west of Iharana (Vohimar) offer excellent opportunities and sightings are common. Forests around Camp Tattersalli are also excellent for other nocturnal species including Daraina Sportive Lemur, Fork-marked Lemur and Aye-aye. Good alternatives are Ankarana Special Reserve, 100m south of Antsiranana, especially forests near both *Campement Anilotra* and *Campement d'Andrafiabe*, and Andrafiamena-Andavakoera Protected Forest (Black Lemur Camp).

MITTERMEIER'S MOUSE LEMUR
Microcebus mittermeieri   Louis *et al.*, 2006a

Malagasy name: Tsidy

MEASUREMENTS
Head-body: *c.*110–125mm.
Tail: *c.*110–160mm.
Weight: mean *c.*45–50g, range *c.*35–60g. (Louis *et al.* 2006a, Rasoloarison *et al.* 2013, Schüßler *et al.* 2020)

HABITAT & DISTRIBUTION Dry deciduous, lowland semi-humid, littoral coastal and gallery forests, including those in limestone canyons, from near sea level to 680m (Rasoloarison *et al.* 2000, Sgarlata *et al.* 2020a). Also recorded in secondary forests and plantations (Hending *et al.* 2018). Range covers most remaining dry forests at northern tip of island, including Analabe, Ankarana, Analamerana and Loky-Manambato Protected Areas (Daraina). Also, likely to occur in Analafiana forest south of Manambato River and several fragments north of Irodo River (Louis *et al.* 2006, 2008, Salmona *et al.* 2015, Sgarlata *et al.* 2018, 2019, 2020a, Le Pors *et al.* 2020). South of Manambato River there is partial overlap with range of Montagne d'Ambre Mouse Lemur. These species may segregate by habitat: Tavaratra Mouse Lemur preferring dry forests, and Montagne d'Ambre Mouse Lemur in relict humid forests (Sgarlata *et al.* 2019). Population density estimates

DESCRIPTION & IDENTIFICATION A diminutive, gracile rainforest mouse lemur. Dorsal pelage reddish-brown to rust with darker brown mid-dorsal stripe. Slightly more orange tints at base of limbs. Ventral fur whitish-grey/brown. Tail similar in colour to body with black tip. Yellowish areas under chin and around neck (Louis *et al.* 2006a). Very similar to Goodman's Mouse Lemur (see below).

In Makira occurs sympatrically with Anjiahely Mouse Lemur: separation in field difficult (for non-expert), although Mittermeier's Mouse Lemur is smaller. Might also be confused with similar-sized

Mittermeier's Mouse Lemur, Marojejy National Park.

Hairy-eared Dwarf Lemur, which is greyish, rather than reddish-brown, and has distinctive and clearly visible ear-tufts.

HABITAT & DISTRIBUTION Lowland, mid-elevation and high-elevation rainforest in far north-east, between c.350m and c.1,760m. Range extends from Marojejy in north to Anjanaharibe-Sud and south to Makira (Louis *et al.* 2006a, Schüßler *et al.* 2020). In highlands west of Marojejy there is an isolated population at Antambato (Randrianambinina *et al.* 2010). Identity of *Microcebus* species on Masoala Peninsula yet to be determined. Molecular research strongly suggests that Mittermeier's Mouse Lemur and Goodman's Mouse Lemur are extremely closely related and constitute variation within the same species (Poelstra *et al.* 2020). Hence, Mittermeier's Mouse Lemur should be regarded as a junior synonym. IUCN Endangered.

BEHAVIOUR Nocturnal and arboreal. No detailed behavioural field studies have been undertaken: aspects probably similar to other rainforest *Microcebus* species.

WHERE TO SEE Best looked for in Marojejy National Park, south-west of Sambava and north of Andapa. Easily seen on night walks around Camp Mantella and Camp Marojejia. Forests close to Camp Marojejia also excellent for Silky Sifaka and several other lemur species.

ANJIAHELY MOUSE LEMUR
Microcebus macarthurii   Radespiel *et al.*, 2008

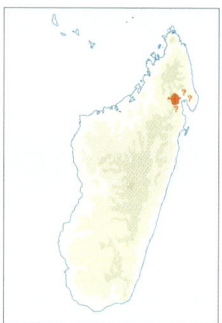

Other name: MacArthur's Mouse Lemur

Malagasy name: Kandrandra

MEASUREMENTS
Head-body: c.120–130mm. Tail: c.140–150mm. Weight: mean 62g, range c.54–68g. (Radespiel *et al.* 2008, Schüßler *et al.* 2020)

DESCRIPTION & IDENTIFICATION A large mouse lemur. Pelage short and dense. Upperparts and tail reddish-brown, with broad dark rufous mid-dorsal line. Head similarly coloured, sometimes more rufous with rufous-orange on cheeks. Dark brownish around eyes. Outer limbs lighter reddish-brown. Tail darker mid-brown towards tip. Underparts pale yellowish-orange, grading to creamy-white under throat and genital region (Radespiel *et al.* 2008). Sympatric with Mittermeier's Mouse Lemur. Anjiahely Mouse Lemur is generally larger, but separation in the field is difficult (for the non-expert).

HABITAT & DISTRIBUTION Lowland rainforest in north-central eastern regions, at c.350–400m. Currently known from Makira region west of Maroantsetra, specifically forests near Anjiahely (Radespiel *et al.* 2008, Schüßler *et al.* 2020) and island of Nosy Mangabe off Maroantsetra in Bay of Antongil (Louis & Lei 2016). Identity of Microcebus species on Masoala Peninsula yet to be determined. IUCN Endangered.

BEHAVIOUR Nocturnal and arboreal. No studies have been undertaken: likely to be similar to other rainforest mouse lemurs.

WHERE TO SEE Best locality is Nosy Mangabe, which is easily reached by boat from Maroantsetra. The Makira region west of Maroantsetra is an alternative, reached by boat upriver from Maroantsetra.

GOODMAN'S MOUSE LEMUR
Microcebus lehilahytsara
Roos & Kappeler, 2005 in Kappeler et al. 2005

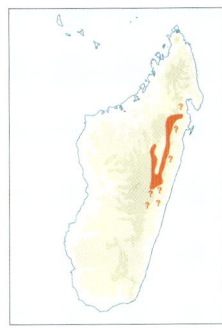

Malagasy name: Tsidy

MEASUREMENTS
Head-body: c.110–120mm.
Tail length: c.100–130mm.
Weight: mean 35–45g,
range c.30–60g. (Kappeler
et al. 2005, Louis et al.
2008, Rasoloarison et al.
2013, Schüßler et al. 2020)

DESCRIPTION & IDENTIFICATION A small rainforest mouse lemur. Pelage short and dense. Upperparts reddish-brown, with rust to orange tones on back, head and tail, and darker brown mid-dorsal stripe. Underparts creamy-white. Distinct off-white stripe from forehead, between eyes and on nose. Tail similar to body colour and uniform. Ears short and round (Kappeler et al. 2005).

Within majority of range, can only be confused with similar-sized Hairy-eared Dwarf Lemur: Goodman's Mouse Lemur is more rufous-brown, rather than greyish, and lacks the distinctive and clearly visible ear-tufts of Hairy-eared Dwarf Lemur. In the Mananara-Nord area (Bay of Antogil) this species is sympatric with Jonah's Mouse Lemur *M. jonahi*; although there is a clear size difference, separation in the field is challenging. Other nocturnal horizontal-posture species in range, e.g., dwarf lemurs, *Cheirogaleus* spp. are appreciably larger.

HABITAT & DISTRIBUTION Lowland, mid-altitude and higher-elevation rainforests: thought to be a 'highland forest specialist', preferring elevations above 825m (Weisrock et al. 2010, Radespiel et al. 2012) but also recorded down to c.235m (Schüßler et al. 2020). Can survive in small relict forest fragments surrounded by grassland and human-altered landscapes in Central Highlands (Yoder et al. 2016, Tiley et al. 2020). Originally described from Andasibe-Mantadia National Park region (Kappeler et al. 2005, Roos & Kappeler 2006) but range now known to extend south to Tsinjoarivo (with Onive/ Mangoro River possible southern boundary) (Dolch et al. 2020), and north to Anjorzorobe and isolated forest fragments such as Ankafobe (Blanco et al. 2016, Yoder et al. 2016) with Ambohitantely as western

limit (Rasoloarison et al. 2013). Currently known northern limits are Marotandrano Special Reserve (Yoder et al. 2016) and Ambavala to south-west of Bay of Antongil (Schüßler et al. 2020). Neighbouring species immediately to south is likely Marohita Mouse Lemur (Rasoloarison et al. 2013), to east in lowlands Gerp's Mouse Lemur (Radespiel et al. 2012) and to north-east in lowlands (Zahamena National Park) Simmons's Mouse Lemur (Roos & Kappeler 2006) and Jonah's Mouse Lemur (Schüßler et al. 2020).

Recent molecular research strongly suggests that mouse lemur populations in highlands further north (Marojejy, Anjanharibe-Sud, Anjiahely, Makira) currently described as Mittermeier's Mouse Lemur are extremely closely related to Goodman's Mouse Lemur and probably represent variation within the same species. Hence, range of Goodman's Mouse Lemur would effectively extend further north, as Mittermeier's Mouse Lemur would become a junior synonym (Poelstra et al. 2020).

In Andasibe population densities estimated at 110 ± 34 individuals/km^2 (when species classified as Brown Mouse Lemur *M. rufus*) (Ganzhorn 1988). In Tsinjoarivo densities estimated at 134 individuals/km^2 in forest fragments and 80 individuals/km^2 in continuous forest (Dolch et al. 2020). Can be more abundant in secondary vegetation than primary forest (Ganzhorn 1995). IUCN Vulnerable.

Goodman's Mouse Lemur, Association Mitsinjo Forest, Andasibe.

BEHAVIOUR Nocturnal, arboreal and generally solitary when foraging. Home range of males generally slightly larger (0.22ha–0.42ha) than those of females (0.21ha–0.30ha) (Randrianambinina 2001). Females moderately dominant over males (Hobenbrink et al. 2016). Prefers forest with dense understorey and feeds on fruits and insects, foraging primarily in shrubs and low trees, and less frequently in canopy (Ganzhorn 1988, 1989).

Both sexes enter bouts of torpor during austral winter (May–September) to avoid unfavourable conditions (low temperatures and food shortages) and so save energy (Karanewsky et al. 2015, Blanco et al. 2017). Bouts of torpor vary in length from c.2–3 days, but only individuals that have laid down sufficient fat reserves (primarily in their tails) are able to do so (Andriambeloson et al. 2020). In highland forests they prefer to hibernate in tree holes, rather than dead leaf nests, as these offer better protection against cold nighttime temperatures (Blanco et al. 2017). In captive populations males and females prefer to sleep in single-sex groups (Jürges et al. 2013). Breeding highly seasonal,

with peak in November (Wrogemann et al. 2001, Randrianambinina et al. 2003).

WHERE TO SEE Most readily seen in and around Andasibe-Mantadia National Park. A local NGO and community-run forest projects offer night walks in the area. The best are Association Mitsinjo (Analamazaotra Forest Station) and V.O.I.M.M.A. Community Reserve (abbreviation of *vondron'olona miaro mitia ala* meaning 'local people love the forest'), both just off the road between Andasibe village and the park entrance. Some 15km from Andasibe lies Vohimanana-Ankera, another community NGO forest with worthwhile night walk options. Alternatively, upland forests in Anjozorobe-Angavo Protected Area are very good: Indri, Diademed Sifaka and Grey Bamboo Lemur can also be seen there.

Jonah's Mouse Lemur.

JONAH'S MOUSE LEMUR
Microcebus jonahi   Schüßler et al., 2020

Malagasy name: Tsidy

MEASUREMENTS
Head-body: c.122–137mm. Tail: c.123–140mm. Weight: c.52–66g. (Schüßler et al. 2020)

DESCRIPTION & IDENTIFICATION A large-bodied and robust mouse lemur. Upperparts including tail reddish-brown, with head more rufous. Darker areas around eyes. Distinctive white patch on nose, between eyes and ending on forehead. Ears relatively small. Cheeks paler than head, becoming whitish around throat. Underparts similarly whitish; abrupt transition with reddish-brown dorsal coloration. Some show a dorsal stripe. Tail densely furred (Schüßler et al. 2020). Named in recognition of Malagasy primatologist Professor Jonah Ratsimbazafy. At Ambavala west of Mananara-Nord National Park, *M. jonahi* sympatric with *M. lehilahytsara*.

HABITAT & DISTRIBUTION Lowland rainforest and adjacent second growth and degraded areas. Recorded at c.40–350m. Currently known only from three sites, Ambavala, Mananara-Nord National Park and Antanambe, in north-east. IUCN Not evaluated.

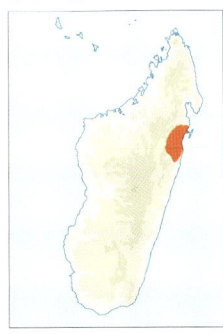

Jonah's Mouse Lemur.

BEHAVIOUR Nocturnal and arboreal. No field studies: aspects of behaviour likely similar to other mouse lemurs in north-eastern regions.

WHERE TO SEE Forest fragments in and around Mananara-Nord National Park, south and inland of Bay of Antongil.

SIMMONS'S MOUSE LEMUR
Microcebus simmonsi   Louis *et al.*, 2006a

Malagasy name: Tsidy

MEASUREMENTS
Head-body: *c.*120–135mm.
Tail: *c.*130–145mm.
Weight: mean 52–65g, range *c.*45–85g. (Louis *et al.* 2006a, 2008, Schüßler *et al.* 2020)

DESCRIPTION & IDENTIFICATION One of several large, robust mouse lemurs in eastern rainforests. Upperparts and head dark reddish-brown to orange-brown, underparts pale grey to off-white. Sometimes mid-dorsal stripe visible. Distinctive white patch between eyes.

HABITAT & DISTRIBUTION Lowland and mid-altitude rainforest in centre-east, between sea level and *c.*950m (Louis *et al.* 2006a, Schüßler *et al.* 2020). Currently known from Betampona Special Reserve, Zahamena National Park, Tampolo coastal forest (Louis *et al.* 2006a, 2008) and Ambodiriana south of Anove River to south of Mananara-Nord (Schüßler *et al.* 2020). Mouse lemurs in marshes and reedbeds around Lac Aloatra may be this species. IUCN Endangered.

BEHAVIOUR Nocturnal and arboreal. No detailed studies have been undertaken: aspects of behaviour and ecology likely similar to other rainforest *Microcebus* species.

Simmons's Mouse Lemur.

DESCRIPTION & IDENTIFICATION A large mouse lemur with relatively long tail and tail tuft. Upperparts reddish, with poorly defined mid-dorsal stripe. Underparts beige with slight reddish tinge. Ears short. This species is regarded as distinct from mainland populations on account of strong genetic differences (Hotaling *et al.* 2016).

HABITAT & DISTRIBUTION Known only from lowland humid forests (Ikalalao Forest, *c.*68m) on island of Ile Sainte Marie off east coast of mainland Madagascar (Hotaling *et al.* 2016). The Malagasy name for the island is Nosy Bohara, hence the species' name. IUCN Data Deficient.

BEHAVIOUR Nocturnal and arboreal. No behavioural field studies have been undertaken: aspects likely similar to other eastern region mouse lemurs.

WHERE TO SEE Ikalalao Forest on Île Sainte Marie.

WHERE TO SEE Tampolo coastal forest, near Fenoarivo and north of Toamasina is the most accessible known location. Betampona Special Reserve, *c.*50km north-west of Toamasina, requires a long strenuous walk. Other lowland, secondary and degraded forest patches in the Toamasina region are all potential haunts for this species.

BORAHA MOUSE LEMUR
Microcebus boraha   Hotaling *et al.*, 2016

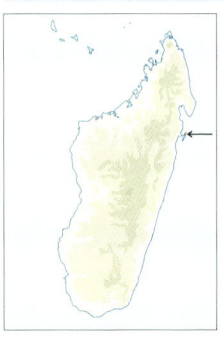

Other name: Ile Saint Marie Mouse Lemur

Malagasy name: Tsidy

MEASUREMENTS
Head-body: *c.*136mm. Tail: *c.*143mm. Weight: *c.*56.5g. (Hotaling *et al.* 2016)

GERP'S MOUSE LEMUR
Microcebus gerpi   Radespiel *et al.*, 2012

Malagasy name: Tsidy

MEASUREMENTS
Head-body: *c.*84mm. Tail: *c.*148mm. Weight: mean *c.*68g. (Radespiel *et al.* 2012)

DESCRIPTION & IDENTIFICATION A large rainforest mouse lemur. Upperparts greyish-brown with broad rufous dorsal line. Extremities of forelimbs and hindlegs noticeably darker than body. Head more rufous-brown, with darker brown around eyes. Underparts vary from creamy-white to pale grey. Ears relatively small. Tail relatively long and brownish-grey (Radespiel *et al.* 2012).

HABITAT & DISTRIBUTION Primary and secondary lowland rainforest at *c.*30–230m. Currently known only from Sahafina Forest in central east, a lowland rainforest fragment of just *c.*15km², lying west of Brickaville and south of Rianila River. Range presumably limited to lowlands (below 700m) between Mongoro River to the south and Rianila

River to the north (Schwitzer *et al.* 2013a). IUCN Critically Endangered.

BEHAVIOUR Nocturnal and arboreal. Yet to be studied: aspects of behaviour and ecology likely similar to other rainforest mouse lemurs.

WHERE TO SEE Sahafina Forest is c.60km east of Andasibe-Mantadia National Park and 12km west of Brickaville.

MAROHITA MOUSE LEMUR

Microcebus marohita   Rasoloarison *et al.*, 2013

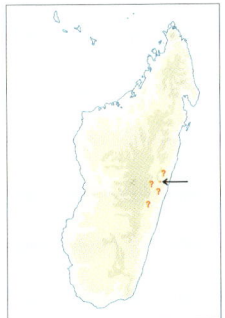

Malagasy name: Tsidy
MEASUREMENTS
Head-body: 132–140mm. Tail: 133–145mm. Weight: 64–89g. (Rasoloarison *et al.* 2013).

DESCRIPTION & IDENTIFICATION A large mouse lemur with a long tail. Upperparts rufous-brown with indistinct mid-dorsal stripe. Distinct transition to greyish-beige underparts. Ears relatively small. Dull white, slightly pinkish patch between eyes. Tail similarly coloured to upperparts, with darker tip (Rasoloarison *et al.* 2013). Named after forest it was discovered in: 'marohita' means 'many views' in Malagasy.

HABITAT & DISTRIBUTION Eastern lowland rainforests. Currently known only from a single location at 695m, Marohita Forest, within the more extensive Marolambo Forest, south of Nosivolo River (Rasoloarison *et al.* 2013). This area of forest is severely fragmented and covers less than 40km². IUCN Critically Endangered.

BEHAVIOUR Nocturnal and arboreal. No formal studies and further information available.

WHERE TO SEE Marolambo Forest is approximately halfway between Antsirabe and Ambodiharina on east coast. This is a very difficult region to reach.

BROWN MOUSE LEMUR

Microcebus rufus   (Lesson, 1840)

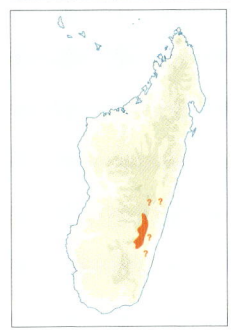

Other name: Rufous Mouse Lemur

Malagasy names: Tsidy, Tsitsihy, Tsitsidy, Antsidy Mena, Pondiky

MEASUREMENTS
Head-body: c.90–115mm. Tail: 110–160mm. Weight: mean 40–45g, seasonal variation 30–55g. (Atsalis *et al.* 1996, Atsalis 1999a,b, Louis *et al.* 2006, Rasoloarison *et al.* 2013)

DESCRIPTION & IDENTIFICATION One of the smaller mouse lemurs of eastern rainforest. Head, upperparts and tail characteristic brown to rufous-brown, with greyish tinges visible on dorsum and tail. Face rounded, often more rufous than body. Dark mid-dorsal line from neck to tail base. Underparts and throat creamy grey-white. Tail relatively long. Ears relatively small, do not protrude. A very small active lemur with long tail. Moves quickly through branches and foliage, and is much smaller than sympatric dwarf lemurs *Cheirogaleus* spp.

HABITAT & DISTRIBUTION Montane rainforests and similar habitat, including secondary vegetation, bamboo forests and adjacent plantations up to c.2,000m (Schwitzer *et al.* 2013b). Observed from ground level to canopy. Formerly thought to occur throughout eastern rainforest belt, from Tolagnaro in south to Tsaratanana Massif in north (Harcourt & Thornback 1990, Garbutt 1999). Description of several new mouse lemur species in eastern regions has prompted significant range re-evaluations for all humid forest mouse lemurs. Brown Mouse Lemur now thought to be restricted to south-east between Ranomafana National Park in north and Andringitra National Park in south (Wright *et al.* 2020c). In Ranomafana National Park population densities estimated at c.24 individuals/km² (Wright *et al.* 2012). IUCN Vulnerable.

BEHAVIOUR Nocturnal and arboreal. Males and females occupy separate home ranges with boundaries scent-marked using urine and faeces. High degree of overlap: male territories larger

than female territories and overlap with those of two or more females, with males regularly moving considerable distances (>1km per night) (Atsalis 2000, Wright *et al.* 2020c).

By day sleeps in groups of 1–4 in tree holes, leaf nests and, sometimes, disused bird nests, occasionally close to ground (Wright & Martin 1995). Largely solitary when foraging at night in understorey and lower canopy; if resources are concentrated, individuals may gather in small numbers. Diet omnivorous: mainly fruit (>40 species eaten), flowers and nectar, young shoots, gum, insects and other arthropods. Composition varies, reflecting seasonal changes in availability. Fruits with high fat value particularly important and feature year-round. Insects, especially beetles (Coleoptera), also taken year-round, but do not feature more during wet season when insect numbers peak (Atsalis 1998a, c, 1999a).

Both sexes lay down fat reserves in tail and other areas of body during wet season. Females and some males become inactive (enter state of torpor) during part of winter (May–September), during which time they may lose 30% of pre-aestivation body weight. Torpor males become active again in August, at least one month before females (Atsalis 1998b, 1999b).

Males prepare for breeding in mid-August when testes enlarge. Females become receptive in October: highly synchronous in proximal females. Mating occurs October and November, births primarily in December to January after gestation of c.60 days. Females construct nest of leaves 1–3m off ground, producing 2–3 young. Young c.5g at birth (Atsalis 1998a, 1999b, Blanco 2008, 2010).

Widely predated: carnivorans like Ring-tailed Vontsira and Fosa able to excavate mouse lemurs from sleep sites (Deppe *et al.* 2008); raptors and other birds, e.g., Hook-billed Vanga also take mouse lemurs (Goodman *et al.* 1993c); some owls and large snakes are significant nocturnal predators (Goodman 2003).

WHERE TO SEE Easily seen in Ranomafana National Park. Walks along road between Ranomafana village and the park entrance at Talatakely, and in higher-elevation forests at Vohiparara can be productive. With patience close views are common.

Brown Mouse Lemur, Ranomafana National Park.

JOLLY'S MOUSE LEMUR
Microcebus jollyae   Louis *et al.*, 2006a

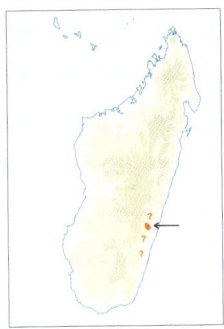

Malagasy name: Tsidy

MEASUREMENTS
Head-body: *c.*120–130mm.
Tail: *c.*120–125mm.
Weight: mean *c.*60g, range
56.8–65.8g. (Louis *et al.*
2006a, 2008)

Jolly's Mouse Lemur.

DESCRIPTION & IDENTIFICATION Upperparts including head uniform reddish-brown. Pale whitish areas between eyes, on nose and under chin grade into grey-white underparts (Louis *et al.* 2006a).

HABITAT & DISTRIBUTION Limited range in south-east, in low-altitude and coastal rainforest, from south of Mananara River to north bank of Mananjary River, including forests of Kianjavato, Vevembe, Mananjary and Karianga (Louis *et al.* 2006a). Forests highly fragmented in this region: total known range <200km². IUCN Endangered.

BEHAVIOUR Nocturnal and arboreal. Consumes fruits of native and non-native species (Borgerson *et al.* 2020), with other elements of diet and aspects of behaviour and ecology probably similar to *M. rufus*.

WHERE TO SEE No sites are readily accessible. Best looked for in forest fragments at Kianjavato to east of Ranomafana and Ifanadiana, and west of Irondro.

ANOSY MOUSE LEMUR
Microcebus tanosi   Rasoloarison *et al.*, 2013

Malagasy names: Tsidy,
Pondiky

MEASUREMENTS
Head-body: 116–140mm.
Tail: 115–150mm. Weight:
48–60g. (Rasoloarison *et al.* 2013)

DESCRIPTION & IDENTIFICATION A relatively large mouse lemur. Upperparts dark brown, more reddish on head, with dark dorsal stripe mainly visible in mid portion to base of tail. Underparts dull beige to dark grey. Noticeable pale whitish patch on nose and between eyes. Tail fur becomes denser towards tip (Rasoloarison *et al.* 2013). Name *tanosi* is derived from Malagasy and means 'from the Anosy Region'.

HABITAT & DISTRIBUTION Lowland and mid-elevation humid and littoral forests in extreme south-east. Precise range limits uncertain: occurs in Tsitongambarika Protected Area, east of Parcel 1 Andohalela National Park and isolated littoral forest fragment of Sainte-Luce Conservation Area (Donati *et al.* 2020b). This range is *c.*1,200 km² within which forests are highly fragmented (Donati *et al.* 2020b). Population densities in Sainte-Luce littoral forests estimated at 0.2–3.2 individuals/ha (Ganzhorn *et al.* 2007), in lowland humid forests of Tsitongambarika 0.7–1.3 individuals/ha (Balestri 2018).

The ranges of several mouse lemur species adjoin in this region. It is thought the western limit of Anosy Mouse Lemur may extend to western edge of humid forest portion of Andohalela National Park (Parcel 1). Previous studies have attributed mouse lemurs here to Brown Mouse Lemur *M. rufus* (Feistner & Schmid 1999). South-eastern limit is Lokaro River, immediately south of which Ganzhorn's Mouse Lemur occurs (Donati *et al.* 2020b). To the south of humid forests of Andohalela, in humid and transitional forests of Lavasoa Mountains is Bemanasy

Anosy Mouse Lemur, Tsitongambarika Forest.

Mouse Lemur (Donati *et al.* 2019). Where clear habitat boundaries (ecotones) occur, there may be partial 'cross-invasion' of these various mouse lemur species with narrow zones of hybridisation (Hapke *et al.* 2011). IUCN Endangered.

BEHAVIOUR Nocturnal and arboreal. Like other mouse lemurs eats small fruits and insects. In littoral forests berries of dwarf shrub *Vaccinium* spp. (Ericaceae) and fruits of endemic *Sarcolaena multiflora* (Sarcolaenaceae) are important foods. Periods of torpor recorded during austral winter, with activity increasing during August and September to coincide with increased forest productivity. In common with some other mouse lemurs, Anosy Mouse Lemur may prefer forest edges (Donati *et al.* 2020b).

WHERE TO SEE Sainte-Luce Reserve, 35km north of Tolagnaro, is the most likely place to see this species.

GANZHORN'S MOUSE LEMUR
Microcebus ganzhorni   Hotaling *et al.*, 2016

Malagasy names: Tsidy, Pondiky

MEASUREMENTS
Head-body: n/a.
Tail: *c*.131–134mm. Weight: 53–83g (seasonal variation). (Data from Lahann *et al.* 2006, as *M. murinus*)

DESCRIPTION & IDENTIFICATION A medium-large grey-coloured mouse lemur, formerly treated as Grey Mouse Lemur *M. murinus*. Molecular analysis has revealed sufficient distinction to warrant species status (Hotaling *et al.* 2016). In appearance and morphology, it is indistinguishable from Grey Mouse Lemur. Also, extremely similar to Grey-brown Mouse Lemur, which has white underparts, rather than the predominantly more greyish underparts of Ganzhorn's Mouse Lemur (Ganzhorn *et al.* 2020b).

HABITAT & DISTRIBUTION Known only from Mandena littoral forest and adjacent corridors composed of native and non-native trees (*Eucalyptus* and *Acacia* spp.) in extreme south-east. Based on current data, presumably limited to littoral forests of Mandena and Lokaro to east of Tolagnaro (Andriamandimbiarisoa *et al.* 2015, Ganzhorn *et al.* 2020b). Formerly these populations were ascribed to Grey Mouse Lemur. West of Tolagnaro (Petriky and Bemanasy Forest) replaced by Bemanasy Mouse Lemur. To the north, where forest is wetter, replaced by Anosy Mouse Lemur. Where these distinct habitat boundaries (ecotones) occur, there may be partial 'cross-invasion' of mouse lemur species with narrow zones of hybridisation (Hapke *et al.* 2011). IUCN Endangered.

BEHAVIOUR All studies during the period when this species was identified as Grey Mouse Lemur. Like all mouse lemurs, nocturnal and arboreal. Home range of females <1ha, but up to 5ha in males, whose ranges overlap. Males sleep alone in open vegetation or clumps of leaves, but females prefer tree holes and sleep with other females, and there is high turnover of sleeping sites (Lahann 2008).

Primarily forages in understorey. Diet comprises a wide variety of small fruits (>60 species recorded) and insects. Consequently, this species plays an important role in seed dispersal (Lahann 2007).

Enters torpor for varying periods during austral winter. May become inactive for up to four weeks: during this time body temperature mirrors that of ambient and may drop to 11.5°C (Schmid & Ganzhorn 2009).

Females are promiscuous and capable of having two perhaps three litters per year. Most offspring are born in either November or January/February, with possible third litters in April/May. Multiple litters probably facilitated by extended wet season and corresponding greater availability of food (Lahann *et al.* 2006).

WHERE TO SEE Mandena littoral forest north-east of Tolagnaro.

BEMANASY MOUSE LEMUR
Microcebus manitatra Hotaling *et al.*, 2016

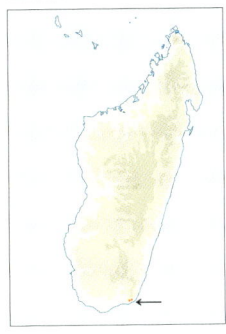

Malagasy names: Tsidy, Pondiky

MEASUREMENTS
Head-body: *c.*125mm. Tail: *c.*150mm. Weight: *c.*58g. (Hotaling *et al.* 2016).

DESCRIPTION & IDENTIFICATION A relatively large mouse lemur with a proportionately long tail. Upperparts and tail uniform greyish-brown. Underparts greyish-beige with darker underfur. Ears relatively long. Similar to Grey Mouse Lemur, but slightly smaller. Also lacks the reddish-cinnamon mid-dorsal stripe often present in Grey Mouse Lemur (Hotaling *et al.* 2016).

HABITAT & DISTRIBUTION Isolated patches of humid forest in far south-east. Described from complex of forest fragments, Ambatotsirongorongo, Grand Lavasoa and Petit Lavasoa (latter also called Bemanasy Forest) that once formed a large forest block at southern extreme of Lavasoa Mountains (Blanco *et al.* 2020c). Bemanasy is a *c.*30ha forest patch (mean elevation 617m). Probably also occurs in neighbouring Petriky Forest east of Ambatotsirongorongo (Hapke *et al.* 2012, Donati *et al.* 2019). Littoral and transition forests there are highly fragmented and under extreme threat (Bollen and Donati 2006). IUCN Critically Endangered.

Apparent distribution of mouse lemur species in extreme south-east is complex. The southern extreme of Lavasoa Mountains is in a zone of pronounced and swift climatic change with correspondingly steep ecological gradients between arid spiny forest to the west, humid lowland forest to the north and littoral humid forest to the east. The vegetation communities and floristic composition reflect influences from both arid and humid zones. Consequently, the ranges of several mouse lemur species juxtapose with Bemanasy Mouse Lemur: to the west in drier habitats, Grey Mouse Lemur and Grey-brown

Mouse lemur, to the north in rainforest Anosy Mouse Lemur and to the immediate east in humid littoral forest Ganzhorn's Mouse Lemur (Donati *et al.* 2019).

BEHAVIOUR Nocturnal and arboreal. Not yet studied. No further information available.

WHERE TO SEE New protected area, Ambato-sirongorongo Special Reserve, 25km west of Tolagnaro.

GIANT MOUSE LEMURS
GENUS *Mirza*

Giant mouse lemurs are small to medium-sized, nocturnal, omnivorous, nest-dwellers restricted to areas in west and north-west. They have an elongate body, pointed snout, long ears and long bushy tail. When first described in 1867, Coquerel's Giant Mouse Lemur *M. coquereli* was included in *Cheirogaleus*, then subsequently moved to *Microcebus*, before finally being placed on its own (at the time) in *Mirza*, largely on the basis of significant genetic, morphological (they are twice the size and 4–5 times the weight of *Microcebus*), behavioural and dentition differences (Tattersall 1982).

In common with other Cheirogaleidae, giant mouse lemurs move quadrupedally but are capable of leaping. Although they remain active during the austral winter, they lay down some fat reserves and may become torpid at times (Dausmann 2014, Dausmann & Warnecke 2016). They are more predatory than other lemurs, taking a wide variety of animal matter, and are unusual in that they periodically hunt mouse lemurs, rodents and other vertebrates (Kappeler in press).

Giant mouse lemurs are opportunistically eaten by various predators, including Fosa and Narrow-striped Boky, large raptors including Madagascar Cuckoo-Hawk *Aviceda madagascariensis*, Madagascar Buzzard *Buteo brachypterus* and Barn Owl, and large snakes, including Western Madagascar Tree Boa *Sanzinia volontany* (Goodman & Ganzhorn in press).

Within the genus *Mirza,* two species are now recognised, although further fieldwork and genetic analysis may prompt future reassessment.

COQUEREL'S GIANT MOUSE LEMUR *Mirza coquereli* (A. Grandidieri, 1867)

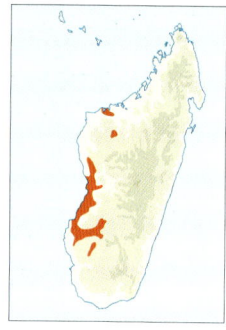

Malagasy names: Tsiba, Tilitilivaha, Siba, Setohy

MEASUREMENTS
Head-body: 235–265mm. Tail: 315–320mm. Weight: mean 310g, range *c.*285–335g. (Kappeler 2003, Kappeler *et al.* 2005)

DESCRIPTION & IDENTIFICATION Fur short and dense. Upperparts and head primarily rich brown to grey-brown, with some reddish, pinkish or yellowish tones. Underparts composed of lighter creamy-grey underfur beneath more rufous or even yellowish tips. Most conspicuous features are relatively long bushy tail (longer than head-body) that darkens towards tip, and large rounded, hairless and very distinctive ears.

Larger than Northern Giant Mouse Lemur (Kappeler *et al.* 2005). A squirrel-sized lemur with long bushy tail and horizontal body posture. Moves quadrupedally, with short leaps and bounds (Kappeler 2003). Movements tend to be rapid, helping distinguish it from similar-sized dwarf lemurs *Cheirogaleus* spp. Ears also very prominent, whereas in dwarf lemurs they are concealed. Might be confused with similar-sized Pale Fork-marked Lemur *Phaner pallescens*, but slightly larger size, characteristic facial markings, small ears and loud vocalisations of fork-marked lemur should differentiate the two.

HABITAT & DISTRIBUTION Dry deciduous forests, coastal moist forest and some secondary forests, from sea level to *c.*700m. In dry forests seems to prefer slightly taller (up to 20m canopy) and thicker vegetation found along riverbanks and by semi-permanent still water.

Range in west apparently discontinuous but exact limits unclear. Probably occurs from north bank of Onilahy River 40km inland from Toliara, east to Zombitse Forest near Sakaraha (Ganzhorn 1994, Goodman *et al.* 1997a) and north to vicinity of Antsalova, including Tsingy de Bemaraha National Park. Another population occurs along west coast, possibly from region of Bemarivo to Mahavavy River, including Tsingy de Namoroka National Park,

Coquerel's Giant Mouse Lemur, Kirindy Forest.

Beanka and Lac Kinkony area (Kappeler *et al.* 2005, Dammhahn *et al.* 2013, LaFleur 2020b). Populations from north bank of Fiherenana River, north-east of Toliara, suggested as a possible new taxon (Gardner & Jasper 2009), but here treated as Coquerel's Giant Mouse Lemur, pending further investigation.

Population density not homogeneous: in dry deciduous forests estimates of 24–286 individuals/km^2 (Ralison 2008) to 30–58 individuals/km^2 (Petter *et al.* 1971, Hladik *et al.* 1980, Schäffler & Kappeler 2014b) or 100–120 individuals/km^2 (Ausilio & Ravelaorinoro 1993, Kappeler 1997), but can reach 210 individuals/km^2 in riverine forests (Petter *et al.* 1971). Populations have been recorded declining rapidly, then recovering (Markolf *et al.* 2008). IUCN Endangered.

BEHAVIOUR Strictly nocturnal, day spent in spherical nest constructed of interwoven lianas, short twigs and leaves chewed off by animals from nearby trees (Kappeler in press). Nests up to 0.5m in diameter and sited in canopy, usually 2–10m above ground, surrounded by tangles of lianas, so difficult for predators to reach (Sarikaya & Kappeler

1997, Kappeler 2003). Individuals utilise up to 10–12 different nests in rotation. Generally occupied by solitary adults, but very occasionally shared by adults and juveniles. Unlike Northern Giant Mouse Lemur, adult males and females never share nests. Nests also utilised by other species like Pale Fork-marked Lemur and introduced Black Rat *Rattus rattus* (Kappeler 2003). In contrast with most other sympatric natural lemurs, Coquerel's Giant Mouse Lemur virtually never uses tree holes as daytime sleep sites (Kappeler in press).

Emerge at dusk, to groom and stretch before foraging. Particularly vocal at this time. Excellent climbers, using all levels of canopy; typically travel and forage between 5m and 10m. Occasionally and briefly, they descend head first down tree trunks to ground level, to hunt insects in leaf litter (Kappeler 2003). Active throughout night, returning to nests just prior to dawn. During second half of night more likely to be involved in social interactions. Active year-round, but less so in austral winter and may enter short bouts of torpor (Dausmann 2014, Dausmann & Warnecke 2016); on colder nights they leave nest later and return earlier (Kappeler 2003).

Coquerel's Giant Mouse Lemur, Kirindy Forest.

Omnivorous; diet varied and includes fruit and flowers, tree gums, especially *Terminalia* spp. and *Adansonia* spp., insects, other invertebrates and bird's eggs (Pagès 1980). Unusual among Cheirogaleidae as small vertebrates (baby birds, frogs, lizards and snakes) are eaten periodically (Hladik *et al.* 1980, Pagès 1980). More remarkably Coquerel's Giant Mouse Lemur is known to hunt small rodents (*Eliurus* sp.) and mouse lemurs, stalking its prey in a 'cat-like' manner in the canopy (Schliehe-Diecks *et al.* 2010). During dry season (June–August), when other resources scarce, sugary excretions from hemipteran nymphs (*Flatida coccinea* and *Phromnia rosea*) may account for 50% of diet (Hladik *et al.*

1980). Further presence of predatory Coquerel's Giant Mouse Lemur influences abundance of Grey Mouse Lemur in non-degraded forests, which in turn reduces competition and promotes niche separation between Grey Mouse Lemur and Madame Berthe's Mouse Lemur (Schäffler *et al.* 2015).

Generally solitary foragers. Males and females occupy overlapping home ranges of 1–4ha; these remain constant throughout year. Male home ranges vary in size between non-breeding and breeding seasons. Most of year they are similar in size to those of females and non-overlapping, but overlap with ranges of 2–3 females. When males and females meet, they interact and remain together for short periods. During October mating season, male ranges quadruple in size (Kappeler 1997). Smell important: saliva, urine and anogenital secretions are used to mark branches within range (Kappeler 2003). Coquerel's Giant Mouse Lemur has pungent distinctive odour that is indicative of its presence.

During breeding season males have dramatically enlarged testes. This, coupled with considerable variance in male home ranges between breeding and non-breeding season, suggests competition between males for receptive females is intense and that Coquerel's Giant Mouse Lemur is promiscuous (Kappeler 1997).

In Kirindy mating is restricted to brief period in September–November, during which oestrous of individual females may be limited to a few hours in a single night (Stanger *et al.* 1995). Female receptiveness indicated by olfactory cues and swelling and reddening of genitals (Markolf & Kappeler 2020). As receptivity approaches, males wait outside female nests, then follow closely and sniff female (Pagès 1980, Kappeler 1997, Markolf & Kappeler 2020). Female produces a distinctive trill 'advertisement' call. Copulation takes places with female grasping by four feet and suspended upside-down beneath branch with male also inverted, 'mounting' from below. After copulation, male grooms female and attempts to 'guard' her, but she soon rejects him aggressively and moves away. Other males then follow female and attempt to mate. Females may mate with 2–5 different males in a two-hour period (Markolf & Kappeler 2020). After successful mating a copulation plug forms in female's vagina, preventing further mating. One or two young are born after gestation of 86–89 days. Young poorly developed, weighing 12–15g, and spend first three weeks in nest (Pagès 1980, Stranger *et al.* 1995, Kappeler 1998). Infants leave nest after 3–4 weeks, initially carried by mother in her mouth. At three months, young forage alone but are vocal and remain in contact with each other during first stages of independence. Sexual maturity reached quickly; females can reproduce at ten months (Kappeler 2003).

Occasionally predated by Fosa and Narrow-striped Boky, and birds such as Madagascar Long-eared Owl and Madagascar Buzzard (Goodman & Ganzhorn in press).

WHERE TO SEE Kirindy Forest, 60km north-east of Morondava, is best locality: nocturnal walks along most of major forest trails should give reasonable chance of success. Five other nocturnal lemurs may also be encountered (Grey Mouse Lemur, Madame Berthe's Mouse Lemur, Fat-tailed Dwarf Lemur, Pale Fork-marked Lemur and Red-tailed Sportive Lemur). Alternatively, Zombitse-Vohibasia National Park can also be rewarding for Coquerel's Giant Mouse Lemur. Encounter rates are probably higher during breeding season when males, in particular, are more active.

NORTHERN GIANT MOUSE LEMUR
Mirza zaza
Kappeler & Roos 2005, in Kappeler *et al.*, 2005

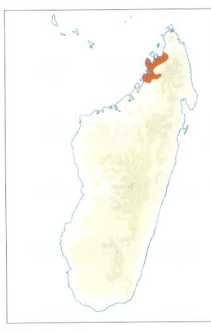

Malagasy names: Tanta, Fitily

MEASUREMENTS
Head-body: 220–250mm. Tail: 260–300mm. Weight: mean *c.*280–290g, range 265–320g. (Kappeler *et al.* 2005, Rode-Margono *et al.* 2016)

DESCRIPTION & IDENTIFICATION Pelage short. Upperparts and head greyish-brown with reddish tinges, underparts greyer. Upperparts in some females blondish-golden (Hending 2021). Tail long, bushy and darkens at tip. Ears rounded and short relative to Coquerel's Giant Mouse Lemur.

A horizontal-posture squirrel-sized lemur with a bushy tail, that moves quadrupedally. Overall, slightly smaller than Coquerel's Giant Mouse Lemur (Kappeler *et al.* 2005, Rode-Margono *et al.* 2016). Northern Giant Mouse Lemur is considerably larger than all sympatric mouse lemur species. Also larger, with conspicuous larger ears and generally faster-

Northern Giant Mouse Lemur.

moving than Fat-tailed Dwarf Lemur. In range, most likely to be confused with Sambirano Fork-marked Lemur *Phaner parienti*, which is similarly active and comparable in size and posture. Sambirano Fork-marked Lemur more obviously vocal and has distinctive facial marking, which giant mouse lemur lacks.

HABITAT & DISTRIBUTION Lowland dry deciduous forests, gallery forest, some humid forests and transitional forests. Also, secondary forests including abandoned cashew, banana, mango and cocoa plantations (Webber *et al.* 2020). Limited to north-west, bordering Sambirano region: range appears restricted by Maevarano River to the south and Mahavavy-Nord River to the north, including Ampasindava and Sahamalaza Peninsulas, and island of Nosy Be (Markolf *et al.* 2008, Reuter & Schwitzer 2020). Densities of 385 individuals/km² recorded in humid secondary forests dominated by mango and cashew trees, and at Ambato reach 1,086 individual/km², probably because the forest is fragmented with many introduced fruit trees (e.g., mangos) and little competition from other lemurs (Kappeler *et al.* 2005). In native forests, densities likely similar to Coquerel's Giant Mouse Lemur, i.e., *c.*100–120 animals/km². In range, native forests extremely fragmented, species occurs over less than 1,700km² (Rode *et al.* 2010, Reuter & Schwitzer 2020). IUCN Vulnerable.

BEHAVIOUR Primarily nocturnal, with some seasonal daytime activity recorded (Hending *et al.* 2021). First half of night spent moving and foraging alone; during second half tends to be more social, vocalising and engaging with other individuals in bouts of grooming, play and mutual resting. When active, prefers areas with tall trees and primarily uses mid to upper storey at heights of 5–10m, but also utilises full range from ground level to canopy top (*c.*25m). Occasionally forages on ground (Rode 2010, Rode-Margono *et al.* 2016).

Diet omnivorous and varied, includes fruits, flowers, buds and gums. In degraded areas, feeds heavily in cashew, banana and mango plantations. During dry season very reliant on sugary excretions of homopteran larvae (Flatidae), which may help facilitate year-round reproduction (Rode 2010).

Nighttime activity patterns do not appear affected by lunar cycle (Rode-Margono *et al.* 2016). Individual home ranges estimated to range from 0.5 to 2.2 ha, whilst group home ranges are between 1ha and 2.4ha (Rode 2010, Rode-Margono *et al.* 2016).

Unlike congener, Northern Giant Mouse Lemur is gregarious when sleeping: groups of 2–8 animals (mean four) including females, unrelated or related males, and juveniles form stable sleep groups, which exclusively share nests (Kappeler *et al.* 2005, Rode *et al* 2013). There is little nest turnover, with 1–3 nests used during a 50-day observation period. This may correlate to high nest preference and/or a scarcity of suitable sites. Spherical nest up to 50 cm diameter, constructed of interwoven lianas, twigs and leaves generally sited in dense part of canopy, near tree trunk, but below crown (Rode *et al.* 2013).

Both males and females are highly vocal and loud, with a considerable repertoire of simple and complex calls (>10 identified), including ultrasound frequencies, all in context of advertisement, alarm, affiliative and agonistic behaviours, and locomotion,

although the likely purpose differs between sexes. Male calls probably territorial and serve to maintain spacing, whereas females call to advertise oestrous and attract a mate. Unusual, as in most other primates, females use visual and olfactory cues to attract males (Seiler *et al.* 2019, Hending *et al.* 2020a).

Unusual among lemurs, reproduction occurs year-round (other examples being Aye-aye and Red-bellied Lemur) (Stanger *et al.* 1995, Rode 2010, Rode-Margono *et al.* 2015). Male Northern Giant Mouse Lemur have largest relative testis size of any primate (8× expected size), which suggests a highly promiscuous mating system with increased sperm competition (Rode-Margono *et al.* 2015).

Also observed mobbing and calling at a potential predator, the colubrid snake *Ithycyphus perineti*, in conjunction with a sportive lemur (Mandl *et al.* 2017).

WHERE TO SEE Often easily seen in secondary forests and abandoned cashew groves near Ambanja. Forests near town of Benavony, near Ambanja, particularly good. Also seen in Ankarafa and Anabohazo Forests, in Sahamalaza-Iles Radama National Park, but these are more difficult to reach.

HAIRY-EARED DWARF LEMUR
GENUS *Allocebus*

When first described in 1875, Hairy-eared Dwarf Lemur was initially assigned to *Cheirogaleus*, but later placed in its own genus, *Allocebus*, on basis of morphological features, mainly aspects of dentition and cranial structure that were considered sufficiently distinctive (Petter-Rousseaux & Petter 1967). The genus is monotypic, with its closest relatives thought to be mouse lemurs (*Microcebus*) and giant mouse lemurs (*Mirza*) (Pastorini *et al.* 2001, Roos *et al.* 2004).

Hairy-eared Dwarf Lemur was originally known from just five museum specimens, four collected in the late 19th century and the fifth in 1965. It was not found again until 1989 (Meier & Albignac 1989, 1991), when it was assumed to be limited to forests near Mananara and was considered possibly the 'rarest of all living primates' (Yoder 1996). However, since 1995 the Hairy-eared Dwarf Lemur has been located at numerous sites in central and northern rainforests (see below). Nonetheless, the species remains elusive and enigmatic, and any sightings should be recorded and conveyed to appropriate lemur authorities.

HAIRY-EARED DWARF LEMUR
Allocebus trichotis (Günther, 1875)

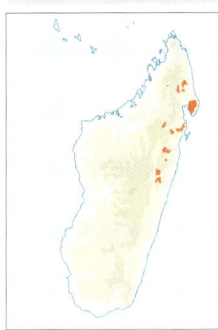

Malagasy names: Tsidiala, Antsidy Mavo

MEASUREMENTS
Head-body: 125–160mm. Tail: 140–200mm. Weight: 65–95g. (Meier & Albinac 1991, Biebouw *et al.* 2009)

Hairy-eared Dwarf Lemur, Association Mitsinjo Forest, Andasibe.

DESCRIPTION & IDENTIFICATION Head and upperparts medium brownish-grey, with slight rosy tinge. Tail similar, but darkens towards tip and becomes bushy. A faint dark dorsal stripe apparent on some individuals. Underparts paler, from light grey to off-white. Narrow dark rings around eyes and sometimes a pale whitish stripe from between eyes to tip of nose. Ears small and largely concealed beneath conspicuous long wavy ear-tufts that extend around cheeks.

A small, active lemur with long tail. Smaller than dwarf lemurs (*Cheirogaleus* spp.), but slightly larger than rainforest mouse lemurs (*Microcebus* spp.). Distinguishing Hairy-eared Dwarf Lemur from mouse lemurs in the field can be problematic and requires close views. Hairy-eared Dwarf Lemur is larger and its distinctive ear-tufts are an obvious feature to look for. Also, more greyish and tends to move in more sporadic, stop-start fashion than mouse lemurs.

HABITAT & DISTRIBUTION Lowland and mid-altitude rainforests from near sea level to *c.* 1,000m, with records up to *c.* 1,600m. Also recorded in partially degraded habitat (Schmid & Smolker 1998). Extent of range now better understood: appears restricted to centre-east and north-east humid forests, including Masoala Peninsula (Louis *et al.* 2020i). From north to south recorded at numerous sites including: Marojejy National Park (Goodman & Raselimanana 2002, Blanco & Tsilanizara 2020); Anjanaharibe-Sud Special Reserve (Schmid & Smolker 1998, Schütz & Goodman 1998); Bemanevika Reserve (Louis *et al.* 2020i); Masoala Peninsula (Sterling *et al.* 1998); Makira Forest (Rasolofoson *et al.* 2007a,b); Mananara-Nord (Meier & Albignac 1989, 1991, Yoder 1996); Antsahanadraitry Forest (Miller *et*

al. 2015); Marotandrano Reserve (Ralison 2006); Zahamena National Park (Rakotoarison 1995); Mantadia National Park (Garbutt 2007); Ambatovy Forest (Louis *et al.* 2020i); Mitsinjo Forest (Biebouw 2009); Analamazaotra Special Reserve (Garbutt 2001); Vohimana Forest (Garbutt 2007, Anania *et al.* 2020); Vohidrazana Forest (Rakotoarison *et al.* 1996, 1997); Maromizaha Forest (Louis *et al.* 2020i) and Ambalafary Forest (Lagadec & Goodman 2010). Throughout range encounters are sporadic and rare, suggesting it occurs either at very low densities and/ or is overlooked as easily confused with mouse lemurs. IUCN Endangered.

BEHAVIOUR Nocturnal and arboreal: active from around dusk until near dawn. Observed both alone and in pairs (probably male/female). Foraging concentrated in dense tangles of vegetation (lianas, vines, twigs) in lower levels of forest). Diet nectar, fruit, tree gum (primarily *Terminalia*), young leaves

*Hairy-eared Dwarf Lemur, Andasibe-Mantadia
National Park.*

and shoots. Insects also an important component
(Biebouw 2009). Very long tongue helps access
nectar in flowers. May remain active year-round,
but can enter short bouts of torpor (Dausmann
2014, Dausmann & Warnecke 2016). Little
seasonal fluctuation in weight, suggesting consistent
resource intake.

Home range is large compared to other
Cheirogaleidae, *c*.5–15ha. This may be linked with
need to forage over larger areas for insect prey or
patchy distribution of preferred gum trees *Terminalia*
spp. (Biebouw 2009).

Sleeps alone (*c*.35% cases) or in groups (*c*.65%),
in holes, primarily in live trees, 1–9m above ground;
holes in larger trees are preferred. Sleep groups
predominantly mixed-sex, consisting of 2–6 animals
(Biebouw *et al*. 2009). Individuals utilise 4–5 different
tree holes, but one to two are preferred and used
more frequently (Meier & Albinac 1991, Biebouw *et
al*. 2009). An individual may use the same hole for
up to eight consecutive days. Tree holes probably
have both a thermoregulatory and anti-predator
function. On occasion also shares tree hole with
an endemic arboreal rodent, White-tailed Tree
Rat *Brachytarsomys albicauda* without any apparent
aggression between them (Biebouw *et al*. 2009).

Females come into oestrous at start of wet season
(November–December), with births following in

January and February (Albignac *et al*. 1991, Meier &
Albinac 1991).

Little is known about vocalisations. Heard to
emit series of short whistles and alternating squeals
similar to those of mouse lemurs. When alarmed
stands erect on hindlimbs to spot danger (Albignac
et al. 1991, Meier & Albinac 1991).

WHERE TO SEE Most accessible localities are
forests in vicinity of Andasibe. The best option is
Mitsinjo Forest Reserve (Analamazaotra Forest
Station), operated by local NGO Association
Mitsinjo. There are known *Terminalia* trees which
these lemurs regularly visit to feed. Alternatively, try
V.O.I.M.M.A. Community Reserve (abbreviation of
vondron'olona miaro mitia ala meaning 'local people
love the forest'), near Andasibe. Some 15km from
Andasibe is Vohimana-Ankera, another community
NGO forest with worthwhile night walk options
where *Allocebus* has been seen. While on any night
walks in Andasibe-Mantadia National Park area, care
should be taken to closely observe any small lemur
resembling a mouse lemur; it could be a Hairy-eared
Dwarf Lemur.

DWARF LEMURS
GENUS *Cheirogaleus*

Dwarf lemurs, genus *Cheirogaleus* (which means
hand-weasel), are unique among primates in that they
are able to variably reduce metabolism for hours,
days or weeks, and are obligate hibernators, i.e., they
are compelled to enter a prolonged period of torpor,
with corresponding suppression of metabolism,
during the cool-dry austral winter (Dausmann 2008,
2014, Blanco in press, Dausmann *et al*. in press). All
currently recognised species hibernate for between
three and seven months per year (Blanco in press).
Duration of torpid period correlates to body size,
with smaller species like Fat-tailed Dwarf Lemur
C. medius being dormant for longer than larger species
like Crossley's Dwarf Lemur *C. crossleyi* (Dausmann
& Blanco 2016, Dausmann & Warnecke 2016). In
preparation, they eat large quantities of highly calorific
foods (fruits and berries with high sugar content)
and accumulate significant amounts of body fat
(hyperphagy), a large proportion of which (*c*.50%) is
stored in the tail. Body mass just prior to hibernation
can be double that when they emerge from torpor.
During dormancy, they hibernate either singly or
in groups of three to five individuals in tree holes,

Crossley's Dwarf Lemur hibernating underground, Tsinjoarivo.

leaf nests or underground. During torpor their body temperature closely tracks ambient temperature: in high-elevation forests, where dormancy chambers are underground and well insulated, ambient temperatures remain stable and low (c.12–14°C), therefore so do body temperatures, whereas in western forests dormancy chambers are in poorly insulated tree holes and daily ambient temperatures may fluctuate from c.10 to 30°C, with corresponding variations in body temperature. They emerge prior to start of warmer rainy season (around October/November) and promptly begin courtship and mating (Fietz 2003, Fietz & Dausmann 2006, Blanco et al. 2013, Schwitzer et al. 2013a).

Being obligate hibernators strongly influences other aspects of dwarf lemur behaviour and ecology. For instance, compared to similarly-sized nocturnal primates that do not enter seasonal torpor, e.g., African bushbabies (*Galago* spp.), dwarf lemurs have shorter gestation and lactation periods, show rapid early (pre-torpor) growth and development, which is subsequently supressed during hibernation, and both delayed attainment of adult size and delayed first breeding (Blanco & Godfrey 2013). Adult females mate shortly after emerging from hibernation (October–December) and generally produce a single litter of one to four offspring per year (Lahann & Dausmann 2011, Blanco & Godfrey

2014). Males generally emerge from hibernation 2–4 weeks prior to females, when their testes enlarge and they secure territories (Müller 1999).

Dwarf lemurs are strictly nocturnal, squirrel-sized lemurs with horizontal body posture that move quadrupedally along branches. They tend to move and forage more deliberately (with a slow creeping gait) and jump less frequently than smaller mouse lemurs. Dwarf lemurs have a well-developed sense of smell, with fruits and slow-moving insects dominating their diet (Lahann 2007a).

Compared to other Cheirogaleidae, e.g., giant mouse lemurs (*Mirza* spp.) and fork-marked lemurs (*Phaner* spp.), dwarf lemurs are not particularly vocal. Recorded vocalisations are beyond human hearing in the ultrasound range (Cherry et al. 1987, Stranger 1995, Zimmermann 2018).

The genus *Cheirogaleus* has been subject to considerable recent taxonomic revision. In the past just two species were recognised: Fat-tailed Dwarf Lemur *C. medius* from drier regions of the north, west and south, and Greater Dwarf Lemur *C. major* in eastern rainforest regions (Harcourt & Thornback 1990). Increased fieldwork and use of molecular techniques, coupled with analysis of museum specimens, has modified the picture significantly and greater species diversity is now acknowledged (Groves 2000, Hapke et al. 2005, Blanco et al. 2009,

Crossley's Dwarf Lemur, Andasibe-Mantadia National Park.

Groeneveld *et al.* 2009, 2010, 2011, Thiele *et al.* 2013, Lei *et al.* 2014). Genetic analysis has revealed four major species clusters or 'subgroups' – the *C. major* subgroup, *C. crossleyi* subgroup, *C. medius* subgroup and *C. sibreei* subgroup – and within these 'cryptic' diversity exists, which will likely lead to descriptions of new species (Thiele *et al.* 2013, Lei *et al.* 2014). Currently nine species are recognised; Fat-tailed Dwarf Lemur *C. medius*, Montagne d'Ambre Dwarf Lemur *C. andysabini*, Ankarana Dwarf Lemur *C. shethi*, Crossley's Dwarf Lemur *C. crossleyi*, Greater Dwarf Lemur *C. major*, Sibree's Dwarf Lemur *C. sibreei*, Haute Matsiatra Dwarf Lemur *C. grovesi*, Lavasoa Dwarf Lemur *C. lavasoensis* and Thomas's Dwarf Lemur *C. thomasi*.

In combination *Cheirogaleus* species are found across most of the island where native forest remains, with representatives in most major forest types, except dry spiny forests in south and south-west (Blanco in press). With so many recent taxonomic revisions, species' distributions have become confused and ill-defined, and those outlined here should be treated cautiously. In many cases, species appear to have non-overlapping ranges, but there are also numerous instances where two or even three species appear to live sympatrically, e.g., at Tsinjoarivo Crossley's Dwarf Lemur and Sibree's Dwarf Lemur (Blanco & Godfrey 2014, Herrera *et al.* 2016), in southern littoral forests Greater Dwarf Lemur and Thomas's Dwarf Lemur (Lahann 2007a,b, 2008) and at Tsihomanaomby (north of Sambava) *C. medius*-like, *C. crossleyi*-like and *C. major*-like populations (Williams *et al.* 2020).

Range boundaries often at ecotones (areas where different habitats adjoin), but species remain separated by subtle differences in behaviour and ecology (Herrera *et al.* 2016).

Dwarf lemurs are eaten by a variety of predators, including the endemic carnivorans, Fosa, Ring-tailed Vontsira and Narrow-striped Boky, various large raptors such as Henst's Goshawk, Madagascar Buzzard and Madagascar Harrier-Hawk, medium to large owls, Madagascar Long-eared Owl and Barn Owl, and a variety of large snakes, including Madagascar Ground Boa *Acrantophis madagascariensis*, Dumeril's Boa *A. dumerili* and Eastern Madagascar Tree Boa *Sanzinia madagascariensis* (Goodman & Ganzhorn in press).

Because they are dormant for 3–7 months, dwarf lemurs are rarely seen during the austral winter (May–August), and in particularly dry or cold areas it can be late September into early October before they emerge and become visible. Identifying dwarf lemurs from other nocturnal species in the field is relatively straightforward (where dwarf lemur species are sympatric, separation of congeners is extremely difficult). Like all nocturnal lemurs they possess a *tapetum lucidum*, a light-reflecting layer on the retina that improves night vision, and shows as bright orange-red eyeshine in a torch beam. However, they are larger and generally slower moving than mouse lemurs and giant mouse lemurs and have distinctive dark eye-rings and pinkish noses. Fork-marked lemurs are similar in size, but faster moving, have more pointed snouts and distinctive facial markings.

FAT-TAILED DWARF LEMUR
Cheirogaleus medius   É. Geoffroy, 1812

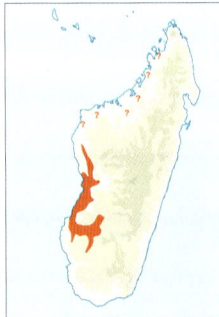

Other names: Lesser Dwarf Lemur, Western Fat-tailed Dwarf Lemur, Spiny Forest Dwarf Lemur

Malagasy names: Tsidihy, Kelybehohy, Matavyrambo, Bodonohy

MEASUREMENTS
Head-body: 200–230mm. Tail: 200–270mm. Weight: seasonal variation 120–275g. (Müller 1999b, Feitz 2003)

DESCRIPTION & IDENTIFICATION Pelage short and dense. Head, dorsal region and broad tail pale fawn-grey, with rufous tones and darker hair bases visible. There is a brown dorsal stripe. Underparts creamy-white with yellowish tinges and distinct partial white collar around throat extending up sides of neck. Face and cheeks pale grey-white with obvious dark rings around eyes and nose is pinkish. Snout and short ears also slightly darker.

Belongs to 'medius' subgroup. A small squirrel-sized lemur with horizontal body posture; a smaller *Cheirogaleus* species. Confusion with giant mouse lemur (*Mirza* spp.) is possible as size and colour are broadly similar; giant mouse lemurs larger, more reddish in colour, have bigger, highly conspicuous ears and a bushier, rather than chunky tail. Movements of giant mouse lemurs are more continuous and rapid than those of Fat-tailed Dwarf Lemur. Distinctive facial markings and more rapid movements of fork-marked lemurs (*Phaner* spp.) serve to separate from Fat-tailed Dwarf Lemur. Larger size and slower movements should prevent confusion with mouse lemurs.

HABITAT & DISTRIBUTION Dry deciduous, transitional and well-established secondary forests in west and north-west, from sea level to 800m. At one time all dwarf lemurs throughout drier regions of north, west and south were assigned to this species. Subsequently, populations in far north (Ankarana Dwarf Lemur *C. shethi*) and far south-east (Thomas's Dwarf Lemur *C. thomasi*) have been described as new species.

As currently understood, this species' southern limit is Onilahy River, including forests of Zombitse and Vohibasia. The range extends north through Kirindy-Mitea and Menabe to at least Tsingy de Bemaraha National Park (Blanco *et al.* 2020f, Webber *et al* 2020). Populations north of Tsingy de Bemaraha to Sahamalaza and Ambanja in north-west, including forests such as Namoroka, Tsiombikibo, Anjamena, Ankarafantsika, Bongolova and Anjajavy, may correspond to this species or an undescribed species (Frasier *et al.* 2016). Isolated population also occurs on island of Nosy Hara (Gardner & Jasper 2015). Dwarf lemurs in far north-west (Sahamalaza and Ambanja) including Nosy Hara are certainly very similar to Fat-tailed Dwarf Lemur, but further fieldwork and genetic investigation may determine they warrant recognition as a distinct species.

Between Onilahy River in south-west and Andohahela region in south-east, dwarf lemur populations have previously been described as a distinct species, *C. adipicaudatus* (Groves 2000), but subsequent work found differentiation from Fat-

Fat-tailed Dwarf Lemur, Kirindy Forest.

tailed Dwarf Lemur was unjustified (Groeneveld *et al.* 2009). The spiny forests of the south-west are highly fragmented and further fieldwork is required to establish if dwarf lemurs still occur in the region and, if so, their specific identity. Subfossil remains of Fat-tailed Dwarf Lemur have been found in Ankilitelo Cave in the south-west (Muldoon *et al.* 2009), which suggests dwarf lemurs were present there within the past 1,000 years and have subsequently experienced substantial range

Dwarf lemurs from Anjajavy Private Reserve may be Fat-tailed Dwarf Lemur or perhaps an undescribed taxon.

contraction, perhaps due to climate change and human-induced deforestation (Blanco in press).

In forests of Menabe Central (Kirindy Forest) density estimates range from 218 individual/km² (Fietz 2003) to 180 individual/km² (Schäffler & Kappeler 2014b). More broadly at sites across centre-west and north-west, estimates range from 50 individuals/km² to 750 individuals/km² (Ralison 2008). Threatened primarily by habitat loss and fragmentation. IUCN Vulnerable.

BEHAVIOUR Nocturnal and arboreal. Lives in small family units consisting of an adult pair and offspring from one or more breeding seasons. Males and females form lifelong pair bonds, broken only if one individual dies (Müller 1999a,b, Fietz 1999b,c). Families occupy territories of 1–2.5ha. By day they sleep communally in tree holes, with up to five animals together (Petter 1978). Territory and sleep sites are defended by faecal marking (Fietz 1999b,c).

An agile climber that forages at all levels in forest (Müller & Thalmann 2002). Diet omnivorous and varied consisting of fruit, flowers, nectar, pollen, leaf buds, gum, insects and other invertebrates, but proportion of animal prey varies seasonally and can comprise up to 20% (Fietz 2003). Prior to dormancy, becomes highly selective with high calorie fruits and berries preferred to maximise accumulation of fat (Fietz & Ganzhorn 1999, Müller & Thalmann 2002).

Fat-tailed Dwarf Lemur is an important agent of small-seed dispersal. During rainy season (November–April) smears faeces onto branches of trees – unique behaviour among primates. Particularly beneficial to germination of parasitic plants like mistletoes (e.g., *Viscum*) thus fills a role normally occupied by birds in other regions (Ganzhorn & Kappeler 1986, Fietz 2003).

Avoids unfavourable cool dry season with long period of hibernation to save energy (Fietz & Dausmann 2006). Duration of torpid period varies with seasonality of location: in highly seasonal western forests (no rain for 5–8 months, food and water in very short supply), torpor lasts 6–7 months, in less seasonal littoral forests it is 4–5 months (Lahann & Dausmann 2011) (populations now regarded as Thomas's Dwarf Lemur). During wet season builds fat reserves (up to c.100g), primarily stored in tail, which results in near doubling of body weight (c.130g to >250g) (Fietz & Ganzhorn 1999, Fietz et al. 2003). At start of dry season (usually April) gradually begins reducing levels of activity (although juveniles remain active up to four weeks longer) and nightly travel distances, eventually becoming inactive and torpid (Dausmann et al. 2004, 2005). Onset of hibernation highly synchronous, occurring over 2–4-week period, likely triggered by change in photoperiod when day length becomes shorter and modulated by body condition, e.g., fat stores (Fietz & Dausmann 2006, Dausmann & Blanco 2016).

Hibernates in a tree hole, where it rolls into a tight ball: mainly solitary, but far less frequently in small groups (Petter et al. 1977, Dausmann & Glos 2015). Individuals may occasionally change tree holes during dormant phase (Müller & Thalmann 2002). During torpor, type and properties of chosen tree hole strongly influence thermoregulation. Holes in large trees provide good insulation, which minimises daily temperature fluctuations, hence lemurs are able to maintain body temperature around 22–25°C, but periodically (weekly) need to emerge from torpor to raise body temperature for short periods (Dausmann & Warnecke 2016). Alternatively, body temperature of lemurs using poorly insulated tree holes, where temperatures fluctuate markedly, instead closely tracks daily ambient temperature (so may fluctuate by 20°C, c.10–30°C) (Dausmann et al. 2004, 2005). This is astonishingly flexible for a mammal and is comparable to thermoregulatory behaviour in some reptiles (Fietz & Dausmann 2006). Most individuals emerge from torpor by October/ November, although considerable regional variation is linked with local climate.

In Kirindy Forest breeds every two years, probably a consequence of energetic demands of extended period of torpor (Dausmann & Blanco 2016); unknown if this is case elsewhere (capable of breeding annually in captivity). Mating occurs November/December shortly after emergence from hibernation (Fietz 1999b). Despite being monogamous, males fight over access to receptive females and c.40% of offspring fathered by extra-pair males (Fietz et al. 2000). Gestation remarkably short, 61–64 days, litter size one to five (average two to three) (Müller & Thalmann 2002, Fietz 2003). Both parents raise young. For first two weeks young remain in nest hole, parents taking turns babysitting so offspring are never alone (Fietz 1999c, Fietz et al. 2000). Seen successfully fending off a snake (Madagascarophis sp.) near a nest (Fietz 2003, Dausmann 2010). When young initially explore outside nest, they are accompanied by one of parents and guided back to sanctuary of nest (Fietz 1999b,c). Offspring from previous season that may still be part of family group do not appear to assist. Juveniles become dormant later in dry season than adults (May/June) and so benefit from a period of reduced competition for food to maximise their fat deposits. Sexual maturity reached after two years, but only after leaving natal group and dispersing from natal area (Fredsted et al. 2007). Individuals do not become 'socially mature' until three years (Müller & Thalmann 2002, Fietz 2003).

Predated by Fosa, which excavates victims from tree holes. Other predators include Narrow-striped Boky, large boas (both Acrantophis spp. and Sanzinia spp.), Malagasy Cat-eyed Snake Madagascariophis colubrinus, Barn Owl and Madagascar Long-eared Owl (Dausmann 2010). Madagascar Harrier-Hawk removes sleeping dwarf lemurs from holes, using its long legs (Goodman 2003).

WHERE TO SEE Best site is Kirindy Forest, north-east of Morondava: here Fat-tailed Dwarf Lemur is often encountered on nocturnal walks during October–April. At this site, five other nocturnal lemurs may also be encountered (Grey Mouse Lemur, Madame Berthe's Mouse Lemur, Coquerel's Giant Mouse Lemur, Pale Fork-marked Lemur and Red-tailed Sportive Lemur). An alternative site is Zombitse-Vohibasia National Park, although arranging night visits to this forest is more difficult. Ampijoroa Forestry Station (part of Ankarafantsika National Park), 120km south-east of Mahajanga, is a further option (although populations in this region may be an undescribed species of Cheirogaleus).

MONTAGNE D'AMBRE DWARF LEMUR
Cheirogaleus andysabini
Lei et al., 2015

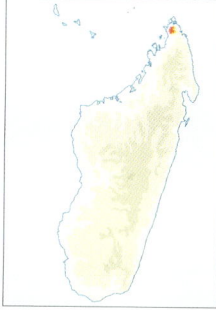

Other name: Sabini's Dwarf Lemur
Malagasy names: Tsidihy, Kelybehohy, Matavyrambo

MEASUREMENTS
Head-body: c.230–240mm.
Tail: c.240–280mm.
Weight: c.220–340g. (Lei et al. 2015)

DESCRIPTION & IDENTIFICATION Upperparts, including head and limbs, rufous-brown; underparts, including under chin, white. Distinct brownish-black eye-rings. Pale grey-whitish patch on nose, between eyes. Belongs to 'crossleyi' subgroup. Very similar to C. crossleyi but generally slightly smaller. Primarily distinguished by molecular differences and isolated distribution.

HABITAT & DISTRIBUTION Lower- and mid-elevation rainforests, c.540m–1,070m. Restricted to isolated rainforests of Montagne d'Ambre and the immediate environs, to north-west of Irodo River in far north (Lei et al. 2015). IUCN Endangered.

BEHAVIOUR No specific studies. Behaviour and ecology assumed to be similar to Crossley's Dwarf Lemur and Greater Dwarf Lemur.

WHERE TO SEE Any dwarf lemur seen during austral summer in and around Montagne d'Ambre National Park and Forêt d'Ambre Special Reserve, including private reserve at Domaine de Fontenay, should correspond to this species.

Ankarana Dwarf Lemur, forests near Daraina.

ANKARANA DWARF LEMUR
Cheirogaleus shethi
Frasier *et al.*, 2016

Other name: Andy Sheth's Dwarf Lemur

Malagasy names: Tsidihy, Kelybehohy, Matavyrambo

MEASUREMENTS
Head-body: *c.*165–175mm.
Tail: *c.*155–165mm.
Weight: *c.*100–125g.
(Fraiser *et al.* 2016)

DESCRIPTION & IDENTIFICATION Smallest of currently described *Cheirogaleus* species. Upperparts, crown, forehead and limbs mid-grey with no dorsal stripe. Underparts, including throat, white. Distinct but narrow brownish eye-rings. Pale grey-whitish patch on nose, between eyes. Ears sparsely furred. Belongs to '*medius*' subgroup. Very similar to Fat-tailed Dwarf Lemur but noticeably smaller. Primarily distinguished by molecular differences. To north occurs Montagne d'Ambre Dwarf Lemur, which is significantly larger and more rufous-brown rather than grey in colour.

HABITAT & DISTRIBUTION Lower-elevation dry deciduous, semi-humid, littoral and savanna scrub forest, including degraded areas and vanilla plantations between *c.*20m and 540m (Hending *et al.* 2017c, 2018a, Sgarlata *et al.* 2020b). Restricted to areas south of Irodo River to around southern extremity of Loky-Manambato Protected Area, but including forest fragments and plantations south of Manambato River near Iharana (Vohemar) (Fraiser *et al.* 2016, Hending *et al.* 2017c, 2018a). May occur sympatrically with congener from '*major*' subgroup, either Crossley's Dwarf Lemur or Greater Dwarf Lemur in parts of Loky-Manambato Protected Area (Hending 2021) and further south at Tsihomanaomby (Williams *et al.* 2020). IUCN Endangered.

BEHAVIOUR No specific studies. Behaviour and ecology probably similar to Fat-tailed Dwarf Lemur, which inhabits similar, seasonally extremely dry forests. Fruits of *Ficus* spp. (Moraeceae) may be particularly important resource prior to torpor: seven individuals seen feeding simultaneously in one tree in Iahaka Forest in March (Hending *et al.* 2017c).

WHERE TO SEE Between November and April, this species can be seen on night walks in vicinity of Ankarana Special Reserve, 100m south of Antsiranana. Also forests around Camp Tattersall, close to Daraina, 45km west of Iharana (Vohimar) provide very good opportunities. This site is also excellent for other nocturnal species including Daraina Sportive Lemur, Northern Fork-marked Lemur and even Aye-aye. The forests of Andrafiamena-Andavakoera Protected Area in the vicinity of Black Lemur Camp are also worth exploring.

CROSSLEY'S DWARF LEMUR
Cheirogaleus crossleyi   A. Grandidier, 1870

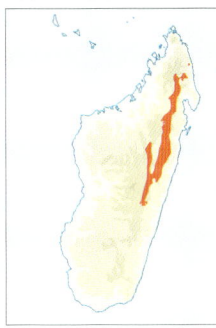

Other name: Furry-eared Dwarf Lemur

Malagasy names: Matavirambo, Tsitsihy

MEASUREMENTS
Head-body: 220–260mm. Tail: 210–270mm. Weight: seasonal range *c.*370–450g. (Schwitzer, *et al.* 2013a)

DESCRIPTION & IDENTIFICATION Upperparts reddish-brown to russet, underparts pale grey becoming creamy around mid-line, more orangey-yellowish around throat. No mid-dorsal line is visible. Mid-face and forehead more yellowish-rufous with clear dark eye-rings. Ears covered inside and out with dark fur.

Belongs to '*crossleyi*' subgroup. Similar to Greater Dwarf Lemur, but slightly smaller and more reddish-rufous colour, darker more pronounced eye-rings and distinctive flattened nose-bridge and pointed snout.

HABITAT & DISTRIBUTION Primarily restricted to mid- and higher-elevation rainforests, up to *c.*1,800m, also recorded in secondary forest and degraded areas, including plantations. Range limits uncertain. Southerly limits appear to be Ambositra area, north through Tsinjoarivo, Andasibe-Mantadia, Anjozorobe, Ankafobe, Zahamena, Ambohitantely and Ambatovy to Andapa region. Precise northern limits uncertain, recorded in Anjanaharibe-Sud, Marojejy, COMATSA forest corridor between these areas, and Tsaratanana and forest fragments of Makirovana-Tsihomanaomby, north-west of Sambava (Lei *et al.* 2014, 2015, Blanco *et al.* 2020a). At some sites, e.g., extreme north-east, sympatric with Greater Dwarf Lemur and perhaps a third congener from '*medius*' subgroup (cf. *C. shethi*) (Williams *et al.* 2020).

In Analamazaotra (now Andasibe-Mantadia National Park) Crossley's Dwarf Lemur (regarded as *C. major* at time of study) recorded at

Crossley's Dwarf Lemur, Andasibe-Mantadia National Park.

Crossley's Dwarf Lemur in underground hibernation chamber, Tsinjoarivo.

densities of 70–110 individuals/km² (Pollock 1979a). Further north in Makira Forest, densities of 15–31 individuals/km² recorded (Rasolofoson *et al.* 2007a) and at sites at northern extreme of range (Marojejy and adjacent forest corridors) densities of *c.*20–60 individuals/km² and *c.*20–90 individuals/km² recorded (Solofoniaina 2018, Tsilanizara 2019). Main threats are habitat loss and some hunting. IUCN Vulnerable.

BEHAVIOUR Nocturnal and arboreal. During austral summer, when active, prefers to sleep in nest-like structures (twigs and dead leaves), less frequently in tree holes. Sleep sites generally high up in canopy. Hibernation period *c.*4–6 months, between April and September, but varies with location and local climate (Blanco & Godfrey 2014, Blanco *et al.* 2018).

Crossley's Dwarf Lemur occupies home ranges that appear variable with location and elevation. In mid-elevation rainforests (Ambatovy, near Andasibe), estimates range from 1–10ha, and at Tsinjoarivo 1–17ha (Blanco in press). Home range smaller post-hibernation versus pre-hibernation (Domoinaharivelo 2014).

In high-elevation forests (1,600–1,800m), e.g., Tsinjoarivo, austral winter temperatures day and night are cold (regularly 5°C, occasionally freezing). At start of hibernation season, a single individual may occupy poorly insulated leaf nests and enter daily bouts of torpor, where body temperature passively tracks ambient temperature. Later moves to occupy excavated underground chamber, *c.*10–40cm deep, often at base of tree below spongy layer of soil, moss, tangled roots and leaf litter (remarkable as dwarf lemurs do not have claws for digging). Within chamber, temperature is fairly constant (no daily high/low extremes) and body temperature is similarly constant around 12–14°C: analogous to mammal hibernation in temperate/arctic areas (Blanco & Rahalinarivo 2010, Blanco *et al.* 2013). At lower elevations, where temperatures higher, Crossley's Dwarf Lemur may also use nest structures during torpid periods, rather than burying themselves underground (Dausmann & Blanco 2016).

Other aspects of behaviour and ecology likely similar to Greater Dwarf Lemur.

WHERE TO SEE Only seen during austral summer/wet season (October–April), when out of dormancy and active. Various forests around Andasibe-Mantadia National Park are best locations. Easily seen on night walks along road near park entrance. Also, Analamazaotra Forest Station, operated by Association Mitsinjo, and V.O.I.M.M.A. Community Reserve near Andasibe village.

GREATER DWARF LEMUR
Cheirogaleus major   É. Geoffroy, 1812

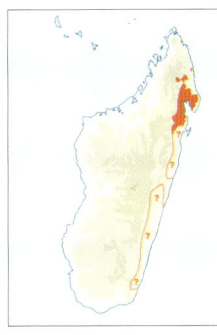

Other name: Geoffroy's Dwarf Lemur

Malagasy names: Tsitsihy, Tsidy, Hataka

MEASUREMENTS
Head-body: *c.*280–290mm.
Tail: *c.*260–300mm.
Weight: seasonal variation *c.*310–640g. (Lahann 2007*, Blanco 2021)

DESCRIPTION & IDENTIFICATION Pelage short and dense. Head, upperparts and broad tail grey-brown to grey-rufous, head may be slightly more rufous-brown. Underparts paler grey to off-white. Vague dark brown mid-dorsal stripe present in some individuals. Dark rings around eyes with whitish-grey areas between eyes and extending down muzzle. Snout dark and slightly pointed with relatively large fleshy nose.

Belongs to '*major*' subgroup. Largest *Cheirogaleus* species. Large squirrel-sized lemur with horizontal posture. Similar to Crossley's Dwarf Lemur, but slightly larger and more greyish, rather than reddish-rufous, with less pronounced eye-rings. *C. ravus* previously was regarded as distinct species (Groves 2000), now a junior synonym of *C. major* (Groeneveld *et al.* 2009).

Greater Dwarf Lemur distinguishable from similar-sized woolly lemurs (*Avahi* spp.) and sportive lemurs (*Lepilemur* spp.) by horizontal rather than vertical posture. Also runs quadrupedally, whereas other genera are vertical leapers. Should not be confused with similar-sized fork-marked lemurs (*Phaner* spp.), which have a more pointed nose and distinctively marked face.

HABITAT & DISTRIBUTION Lowland and mid-altitude rainforest, littoral forest and older secondary forest. With recent taxonomic revisions, precise limits unclear. Occurs in lowland centre-east regions from Brickaville/Andevoranto area to north-east regions including Mananara-Nord, Tampolo, Makira Forest and Masoala Peninsula. Also recorded in Marojejy and adjacent forest corridors, and forest fragments (Makirovana-Tsihomanaomby) between Sambava and Iharana (Vohemar) where sympatric with Crossley's Dwarf Lemur and perhaps a third species (Williams *et al.* 2020, Blanco *et al.* 2020b). Throughout range appears to prefer lower elevations, whereas Crossley's Dwarf Lemur prefers higher elevations.

Greater Dwarf Lemur, Farankaraina Reserve.

Greater Dwarf Lemur, Masoala National Park.

Range may extend to south-east rainforests, but populations in littoral forests near Tolagnaro that were previously designated as *C. major* (Hapke *et al.* 2005, Lehann 2007b), are now recognised as a distinct species, Lavasoa Dwarf Lemur *C. lavasoensis* (Thiele *et al.* 2013), whilst populations in nearby Anosyennes Mountains are possibly a species awaiting formal description (Lei *et al.* 2014). Densities of *c.*100 individuals/km² in Makira Forest (Rasolofoson *et al.* 2007a) and 25–100 individuals/km² in Marotandrano Reserve (Ralison 2006). Main threats are habitat loss and fragmentation, and subsistence hunting (Golden 2009, Borgerson *et al.* 2018, Brook *et al.* 2019). IUCN Vulnerable.

BEHAVIOUR Strictly nocturnal and arboreal. Likely monogamous, with adult male and female forming stable pair in a home range of *c.*4–4.5ha. Offspring from previous year remain with parents and all share sleep sites, either tree holes high in canopy or dense clumps of 'nest-like' tangled vegetation (Lahann, 2007b*, Blanco in press). Forages, either alone or socially (*c.*50% of time) at all levels from understorey to canopy. Also descends to ground to search for insects in leaf litter (Wright & Martin 1995**).

Dormant during austral winter (normally May/June–September). Prior to hibernation, fat reserves laid down in tail, which swells and constitutes up to *c.*30% of body weight. Before torpor, after prolonged feeding body weight can be *c.*640g, post-dormancy reduces to *c.*350g (Blanco 2021). Hibernating animals conceal themselves underground in leaf litter at base of large trees, or in tree holes (Wright & Martin 1995**).

Feeds on nectar, ripe fruit, flowers and, to lesser extent, young leaves and buds. Insects also taken but form small part of diet (Ganzhorn 1988***, Wright & Martin 1995**). After emergence from hibernation (November–December), pollen and nectar of lianas, e.g., *Strongylodon* spp. (Fabaceae) particularly important. Greater Dwarf Lemur thought to be primarily a pollinating species. When feeding it does not to destroy inflorescence, instead parts petals dextrously with hands and licks nectar for several minutes, and so triggers flower's pollen removal/receipt mechanism. Pollen is deposited on lemur's forehead and transferred to another flower when it feeds elsewhere (Nilsson *et al.* 1993**).

Mating takes place shortly after emerging from hibernation in late September–October. Copulation bouts last 2–3 minutes, at intervals of ten minutes. During courtship larger aggregations up to 14 individuals occur. Such groups are very noisy, highly active and agitated; may be groups of males competing for females. Female constructs nest of leaves and vines 6–12m above ground. Litter of two or three born January/February: gestation *c.*70 days (Petter *et al.* 1977***). Initially mother carries young in her mouth, but later they cling to her back. At one month infants are able to follow mother and begin eating soft fruit, but lactation lasts up to six weeks (Petter-Rousseaux 1964***).

Known predators include Ring-tailed Vontsira and Fosa which are both able to remove lemurs from sleep sites, and large snakes such as Madagascar Tree Boa (Goodman 2003).

Note: The studies on which this information is based were carried out when all rainforest *Cheirogaleus* species were regarded as *C. major*. Under current taxonomic arrangements, studies marked [1] are now likely *C. lavasoensis* (littoral forest, Mandena), those marked [2] are *C. grovesi* (Ranomafana) and studies marked [3] are *C. crossleyi* (Andasibe-Mantadia).

WHERE TO SEE During austral summer/wet season (October–April) can potentially be seen in most forests in range. Often seen in Farankareina Reserve near Maroantsetra or nearby Masoala National Park. An alternative location is Marojejy National Park, around both Camp 1 (Mantella) and Camp 2 (Marojejia), although very difficult to distinguish from sympatric Crossley's Dwarf Lemur. Dwarf lemurs in other accessible rainforest reserves like Ranomafana and Andasibe-Mantadia National Parks have previously been designated *C. major*, but with current taxonomic arrangements are now regarded as different species – see *C. crossleyi* and *C. grovesi*.

SIBREE'S DWARF LEMUR
Cheirogaleus sibreei (Forsyth Major, 1896)

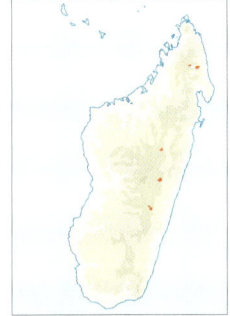

Malagasy name: Tsitsihy

MEASUREMENTS
Head-body: *c.*205–235mm.
Tail: *c.*220–265mm.
Weight: seasonal variation *c.*220–360g. (Blanco *et al.* 2009)

DESCRIPTION & IDENTIFICATION Pelage short and dense. Upperparts, including head and crown and limbs, are grey-fawn. Underparts, including throat and bands extending on side of neck, light grey to cream. Some individuals show a mid-dorsal line. Large and very dark rings around and below eyes, contrast sharply with paler grey-brown muzzle. Ears prominent and sparsely furred (Blanco *et al.* 2009). Belongs to '*sibreei*' subgroup. A relatively small dwarf lemur, squirrel-sized with horizontal posture. Smaller than Crossley's Dwarf Lemur, and more grey (not rufous-brown), with very pronounced eye-rings.

HABITAT & DISTRIBUTION Restricted to higher-elevation montane forests (above 1,300m). These forests now highly fragmented. Known from several dispersed montane sites on east side of central plateau, including: Mount Maharira (1,470m) in Ranomafana National Park; Andasivodihazo and Ankadivory forests (1,600–1,700m), fragment in the Tsinjoarivo protected area; Anjozorobe-Angave corridor; Marojejy National Park around Camp 4 (*c.*1,600m) and the COMATSA forest corridor between Marojejy and Tsaratanana (Blanco *et al.* 2020e).

Sympatric with Crossley's Dwarf Lemur in some locations, e.g., Tsinjoarivo and Marojejy National Park. Sympatric with Haute Matsiatra Dwarf Lemur

Sibree's Dwarf Lemur, Tsinjoarivo.

Haute Matsiatra Dwarf Lemur, Ranomafana National Park.

chamber. Hibernation lasts *c*.4.5–7 months, between April and September, and exceptionally has been recorded from end of February / early March (Blanco & Godfrey 2014, Dausmann & Blanco 2016).

In Tsinjoarivo, Sibree's Dwarf Lemur, occupies home ranges of 2–5ha (smaller than sympatric Crossley's Dwarf Lemur) (Domoinaharivelo 2014, Blanco in press). Both species have smaller home ranges post-hibernation compared to pre-hibernation season (Domoinaharivelo 2014).

Genetics suggest Sibree's Dwarf Lemur was earliest evolving of extant *Cheirogaleus* species and is therefore likely the closest living relative to ancestral dwarf lemur. It may be that obligatory hibernation evolved in dwarf lemurs inhabiting high-altitude forests and the behaviour has been 'inherited' and 'modified' by subsequently evolving species at lower elevations (Blanco & Godfrey 2014).

WHERE TO SEE For the adventurous, areas of forest around Camp 4 in Marojejy National Park. Also forests in Anjozorobe-Angave corridor or alternatively Andasivodihazo forest in Tsinjoarivo, although the latter location is more challenging to reach.

HAUTE MATSIATRA DWARF LEMUR
Cheirogaleus grovesi　　　　McLain *et al.*, 2017

Other name: Groves' Dwarf Lemur

Malagasy name: Tsitsihy

MEASUREMENTS*
Head-body: *c*.260mm. Tail: *c*.290mm. Weight: mean *c*.450g. (McLain *et al*. 2017) (*based on three specimens)

DESCRIPTION & IDENTIFICATION Pelage short and dense. Upperparts, head and limbs rufous-brown. Underparts rufous-grey, grading to near white below muzzle. Distinct broad blackish-brown eye-rings and slight tear mark. Clearly defined pale whitish patch on nose and upper muzzle. Ears dark. Largest of '*crossleyi*' subgroup. Very similar to Crossley's Dwarf Lemur but larger. Primarily distinguished from Crossley's Dwarf Lemur, Montagne d'Ambre Dwarf Lemur and Lavasoa Dwarf Lemur by genetic

in Ranomafana National Park. Extremely limited and fragmented habitat. IUCN Critically Endangered.

BEHAVIOUR Strictly nocturnal and arboreal. Like other dwarf lemurs, diet includes fruits, flowers, buds and insects. During austral summer when active, sleeps exclusively in tree holes. However, hibernation during austral winter always in underground differences. A large dwarf lemur, squirrel-sized with horizontal posture. Larger than sympatric Sibree's Dwarf Lemur, and more rufous-brown (not greyish), with very pronounced eye-rings.

HABITAT & DISTRIBUTION Restricted to remaining mid-elevation humid forests in Haute Matsiatra region (c.750–1,000m). Primarily known from Andringitra and Ranomafana National Parks and possibly Vohibola II Classified Forest, where it appears to avoid forest edge. Density estimates in Vohibola II Classified Forest, 15–59 individuals/km² (Lehman et al. 2006a). Sympatric with Sibree's Dwarf Lemur in Ranomafana National Park, although in this area species probably segregated elevationally, with Haute Matsiatra Dwarf Lemur below c.1,100m and Sibree's Dwarf Lemur above c.1,300m. IUCN Data Deficient.

BEHAVIOUR Strictly nocturnal and arboreal. Like other dwarf lemurs, Haute Matsiatra Dwarf Lemur hibernates during austral winter. In Ranomafana, not seen from mid-April until September. Records of Fosa predation in Andringitra National Park (Goodman et al. 1997b). No specific studies have been undertaken, but other aspects of behaviour probably similar to Crossley's Dwarf Lemur and Greater Dwarf Lemur.

WHERE TO SEE Best accessible location is Ranomafana National Park. During austral summer night walks along road in vicinity of Talatakely are productive. Dwarf lemurs seen in higher-elevation forest at Vohiparara are also likely to be this species.

LAVASOA DWARF LEMUR
Cheirogaleus lavasoensis   Thiele et al., 2013

Malagasy names: Tsitsihy, Kelybehohy

MEASUREMENTS
Head-body: c.220–285mm.
Tail: c.210–280mm.
Weight: c.250–300g,
seasonal variation up to c.420g. (Lahann 2007b, Thiele et al. 2013)

DESCRIPTION & IDENTIFICATION Head and upper neck reddish-brown, grading to grey-brown upperparts and tail. Underparts pale creamy-white with sharp delineation from upperparts. Pale ventral

coloration partially extends around lower neck forming distinct band. Forehead reddish-brown, with very distinct broad blackish-brown eye-rings and dark coloration extending around muzzle and upper jaw. Clearly defined pale whitish patch on nose. Ears well furred and dark.

Belongs to '*crossleyi*' subgroup and very similar to Crossley's Dwarf Lemur. Populations initially assigned to *C. major/C. crossleyi* (Hapke et al. 2005) later given species status largely due to genetic differences (Thiele et al. 2013). Possibly sympatric with smaller more greyish Thomas's Dwarf Lemur in some littoral forests. A relatively small, squirrel-sized dwarf lemur with horizontal posture. Considerably larger than sympatric mouse lemurs (*Microcebus* spp.) and smaller than vertically clinging and leaping woolly lemurs (*Avahi* spp.).

HABITAT & DISTRIBUTION Restricted to transitional and mid-elevation humid forests. Currently known only from very small fragments of transitional forest in Lavasoa-Ambatotsirongorongo Mountains (c.300–800m) in far south-east, and mid-elevation humid forest (c.1,200m) in Kalambatritra, 170km further north. Kalambatritra is westernmost extent of rainforest region, lying directly on Madagascar's west–east drainage divide. IUCN Endangered.

BEHAVIOUR Strictly nocturnal and arboreal. Studied in Mandena when regarded as *C. major* (Lahann 2007a,b, 2008). Lives in monogamous pairs with dependent offspring that occupy home ranges of c.4ha. Males and females sleep together, preferring holes in larger trees high in canopy (Lahann 2007b). Hibernates between early May and late September.

Males, females and offspring often forage together (Lahann 2007b). Diet dominated by fruits (>60 species eaten) and reduces competition with sympatric species (Ganzhorn's Mouse Lemur and Thomas's Dwarf Lemur) by primarily utilising upper storey (Lahann 2008). Seeds pass through digestive system unharmed, so species plays important role in seed dispersal (Lahann 2007a). Further aspects of behaviour are likely similar to other humid forest *Cheirogaleus* species.

WHERE TO SEE Complex of forest fragments, Ambatotsirongorongo, Grand Lavasoa and Petit Lavasoa (latter also called Bemanasy Forest) forming new protected area, Ambatotsirongorongo Special Reserve, 25km west of Tolagnaro.

THOMAS'S DWARF LEMUR
Cheirogaleus thomasi   (Forsyth Major, 1894)

Malagasy names: Tsitsihy, Kelybehohy

MEASUREMENTS
Head-body: *c.*200–220mm.
Tail: *c.*190–250mm.
Weight: mean *c.*180g,
seasonal variation *c.*160–260g. (Lahann & Dausmann 2011*) (*regarded as *C. medius* at time of study)

DESCRIPTION & IDENTIFICATION Pelage short and dense. Upperparts darker fawn-grey with faint mid-dorsal stripe. Underparts pale grey with creamy-yellow midline. Moderately distinct whitish collar. Prominent brown-black eye-rings. Darker fur around ears. Belongs to '*medius*' subgroup. Originally described as *Opolemur thomasi* (Forsyth Major 1894) and specific name recently resurrected (Lei *et al.* 2014), and likely to correspond in part to *C. adipicaudatus* (Groves 2000), a taxon no longer recognised (Lei *et al.* 2014). A smallish, squirrel-sized dwarf lemur with horizontal posture. Appreciably larger than sympatric mouse lemurs (*Microcebus* spp.) and smaller than woolly lemurs (*Avahi* spp.), which adopts a vertical, rather than horizontal posture.

HABITAT & DISTRIBUTION Probably restricted to littoral forest fragments below *c.*50m on coastal plain, in far south-east between Saint Luce and Petriky. This is transitional zone, ranging from humid forests in east at Saint Luce (*c.*2,400mm rainfall per year) to dry forests in west at Petriky (*c.*800mm) (Ganzhorn *et al* 2020c). Probably sympatric in some localities with Lavasoa Dwarf Lemur, which is larger and more reddish-brown. IUCN Endangered.

BEHAVIOUR Strictly nocturnal and arboreal. All studies undertaken when populations were thought to be *C. medius* (Lahann 2007a, 2008, Lahann & Dausmann 2011). Lives in mated pairs with dependent offspring in home ranges of *c.*1.8–2.4ha. Produces a single litter of 2–4 offspring per year, with young born late November to early December (Lahann & Dausmann 2011). Diet dominated by smaller fruits (>60 species eaten) and reduces competition with sympatric species (Ganzhorn's Mouse Lemur and Lavasoa Dwarf Lemur) by primarily utilising

midstorey (Lahann 2008). Seeds pass through digestive system unharmed, so species plays important role in seed dispersal (Lahann 2007a).

Hibernates for several months between late April/early May and September (Lahann & Dausmann 2011), preferring to use large tree holes. Duration of hibernation shorter than in similar Fat-tailed Dwarf Lemur that lives in much drier western forests, but life expectancy shorter due to higher mortality rates. Consequently, reproductive rates are higher (larger litter sizes, greater number of litters) (Lahann & Dausmann 2011).

WHERE TO SEE Littoral forests near Mandena, north-east of Tolagnaro.

FORK-MARKED LEMURS
GENUS *Phaner*

In size, fork-marked lemurs, genus *Phaner*, are the second largest in the family Cheirogaleidae (some *Cheirogaleus* spp. are larger) and are the only non-heterothermic members of the family (Blanco in press). They are perhaps the most enigmatic of the Cheirogaleidae, and have been relatively under-studied and remain the least well-known members of the group.

When initially described, the fork-marked lemur was placed in the genus *Lemur* (as *L. furcifer*), but later reclassification resulted in the creation of genus *Phaner*. Until relatively recently, *Phaner* contained a single species; further study recognised four distinct subspecies (Groves & Tattersall 1991), which were subsequently but contentiously elevated to species (Groves 2001, Tattersall 2007). Some known populations may represent new species yet to be formally described, e.g., those between Loky and Manambato Rivers (e.g., around Daraina) and south of Manambato River to Fanambana River (Salmona *et al.* 2018, Hending *et al.* 2020). It is likely the taxonomy of *Phaner* will be amended as research continues.

The four currently recognised species are very similar in appearance (with subtle variations in pelage coloration) and occupy apparently discontinuous ranges in the west, north and north-east, the limits of which are poorly understood in some cases. Furthermore, several recently discovered populations have yet to be specifically identified. Fork-marked lemurs are absent from dry regions of the south and south-west. The presence of fork-marked lemurs inferred by vocalisations, from Andohahela region in extreme south-east (Feistner & Schmid 1999), requires further investigation as this is outside any known species' range. Nonetheless, location is the most reliable guide for identifying each species.

All *Phaner* species are broadly similar in size (head-body *c.*230–285mm, tail length *c.*300–400mm, mean weight *c.*300–400g), habits and general markings, but some subtle differences in coloration exist. Upperparts various shades of brown to brownish-grey, underparts and bushy tail lighter shades of brown and grey. Tail darkens towards tip, or tip may be white. Tail longer than head-body length. Their obvious characteristic is broad dark stripe from base of tail that runs along dorsal ridge, back of neck to crown, where it divides, with two stripes continuing down face, around eyes and sometimes muzzle: hence 'fork-marked'.

Fork-marked lemurs adopt horizontal postures and move quadrupedally in characteristic manner. Very agile, running at speed along branches and leaping from one branch to next (covering gaps of 4m+) without pause, therefore difficult to follow. The hands and feet have large pads on the digits, with long nails (except on thumb) which allow individuals to grip to and traverse the smooth surface of large tree trunks when feeding. At rest bobs head up and down, or wags it sideways in distinctive manner (which can be detected by eyeshine).

Fork-marked lemurs are specialist exudate feeders (exhibiting several convergent adaptations with Central-West African needle-clawed bushbabies, *Euoticus* spp.), primarily subsisting on tree gums and saps, supplemented with fruits, nectar, flowers and occasionally insects and their excretions. They are adapted to a gummivorous diet by virtue of a unique dental arrangement: incisors and canines on lower jaw are almost forward-facing horizontal and act like a scraper that helps strip away tree bark. Any exposed wood-boring invertebrates are quickly consumed and oozing sap is then lapped with a particularly long, narrow tongue. The gut also contains specialised bacteria to help digest this diet (Schwitzer *et al.* 2013b). There is no evidence to suggest fork-marked lemurs accumulate fat reserves, nor enter periods of torpor.

Fork-marked lemurs appear to form family units: an adult pair and their offspring in a territory that is defended, but male and female spend little time together. They sleep in tree holes or abandoned lemur nests (Schülke 2003a,b, 2005) and are extremely vocal: their high-pitched calls and whistles shortly after dusk and around dawn often being the first indicators of presence. Communication is both

between individuals in a group and between groups. Several different call types have been identified, the piercing long-distance territorial call of males being the most distinctive with several males often calling simultaneously. Vocalisations have low-frequency elements that travel greater distances in densely foliated forest and the calls of each species is subtly different and elicits species-specific responses (Forbanka 2020).

Fork-marked lemurs appear to prefer the upper strata and forest canopy, which often makes them difficult to locate and observe clearly. This may be a contributory factor in the genus being the least studied in Cheirogaleidae.

MASOALA FORK-MARKED LEMUR
Phaner furcifer (de Blainville, 1839)

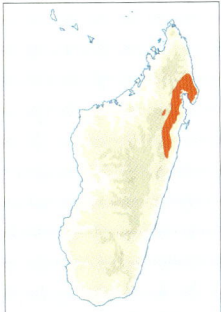

Other name: Eastern Fork-marked Lemur

Malagasy names: Tantana, Tanta, Vakiandrina

MEASUREMENTS
No specific data available.

DESCRIPTION & IDENTIFICATION Largest-bodied and darkest fork-marked lemur with long dense pelage. Upperparts dark brown, underparts buffy-grey to creamy-buff. Ears large and unfurred. Facial fork and dorsal stripe very dark, broad and pronounced; stripe does not reach base of tail. Tail bushy and relatively long: up to 150% of head-body length. Tail end dark but extreme tip sometimes pale grey.

Distinctive markings and rapid movements should prevent confusion with dwarf lemurs (*Cheirogaleus* spp.) that are similar in size, colour and posture.

HABITAT & DISTRIBUTION Lowland and mid-altitude rainforest up to *c.*1,000m (Mittermeier *et al.* 2010). Current extent of range uncertain. Historic range from Zahamena in centre-east to Marojejy in north-east (Goodman & Raselimanana 2002, Mittermeier, *et al* 2010), including Makira Forest and Masoala Peninsula. Record from Zahamena National Park in 2004 and Makira Forest (Rasolofoson 2007a), but subsequent expeditions over next decade to numerous sites within historic range failed to confirm presence (Louis *et al.* 2020m). Densities in Makira Forest, 3–12 individuals/km² (Rasolofoson 2007a) and in Zahamena, in 2014, 2.3 individuals/km² (Forbanka 2018). IUCN Endangered.

BEHAVIOUR Nocturnal and arboreal. Brief studies in Zahamena National Park suggest a preference for forests dominated by trees with rough rather than smooth bark, which aids climbing up and down large trunks. These trees produce crucial exudates on which this species feeds (Forbanka 2018b). No other specific wild studies.

WHERE TO SEE Potentially encountered at localities on Masoala Peninsula, including Lohatrozona and Tampolo, and in Mikira Natural Park (Andranomenahely, Antaka and Mangabe), west of Maroantsetra. Zahamena National Park is another option, but challenging to reach requiring expedition-level logistics. Any sightings or inferred presence from distinctive vocalisations should be reported in as much detail as possible to appropriate lemur authorities.

SAMBIRANO FORK-MARKED LEMUR
Phaner parienti   Groves & Tattersall, 1991

Other name: Pariente's Fork-marked Lemur

Malagasy names: Tanta, Valvihy

MEASUREMENTS
Head-body: *c.*230–240mm.
Tail: *c.*390–410mm.
Weight: *c.*320–400g.
(Mittermeier *et al.* 2010, Salmona *et al.* 2018)

DESCRIPTION & IDENTIFICATION Pelage thick and dense. Upperparts mid-brown to dark brown, underparts buff with reddish tinges. Crown/facial fork dark, broad and prominent, dorsal mid-stripe extends to base of tail. Distinct darkening of distal third of tail, but extreme tip white (Groves 2001). Darker than Northern Fork-marked Lemur *P. electromontis* to north, and Pale Fork-marked Lemur *P. pallescens* that occurs to south. Two sympatric nocturnal lemurs, dwarf lemurs (*Cheirogaleus* spp.) and Northern Giant Mouse Lemurs, are similar in size, colour and posture: the distinctive facial markings, characteristic rapid

movements and loud vocalisations of Sambirano Fork-marked Lemur should prevent misidentification.

HABITAT & DISTRIBUTION Lowland moist forest and secondary forest up to 800m. Also recorded in shade-grown coffee plantations and cacao plantations around villages (Forbanka 2018b, Webber *et al.* 2020). Restricted to Sambirano region in north-west, south of Ambanja between Andranomalaza River to south and Sambirano River to north, including Ampasindava Peninsula to west and western side of Tsaratanana Massif at eastern extreme of Sambirano region (Louis *et al.* 2020k). May previously have occurred on Sahamalaza Peninsula but now presumably extirpated (Hending 2021). Encounter rates of 2.4 individuals/km on night walks in Manongarivo (Forbanka 2018a) and 0.2–2.4 individuals/km at Ampopo and Ambaliha and cocoa plantations near Ambanja (Ralantoharijaona *et al.* 2014, Webber *et al.* 2020). IUCN Endangered.

BEHAVIOUR Nocturnal and arboreal. Like all fork-marked lemurs, diet likely dominated by saps and gums. Similar to other fork-marked lemurs, anecdotal observations suggest preference for tall trees. Brief observations in Manongarivo suggest nightly division of activity split between moving, feeding, grooming and resting (Forbanka 2018a). No further specific wild studies. See Pale Fork-marked Lemur below for more details of fork-marked lemur behaviour and ecology.

WHERE TO SEE Well-preserved forest on Ampasindava Peninsula. Also forests around village of Beraty at western edge of Manongarivo Special Reserve, 45km south of Ambanja, can be productive (Mittermeier *et al.* 2010).

PALE FORK-MARKED LEMUR
Phaner pallescens   Groves & Tattersall, 1991

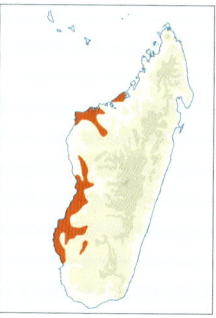

Malagasy names: Tanta, Tantaraolana, Vakivoho

MEASUREMENTS
Head-body: *c.*235–275mm.
Tail: *c.*300–340mm.
Weight: *c.*310–370g.
(Schülke 2003, Mittermeier *et al.* 2010)

Pale Fork-marked Lemur, Kirindy Forest.

DESCRIPTION & IDENTIFICATION Palest and probably smallest *Phaner*. Upperparts light grey to grey-fawn with silvery sheen. Underparts pale yellowish-white. Crown/facial fork and mid-dorsal stripe less distinct than in other fork-marked lemurs. Distal two-thirds of tail slightly darker than body, but unlike congeners, tail lacks dark tip. Occurs sympatrically with several other nocturnal lemurs, including Fat-tailed Dwarf Lemur and Coquerel's Dwarf Lemur, which are broadly similar in size, colour and posture: distinctive markings and rapid movements of Pale Fork-marked Lemur should prevent confusion.

HABITAT & DISTRIBUTION Utilises variety of habitats, including dry deciduous forest, transitional forest (between deciduous forest and spiny forest) and secondary forest, from sea level to c.800m. Confined to western Madagascar, but has largest range of any fork-marked lemurs species. Main population occurs north from Fiherenana River, including Mikea Forest, Zombitse and Isalo National Parks to around Antsalova, north of Tsiribihina River (Hawkins 1999, Borgerson 2020). A second subpopulation is centred on Namoroka National Park and Baly Bay/Soalala area. Further north,

between Betsiboka and Mahajambe Deltas, is a third subpopulation. Both northerly populations appear isolated and may constitute distinct taxa (Borgerson 2020). In Menabe Central (Kirindy Forest) previous density estimates 50–70 individuals/km² (Schülke 2003) and more recently 32 individuals/km² (Schäffler & Kappeler 2014b). Recent estimates from Zombitse just 3.3 individuals/km² (Forbanka 2018a). Primary threats are habitat destruction and fragmentation. IUCN Endangered.

BEHAVIOUR Most information on behaviour and ecology in fork-marked lemurs is based on studies of this species in dry deciduous forests north of Morondava.

Nocturnal, although activity starts at dusk before complete darkness. Can be extremely vocal, frequently answering calls of neighbours; bouts of calling are a distinctive twilight feature in forests the species inhabits. Further bouts of calling occur as individuals return to sleep sites prior to dawn (Schülke 2003a).

Apparently monogamous (van Schaik & Kappeler 1993) and pairs remain together for several breeding seasons (Schülke 2003a, 2005). Territories of 3–10ha are defended and remain constant in

163

size year-round. When active, males and females spend little time together, with very brief bouts of grooming and contact that occur face-to-face, often while hanging upside-down. Majority of time is spent apart (regularly >100m away), foraging alone to avoid competition, but maintaining vocal contact (Schülke & Kappeler 2003).

Little territorial overlap; interactions between pairs at boundaries are frequent, but not necessarily antagonistic as neighbouring females and juveniles may groom one another. Males are aggressive towards other males and females, but females are invariably dominant (Schülke 2003a).

By day, sleeps alone or in pairs in hollows and holes near tops of large trees (Schülke 2003a), including baobabs *Adansonia* spp. (Du Puy 1996) and occasionally in abandoned nests of Coquerel's Dwarf Lemur (Kappeler 2003). Sleep sites lined with leaves. Individuals can use >30 different sites per year. Pale Fork-marked Lemur is sympatric with several other nocturnal lemurs; its preference for sleep sites towards the canopy top may reduce competition (Schülke 2003a).

Mating highly seasonal: occurs late October/ early November, births in February and March. Single young born in tree hole and remains in nest for first couple of weeks. Offspring stay with parents in family unit for up to three years (Schülke 2003a). Despite being highly territorial, parents tolerate subadult offspring in natal area, because of scarcity and patchy distribution of exudate resources (Schülke 2005, Schülke & Ostner 2005).

Dentition and diet specialised; lower incisors modified into inclined 'dental comb', used to scrape and gouge tree bark to stimulate flow of sap and gums. These rich sources of sugar and protein are primary dietary constituents throughout year: *c.*75% of monthly feeding time is directed at their consumption, primarily from baobabs and taly trees (*Terminalia* spp.) (Schülke 2003a). Such resources are scarce, patchily distributed and can be utilised by only one individual at a time, so are aggressively defended, with females dominant over males (Schülke 2003b). Other foods include flowers and nectar, so fork-marked lemurs are responsible for pollinating some baobabs (Baum 1996), insects (adults, larvae and pupae) and their excretions. Pale Fork-marked Lemur has a long extensible tongue, which helps exploit its diet (Schülke 2003a). Two species of needle-clawed bushbaby (*Euoticus* spp.: Galagonidae) from rainforests of Central-West Africa, have very similar adaptations, and also feed on tree gums and sap.

As a specialist gum feeder, Pale Fork-marked Lemur can remain active throughout year, without physiological adaptations to low temperatures and low food availability (Hladik *et al.* 1980); however, during austral winter experiences weight loss up to 20% (Schülke 2003a).

Predated by Fosa and several raptors, e.g., Madagascar Buzzard, Madagascar Harrier-Hawk and Madagascar Cuckoo-Hawk (Goodman 2003).

WHERE TO SEE Kirindy Forest, north of Morondava, is the best locality. Also, Kirindy-Mitea National Park south of Morondava, and Zombitse-Vohibasia National Park east of Sakaraha, offer potential opportunities. The species' loud vocalisations at dusk are unmistakable, but their tendency to be high in the canopy makes close observation challenging. Their habit of head-bobbing helps make eyeshine distinctive.

NORTHERN FORK-MARKED LEMUR
Phaner electromontis   Groves & Tattersall, 1991

Other names: Amber Mountain Fork-marked Lemur, Montagne d'Ambre Fork-marked Lemur

Malagasy name: Tanta

MEASUREMENTS*
Head-body: *c.*265–275mm. Tail: *c.*340–350mm. Weight: *c.*350–420g. (Mittermeier *et al.* 2010) (*based on three specimens)

DESCRIPTION & IDENTIFICATION Relatively large *Phaner*. Upperparts light grey to silver-grey with dark, well-defined crown fork and broad mid-dorsal stripe extending to base of tail. Tip of tail (distal third) also very dark. Hands and feet slightly darker than upperparts. Distinctive markings and rapid movements should prevent confusion with *Cheirogaleus*, the only sympatric nocturnal species of similar size and posture.

HABITAT & DISTRIBUTION Occurs in rainforest, moist tropical forest, dry deciduous forest and secondary forest between *c.*20m and *c.*1,300m. Restricted to far north: centred on humid forests of Montagne d'Ambre complex of protected areas,

Currently regarded as Northern Fork-marked Lemur, populations from forests in Daraina region may be re-classified as a new species.

and drier forests of Ankarana and Analamerana; also, Andrafiamena-Andavakoera Protected Area, Ankarongana and Sahafary Classified Forests (Salmona 2014, Sgarlata *et al.* 2020c).

Fork-marked lemurs also recorded in Loky-Manambato Protected Area, e.g., Daraina, and sites further south between Manambato and Manambery Rivers (Analafiana, Analamanara and Salafaina) and south of Manambery River at Bezavona Ankirendrina, meaning southern limit possibly Fanambana River (Hending *et al.* 2020b). These populations currently regarded as *P. electromontis*, but might correspond to a new species (Mittermeier *et al.* 2010, Salmona *et al.* 2018) or indeed two species (Hending 2021): pending further fieldwork and genetic analysis, their taxonomic status remains undetermined.

Within range, Northern Fork-marked Lemur appears patchily distributed: a survey of >60 sites covering all habitat types across its range, revealed presence in just 33% of sites, with no obvious correlation to habitat type or quality (Hending *et al.* 2020b). IUCN Endangered.

BEHAVIOUR Nocturnal and arboreal. Like other fork-marked lemurs, diet likely dominated by saps and gums. In common with congeners, Northern Fork-marked Lemur is very vocal year-round, but especially in June/July and November. Calling most apparent during first four hours of darkness: peak vocalisations at onset of dusk (c.17.30–18.00) and early night (c.19.00–20.00) (Hending *et al.* 2020b). No further specific wild studies. See Pale Fork-marked Lemur for more details of likely behaviour and ecology.

WHERE TO SEE Montagne d'Ambre National Park is most accessible location: Botanical Garden trail and areas around camp site and *Petite Cascade* are good places to concentrate efforts. As trees are tall, views are often distant. Fork-marked lemurs that may correspond to Northern Fork-marked Lemur, or an as yet undescribed species, are relatively easy to see near Daraina: forest around Camp Tattersalli provides excellent opportunities. This site is also very good for other nocturnal species including Daraina Sportive Lemur and even Aye-aye, and by day Golden-crowned Sifaka *Propithecus tattersalli* and Crowned Lemur *Eulemur coronatus*. The more adventurous might try nearby Analamanara Forest, a two-hour climb from village of Mahasoa on *Route Nationale* 5a between Iharana and Daraina (Hending 2021).

Sportive Lemurs Family Lepilemuridae

The term 'sportive' was perhaps first coined in reference to the agility of these lemurs and may also be a consequence of them adopting a 'boxer-like' posture when threatened (Schwitzer *et al.* 2013c). Sportive lemurs form their own family Lepilemuridae, although some authorities have previously suggested dental features suggest alignment with extinct taxa in the family Megaladapidae (Tattersall & Schwartz 1991, Groves 2001). Molecular analysis fails to support this (Yoder *et al.* 1999, Karanth *et al.* 2005). The evolutionary divergence of Lepilemuridae from Lemuridae is estimated to be *c.*30 mya (Lei *et al.* 2017).

SPORTIVE LEMURS
GENUS *Lepilemur*

All sportive lemurs are assigned to the genus *Lepilemur*, derived from the Latin *lepidus* meaning 'agreeable, pleasant, pretty' in combination with *Lemur*. Taxonomy of this group has been subject to frequent revision. At one time, all recognised forms were regarded as subspecies of *L. mustelinus* (Tattersall 1982), with later authors elevating these taxa to seven species, each occupying a distinct geographical range (Jenkins 1987, Harcourt & Thornback 1990).

The past 20 years has seen significant increase in research into this group, much involving the exploration of many forests previously not surveyed. This, combined with the application of genetic techniques, has revealed significant 'cryptic' diversity (Lei *et al.* 2017), which has resulted in a considerable proliferation in described forms: currently 26 species are recognised (Andriaholinirina *et al.* 2006a,b, Louis *et al.* 2006b, Rabarivola *et al.* 2006, Craul *et al.* 2007, Lei *et al.* 2008, 2017, Ramaromilanto *et al.* 2009), although the validity of many has been questioned (Tattersall 2007) (See 'The Species Conundrum').

Sportive lemurs occur throughout the various native forest types on the island. In broad terms, there are discernible species differences from east to west across the Central Highlands divide, where evolutionary divergence occurred *c.*15 mya, and north to south along a climatic cline (Lei *et al.* 2017) and across major rivers, which appear to act as barriers to dispersal (Craul *et al.* 2007). Yet in some regions, where several species have been described in a relatively small geographic area, the situation is unclear. This is especially true in the north-west, where eight different *Lepilemur* species

are recognised, with some described from localities very close together, and where, in some cases, there are no obvious biogeographic barriers separating populations that might lead to speciation (Zinner *et al.* 2007).

Each species apparently occupies a distinct range (very occasional examples of species in sympatry, e.g., Ankarana Sportive Lemur *L. ankaranensis* and Daraina Sportive Lemur *L. milanoii* in Andrafiamena Classified Forest), with major rivers often appearing to be boundaries between species (Goodman & Ganzhorn 2004, Craul *et al.* 2007, Lei *et al.* 2017). Where a single species occurs both sides of a major river, it is likely that colonisation occurs via migration around the headwaters (Craul *et al.* 2008). Nonetheless, boundaries between ranges of some species are unclear and it is likely that ongoing research will prompt amendment and further taxonomic revision, including the recognition of further new taxa (Louis *et al.* in press).

Most reasonably sized patches of remaining native forest are home to sportive lemurs, although ranges are extremely fragmented, a consequence of massive deforestation (like all lemurs). They also occur in secondary and degraded forests, but population densities are appreciably lower: larger trees with tree holes are reduced or absent in such forests, and paucity of sleep sites may be a significant limiting factor (Schwitzer *et al.* 2013c). The smallest recorded forest patch containing the genus is 6ha, for Red-tailed Sportive Lemur *L. ruficaudatus* in western deciduous forest (Schwitzer *et al.* 2013c).

Sportive lemurs are primarily nocturnal, although some species do show low levels of daylight activity, e.g., Sahamalaza Sportive Lemur *L. sahamalazensis* (Seiler *et al.* 2013b), medium-small to medium-sized lemurs (*c.*500–1,100g) with tails generally slightly shorter than head/body length.

They are primarily 'vertical clingers and leapers', grasping vertical tree trunks and leaping between them, sometimes considerable distances, using powerful hindlegs; however some species do exploit horizontal and oblique branches at times. Unlike other lemurs, adults have only 32 teeth, as they lose their upper incisors when juvenile. The lower incisors and canines form a forward-facing toothcomb (Schwitzer *et al.* 2013c).

Sportive lemurs spend the day sleeping in tree holes or dense tangles of vegetation (preference varies between species and with habitat), which may provide thermoregulatory benefits and also avoid detection by predators (Radespiel *et al.* in press).

Social structure ranges from primarily solitary, to dispersed pairs, to moderately cohesive pairs, depending on species (Kappeler 2014, Seiler *et al.* 2015a, Mandl *et al.* 2019a), but when foraging they are generally solitary. During daylight they usually sleep alone, or more occasionally in pairs or groups of three (pair with offspring), in favoured tree holes, tight crevices between branches, or less frequently dense tangles of twigs and leaves. Leaves dominate their diet, with flowers, buds and occasionally fruit constituting a minor proportion; they are the smallest folivorous primates (Ganzhorn 1993).

Folivory is rare in nocturnal mammals, principally because leaves are low in sugars during hours of darkness, when photosynthesis stops. Sportive lemur diets are among the most nutritionally poor of any primate, hence an adaptation is the enlarged ceacum and proximal colon, where both these compartments host bacteria and other microbes capable of breaking down cellulose (the major constituent of plant cell walls) and thus act as fermentation chambers that produce volatile fatty acids, the principal energy source for sportive lemurs. The metabolic rates of sportive lemurs are among the lowest in folivorous mammals (Schmid & Ganzhorn 1996, Bethge *et al.* 2017). Consequently, they are characterised by behaviour that minimises energetic costs, e.g., small home ranges, short nightly travel distances and increased resting periods.

In general, sportive lemurs in drier western and southern forests tend to be quite vocal; however, species in eastern humid forests appear comparatively silent, although data on eastern species are scanter. In many western forests, the single-note, high-pitched, scream-like calls of sportive lemurs are characteristic of initial periods of darkness and on nights with more intense moonlight (Schwitzer *et al.* 2013c).

Red-tailed Sportive Lemur, Kirindy Forest.

There are records of several mammalian and avian predators taking sportive lemurs, including Fosa, Narrow-striped Boky, Henst's Goshawk, Madagascar Buzzard, Madagascar Harrier-Hawk, Madagascar Long-eared Owl and Barn Owl (Goodman & Ganzhorn in press). Large boid snakes (*Acrantophis* and *Sanzinia* spp.) are also likely predators.

Overall sportive lemur morphology, appearance and coloration are broadly very similar across all species and most cannot easily be distinguished

167

from one another visually (Radespiel *et al.* in press), hence geographic location is often the most reliable factor for identification. In the field, sportive lemurs are most likely to be confused with nocturnal woolly lemurs (*Avahi* spp.), which adopt a similar vertical posture, although woolly lemurs tend to be slightly larger and have prominent, highly visible white patches on the thighs. Confusion with other nocturnal species, like dwarf lemurs (*Cheirogaleus* spp.), is less likely, as they adopt a horizontal posture and run quadrupedally along branches, in contrast to the vertical clinging and leaping of sportive lemurs. The generic name *Lepilemur* is often used as the vernacular.

Young Weasel Sportive Lemur, Andasibe-Mantadia National Park.

WEASEL SPORTIVE LEMUR
Lepilemur mustelinus

I. Geoffroy Saint-Hilaire, 1851

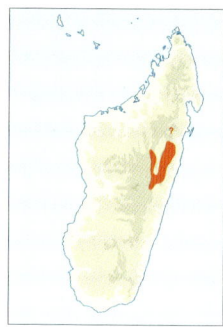

Malagasy names: Hataka, Trangalava, Kotrika, Fitiliky, Kotriana

MEASUREMENTS
Head-body: 260–300mm. Tail: 250–290mm. Weight: 900–1,100g. (Louis *et al.* 2006b, Lei *et al.* 2008, Schwitzer *et al.* 2013b)

DESCRIPTION & IDENTIFICATION A large sportive lemur. Pelage long and dense. Upperparts and head grey-brown to chestnut-brown with faintly darker crown and dorsal stripe sometimes apparent. Tail similar colour, but darkens towards tip; relatively short. Face, throat and underparts paler grey-brown, muzzle darker. Ears naked and clearly visible. In Andasibe region, a highly unusual bright orange colour variant has been recorded (Mittermeier *et al.* 2010).

A medium-sized, vertically clinging lemur with tail equivalent to, or slightly shorter than, head/body length. Potentially confused with Eastern Woolly Lemur *Avahi laniger* of similar size, coloration and posture. Weasel Sportive Lemur has prominent ears, more pointed muzzle and uniform coloration, whereas woolly lemur has very visible white thigh and eyebrow patches, and distinctive pale 'facial mask'. Easily distinguished from Crossley's Dwarf Lemur, which is smaller, adopts a horizontal posture and moves along branches quadrupedally.

HABITAT & DISTRIBUTION Primary and secondary rainforest. Weasel Sportive Lemur is found over central-northern portion of eastern rainforest belt, approximately from line formed by Onive and Mangoro Rivers in south (Andriaholinirina *et al.* 2006a) to at least Maningory River, including higher-elevation forests around Anjozorobe (LaFleur 2020a). At southern extreme of range, boundary with Betsileo Sportive Lemur *L. betsileo* is unclear; similarly, at northern extreme the identity of sportive lemurs immediately north of Maningory River is unclear. Further north still is Mananara-Nord Sportive Lemur *L. hollandorum*. IUCN Vulnerable.

BEHAVIOUR Strictly nocturnal and arboreal. Both sexes generally sleep alone (females may sleep with current or previous years' offspring) in tree holes (c.80% of sites), 6–12m above ground. Less frequently in nests of densely tangled vines and leaves, constructed in hanging lianas (c.20% of sites), the latter more frequently during wet season (November–March) (Ratsirarson & Rumpler 1988).

At dusk emerges and rests at edge of tree hole becoming active at onset of darkness. Sometimes vocal at dusk; calling soon subsides and stops within an hour of nightfall. Occupies territories of 1.5ha (Schwitzer *et al.* 2013b). Solitary when foraging and feeds on leaves, but also some fruit and flowers. Appears to be outcompeted by Eastern Woolly Lemur for leaves of higher nutritional value (Ganzhorn 1988). Grooming recorded between mother and young, but never between adults.

Known predators include Fosa, Madagascar Long-eared Owl and Madagascar Harrier-Hawk (Goodman 2003, Goodman & Ganzhorn in press). Also reports of Madagascar Tree Boa taking sportive lemurs from nests or tree holes.

WHERE TO SEE Forest patches in vicinity of Andasibe-Mantadia National Park are best. Also, Analamazaotra Forest Station, run by Association Mitsinjo, and V.O.I.M.M.A. Community Reserve near Andasibe village. Some 15km away is Vohimanana-Ankera, another community NGO forest with rewarding night walk possibilities where *L. mustelinus* can be seen.

BETSILEO SPORTIVE LEMUR
Lepilemur betsileo   Louis *et al.*, 2006

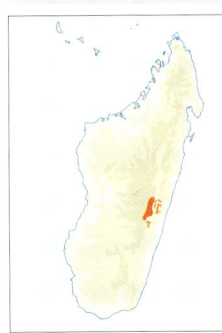

Malagasy names: Hataka, Trangalava, Kotrika, Fitiliky, Kotriana

MEASUREMENTS
Head-body: c.320–330mm.
Tail: c.265–290mm.
Weight: mean 1,150g, range 1,020–1,280g. (Louis *et al.* 2006b)

DESCRIPTION & IDENTIFICATION A noticeably large sportive lemur. Dense fur mainly greyish to reddish-brown dorsally, including limbs, but paler ventrally. Head more greyish. Fur at edges of ears

Betsileo Sportive Lemur, Vohiparara, Ranomafana National Park.

noticeably darker than head. Tail black, contrasting obviously with body. Face and muzzle grey, lower mandible and around mouth white (Louis *et al.* 2006b). Within range, most likely to be confused with similar-sized and -coloured Peyrieras's Woolly Lemur *Avahi peyrierasi*: more obvious ears, russet forelimbs and shoulders (rather than grey-brown), and absence of white thighs and tail base are main distinguishing features.

HABITAT & DISTRIBUTION Primary and secondary rainforest. Type locality Fandriana Classified Forest. Range thought to extend between Namorona River in south and Mangoro/Onive Rivers in north (Louis *et al.* 2020j), but further work required to confirm precise limits. As currently known, total range may only be *c*.1,150km² (Herrera *et al.* 2018). IUCN Endangered.

BEHAVIOUR Nocturnal. Yet to be studied in the wild. Aspects of behaviour probably similar to other rainforest *Lepilemur* species.

WHERE TO SEE Sportive lemurs in higher-elevation forests of Ranomafana National Park, e.g., Vohiparara, may be this species. Otherwise, any *Lepilemur* in forests between Namorona and Mangoro Rivers are likely this species.

Small-toothed Sportive Lemur, Talatakely, Ranomafana National Park.

SMALL-TOOTHED SPORTIVE LEMUR
Lepilemur microdon (Forsyth Major, 1894)

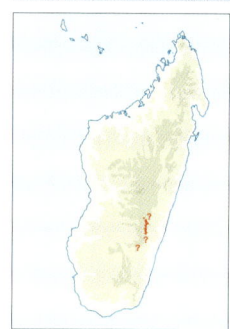

Malagasy names: Fitiliky, Hataka, Trangalava, Kotrika

MEASUREMENTS
Head-body: 270–320mm. Tail: 250–290mm. Weight: 900–1,200g.

DESCRIPTION & IDENTIFICATION A large sportive lemur. Pelage dense and red-brown on head and dorsal region with darkish dorsal stripe. Forelimbs and shoulders particularly rich chestnut. Tail similar to body and darker towards tip. Underparts, face and throat pale grey-brown. Sometimes yellowish-buff wash on belly/abdomen. Eyes pale orange/yellow. As compared to similar congeners, molars relatively small, hence differentiation from very similar Weasel Sportive Lemur, which has been confirmed by molecular analysis (Andriaholinirina *et al.* 2005). A medium-sized vertically clinging lemur with medium-long tail. In known range, most likely confused with Peyrieras's Woolly Lemur of similar size, general coloration and posture. More obvious ears, absence of facial disc, russet forelimbs and shoulders, and absence of white thighs are main separating features.

HABITAT & DISTRIBUTION Primary and secondary rainforest. As currently understood, range extends from vicinity of Ranomafana National Park south-west to Andringitra National Park. In Ranomafana area, boundary and relationship with Betsileo Sportive Lemur unclear. Similarly, range boundaries to south with other sportive lemurs requires further fieldwork to clarify. IUCN Endangered.

BEHAVIOUR Nocturnal and arboreal. Prefers to sleep in holes of larger trees in groups of 1–3 individuals (two adults, one juvenile) (Wright *et al.* 2020a). Solitary when active and feeds on leaves, fruit and flowers. Like other eastern sportive lemurs, may compete for resources with woolly lemurs (*Avahi* spp.). Home range size of less than 1ha recorded (Porter 1998). Records of predation by Henst's Goshawk (Goodman *et al.* 1998)

WHERE TO SEE Ranomafana National Park is only accessible site. However, sportive lemurs are encountered only rarely in main tourist area of Talatakely. Sportive lemurs are seen more often in higher-altitude forests of Vohiparara, although these may correspond to Betsileo Sportive Lemur. The rainforest in Andringitra National Park is considerably more challenging to visit. Any reasonable-sized patches of native humid forest between these two parks may contain Small-toothed Sportive Lemur.

MANOMBO SPORTIVE LEMUR
Lepilemur jamesorum   Louis *et al.*, 2006

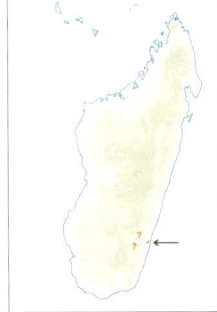

Other name: James's Sportive Lemur
Malagasy names: Hataka, Trangalava, Kotrika, Fitiliky, Kotriana

MEASUREMENTS
Head-body: *c*.260–300mm. Tail: 290–310mm. Weight: *c*.800–1,000g. (Louis *et al.* 2006b)

DESCRIPTION & IDENTIFICATION A relatively large sportive lemur. Dorsal pelage short and smooth; mainly brown dorsally with distinctive black midline from head along spinal ridge to base of tail. Underparts paler greyish-brown than upperparts. Tail brown, becoming progressively darker to black at tip. Head and crown mid-grey. Ears large, grey and edged black, some cream patches below ears. Distinctive whitish-grey mask on face (Louis *et al.* 2006b). Molecular study has confirmed taxonomic differentiation from similar species (Andriaholinirina *et al.* 2006b). Shares limited range with broadly similar Manombo Woolly Lemur *Avahi ramanantsoavanai*: more obvious ears, brown (rather than grey-brown) upperparts, absence of white thighs, and brown rather than russet tail are main distinguishing features.

HABITAT & DISTRIBUTION Primary and secondary coastal rainforest. Described from Manombo Special Reserve near south-east coast. Also known from Vevembe Classified Forest (Andriamisedra *et al.* 2019). Range limits unclear; may occur from Manampatrana River north of Farafangana, south to Mananara River near Vangaindrano (Louis *et al.* 2006b). Population densities *c*.45 individuals/km²

Manombo Sportive Lemur.

(Manombo Special Reserve) and 23 individuals/km² (Manombo Classified Forest) with total population fewer than 1,400 individuals (Ingraldi 2010). Total range may be no more than 70km². Hunting for bushmeat is a significant threat, along with deforestation.

BEHAVIOUR Like all sportive lemurs nocturnal and arboreal. Presumed to be folivorous. Brief studies indicate as few as two individuals occupy 800ha study plot (Andriamisedra *et al.* 2019).

WHERE TO SEE Manombo Special Reserve is currently the only known site to see this species. Access is difficult in wet season when local rivers swell and cut off portions of forest.

KALAMBATRITRA SPORTIVE LEMUR *Lepilemur wrightae* Louis *et al.*, 2006

Other name: Wright's Sportive Lemur
Malagasy names: Hataka, Trangalava, Kotrika, Fitiliky, Kotriana

MEASUREMENTS
Head-body: *c.*250–270mm. Tail: *c.*250–260mm. Weight: *c.*1,000–1,900g. (Louis *et al.* 2006b, Rabeson *et al.* 2006)

DESCRIPTION & IDENTIFICATION Very large with distinctive appearance: probably largest sportive lemur and only one known to exhibit sexual dimorphism. In both sexes, upperparts reddish-brown to greyish-brown, underparts paler greyish-brown. Females have uniform grey head, which sharply contrasts with body coloration. Head of males reddish-brown to grey, similar to body colour with little contrast. Some females also have paler coloration around face (faint mask). Ears in both sexes mainly bare (very lightly furred) and pale (Louis *et al.* 2006b). Within range, potentially confused with Peyrieras's Woolly Lemur: Kalambatritra Sportive Lemur has more obvious ears, lacks a facial disc and has no white thigh patches.

HABITAT & DISTRIBUTION Primary and secondary mid-altitude rainforest, between 1,000m and 1,750m. Described from Kalambatritra Special Reserve in south-east. Also known from unprotected Beakora Forest slightly further south-east (Irwin *et al.* 2001, Rabeson *et al.* 2006). In Kalambatritra population densities of *c.*72 individuals/km² (Irwin *et al.* 2001) and *c.*41 individuals/km² (Rakotonirina *et al.* 2017) calculated. In Beakora Forest estimates considerably lower, *c.*6 individuals/km² (Rabeson *et al.* 2006). Range presumed to extend north of Mandrare River and west of Mananara River (Louis *et al.* 2006b), but forests extremely fragmented, covering less than 1,000km². Further field studies needed to determine limits of distribution and range boundaries with genus *Lepilemur* to north (Andringitra), east (Midongy du Sud) and south (Andohahela). IUCN Endangered.

BEHAVIOUR Nocturnal and arboreal. Yet to be studied in detail. Initial evidence suggests individuals regularly use specific latrines and scent-mark immediate vicinity; latrines may serve a role in territorial demarcation and resource defence (Irwin *et al.* 2004). Population densities in Kalambatritra are highest yet recorded for a rainforest sportive lemur. This may be linked to apparently reduced competition due to absence of other folivorous lemurs, like woolly lemurs (*Avahi* spp.) and sifakas (*Propithecus* spp.) (Irwin *et al.* 2004).

WHERE TO SEE Kalambatritra Special Reserve in south-east Madagascar, *c.*100km south of Ihosy, is very difficult to reach. There are no facilities.

Kalambatritra Sportive Lemur.

ANDOHAHELA SPORTIVE LEMUR
Lepilemur fleuretae   Louis *et al.*, 2006

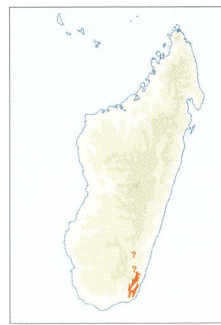

Other name: Fleurete's
Sportive Lemur
Malagasy names: Hataka,
Trangalava, Kotrika, Fitiliky,
Kotriana

MEASUREMENTS
Head-body: *c.*240–260mm.
Tail: *c.*270–320mm. Weight:
*c.*800–1,300g. (Louis *et al.*
2006b, Campera 2018)

DESCRIPTION & IDENTIFICATION Medium-sized sportive lemur. Upperparts mainly grey with greyish-brown limbs and extremities. Diffuse midline from crown to around halfway down spine. Underparts brownish-grey with light brown on sides of belly. Tail reddish-grey, becoming darker at tip (Louis *et al.* 2006b). Shares range with similarly-sized Southern Woolly Lemur *Avahi meridionalis*: more visible ears, absence of facial disc, brown rather than grey-brown upperparts, absence of white thighs, and brown rather than russet tail are main distinguishing features.

HABITAT & DISTRIBUTION Restricted to primary and secondary lowland and mid-elevation rainforest (below *c.*1,300m) in extreme south-east. Originally described from Manangotry, in parcel 1, Andohahela National Park. Also recorded to east in Tsitongambarika Protected Area (Campera 2018). Range probably covers rainforest areas between Mandrare River in west and Mananara River to north centred on Anosyennes Mountains, Vohimena Mountains further east, and connecting Manangotry corridor. Surveys of undisturbed lowland forests in Ampasy Valley, Tsitongambarika Protected Area, recorded very high population density of 81 individuals/km² (Campera *et al.* 2019c). Density much lower in degraded and secondary forests. Further fieldwork required to determine northern range limits, and boundaries with congeners to north. In degraded areas extremely susceptible to human hunting (Campera *et al.* 2019b, Campera 2018). IUCN Endangered.

BEHAVIOUR Strictly nocturnal, with likely 'dispersed pair' social system; males and females solitary while foraging and do not share sleep sites (Campara 2018, Campera *et al.* 2019a). Andohahela Sportive Lemur is largely folivorous (*c.*26% young leaves, *c.*38% mature

Andohahela Sportive Lemur, Tsitongambarika Forest.

leaves), but compared with congeners, a far greater proportion of fruits and flowers (*c*.35%) comprise diet, particularly during wet season; potentially linked to low diversity of diurnal frugivorous lemurs (*Eulemur* species occur at very low densities in region, ruffed lemurs *Varecia* spp. are absent) and therefore reduced competition. In addition, helps reduce competition with nocturnal Southern Woolly Lemur *Avahi meridionalis* (Campera 2018). The most important dietary plant species include *Albizia* spp., *Brochoneura acuminata*, *Cynometra* spp., *Syzygium* spp., *Uapaca thouarsii* and *Humbertia madagascariensis* (wet season alone) (Campera 2018).

Consequently, home ranges are larger than other sportive lemurs, and individuals more active and travel further per night, spending proportionately less time resting than congeners (Campera 2018). Female home ranges vary between *c*.2.6 and 5.3ha, while male home ranges are 7.6–7.9ha, with seasonal variation. Male ranges increase in size during wet season when food more abundant, whilst female ranges decrease in size. In wet season, males may range further in search of patchily clumped and valued resources (certain fruits and flowers), whilst females minimise their energy expenditure when lactating (Campera 2018, Campera *et al*. 2019a).

Appears to avoid direct competition for resources with similar-sized nocturnal Southern Woolly Lemur by being active at different times of night; Southern Woolly Lemur is primarily active during twilight and early hours of darkness, whereas Andohahela Sportive Lemur concentrates activity in central hours of darkness. Southern Woolly Lemur also active on moonlit nights, whereas Andohahela Sportive Lemur shows significantly reduced activity at such times (Campara 2018, Campera *et al*. 2019a).

WHERE TO SEE Best looked for at Malio, the rainforest area of Andohahela National Park (parcel 1). This area should become more accessible as ecotourism is developed.

MANANARA-NORD SPORTIVE LEMUR *Lepilemur hollandorum*

Ramaromilanto *et al*., 2009

Other name: Holland's Sportive Lemur
Malagasy names: Hataka, Trangalava, Kotrika, Fitiliky, Kotriana

MEASUREMENTS
Head-body: *c*.290–335mm. Tail: *c*.270–295mm. Weight: *c*.1,000g. (Ramaromilanto *et al*. 2009)

DESCRIPTION & IDENTIFICATION A large sportive lemur. Pelage short and dense. Head, shoulders and upper dorsal area mottled reddish-grey, becoming paler greyish-brown in lower dorsal area and pygal region to tail. Hands and feet also paler greyish-brown. Underparts pale grey. Tail dark brown, black towards tip. A faint dark brown to black midline stripe or inverted Y-shape may be visible on head extending down back. Face generally pale grey. Ears fleshy and prominent (Ramaromilanto *et al*. 2009).

A medium-sized, vertically clinging lemur with tail approximately equal to head-body length. Potentially confused with Eastern Woolly Lemur of similar size, coloration and posture. Mananara-Nord Sportive

Mananara-Nord Sportive Lemur.

Lemur has prominent ears and uniform coloration, whereas Eastern Woolly Lemur has very visible white thighs and eyebrow patches, and distinctive pale facial disc.

HABITAT & DISTRIBUTION Primary and secondary lowland rainforest to c.375m (Louis *et al.* 2020g). Currently known only from Mananara-Nord region, more specifically Ivontaka-Sud and Verezanantsoro (Ambinanibeorana) parcels of Mananara-Nord Biosphere Reserve. Northern and southern limits of range ill-defined, but tentatively thought to be south of Fahambahy or Mananara Rivers and north of Simianona, Sandratsio and Maningory Rivers (Ramaromilanto *et al.* 2009). Additional surveys required to determine full extent of distribution. Known range covers <400 km^2, where habitat severely fragmented and bushmeat hunting widespread. IUCN Critically Endangered.

BEHAVIOUR Nocturnal. Like other sportive lemurs, presumably folivorous. Yet to be studied in the wild. No further specific behavioural details available.

WHERE TO SEE Any forest areas or fragments in and around Mananara-Nord National Park are potentially places to see this lemur.

SEAL'S SPORTIVE LEMUR
Lepilemur seali   Louis *et al.*, 2006

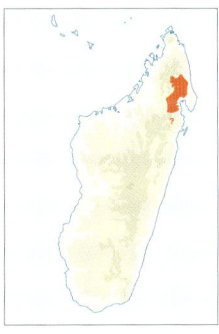

Malagasy names: Tsitsihy

MEASUREMENTS
Head-body: 260–270mm.
Tail: 250–260mm. Weight:
c. 900–1,000.g (Louis *et al.*
2006b)

DESCRIPTION & IDENTIFICATION A largish sportive lemur. Pelage dense, long and luxuriant. Upperparts uniform light chocolate-brown to reddish-brown. Underparts pale brownish-grey. Face uniform light brownish-grey, hands and feet paler. Tail also brownish-grey but contrasts with body (Louis *et al.* 2006b). Potentially confused with similar-sized Eastern Woolly Lemur. Seal's Sportive Lemur has prominent ears and uniform coloration, whereas

Seal's Sportive Lemur, Marojejy National Park.

Eastern Woolly Lemur has very visible white thighs and eyebrow patches, distinctive pale facial disc and a reddish tail.

HABITAT & DISTRIBUTION Primary and secondary mid-altitude rainforest to c.1,100m (Louis *et al.* 2020a). However, probably tolerates higher elevations, allowing it to migrate and colonise around headwaters of major rivers (Craul *et al.* 2008). Seal's Sportive Lemur is found at north-eastern extreme of rainforest belt. Originally described from Anjanaharibe-Sud south of Andapa (Louis *et al.* 2006b), range now known to extend south to Makira region where Fahambahy River or Mananara River may form boundary with Mananara-Nord Sportive Lemur (Craul *et al.* 2008). To north, range probably extends to Marojejy National Park and perhaps COMATSA forest corridor. To the east, boundary with Masoala Sportive Lemur may be Andranofotsy River or Sakafihitra River, but further study required to verify this. The species' range remains relatively extensive, including some of most remote blocks of remaining rainforest, but bushmeat hunting in them is high (Golden 2005, 2009). IUCN Vulnerable.

BEHAVIOUR Nocturnal and presumably folivorous like all congeners. Yet to be studied in the wild. No further information available.

WHERE TO SEE In all likelihood, sportive lemurs in Marojejy National Park, in forests around Camp Marojejia (Camp 2), probably correspond to this species. Otherwise Anjanaharibe-Sud Reserve south of Andapa, and the more difficult-to-reach forests of Makira Natural Park are worth exploring to look for this species.

Masoala Sportive Lemur, Masola National Park.

MASOALA SPORTIVE LEMUR
Lepilemur scottorum   Lei *et al.*, 2008

Other name: Scott's Sportive Lemur
Malagasy names: Hataka, Trangalava, Kotrika, Fitiliky, Kotriana

MEASUREMENTS
Head-body: *c.*250–280mm.
Tail: *c.*250–290mm.
Weight: *c.*700–880g. (Lei *et al.* 2008)

DESCRIPTION & IDENTIFICATION A medium-sized sportive lemur. Pelage long and dense. Upperparts uniform reddish-brown, with diffuse dark midline stripe from crown continuing partially on back. Head mid-grey, face pale grey. Underparts pale reddish-brown. Limbs mainly reddish-brown. Tail reddish-brown at base, becoming more brownish-grey, then black at tip (Lei *et al.* 2008).

A medium-sized vertically clinging lemur with long tail. Within range, potentially confused with Masoala Woolly Lemur *Avahi mooreorum*, which is similar in size, general coloration and posture. More obvious ears, russet forelimbs and shoulders (rather than chocolate-brown), absence of white thighs and tail base, and brown rather than russet tail, are main distinguishing features.

HABITAT & DISTRIBUTION Littoral, lowland and mid-elevation primary and secondary rainforest. Described from Masiaposa Forest (Masoala National Park) in far north-east. Range presumably restricted to Masoala Peninsula, with western limit Sakafihitra River or Andranofotsy River (boundary with Seal's Sportive Lemur), and northern limit perhaps Ankavia River. Population densities in primary forest *c.*25 individuals/km² (Sawyer *et al.* 2017) and 33 individuals/km² (Sterling & Rakotoarison 1998). Suffers some negative effects of habitat alteration and disturbance (but to lesser extent than other nocturnal lemurs) and found at lower densities in selectively logged forest (Sawyer *et al.* 2017, Fenosoa *et al.* 2018). Total range covers *c.*2,500km². Rosewood (*Dalbergia* spp.) is a key dietary resource for sportive lemurs (Ganzhorn *et al.* 2001) and its continued illegal extraction on the Masoala Peninsula negatively impacts Masoala Sportive Lemur abundance (Sawyer *et al.* 2017). IUCN Endangered.

BEHAVIOUR Nocturnal and presumed to be largely folivorous. In primary forest uses holes and larger trees as sleep sites. Apparent abundance is highest in primary lowland forest where endemic tree *Garcinia commersonii* (Clusiaceae) is common (Eppley *et al.* 2020a).

WHERE TO SEE Relatively easy to see on Masoala Peninsula. Forests around village of Ambodiforaha are generally productive, particularly adjacent to Arol Lodge or Masoala Forest Lodge. Also possible to see further along coast near Tampolo.

DARAINA SPORTIVE LEMUR
Lepilemur milanoii   Louis *et al.*, 2006

Malagasy names: Hataka, Trangalava, Kotrika, Fitiliky, Kotriana

MEASUREMENTS
Head-body: *c.*220–240mm.
Tail: *c.*250–260mm.
Weight: *c.*650–750g. (Louis *et al.* 2006b)

DESCRIPTION & IDENTIFICATION A medium-sized sportive lemur. Fur dense. Upperparts reddish-brown with faint dark midline stripe from crown to part way down back. Head similarly reddish-brown, face more grey-brown forming noticeable 'mask'. Underparts greyish-white. Hands and feet mainly grey, some reddish-brown on upper thighs. Tail uniformly reddish-brown (Louis *et al.* 2006b).

Should not be confused with any other species in its range (other than extremely similar Ankarana Sportive Lemur, in Andrafiamena Forest). Sympatric Ankarana Dwarf Lemur is significantly smaller, adopts a horizontal body posture, and moves quadrupedally.

HABITAT & DISTRIBUTION Dry deciduous, gallery and humid semi-evergreen forests (found at some higher elevations), from close to sea level to *c.*380m. Restricted to small area in far north-east centred on Loky-Manambato region. Originally described from Andranotsimaty, near Daraina. Also recorded further north from Andrafiamena Classified Forest, where sympatric with Ankarana Sportive Lemur (Louis *et al.* 2020h); currently, the only known example of two

sympatric sportive lemur species. Southern range limit perhaps Fanambana River. Further field research required to clarify precise distribution. Current range covers *c.*2,600km², but forests extremely fragmented (ten major fragments). Also, significant differences in sportive lemur densities between forest fragments: in Ambilondambo *c.*40 individuals/km², in Antsaharaingy *c.*50 individuals/km², whereas in nearby Ampondrabe figure is *c.*590 individuals/km² (Salmona *et al.* 2014). Most forests in species' range have limited protection and sportive lemurs and other lemur species are hunted for bushmeat (Salmona *et al.* 2014). There is considerable illicit

Daraina Sportive Lemur, forests near Daraina.

gold mining in region, with workers adding to hunting pressure. IUCN Endangered.

BEHAVIOUR Nocturnal and presumably largely folivorous. Behaviour and ecology yet to be studied in the wild.

WHERE TO SEE Relatively easy to see near Daraina and environs: forests at Camp Tattersalli run by FANAMBY provide excellent opportunities, as do areas near Andranotsimaty, 5km north-east of Daraina. This site is also very good for other nocturnal species including Tavaratra Mouse Lemur, Northern Fork-marked Lemur and Aye-aye. During day, Golden-crowned Sifaka and Crowned Lemur are easily seen. Daraina Sportive Lemur has also been recorded in Andrafiamena-Andavakoera Protected Forest, south of Analamerana, where it occurs alongside Ankarana Sportive Lemur. Visiting Black Lemur Camp potentially offers the opportunity to see both species, although differentiating them in the field is challenging.

ANKARANA SPORTIVE LEMUR
Lepilemur ankaranensis
Rumpler & Albignac, 1975

Malagasy name: Valiha

MEASUREMENTS
Head-body: *c.*220–230mm.
Tail: *c.*250–270mm.
Weight: *c.*700–800g.

DESCRIPTION & IDENTIFICATION Medium-small sportive lemurs. Upperparts pale grey-brown, becoming darker towards and on tail, especially at tip. Brown tones apparent on crown and shoulders. Darkish dorsal stripe from head to part way down back. Underparts pale grey. Ears small and less prominent than in congeners.

Ankarana Sportive Lemur potentially confused with woolly lemurs *Avahi* spp., which have obvious white thighs, a distinctive pale facial disc and reddish tail, all of which are absent in Ankarana Sportive Lemur. In Andrafiamena Forest this species is sympatric with Daraina Sportive Lemur;

differentiation in the field is extremely difficult. Sympatric Ankarana Dwarf Lemur is appreciably smaller than Ankarana Sportive Lemur, adopts a horizontal body posture and moves quadrupedally.

HABITAT & DISTRIBUTION Restricted to far north of island, in dry deciduous and some humid evergreen forests below *c.*430m. As currently understood, range covers forests of Ankarana Reserve, Analamerana Reserve and Andrafiamena Protected Area (previously, these populations were assigned to *L. septentrionalis*). In Andrafiamena, occurs with Daraina Sportive Lemur, the only known instance of sportive lemur sympatry (Louis *et al.* 2006b, Salmona *et al.* 2014). North and north-west of Ankarana, Ankarana Sportive Lemur may also occur in dry forests of Analabe and mid-elevation humid forests of Montagne d'Ambre National Park and Forêt d'Ambre Reserve; which sportive lemur species occurs in these forests has yet to be verified. Entire range covers <2,500km². IUCN Endangered.

BEHAVIOUR Most information is from studies conducted when populations were assigned to *L. septentrionalis*. Like other sportive lemurs, leaves form bulk of diet in dry season, but in wet season fruits also comprise a significant element. Preliminary observations in patches of degraded forest show fruit may constitute up to *c.*30% of diet in wet season (Lowin 2010), a far higher proportion than seen in any other sportive lemur. Also recorded eating sap and tree exudates in dry season (Lowin 2010, Schwitzer *et al.* 2013b). Forages alone, with home ranges covering *c.*1ha. Adults rarely associate, but mothers and infants have been observed together. Spends day in tree holes or dense tangles of vines. Most sleep sites are 6–8m above ground, but some as low as 1m. Large boas (*Sanzinia* and *Acrantophis* spp.) have been reported to wait in ambush outside sleep sites and take lemurs from within tree holes (Mittermeier *et al.* 2010).

WHERE TO SEE Easily seen in Ankarana Reserve, particularly *Canyon Forestière* and forests around *Campement Anilotra* (*Camp des Anglais*) and *Camps des Princes* near Mahamasina (East) entrance. By day, can often be found sitting on edge of tree holes. Ankarana Sportive Lemur is also seen in Andrafiamena-Andavakoera Protected Forest south of Analamerana, where it is sympatric with Daraina Sportive Lemur. Visiting Black Lemur Camp offers the chance to see both species.

Ankarana Sportive Lemur, Ankarana Special Reserve.

Sahafary Sportive Lemur, Montagne des Français.

SAHAFARY SPORTIVE LEMUR
Lepilemur septentrionalis
Rumpler & Albignac, 1975

Other name: Northern
Sportive Lemur
Malagasy name: Fitsidika

MEASUREMENTS
Head-body: 180–200mm.
Tail: 250mm. Weight:
600–750g. (Louis *et al.*
2006b)

DESCRIPTION & IDENTIFICATION Previous descriptions of *L. septentrionalis* relate to populations
now assigned to *L. ankaranensis* (Rumpler *et al.* 2001,
Rumpler 2004); these two are morphologically very
similar, but genetic differences confirm species status

(Ravoarimanana *et al.* 2004, Andriaholinirina *et al.*
2006a). A small sportive lemur. Head and upperparts
greyish-brown, underparts grey. Shoulders may have
brownish tones. Dark medial stripe often extends
from crown down spine. Tail pale brown, darker at
tip. Ears less prominent than other sportive lemur
species (Schwitzer *et al.* 2013c).

HABITAT & DISTRIBUTION Dry deciduous forest
to *c.*600m. Sahafary Sportive Lemur occurs only at
extreme north tip of island, where it is now believed
to be restricted to remaining forest fragments around
Montagne des Français (Louis & Zaonarivelo 2015,
Bailey *et al.* 2020). Previously regarded as being
distributed over more extensive areas of northern
Madagascar, from Montagne d'Ambre south to
Mahavavy River near Ambilobe (Hawkins *et al.*
1990, Rumpler *et al.* 2001). Populations in Ankarana,
Analamerana and Andrafiamena later reclassified as
L. ankaranensis, which effectively reduced range of
Sahafary Sportive Lemur to a few forest fragments in

Sahafary region (Rumpler *et al.* 2001, Ravaorimanana *et al.* 2004, Wilmet *et al.* 2014). Sportive lemurs in Analabe forest fragments recorded as this species (Le Pors *et al.* 2020), but are more likely Ankarana Sportive Lemur. Surveys over last decade have indicated significant declines, as anthropogenic pressures have increased (charcoal production, tree felling, cattle grazing and slash-and-burn agriculture), including its apparent disappearance from some forest fragments (Sabel *et al.* 2009, Ranaivoarisoa *et al.* 2013). There are probably fewer than 100 individuals remaining, and the species teeters on cusp of extinction (Bailey *et al.* 2020). IUCN Critically Endangered.

BEHAVIOUR As now defined, there have been no long-term wild studies of Sahafary Sportive Lemur. Preliminary studies show individuals are able to survive in forest patches with varying degrees of human disturbance and suggest home ranges of *c*.1.3–2.9ha (based on four individuals). Diet primarily foliage, especially new-growth leaves with small amounts of fruit (Dinsmore *et al.* 2016). Known to sleep in tree holes. Most other aspects of behaviour and ecology probably similar to Ankarana Sportive Lemur.

WHERE TO SEE Forest fragments at Montagne des Français are the only localities where this species might be seen.

GREY-BACKED SPORTIVE LEMUR
Lepilemur dorsalis
J.E. Gray, 1871

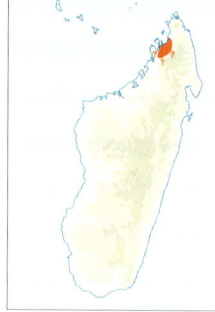

Other name: Gray's Sportive Lemur
Malagasy names: Fitsidika, Apongy

MEASUREMENTS
Head-body: 230–260mm.
Tail: 260–280mm.
Weight: *c*.600–770g.
(Andriaholinirina *et al.* 2006, Louis *et al.* 2006b)

DESCRIPTION & IDENTIFICATION Relatively small sportive lemur. Upperparts and tail medium-brown to grey-brown with faint dark brown dorsal stripe. Underparts paler grey-brown. Throat appears paler grey. Head more greyish, face darker grey-brown with blunt muzzle. Ears rounded, relatively small, appearing almost hidden in fur. Eyes deep fiery-orange.

A medium-small vertically clinging lemur; one of smallest sportive lemurs. In Manongarivo region may be confused with Sambirano Woolly Lemur *Avahi unicolor*, which has similar posture but is noticeably larger, with smaller concealed ears, a rounded

Grey-backed Sportive Lemur, forests near Ambanja.

'owl-like' face and distinctive white thigh patches. Less likely to be confused with either dwarf lemurs (*Cheirogaelus* spp). or giant mouse lemurs (*Mirza* spp.), as these adopt horizontal, rather than vertical posture, and run quadrupedally.

Taxonomic status of various sportive lemurs in north-west is contentious and unclear. Previously all *Lepilemur* in Sambirano region, including island of Nosy Be, assigned to *L. dorsalis* (Harcourt & Thornback 1990). More recently, population on Nosy Be described as *L. tymerlachsoni* (Louis *et al.* 2006b), whilst sportive lemurs on Ampasindava Peninsula are now *L. mittermeieri* (Rabarivola *et al.* 2006) and Sahamalaza Peninsula are *L. sahamalaza* (Andriaholinirina *et al.* 2006a).

HABITAT & DISTRIBUTION Humid lowland forest, subhumid forest and some secondary and degraded forests in areas subject to short dry season. As currently understood, restricted to Sambirano region in north-west, including Manongarivo Massif (Goodman & Schütz 2000, Louis *et al.* 2020e). Precise range limits unclear. Northern limit may be Mahavavy River (Ambilobe) or Sambirano River (Ambanja). Specific identity of sportive lemurs in Tsaratanana Reserve undetermined. Southern and western range limits also ill-defined and how these relate to boundaries with Mittermeier's Sportive Lemur *L. mittermeieri*, Sahamalaza Sportive Lemur *L. sahamalaza* and Anjiamangirana Sportive Lemur *L. grewcockorum* remain unclear. Range of Grey-backed Sportive Lemur may cover <5,000km^2. IUCN Endangered. Sportive lemur populations on island of Nosy Be, once assigned to *L. dorsalis*, now classified as Nosy Be Sportive Lemur *L. tymerlachsoni* (Louis *et al.* 2006b).

BEHAVIOUR Nocturnal and folivorous. Anecdotal observations suggest that in primary humid forest appears to prefer sleeping in tree holes, but in more open drier forests will sleep in dense vegetation tangles. Births observed August–November (Louis *et al* 2020abc). Previous descriptions of *L. dorsalis* behaviour relate to populations on Nosy Be, now described as *L. tymerlachsoni*. Known predators include Henst's Goshawk (Goodman *et al.* 2014b).

WHERE TO SEE No easily visited sites. Any decent-sized forest patch in the Ambanja area is worth investigating. This species can be found in Manongarivo Reserve, but this is a challenging site to reach.

NOSY BE SPORTIVE LEMUR
Lepilemur tymerlachsoni   Louis *et al.*, 2006

Other name: Hawk's Sportive Lemur
Malagasy names: Fitsidika, Apongy

MEASUREMENTS
Head-body: c.230–240mm.
Tail: c.240–250mm.
Weight: c.780–980g. (Louis *et al.* 2006b)

DESCRIPTION & IDENTIFICATION Medium-sized sportive lemur. Upperparts mainly light brownish-grey, with paler reddish-brown on upper half of back. Underparts light greyish-white. Front of thighs and edges of limbs diffuse reddish-brown. Dark brown midline stripe from top of head to lower half of back. Facial mask grey. Tail uniform reddish-grey to brown (Louis *et al.* 2006b). Taxonomic validity and distinction from mainland populations assigned to *L. dorsalis* has been questioned (Zinner *et al.* 2007), but later studies have confirmed species status

Nosy Be Sportive Lemur, Lokobe Reserve.

(Lei *et al*. 2017). Originally named *L. tymerlachsoni* (Louis *et al*. 2006a), sometimes incorrectly referred to as *L. tymerlachsonorum* (e.g., Hoffman *et al*. 2009, Mittermeier *et al*. 2010).

HABITAT & DISTRIBUTION Moist tropical lowland forest subject to short dry season and some modified secondary habitats, including coffee, vanilla and cashew plantations below 35m. Described from Lokobe National Park on island of Nosy Be, off north-west coast, its only known locality. Sportive lemurs have previously been recorded on neighbouring island of Nosy Komba, but forest there now very degraded and no evidence has been found recently (Hyde Roberts & Daly 2014). Range as currently understood <75km². IUCN Critically Endangered.

BEHAVIOUR Past anecdotal studies of sportive lemurs on Nosy Be, when species was assigned to *L. dorsalis* (Andrews *et al*. 1998). More recent studies confirm diet dominated by foliage, buds and stems, with fruits constituting a small but important component. This species may play a role in seed dispersal (Sawyer *et al*. 2015).

Appears to prefer dense tangles of vegetation as daytime sleep sites, rather than tree holes; this may be a bias due to reduced numbers of large trees with holes (Sawyer *et al*. 2015). Shows an affinity for favoured sleep sites; individuals may return to same site for 14 or more consecutive nights. It appears more common in secondary forest dominated by smaller trees, but this may be an artefact of individuals being easier to find in such forest compared to denser primary forest (Bederu 2014). In marginal areas at forest edges, utilises purpose-built nest boxes fixed to trees (Andrews *et al*. 1998).

Little is known about breeding; single infant born August–November. In secondary habitat, predated by feral dogs and also Madagascar Buzzard and Madagascar Harrier-Hawk (Andrews *et al*. 1998).

WHERE TO SEE Easily seen in buffer zone at north-east edge of Lokobe Reserve. Access is via the village of Ampasipohy. Nosy Be Sportive Lemur is quite common in nearby forests. Be warned, the ecotourist experience 'created' here is contrived and unfulfilling, and not the type of development that should be encouraged.

MITTERMEIER'S SPORTIVE LEMUR
Lepilemur mittermeieri   Rabarivola *et al*., 2006

Malagasy names: Fitsidika, Apongy

MEASUREMENTS
Head-body: *c*.250–300mm.
Tail: *c*.230–280mm.
Weight: 630–810g, mean *c*.700g. (Wilmet *et al*. 2019)

DESCRIPTION & IDENTIFICATION Medium-small sportive lemur. Upperparts mainly brown to reddish-grey. Underparts light brownish-grey. Some individuals show dark brown midline stripe from top of head to lower back. Tail colour as lower body, darkening towards tip. Facial mask dark grey, with white fur beneath eyes and mandibles (Schwitzer *et al*. 2013b, Wilmet *et al*. 2017b, 2019).

HABITAT & DISTRIBUTION Primary and secondary subhumid and humid forest to *c*.550m, in southern Sambirano region (Wilmet *et al*. 2017a,b). Described from Ampasindava Peninsula in north-west, which is subject to more humid microclimate than other western forests and currently constitutes entire extent of species' range, which covers *c*.1,000km². Forests extremely fragmented; largest patches of primary forest cover two hills, Andranomatavy and Ambohimirahavavy. Also found in small patches (Wilmet *et al*. 2017a,b). In intact forest density estimate 133 individuals/km² (Ralantoharijaona *et al*. 2014). Ampasindava established as protected area in 2014, but now threatened by a rare earth elements mining project (Wilmet *et al*. 2017a). IUCN Critically Endangered.

Sportive lemurs on Ampasindava Peninsula previously regarded as *L. dorsalis*. Further research is required to clarify extent of range beyond the peninsula and evolutionary relationships with other sportive lemur taxa in region (*L. dorsalis*, *L. tymerlachsoni* and *L. sahamalaza*).

BEHAVIOUR Nocturnal and forages alone, although pairs have been observed together (Wilmet *et al*. 2017b). Like congeners, primarily folivorous; diet diverse with leaves of >70 species recorded. Also eats small amounts fruit (Wilmet *et al*. 2020). Males

Mittermeier's Sportive Lemur, Ampasindava Peninsula.

and females each occupy home ranges of c.2ha, with no evidence of territoriality and some overlap, even among males (Wilmet *et al.* 2019). Mittermeier's Sportive Lemur also recorded in edge areas of forest. Individuals utilise a number of different sleep sites (up to c.4–5), most of them tangles of branches and leaves, rather than tree holes. A single sleep site may be used repeatedly over several consecutive nights (Wilmet *et al.* 2017b).

WHERE TO SEE Ampasindava Peninsula lies c.45km south of Ambanja and is challenging to reach. Any reasonably sized fragment of forest on the peninsula is worth exploring to find this species.

SAHAMALAZA SPORTIVE LEMUR
Lepilemur sahamalaza
Andriaholinirina *et al.*, 2017

Malagasy name: Fitsidiky

MEASUREMENTS
Head-body: 200–285mm. Tail: 240–290mm. Weight: c.550–850g. (Louis *et al.* 2006b, Seiler *et al.* 2015a, Mandl 2018)

Sahamalaza Sportive Lemur, Ankarafa Forest, Sahamalaza Peninsula.

DESCRIPTION & IDENTIFICATION A medium-small sportive lemur. Pelage apparently varies with age. Upperparts, including shoulders and arms, reddish-brown. Thighs and lower limbs less reddish, more greyish. Underparts pale grey to creamy. Face grey, forehead and ear bases red-brown. Diffuse mid-dorsal line from crown to mid or lower back, but no further; this is most distinct on mid-back. Tail reddish-brown to deep brown (Andriaholinirina *et al.* 2006a). Some individuals have white tail tip (cf. Anjiamangirana Sportive Lemur *L. grewcockorum*, Milne-Edwards's Sportive Lemur *L. edwardsi*, Ambodimahabibo Sportive Lemur *L. otto*) (Mandl *et al.* 2020). Originally named *L. sahamalazensis* (Andriaholinirina *et al.* 2006a), amended to *L. sahamalaza* (Andriaholinirina *et al.* 2017). Within known range, might only be mistaken for dwarf lemurs *Cheirogaleus* spp. or Northern Giant Mouse Lemur; these species are smaller and adopt a horizontal rather than vertical posture.

HABITAT & DISTRIBUTION Subhumid primary and older secondary forests with trees up to 25m tall, in transition zone between Sambirano region to north and western deciduous forests to south. Described from, and apparently restricted to, Sahamalaza Peninsula in north-west (Wilmet *et al.* 2017, Randriatahina *et al.* 2020). Surveys of forests outside peninsula to east failed to record the genus. (Mandl *et al.* 2020). Only three significant blocks remain (Ankarafa, Anabohazo and Analavory); these are extremely degraded, fragmented and highly

variable in structure (Seiler *et al.* 2013a). Despite official protection since 2007 (Sahamalaza-Iles Radama National Park), degradation through bush fires, tree felling and hunting continues (Seiler *et al.* 2010, 2012, Randriatahina *et al.* 2014). In Ankarafa fragment, population density estimated at *c.*1.5–1.9 individuals/ha (Mandl *et al.* 2020), and appears unaffected by increased 'edge effects' (Mandl 2018). Remaining area of forest supporting Sahamalaza Sportive Lemur may be only 2,500ha, with a total population of *c.*3,800–4,700 individuals (Mandl *et al.* 2020). IUCN Critically Endangered.

BEHAVIOUR Sahamalaza Sportive Lemur is primarily nocturnal (with low levels of daytime activity) and solitary. Adults exhibit minimal interaction and social complexity; only interactions observed are between mother and offspring (Seiler 2012, Seiler *et al.* 2014, Mandl *et al.* 2019a). Day spent resting alone in tree hole or dense tangle of twigs and leaves, except mother-infant pairs that co-habit sleep sites (Seiler *et al.* 2013b). Tree holes preferred in dry, colder season (May–September), probably due to better insulation and also facilitates bouts of sunbathing, where individuals rest at entrance of tree hole, maintain vigilance and occasionally groom (Ruperti 2007, Mandl *et al.* 2018).

Tree hole sleep sites usually in larger trees (especially *Bridelia pervilleana*), whilst twig and leaf sleep sites mostly in *Sorindeia madagascariensis* (Seiler *et al.* 2013b). Individuals resting by day in degraded forest and vegetation tangles (rather than intact forest and tree holes) tend to spend longer periods awake and remain more vigilant, presumably due to increased likelihood of predation (Seiler *et al.* 2013c).

During nighttime active foraging period, around half of time is spent resting while remaining vigilant, with total nightly ranges covering *c.*0.5ha, but during dry, colder season (May–September) they rest more and travel less each night (Seiler *et al.* 2014, Mandl *et al.* 2018). Home range varies between *c.*0.2ha and *c.*1.8ha (mean *c.*0.7ha) with some degree of overlap between neighbouring ranges (mean *c.*20% overlap), but home range size does not vary seasonally (Mandl *et al.* 2018, 2019a, 2020). Availability of suitable sleep sites, density of preferred feeding trees and canopy cover all influence home range choice (Seiler *et al.* 2014). Home ranges of Sahamalaza Sportive Lemur are smaller, with greater overlap than most other congeners; this may correlate to limited dispersal options in highly fragmented habitat (Mandl *et al.* 2020).

Female Sahamalaza Sportive Lemur with infant, Sahamalaza Peninsula.

Sahamalaza Sportive Lemur is primarily folivorous; recorded feeding on leaves of >40 different trees. Tends to prefer more abundant species like *Clitoria lasciva*, *Mangifera indica*, *Garcinia pauciflora* and *Sorindeia madagascariensis*. Occasionally fruits of *Ficus* spp. (Moraeceae), spiders, insects and their larvae are also eaten (Seiler *et al*. 2014).

Sahamalaza Sportive Lemur is highly vocal: at least 12 distinct loud calls have been identified. These vary between the sexes, and seasonally, with likely functions relating to mate advertisement and sexual receptiveness, communicating with offspring and territory defence (Seiler *et al*. 2015b, Mandl *et al*. 2019b). Playback experiments suggest resting Sahamalaza Sportive Lemurs differentiate between vocalisations of aerial predators (raptors) and ground predators (Fosa) and consequently engage different anti-predator behaviours (Seiler *et al*. 2013c). They may also recognise the anti-predator calls of other sympatric species (birds and lemurs), which leads to increased vigilance (Seiler *et al*. 2013d,e). Has also been observed calling at and mobbing a potential predator, the colubrid snake *Ithycyphus perineti* (Mandl *et al*. 2017).

Females give birth to single offspring in September–October (Mandl 2021). Predation by Madagascar Buzzard recorded (Randriatahina & Volampeno 2013).

WHERE TO SEE Ankarafa forest, part of Sahamalaza-Iles Radama National Park, is the best place to see this species. There is a research station, and adjacent campsite with covered pitches. This is a challenging site to reach, in the dry season a drive of *c*.4 hours, and in wet season involving a boat trip from Analalava and a walk of *c*.2 hours.

ANJIAMANGIRANA SPORTIVE LEMUR
Lepilemur grewcockorum   Louis *et al*., 2006

Other name: Grewcock's Sportive Lemur
Malagasy names: Repahaka, Boenga, Boengy, Kitrontro, Kitanta

MEASUREMENTS
Head-body: *c*.250–265mm. Tail: *c*.265–290mm. Weight: *c*.780–950g. (Louis *et al*. 2006b, Craul *et al*. 2007)

DESCRIPTION & IDENTIFICATION Medium-sized sportive lemur. Upperparts grey to grey-brown, including shoulders and arms. Underparts light grey to creamy-white. Diffuse dark stripe from head down

Anjiamangirana Sportive Lemur, Anjajavy Private Reserve.

dorsal ridge to lower back. Face and forehead greyer than body. Ears large and obvious, snout relatively long. Tail relatively long, grey-brown, sometimes browner. White tail tip apparent in some individuals (cf. Sahamalaza Sportive Lemur, Milne-Edwards's Sportive Lemur, Ambodimahahibo Sportive Lemur).

Originally named *L. grewcocki* in error (Louis *et al.* 2006b), then amended to *L. grewcockorum* (Hoffman *et al.* 2009). *L. manasamody* also described from Anjiamangirana and Ambongabe Forests (Craul *et al.* 2007) is considered a junior synonym (Zinner *et al.* 2007, Lei *et al.* 2008, 2017).

Within limited range, might be mistaken for Western Woolly Lemur *Avahi occidentalis*, which adopts similar vertical posture, but has clearly visible white thighs, a white face and less obvious ears. Other nocturnal species like dwarf lemurs *Cheirogaleus* spp. are smaller and adopt horizontal rather than vertical posture.

HABITAT & DISTRIBUTION Primary, secondary and degraded deciduous forests. Described from Anjiamangirana Classified Forest. Known only from forest fragments in Antsohihy region in north-west, between Maevarano River to north and Sofia River to south (boundary with Ambodimahahibo Sportive Lemur), from sea level to *c.*360m (Louis *et al.* 2006b,

Craul *et al.* 2007, Randrianambinina *et al.* 2010). Sportive lemur populations in Antsohihy region were previously classified as *L. edwardsi.* Most records from larger forest areas like Anjiamangirana, Ambongabe, Anjajavy, Ambarijeby and Bekofafa. Appears absent from smaller forest fragments, probably linked to lack of large trees, sufficient food and suitable sleep sites (Randrianambinina *et al.* 2010). This species suffers a high degree of hunting pressure (Olivieri *et al.* 2005). IUCN Critically Endangered.

BEHAVIOUR Nocturnal and folivorous. Behaviour yet to be studied, but probably similar to other western sportive lemurs, cf. Sahamalaza Sportive Lemur, Milne-Edwards's Sportive Lemur, Ambodimahabibo Sportive Lemur.

WHERE TO SEE Anjajavy Private Reserve on north-west coast is the most accessible location. This species can be encountered with careful searching on night walks in forests with larger trees. Occasionally also found resting at edge of sleep hole during daytime.

AMBODIMAHABIBO SPORTIVE LEMUR *Lepilemur otto*   Craul *et al.*, 2007

Other name: Otto's Sportive Lemur
Malagasy names: Repahaka, Boenga, Boengy, Kitrontro, Kitanta

MEASUREMENTS
Head-body: *c.*280–300mm.
Tail: *c.*250–280mm.
Weight: *c.*850–950g. (Craul *et al.* 2007, Mittermeier *et al.* 2010)

DESCRIPTION & IDENTIFICATION A medium-sized sportive lemur. Upperparts, including shoulders, upper and lower arms. predominantly grey-brown; shoulders and arms may be more mid-brown. Underparts grey to creamy. Diffuse dark midline from crown down spine, ending on mid- or lower back. Face, forehead and crown grey, whitish areas around lower jaw and throat. Snout relatively long. Tail relatively short and greyish-brown to deep brown, sometimes with a white tip, cf. Anjiamangirana Sportive Lemur, Sahamalaza Sportive Lemur, Milne-Edwards's Sportive Lemur (Craul *et al.* 2007).

Ambodimahabibo Sportive Lemur.

HABITAT & DISTRIBUTION Patches of dry decidu-ous forest. Described from Ambodimahabibo Forest and, as currently understood, known only from there and three other sites (Marosely, Antsahonjo and Ankarahara) in Bongolava region, north-west Madagascar (Craul *et al.* 2007, Mahaboubi *et al.* 2015). Apparent distribution constrained by Sofia River to north and Mahajamba River to west and south (Craul *et al.* 2007). North of Sofia River is Anjiamangirana Sportive Lemur, whilst Milne-Edwards's Sportive Lemur occurs on opposite side of Mahajamba River. Within limited range, tree felling and forest burning continues. IUCN Endangered.

BEHAVIOUR Nocturnal and folivorous. In common with other western sportive lemurs, a highly vocal species. Peak vocalisations occur around and after dusk (18.00–19.00) and seven different calls have been identified, with similar structure to those o *L. edwardsi* (Mahaboubi *et al.* 2015). Likely similar to other western sportive lemurs, cf. Anjiamangirana Sportive Lemur, Sahamalaza Sportive Lemur, Milne-Edwards's Sportive Lemur.

WHERE TO SEE No sites are accessible. Anyone with an adventurous spirit should search Ambodimahabibo Forest and any neighbouring forest areas.

MILNE-EDWARDS'S SPORTIVE LEMUR
Lepilemur edwardsi (Forsyth Major, 1894)

Malagasy names: Repahaka, Boenga, Boengy, Kitrontro, Kitanta

MEASUREMENTS
Head-body: 270–300mm. Tail: 265–300mm. Weight: c.800–1,100g. (Warren 1997, Warren & Crompton 1997, Thalmann 2001, Louis *et al* 2006b)

DESCRIPTION & IDENTIFICATION One of largest western sportive lemur species. Upperparts grey-brown or beige-grey, with noticeable chestnut-

Milne-Edwards's Sportive Lemur, Ankarafantsika National Park.

HABITAT & DISTRIBUTION Restricted to dry deciduous forests below 450m (Ankarafantsika). As now understood, range covers area between Betsiboka River to west and south, and Mahajamba River to east and north. On opposite side of Betsiboka River is Antafia Sportive Lemur *L. aeeclis*, and opposite side of Mahajamba River is Ambodimahabibo Sportive Lemur (Craul *et al.* 2007, Louis *et al.* 2020f). Prior to recognition of these other species, range of Milne-Edwards's Sportive Lemur also incorporated these areas. In primary forests like Ankarafantsika National Park, especially around Ampijoroa Forest Station, Milne-Edwards's Sportive Lemur appears relatively common: densities up to 60 animals/km² have been recorded (Warren & Compton 1997a). However, ongoing significant reductions in forest extent and quality over large areas mean densities at most locations in Ankarafantsika are significantly lower (Rabesandratana *et al.* 2012). Where larger trees removed, paucity of tree holes/sleep sites likely to be a limiting factor. More broadly, recent studies have suggested possible population collapse of two orders of magnitude, even complete disappearance, at some sites (Craul *et al.* 2009). Hunting pressure, even inside protected areas, is high (Rabesandratana 2006). IUCN Endangered.

BEHAVIOUR Exclusively nocturnal. Feeds on leaves, some fruits, seeds and flowers (Thalmann & Ganzhorn 2003). During dry season, when foliage reduced, several individuals may feed in same tree, without apparent aggression. In Ankarafantsika Milne-Edwards's Sportive Lemur is sympatric with similar-sized, nocturnal Western Woolly Lemur. To avoid competition woolly lemurs are more active, prefer higher forest strata, range over larger areas and feed on leaves of high nutritional value, whilst Milne-Edwards's Sportive Lemur minimises its movements and subsists on vegetation of poorer quality (Ganzhorn 1993, Warren & Compton 1997a,b, Thalmann 2001, 2002, 2006).

Lives in dispersed pairs, where males and females share sleep sites and share a home range *c.*1ha (Warren & Compton 1997a, Rasoloharijaona *et al.* 2003). Pair spends majority of time in core area of *c.*0.5ha (Warren 1997). Exhibits highest degree of male-female cohesion of any sportive lemur (Méndez-Cárdenas & Zimmermann 2009). Mainly solitary foragers when active at night, but individuals regularly meet with partner throughout night, and encounters often include sessions of allo-grooming (Thalmann & Ganzhorn 2003).

reddish areas around shoulders, forelimbs and upper thighs. Tail is similar to body colour, often with white tip. Underparts paler grey with creamy areas. An indistinct darker mid-dorsal stripe from nape down back. Head, crown and face grey, and muzzle grey-brown, sometimes more chestnut. Ears large and prominent.

A medium-sized, vertically-clinging lemur. Potentially confused with Western Woolly Lemur of similar size and posture. Woolly lemur has smaller concealed ears, a rounded 'owl-like' face with larger eyes, and distinctive white patches on thighs. Also, Western Woolly Lemur is more likely to be seen in small groups, whereas Milne-Edwards's Sportive Lemur is generally solitary.

There is some overlap between adjacent territories, which are vociferously defended with loud calls (nine different call types identified) (Rasoloharijaona *et al.* 2006). Pairs regularly duet. Vocalisations synchronous, with males and females contributing different elements; thought to help pair coordinate activity when dispersed (Méndez-Cárdenas & Zimmermann 2009). Appears most vocal towards end of dry season and start of rainy season, perhaps when food resources are at a premium (Thalmann & Ganzhorn 2003).

Tree holes are preferred sleep sites, with dense leaf tangle used very rarely; most individuals sleep alone, but pairs with offspring may share larger holes. Majority of sleep holes 4–5m above ground, some as low as 1m; individuals utilise several different holes in their territory, but often return to same hole for several consecutive nights (Warren 1994, 1997, Warren & Compton 1997a, 1998a,b, Rasoloharijaona *et al.* 2006).

Activity begins around dusk and ends just before daybreak. After emerging, individuals sit at tree hole entrance before moving into forest. Most active for first two hours, thereafter foraging bouts punctuated by extended periods of rest and grooming on favoured branches (Warren & Crompton 1997a). When leaping, seems to prefer vertical supports greater than 5cm diameter (Blanchard *et al.* 2015). Nightly distance travelled varies seasonally; in austral winter (May–September) may move 280–680m per night. In summer months (December–March) distances increase to 400–1,200m (Thalmann & Ganzhorn 2003).

Mating occurs May–June, when males' testes size increases and females enter oestrous for a short period. Gestation lasts 4–5 months, with single offspring born October–November (Randrianambinina *et al.* 2007). Infanticide has been recorded, where a male newcomer killed infant of female whose partner had left (Rasoloharijaona *et al.* 2000).

Known predator is Fosa, which is capable of excavating sportive lemurs from sleep sites (Goodman 2003, Dollar *et al.* 2007).

WHERE TO SEE Ampijoroa Forestry Station, part of Ankarafantsika National Park, is the best place to see this species. Forests around campsite provide good opportunities. An alternative location is Mariarano Classified Forest near the coast, north-east of Mahajanga.

ANTAFIA SPORTIVE LEMUR
Lepilemur aeeclis   Andriaholinirina *et al.*, 2017

Other names: Red-shouldered Sportive Lemur, AEECL's Sportive Lemur
Malagasy name: Boengy

MEASUREMENTS
Head-body: 210–240mm.
Tail: *c.*240–260mm.
Weight: *c.*650–850g.
(Andriaholinirina *et al.* 2006a, Louis *et al.* 2006b)

Antafia Sportive Lemur, Antrema Reserve, Katsepy.

DESCRIPTION & IDENTIFICATION A medium-sized sportive lemur. Pelage variable in coloration (perhaps age-related): generally, upperparts greyish-brown to reddish-grey, head, shoulders and arms more reddish-brown, thighs and lower limbs less reddish. Underparts pale grey with darker areas. Face grey, ears rounded and protruding. Above eyes darker diffuse stripes join on top of head and continue down neck and dorsal ridge, becoming darker and distinct on back (Andriaholinirina *et al.* 2006a). Tail varies from deep rusty-red to grey; when grey becomes more reddish towards tip.

Smaller than Milne-Edwards's Sportive Lemur, but larger than Tsiombikibo Sportive Lemur *Lepilemur ahmansoni*. Within range, might only be mistaken for dwarf lemurs *Cheirogaleus* spp., which are smaller and adopt horizontal rather than vertical posture.

Tsiombikibo Sportive Lemur, Tsiombikibo Forest.

HABITAT & DISTRIBUTION Confined to western dry deciduous forest. Described from Antafia Forest on north bank of Mahavavy River. As currently understood, range probably extends between Betsiboka River in north and east, and Mahavavy River in south and west. Opposite bank of Betsiboka River is Milne-Edwards's Sportive Lemur, while opposite bank of Mahavavy River is presumed to be range of Tsiombikibo Sportive Lemur. The taxonomic validity of several apparently very similar sportive lemur species in western regions has been questioned (Tattersall 2007), but molecular analysis supports their recognition (Lei *et al.* 2017). Within its range, forests extremely fragmented and subject to ongoing encroachment and destruction. IUCN Endangered.

BEHAVIOUR Nocturnal and folivorous. No specific studies. Most aspects of behaviour likely similar to Milne-Edwards's Sportive Lemur.

WHERE TO SEE Most accessible location is forests near Katsepy (Antrema Reserve) across Betsiboka River from Mahajanga. Alternatively, forests near Anjahamena on north/east bank of Mahavavy River and potentially any reasonable-sized patches of forest within range.

TSIOMBIKIBO SPORTIVE LEMUR
Lepilemur ahmansoni   Louis *et al.* 2006

Other name: Ahmanson's Sportive Lemur
Malagasy name: Boengy

MEASUREMENTS
Head-body: 240–290mm. Tail: 230–300mm. Weight: c.460–760g. (Louis *et al.* 2006b)

DESCRIPTION & IDENTIFICATION Relatively small sportive lemur. Upperparts primarily dark grey, with reddish-brown tones around dorsal extremities. Lacks prominent dorsal midline of Antafia Sportive Lemur. Faint black stripe on crown. Underparts dark grey near midline becoming light grey below. Tail relatively long, reddish-brown, darker dorsally and paler ventrally (Louis *et al.* 2006b). Due to errors in previous publications and their perpetuation, sometimes incorrectly referred to as *L. ahmansonorum*, *L. ahmansorum* or *L. ahmansonori* (Louis *et al.* 2020d).

HABITAT & DISTRIBUTION Dry deciduous western forest. Described from Tsiombikibo Forest, west of Mahavavy River, near Mitsinjo. Range may also incorporate forests in Mahavavy Kinkony Wetland Complex and Namoroka National Park (Rylands *et al.* 2018). Mahavavy River probably forms eastern limit and boundary with Antafia Sportive Lemur. Southern range limit and relationship to distribution of Bemaraha Sportive Lemur *L. randrianasoloi*, yet to be determined: it may be either the Maningoza River or Mananbaho River. This requires further investigation. IUCN Critically Endangered.

BEHAVIOUR Nocturnal and folivorous. Yet to be studied in wild. Most aspects of behaviour probably similar to Milne-Edwards's Sportive Lemur or Red-tailed Sportive Lemur *L. ruficaudatus*.

WHERE TO SEE Best looked for in Tsiombikibo Forest, near Mitsinjo west of the Mahavavy River.

BEMARAHA SPORTIVE LEMUR
Lepilemur randrianasoloi

Andriaholinirina *et al.*, 2017

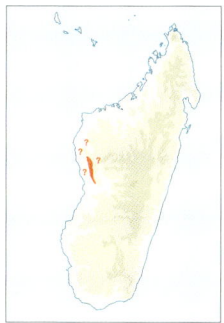

Other name: Randrianasolo's Sportive Lemur
Malagasy name: Boengy

MEASUREMENTS
Head-body: *c.*280–300mm.
Tail: *c.*210–275mm.
Weight: *c.*700–880g.
(Andriaholinirina *et al.* 2006a, Schwitzer *et al.* 2013b)

DESCRIPTION & IDENTIFICATION A medium-small sportive lemur. Upperparts pale grey, with areas of rufous-brown and darker grey on forearms, shoulders, back and hindlimbs. Head mainly grey, face pale grey forming faint mask. A dark mid-dorsal line from head to base of back. Tail more reddish than body. Broadly similar to Red-tailed Sportive Lemur, Tsiombikibo Sportive Lemur and Antafia Sportive Lemur with narrower and longer head, which is more apparent in males (Andriaholinirina *et al.* 2006a). Originally named *L. randrianasoli* in error

Bemaraha Sportive Lemur, Tsingy de Bemaraha National Park.

191

(Andriaholinirina *et al.* 2006b), later corrected to *L. randrianasoloi* (Hoffman *et al.* 2009, Andriaholinirina *et al.* 2017).

Within range, similar in size and posture to Bemaraha Woolly Lemur *Avahi cleesei*, but lacks obvious owl-like 'facial disc' or white thigh patches that are clearly visible in woolly lemurs. Unlike dwarf lemurs *Cheirogaleus* spp. or fork-marked lemurs *Phaner* spp. adopts vertical rather than horizontal posture.

HABITAT & DISTRIBUTION Dry deciduous western forest from sea level to *c.*140m, often in forests within and adjacent to limestone pinnacle karst formations. Described from Andramasay, Bemaraha National Park, and adjacent forests. Range presumed to extend from Tsiribihina River in south to at least Manambaho River and possibly Maningoza or Mahavavy Rivers to north. At northern limits, boundary with Tsiombikibo Sportive Lemur yet to be clarified (Andriaholinirina *et al.* 2006). Forests in range are very fragmented and suffer ongoing encroachment and destruction through tree felling and annual cycles of burning. IUCN Endangered.

BEHAVIOUR Nocturnal and folivorous. No specific wild studies. Most aspects of behaviour probably similar to other western sportive lemurs cf. Milne-Edwards's Sportive Lemur and Red-tailed Sportive Lemur.

WHERE TO SEE Tsingy de Bemaraha National Park, near Bekopaka on north bank of Manambolo River; relatively easy to see at daytime sleep sites in forests surrounding both *Grand Tsingy* and *Petit Tsingy*. Alternatively, explore any similar large forest areas north of Tsiribihina River.

Red-tailed Sportive Lemur, Kirindy Forest.

RED-TAILED SPORTIVE LEMUR
Lepilemur ruficaudatus　　A. Grandidier, 1867

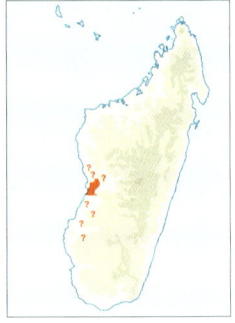

Malagasy name: Boengy

MEASUREMENTS
Head-body:
*c.*260–300mm. Tail:
*c.*240–280mm. Weight:
*c.*650–900g.

DESCRIPTION & IDENTIFICATION A medium-sized sportive lemur. Upperparts grey-brown with chestnut-rufous tones on shoulders and forelimbs, underparts paler grey, throat creamy-white. Tail distinctly reddish. Head grey with darker grey around muzzle. Ears large, rounded and prominent. Eyes pale orange to fiery orange. Within its range, all other small to medium-sized nocturnal lemurs, dwarf lemurs *Cheirogaleus* spp., Coquerel's Giant Mouse Lemur and Pale Fork-marked Lemur adopt horizontal posture and run and jump quadrupedally, rather than vertical posture and leaping behaviour of Red-tailed Sportive Lemur.

HABITAT & DISTRIBUTION Dry deciduous forest of centre-west from sea level to *c.*900m. Range as currently understood centred on Central Menabe region. Northern limit probably Tsiribihina River, with Bemaraha Sportive Lemur occurring north of river. Southern limits much less clear; probably Mangoky River north of Morombe. In intact dry deciduous forests recorded at very high densities; estimates vary between *c.*90–160 individuals/km² in Kirindy Forest and 180–350 individuals/km² in Marosalaza Forest (Zinner *et al.* 2003). In degraded and fragmented forests densities much lower. Although range extends over relatively large area, levels of deforestation and habitat destruction are massive. Areas of protected forest within range cover *c.*1,600km². IUCN Critically Endangered.

BEHAVIOUR As with other sportive lemurs, nocturnal and folivorous. Males and females live as a 'dispersed' pair, with minimal social interaction and have directly overlapping home ranges; male home ranges larger than partner (male *c.*1.6ha, female *c.*1.1ha) (Ganzhorn & Kappeler 1996, Pietsch 1998, Zinner *et al.* 2003, Hilgartner *et al.* 2012). Home ranges may be stable over long periods (Zinner *et al.* 2003). Foraging bouts solitary, with activity only synchronised between pair for *c.*15% of time, when pair in visual contact (Fichtel *et al.* 2011). Foraging punctuated by a single long period of inactivity. Nightly distances travelled *c.*100–1,000m (Schwitzer *et al.* 2013c) and similar for males and females (Ganzhorn *et al.* 2004). When leaping, seems to prefer vertical supports no greater than 5cm diameter (Blanchard *et al.* 2015). When aggressive, individuals bang their hindfeet on branches and shake the tree (Ganzhorn 1993).

Within home range, uses several different sleep sites, up to six over three-month period. Pairs sleep together in same tree hole approximately every fourth night (Zinner *et al.* 2003). Often spends part of day at entrance to tree hole sunbathing. In Kirindy-Mitea National Park (sometimes Kirindy Mité), sleep holes in tall intact living trees, particularly *Strychnos madagascariensis*, are preferred, with holes generally at least 4m above ground. Presumed to help reduce predation and may serve thermoregulatory function (Rakotomalala *et al.* 2017b).

Primarily eats leaves, although fruits of some trees, especially *Diospyros* spp., also eaten seasonally. No obvious dietary differences between sexes (Ganzhorn *et al.* 2004). Able to tolerate leaves with high toxin concentrations and during dry season (May–November) can subsist on dry leaves (Hladik *et al.* 1980). Red-tailed Sportive Lemur is nowhere sympatric with woolly lemurs *Avahi* spp., consequently feeds on higher quality foliage

Red-tailed Sportive Lemurs, probably a pair with offspring (middle), Kirindy Forest.

than western congeners that are sympatric with woolly lemurs cf. Milne-Edwards's Sportive Lemur (Ganzhorn 1993).

During daytime rest periods has one of lowest metabolic rates of any mammal. Raised substantially (doubled) prior to nighttime activity; probably an adaptation to survive at relatively high densities on a poor-quality diet (Schmid & Ganzhorn 1996). During cold dry season when food resources more limited, reduces nightly travel distances (Schwitzer *et al.* 2013b).

Mates between May and June, when males guard females from potential competitors and become more aggressive (Hilgartner *et al.* 2008 & 2012). Single young, generally born in November, after gestation of *c*.5 months (longer than expected for lemur of this size) (Hilgartner *et al.* 2008). Infants regularly groomed by mother. When moving initially carried by mother in her mouth; later, infant clings to fur on her back. When older, mother may leave or 'park' infant in tree hole or similar 'safe' site while she forages. Lactation and suckling last *c*.50 days. Males seen rarely in close proximity to infants and appear to take no direct part in parental care (Hilgartner *et al.* 2008). Independence reached around one year (Petter-Rousseaux 1964).

Red-tailed Sportive Lemurs lack specific responses to potential aerial or terrestrial predators, but they do respond to general alarm calls (barks) of conspecifics by fleeing (Fichtel 2007). Predation rates can be high, up to 40% in some years, but generally much lower (Hilgartner *et al.* 2008). In Kirindy Forest, known predators include Madagascar Long-eared Owl, Madagascar Harrier-Hawk, Narrow-striped Boky and Fosa: Madagascar Harrier-Hawk and Fosa are capable of excavating the lemurs from tree holes (Goodman 2003, Goodman & Ganzhorn in press).

WHERE TO SEE The best locality is Kirindy Forest in Central Menabe, north of Morondava. Here relatively common and with help of guides from camp, sportive lemurs can be found on most nights. Kirindy Forest is one of best sites in Madagascar to see a number of nocturnal lemur species (six in all, others being Grey Mouse Lemur, Madame Berthe's Mouse Lemur, Fat-tailed Dwarf Lemur, Coquerel's Giant Mouse Lemur and Pale Fork-marked Lemur) and several other mammals. The more adventurous may wish to visit Kirindy-Mitea National Park, *c*.100km south of Morondava.

ZOMBITSE SPORTIVE LEMUR
Lepilemur hubbardorum   Louis *et al.*, 2006

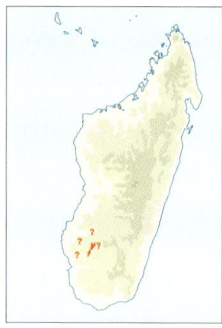

Other name: Hubbard's Sportive Lemur
Malagasy names: Apongy, Fitsidika

MEASUREMENTS
Head-body: *c*.230–260mm.
Tail: *c*.230–250mm.
Weight: *c*.850–1,100g.
(Louis *et al.* 2006b)

DESCRIPTION & IDENTIFICATION A medium-large sportive lemur. Upperparts dark reddish-brown on shoulders and upper back, becoming paler reddish-grey towards lower body and tail base. Underparts entirely creamy-white to white. Face greyish-brown, crown and nape reddish-brown, fur around neck paler, forming reddish-blonde collar. Eyes pale orange to yellowish. Tail uniformly reddish-blonde (Louis *et al.* 2006b). Originally named *L. hubbardi* in error (Louis *et al.* 2006b), later corrected to *L. hubbardorum* (Hoffman *et al.* 2009).

HABITAT & DISTRIBUTION Dry deciduous and dry transitional forest at *c*.300–500m. Described from Zombitse National Park. Currently known only from Zombitse-Vohibasia National Park region, north of Onilahy River, and south of Fiherenana River in south-west. Further field surveys needed to establish boundaries of species' range. Population density of *c*.145 individuals/km² recorded, which extrapolates to tentative total population of 16,500–18,000 in Zombitse portion (Martin *et al.* 2021). Virtually all forest in region has disappeared; Zombitse-Vohibasia National Park covers less than 275km². IUCN Endangered.

BEHAVIOUR Nocturnal and presumed folivorous. No detailed wild studies.

WHERE TO SEE Relatively easy to see in Zombitse-Vohibasia National Park during day, resting in sleep sites. Local guides can be arranged in nearby town of Sakaraha. There are several trails and circuits, where this species, as well as Verreaux's Sifaka and Red-fronted Brown Lemur can be seen.

Zombitse Sportive Lemur, Zombitse National Park.

Petter's Sportive Lemur, Bezà-Mahafaly Special Reserve.

PETTER'S SPORTIVE LEMUR
Lepilemur petteri   Louis *et al.*, 2006

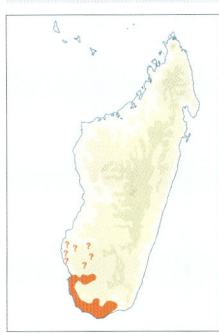

Malagasy names: Apongy, Fitsidika

MEASUREMENTS
Head-body: *c.*230–240mm.
Tail: *c.*220–250mm.
Weight: *c.*620–650g. (Louis *et al.* 2006b)

DESCRIPTION & IDENTIFICATION A small sportive lemur. Upperparts grey to greyish-brown, more brownish-grey on tops of thighs and midline of back. Underparts whitish-grey. Head and face grey, whitish areas around eyes, sides of muzzle and under chin. Ears dark, with light grey to white fur at edges. Eyes deep orange. Coastal forests at Ifaty lie north of Onilahy River (and Fiherenana River) and outside currently accepted range of Petter's Sportive Lemur.

Sportive lemurs at Ifaty are outwardly similar to Petter's Sportive Lemur, but some have highly distinctive rich chocolate-brown to black patches on thighs, forearms and across shoulders, with very white areas around eyes, on cheeks and throat. The specific identity of these populations has yet to be determined.

The validity of many sportive lemur species, including differentiation of Petter's Sportive Lemur and White-footed Sportive Lemur *L. leucopus*, has been questioned (Tattersall 2007), but molecular analysis suggests these distinctions are robust (Lei *et al.* 2017).

HABITAT & DISTRIBUTION Confined to dry habitats, principally spiny forest, thorn scrub and some gallery forest, at low elevations, to *c.*200m. Described from Bezà-Mahafaly Special Reserve. As currently understood, range extends across south-west and southern areas, from Onilahy River in south-west to Mandrare River in south-east, and includes Mahafaly Plateau (Louis *et al.* 2020o). Occurs at higher densities within intact forest and lower densities in degraded forest (Nash 2000).

An unusually marked sportive lemur from Ifaty that may correspond to Petter's Sportive Lemur or possibly a new taxa.

Prior to description of Petter's Sportive Lemur this range was presumed to be occupied by White-footed Sportive Lemur, whose range is now restricted to areas east of Mandrare River. Although range of Petter's Sportive Lemur is apparently extensive, spiny forest areas in south and south-west have suffered greater rates of deforestation over last 30 years than any other native forest type (Harper *et al.* 2007). Remaining forest severely fragmented, with few large tracts remaining. IUCN Endangered.

BEHAVIOUR Nocturnal and folivorous. Petter's Sportive Lemur has been studied in Bezà Mahafaly Special Reserve and Berenty Private Reserve, when populations were assigned to *L. leucopus*.

By day tree holes, concealed spaces between tightly clustered branches or tangles of lianas are used as sleep sites; males and females sometimes share sites, but often sleep separately. Can be very vocal, especially just after dusk when they emerge.

Petter's Sportive Lemur using an artificial 'nest' box, Berenty Private Reserve.

During nighttime wake period, *c*.10–15% of time is spent moving between forage sites, 30% actually feeding and *c*.50% resting and/or self-grooming. Social behaviour, vocalising and other behaviours comprise 5% (Nash 1998). In cooler dry season periods of rest increase and shorter distances are travelled when foraging (as little as 200–300m per night), although total time spent foraging per night remains constant year-round (Nash 1998).

Diet consists of leaves, primarily Didiereaceae spp. in spiny forest, while in gallery forests leaves of *Tamarindus* and *Euphorbia* together with various vines dominate (Nash 1998). In dry season (May–October), flowers of Didiereaceae also eaten.

Phases of moon apparently have little influence on activity patterns, Petter's Sportive Lemur being equally active (or inactive) on moonlit or dark nights. However, during periods with intense moonlight, uses higher levels in forest far less frequently (Nash 2000, 2007). Probably linked to predator avoidance as it is predated by Barn Owl and White-browed Owl *Ninox superciliaris*.

WHERE TO SEE Lac Tsimanampetsotse National Park, 80km south of Toliara. Best looked for in spiny forest on top of escarpment on west edge of Mahafaly Plateau. An alternative location is Bezà-Mahafaly, south-east of Toliara, although this can be challenging to reach. Furthermore, under current classification populations at Berenty Private Reserve on west bank of Mandrare River are now regarded as Petter's Sportive Lemur (previously White-footed Sportive Lemur). Sportive lemurs are sometimes seen in coastal spiny forest areas near Ifaty, 25km north of Toliara; similar to Petter's Sportive Lemur, but specific identity requires clarification.

WHITE-FOOTED SPORTIVE LEMUR
Lepilemur leucopus　　　　(Forsyth Major, 1894)

Malagasy name: Songiky

MEASUREMENTS
Head-body: 190–260mm.
Tail: 215–260mm. Weight: *c*.500–650g.

DESCRIPTION & IDENTIFICATION A small sportive lemur. Head and upperparts principally pale grey with brownish tinges around shoulders, upper forelimbs and upper thighs. Tail more brownish-grey. Underparts and throat very pale grey to off-white, extending part way up flanks. Face grey-brown with whitish 'spectacles' around eyes. Ears relatively large and rounded with whitish tufts at bases. Eyes rich orange. Probably the smallest sportive lemur species. Confusion with other nocturnal species in range unlikely: dwarf lemurs *Cheirogaleus* spp., and mouse lemurs *Microcebus* spp., are smaller than White-footed Sportive Lemur, adopt a horizontal posture and move quadrupedally.

HABITAT & DISTRIBUTION Spiny forest dominated by Didiereaceae species thorn scrub, gallery forest and subtropical dry lowland forest to *c*.300m. Range extends from west base of Anosyennes Mountains to Mandrare River in south-east. This includes spiny forest and transitional forest parcels of Andohahela National Park and Nord-Ifotaka, Ankodida, Behara-Tranomaro and Vohidava-Betsimalaho protected areas. Anosyennes Mountains create major rain shadow, resulting in aridity of region. Anosyennes Mountains themselves incorporate rainforest parcel of Andohahela National Park (range of Andohahela Sportive Lemur). Range covers *c*.2,300km^2, but is highly fragmented with few large areas of forest remaining. IUCN Endangered.

BEHAVIOUR Note: information presented here is derived from studies undertaken at Berenty Private Reserve, on west bank of Mandrare River. All studies list species as White-footed Sportive Lemur; however, based on now accepted distribution and range outlined above, species at Berenty is Petter's Sportive Lemur. Sportive lemur populations east of Mandrare River are White-footed Sportive Lemur.

Nocturnal and folivorous. In spiny forest, diet dominated by most abundant species, leaves of *Alluaudia procera* (Didiereaceae) and the liana *Metaporana parvifolia*. However, leaves of >30 tree species and 16 liana species have been recorded. Flowers, fruits and shoots are eaten infrequently. During early wet season young leaves dominate, mature leaves eaten primarily at other times (Dröscher & Kappeler 2014a). Food selection year-round appears to prioritise protein intake over non-proteins, with no apparent differences between sexes. In females, relative proportion of protein in diet increases during lactation (Dröscher et al. 2016).

White-footed Sportive Lemur, Ihazofotsy, Andohahela National Park.

White-footed Sportive Lemur has one of lowest metabolic rates recorded for a mammal; rates are lower in warm, wet season and higher in cold, dry season (presumed adaptations to extreme habitat and nutritionally poor diet) (Bethge *et al.* 2017). Also is able to reduce energy and water requirements in hot dry conditions, when extreme temperatures limit movement and they seek more sheltered cooler resting locations (Stalenberg 2019).

Probably most asocial sportive lemur. Males and females have overlapping home ranges and live as 'dispersed pairs', meaning the sexes co-habit common areas, but within them are largely solitary and show signs of active avoidance (Dröscher &

Kappeler 2013). Occasionally one male may be associated with two females. Home range small, 0.2–0.4ha, with male ranges larger (Russell 1980, Dröscher & Kappeler 2013, 2014). Home range overlap especially with neighbouring females is minimal (Dröscher & Kappeler 2013). Within shared area, male and females forage alone and there is little competition for food. When resource competition occurs, females appear dominant (Dröscher & Kappeler 2013).

White-footed Sportive Lemur becomes active around dusk and can be very vocal. At night *c.*40% of time spent foraging, interspersed with prolonged periods of rest and/or self-grooming (Russell 1980). By day, individuals sleep singly, never as a pair, in tree holes (often in *Alluaudia procera*), crevices between tightly clustered branches, or tangles of lianas. Individuals may use partner's sleep site on consecutive days. At beginning of sleep period (early morning) female may sometimes displace male from a favoured sleep site. Females share sleep site with offspring (Dröscher & Kappeler 2013).

Males and females in a pair utilise a latrine site for defecation and urination, located near core of home range. This appears to facilitate familiarity between individuals and stimulate a degree of social bonding. Urine may be a more important component in olfactory communication (Dröscher & Kappeler 2014b).

Mating occurs from May to July; gestation *c.*130 days. Single infant born between October and November, and attains sexual maturity by 18 months (Petter *et al.* 1977).

White-footed Sportive Lemurs suffer high rates of predation: over a two-year period, *c.*20–30% of population may be taken (Dröscher & Kappeler 2014a). Fosa is extremely uncommon in spiny forest areas, but Barn Owl, White-browed Owl and Madagascar Buzzard are significant predators (Goodman 2003). In localities in near human habitation, feral cat *Felis silvestris* is also a major predator (Dröscher & Kappeler 2014a).

WHERE TO SEE White-footed Sportive Lemur can be seen in spiny forest in Andohahela National Park (Mangatsiaka and Ihazofotsy). Sportive lemurs in Berenty Private Reserve, and forests further north at Ifotaka, on west side of Mandrare River, were previously listed as White-footed Sportive Lemur but are now assigned to Petter's Sportive Lemur.

True Lemurs

True Lemurs Family Lemuridae

The most recognisable and well known of Madagascar's primates reside in the family Lemuridae; indeed, some are arguably the island's most iconic mammals. The family contains five genera, *Lemur*, *Eulemur*, *Varecia*, *Hapalemur* and *Prolemur* that divide into two informal subgroupings: the genera *Hapalemur* and *Prolemur* are specialist bamboo feeders (although other foods are also eaten), whilst the genera *Lemur*, *Eulemur* and *Varecia*, often considered the 'true lemurs', have diets dominated by fruits and leaves. Some previous classifications have formalised these broad divisions as subfamilies, Lemurinae and Hapalemurinae, which would, by definition, require each to be monophyletic. However, molecular analysis suggests the evolutionary relationships between *Hapalemur* and *Lemur* are closer than between *Hapalemur* and *Prolemur*, precluding this (Simons & Rumpler 1988, Groves & Eaglen 1988).

Members of Lemuridae are found across all major native forest types; southern spiny forest and thorn scrub, western deciduous forest and eastern humid forests. They range in size from *c*.750g (*Hapalemur* spp.) to *c*.4kg (*Varecia* spp.). All mainly adopt a horizontal body posture and are primarily quadrupedal, with forelimbs being slightly shorter than hindlimbs. They move through the canopy with a combination of walking and running along branches and leaping between them.

They are all gregarious, typically living in groups that vary considerably in size between species. Most Lemuridae, i.e., the genera *Hapalemur*, *Prolemur*, *Lemur* and *Eulemur* are cathemeral (can be active anytime over 24-hour cycle), whilst *Varecia* exhibits only limited nighttime behaviour (Bray *et al.* 2017). Cathemeral behaviour is very unusual in primates and has been recorded in just three species outside Madagascar, Azara's Night Monkey *Aotus azarae*, Black Howler Monkey *Alouatta pigra* and Mantled Howler Monkey *A. palliata*, all from the Neotropics (Donati *et al.* 2009).

Unsurprisingly, being social, all Lemuridae communicate with a variety of calls, with some species among the most vocal lemurs. All have 36 teeth and a dental comb, where the narrow incisors and canines of the lower jaw jut straight forward. This is thought to aid grooming (Schwitzer *et al.* 2013d).

Bamboo Lemurs Genera *Hapalemur* and *Prolemur*

Bamboo lemurs are the smallest diurnal lemurs (*c*.750g to 3.5kg) and are best known for their dietary preference for various species of woody bamboo (grass family Poaceae), with some species spending *c*.80–95% of feeding time consuming these plants (Tan 2006). They are Madagascar's ecological equivalents of the Giant Panda *Ailuropoda melanoleuca* and Red Panda *Ailurus fulgens* (and even share some dental traits) (Eronen *et al.* 2017, Tan *et al.* in press). Although other primates opportunistically or seasonally feed on bamboo in small quantities, bamboo lemurs are unique among primates in specialising in this diet (Mutschler & Tan 2003, Tan 2006). The evolution of this behaviour has allowed these species to 'corner the market' in bamboo and reduce competition with other lemurs. However, there is a price to pay: having 'driven into an evolutionary cul-de-sac,' they are closely tied to their chosen food sources (especially *Prolemur simus*, but to a lesser extent some *Hapalemur* spp. that exhibit some dietary diversity) (Eronen *et al.* 2017), and therefore more vulnerable to any changes that might impact these.

Bamboo is unpalatable to most folivores as it is very fibrous and some species produce potent toxins to deter leaf predators like bamboo lemurs. The two sides are embroiled in an evolutionary arms race, with bamboos producing more toxins to dissuade the lemurs, and the lemurs evolving more sophisticated mechanisms for detoxification (Ballhorn *et al* 2016). An extreme example is Madagascar Giant Bamboo *Cathariostachys madagascariensis* that produces very high concentrations of cyanide toxins in its ground and branch shoots (Ballhorn *et al.* 2009). In Ranomafana National Park, three species of bamboo lemur feed on this plant, each in differing ways: Eastern Grey Bamboo Lemur *Hapalemur griseus ranomafanensis* is a relative generalist and mainly eats parts of the plant with relatively low levels of toxins (in small quantities), whereas Golden Bamboo Lemur *H. aureus* targets young leaves and branch shoots, and Greater Bamboo Lemur *Prolemur simus* specialises in ground shoots, which are all laced with high toxin content, thereby reducing competition with congeners (Wright 1989, Wright & Randraimamamtena 1989, Tan 2006, Eppley *et al.* 2017b).

The teeth of bamboo lemurs are also appropriately specialised: the lower incisors and canines protrude straight forward and are used in chisel-like fashion to strip away the hard outer layers of bamboo stems. All other teeth, except the molars, have a serrated cutting edge, undoubtedly an adaptation to feeding on coarse and fibrous bamboo leaves (Schwitzer *et al.* 2013d). Given the high fibre content in their diets, digestion times in bamboo lemurs are longer that other similar-sized lemur species: typically, 18–36 hours for *Hapalemur* spp. (Campbell *et al.* 2004a,b, Perrin 2013) and *c*.8 hours for Greater Bamboo Lemur (Tan 2000).

Formerly all species were placed in the genus *Hapalemur*. Primarily on basis of distinctive chromosomal characteristics, but also morphological (cranial and dental) and behavioural aspects, Greater Bamboo Lemur has subsequently been assigned its own monotypic genus *Prolemur* (Groves 2001, Hawkins *et al.* 2018).

Compared to 'true lemurs' (*Lemur*, *Eulemur* and *Varecia*), bamboo lemurs (both *Hapalemur* and *Prolemur*) have a broadened and relatively short muzzle, with a more 'rounded' head and 'blunt' face and long tail, all features that aid their recognition in the field.

Bamboo or Gentle Lemurs
genus *Hapalemur*

Currently five species are placed in the genus *Hapalemur*: all are characterised by medium-small to medium-sized bodies, long tails and moderately long hindlegs for leaping. Hands and feet have large finger and toe pads, and the big toe (hallux) is very large, with the thumb (pollex) semi-opposable. Also characteristic is a broad, short muzzle, with 'rounded' head and 'blunt' face, and relatively small, largely hidden ears. There is no significant sexual dimorphism.

Activity patterns are diurnal and often cathemeral. Because of their penchant for feeding in stands of bamboo, they prefer to adopt a vertical posture and are capable of quick bounding leaps between close upright stems of bamboo and similar vegetation. One exception is Lac Alaotra Reed Lemur *H. alaotrensis*, which has evolved a specialised form of locomotion associated with its unique habitat.

The use of latrine sites is rare in primates, but has been recorded in some lemurs, e.g., dwarf lemurs *Cheirogaleus* spp. (Ganzhorn & Kappeler 1996) and sportive lemurs *Lepilemur* spp. (Irwin *et al.* 2004); the behaviour appears more common in bamboo lemurs, especially Southern Grey Bamboo

Lemur *H. meridionalis* (Irwin *et al.* 2004, Eppley *et al.* 2016b).

Bamboo lemurs have a number of vocalisations that serve as within-group contact and cohesion calls, more distant communication calls, mating calls, and distress and predator alarm calls.

The taxonomy and the phylogenetic relationships within this group remain contentious and there have been several recent changes (Fausser *et al.* 2002a,b, Pastorini *et al.* 2002a,b, Rabarivola *et al.* 2007). In addition to Greater Bamboo Lemur being removed to the genus *Prolemur*, the Golden Bamboo Lemur *H. aureus* was discovered in 1985 and formally described two years later (Meier *et al.* 1987). Taxa previously regarded as subspecies of Grey Bamboo Lemur have now been elevated to species, *H. griseus*, *H. occidentalis* and *H. alaotrensis* (Groves 2001,) and bamboo lemurs at the southern extremity of the rainforest region are now recognised as a distinct species, Southern Grey Bamboo Lemur *H. meridionalis* (Fausser *et al.* 2002). Furthermore, two new subspecies of Grey Bamboo Lemur have been proposed, *H. griseus gilberti* and *H. g. ranomafanensis* (Rabarivola *et al.* 2007).

Beyond the genus, *Hapalemur* may share closest relationships with genus *Lemur* (Ring-tailed Lemur); molecular DNA evidence suggests close evolutionary affinities, and wrist glands are a trait shared by both.

EASTERN GREY BAMBOO LEMUR
Hapalemur griseus   (Link, 1795)

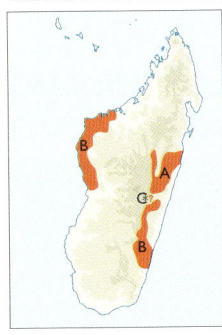

Other names: Eastern Grey Gentle Lemur, Eastern Lesser Bamboo Lemur
Malagasy names: Bokombolo, Kotrika, Alokoteha, Konté

MEASUREMENTS
Head-body: *c.*240–300mm.
Tail: *c.*320–400mm. Weight: *c.*750–1,000g. (Tan 1999, 2000, Rabarivola *et al.* 2007)

DESCRIPTION & IDENTIFICATION Medium-sized active species; smallest diurnal lemur. Upperparts medium-grey to olive-grey with chestnut-russet tinges over head, around shoulders and sometimes extending on dorsum; these areas are more noticeable on individuals in south of species'

Eastern Grey Bamboo Lemur, Andasibe-Mantadia National Park.

range. Tail darker grey and relatively long (*c.*120% head-body). Face and underparts slightly paler grey, gradually becoming creamy-grey on belly. Ears small and rounded, muzzle much shorter than other diurnal lemurs.

Three subspecies proposed on basis of slight chromosomal differences, *H. g. griseus*, *H. g. ranomafanensis* and *H. g. gilberti* (see below), all broadly similar in appearance. *H. g. gilberti* is considered slightly larger (mean *c.*1,000g), with darker, more reddish, areas around head and

Eastern Grey Bamboo Lemur, Ranomafana National Park.

shoulders, and underparts grey anteriorly grading to reddish posteriorly (Rabarivola *et al.* 2007).

Most often seen in vertical clinging posture. Small size, blunt muzzle and rounded face should prevent confusion with other species. In areas of sympatry (e.g., Ranomafana National Park), confusion with Golden Bamboo Lemur or Greater Bamboo Lemur is possible. Distinguished from Golden Bamboo Lemur by smaller size and grey, rather than golden-brown, coloration. Greater Bamboo Lemur broadly similar in colour, but significantly larger with distinctive whitish ear-tufts.

HABITAT & DISTRIBUTION In eastern regions occurs in primary and secondary lowland and montane humid forests with bamboo and bamboo vines, from sea level to *c.*1,600m (Andringitra) (Sterling & Ramaroson 1996). In western regions in dry and humid deciduous forests with bamboo and bamboo vines. Able to tolerate forest edge and human-altered habitats (Arrigo-Nelson & Wright 2004, Lehman *et al.* 2006b). Broadly, range covers eastern rainforest regions from Onibe River in north, to at least Andringitra region, and possibly further south to Mananara River (boundary and relationship with Southern Grey Bamboo Lemur requires clarification). In western areas probably occurs between Betsiboka and Manambolo Rivers (Hawkins *et al.* 1998, Rakotoarison *et al.* 1993, Curtis *et al.* 1995). Previously, grey bamboo lemurs north of Onibe River were described as *H. griseus* (Harcourt & Thornback 1990, Garbutt 1999). These populations are now assigned to *H. occidentalis* (Rumpler 2004). Precise ranges of three subspecies remain uncertain and require clarification; as currently recognised, tentatively:

- *H. g. griseus* (distribution A) occurs in central-eastern regions, with Onibe River as northern limit and Onive River and Mangoro River forming southern limit (Rabarivola *et al.* 2007).

- *H. g. ranomafanensis* (distribution B) has disjunct distribution, occurring in both eastern and western regions. In east, occurs from Nosivolo and Mangoro Rivers area in north, to Andringitra region and perhaps as far south as Mananara River. South of this is Southern Grey Bamboo Lemur and range limits of these taxa require clarification. In western regions has patchy distribution between Betsiboka River in north and Manambolo River in south (Rabarivola *et al.* 2007).

• *H. g. gilberti* (distribution C) was described from Beanamalao area and is thought to occupy limited range between Nosivolo River (southern limit) and Onive and Mangoro Rivers (northern limit) (Rabarivola *et al.* 2007).

Within forests, patchily distributed, depending on prevalence of bamboo. Population density of *c.*2.5 individuals/km^2 recorded in Ranomafana National Park; a significant reduction in population recorded at this site over last 30 years (Wright *et al.* 2012). In undisturbed forests, e.g., Tsinjoarivo, density of *c.*31.5 individuals/km^2 recorded (Rakotomalala *et al.* 2017a). Eastern Grey Bamboo Lemur is threatened by habitat loss and fragmentation (Eppley *et al.* 2020d), especially in west. Also one of most popular lemurs in illegal pet trade (Reuter *et al.* 2019). IUCN Vulnerable.

BEHAVIOUR Previously regarded as principally diurnal (Overdorff *et al.* 1997, Tan 2000) but generally more active around dawn and dusk, so better described as crepuscular. Not active after dark (Tan *et al.* in press). Vocalisations begin *c.*2 hours before dawn and occasionally given at night. Around half day spent foraging at all levels in forest (Wright 1986, 1989).

Diet dominated by various species of bamboo, *c.*50–80%, with marked preference for bases of young leaves, branch shoots, some ground shoots and stem pith (Tan *et al.* in press). Demonstrates considerable dexterity, especially when feeding on bamboo; leaves pulled from end of branch, soft leaf bases bitten off, tough leaf blades discarded. Some preferred bamboos (e.g., *Panicum* spp.) that may comprise *c.*50% of diet in dry season, contain potentially harmful amounts of cyanogenic toxins, thus lemurs descend on daily basis to eat soil, to nullify detrimental effects of accumulating toxins (Andrianandrasana *et al.* 2020). Remainder of diet comprises other grasses, fruit, flowers and some fungi, the proportion of which varies seasonally, e.g., in Ranomafana National Park during March–April ripening non-native, invasive Strawberry Guava *Psidium cattleianum* is eaten and fruit can constitute *c.*25% of diet during wet season (Tan *et al.* in press). At other times in Ranomafana, *c.*90% of diet consists of two species of bamboo vine (*Cephalostachyum* spp.). Giant bamboo (*Cathariostachys* spp.) also eaten in smaller quantities (Wright 1986, Overdorff *et al.* 1997, Tan 1999, 2000). Sometimes defecates at latrine sites (Irwin *et al.* 2004).

Young Eastern Grey Bamboo Lemur eating bamboo shoot, Andasibe-Mantadia National Park.

Eastern Grey Bamboo Lemur is territorial: both sexes use scent-marking, vocalisations and direct chasing to defend home range from conspecifics (Overdorff *et al.* 1997, Tan 1999, 2000, Grassi 2002). Other species of bamboo lemur i.e. Golden Bamboo Lemur and Greater Bamboo Lemur, appear to be tolerated in range (Tan 2000, 2006). Home range sizes of 6—10ha (Wright 1986) and 15–20ha (Tan 1999) recorded; differences correlate to amount of bamboo within occupied area. In disturbed forest fragments where bamboo is abnormally abundant home ranges as small as 1–2ha recorded (Irwin *et al.* 2020). During a day's foraging, groups move up to c.450m (Wright 1986). Prefers to sleep in taller trees.

Group size 2–7 (average 4–5), although larger groups up to 11 are seen occasionally (Pollock 1986, Overdorff *et al.* 1997, Tan 1999, 2000). Smaller groups contain breeding pairs with offspring, larger groups one breeding male with multiple breeding females plus their offspring. Females dominant over males when feeding and in other social contexts (Tan *et al.* in press).

Mating occurs in June–July, gestation 137–140 days, and females give birth to single young between late October and November (Wright 1990, Tan 2000), although births have been recorded in January (Tan 2006). Infant carried by mother in her mouth for first two weeks, later clings to fur and rides on her back. When foraging, infant sometimes 'parked' by mother for brief periods. Young take solid food at three weeks, eat bamboo readily by six weeks and fully weaned at four months (Tan 2000, 2006). Both sexes disperse from natal area (Tan 2006).

Predated by Fosa, Madagascar Long-eared Owl, large raptors, e.g., Madagascar Harrier-Hawk, Henst's Goshawk and Madagascar Buzzard (Karpanty 2006, Karpanty & Wright 2007, Andrianandrasana *et al.* 2018) and large snakes such as Madagascar Tree Boa (Goodman & Ganzhorn in press).

WHERE TO SEE Both Andasibe-Mantadia and Ranomafana National Parks are very good places to find this species. In Andasibe, the areas either side of road around Orchid Garden or by bridge over Analamazaotra River (adjacent to old fish ponds) on trail into main reserve can be rewarding. In Ranomafana, the main trails in Talatakely are reliable. In western regions this species is harder to see: search Tsingy de Bemaraha National Park, in areas of dense bamboo near village of Bekopaka, or Tsiombikibo Forest, near Mitsinjo.

NORTHERN GREY BAMBOO LEMUR
Hapalemur occidentalis   Rumpler, 1975

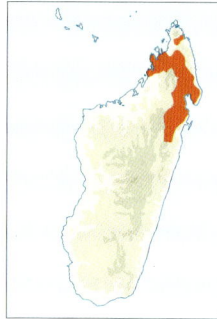

Other names: Northern Grey Gentle Lemur, Northern Lesser Bamboo Lemur, Western Grey Bamboo Lemur, Sambirano Grey Bamboo Lemur Malagasy names: Akomba'valiha, Bekola, Kofiy, Kintrontro

MEASUREMENTS
Head-body: 270–280mm. Tail: 360–390mm. Weight: c.800–1,200g; males c.800–900g, females 900–1,200g. (Schwitzer *et al.* 2013c)

DESCRIPTION & IDENTIFICATION Medium-sized active lemur. Upperparts medium to pale grey-brown, face and underparts noticeably paler than congeners. Ears small and snout short. Overall, very similar to *H. griseus* but larger with more uniform coloration. Only *Hapalemur* with apparent sexual dimorphism: females slightly larger than males. Sympatric with several *Eulemur* species; small size, uniform colour, short muzzle and round face should prevent confusion.

HABITAT & DISTRIBUTION Broad habitat tolerance: dry deciduous, semi-humid transitional and humid forests with stands of bamboo; also, secondary, degraded and human-altered forests and cacao plantations (Mutschler & Tan 2003, Martinez 2008). Range poorly understood and apparently perplexing: name '*occidentalis*' means 'coming from the west'. Previously, distribution of *H. occidentalis* was assumed to be confined to western regions (Harcourt & Thornback 1990). Later genetic analysis prompted re-evaluation (Rumpler *et al.* 2002), with populations in north-east rainforest now regarded as *H. occidentalis* (formerly *H. griseus*) and populations of *Hapalemur*, previously regarded as '*occidentalis*', in west between Betsiboka and Manambolo Rivers, now assigned to *H. griseus ranomafanensis*.

As currently understood, range of Northern Grey Bamboo Lemur includes Ampasindava Peninsula, Sahamalaza region, Ambato Massif and Sambirano region in north-west, from Andranomalaza River

Northern Grey Bamboo Lemur feeding on fruit, Marojejy National Park.

in south to Mahavavy du Nord River in north (Raxworthy & Rakotondraparany 1988, Goodman & Schütz 2000, Mutschler 2000, Rakotoarinivo *et al.* 2011). Also occurs in Ankarana and Analamerana (Hawkins *et al.* 1990, Rakotonirina and King 2014). In eastern areas recorded in Marojejy, Anjanaharibe-Sud, COMATSA forest corridor between these areas and Tsaratanana, onto Masoala Peninsula, and south in eastern rainforests to Onibe River (Zahamena region) (Eppley *et al.* 2020c). South of Onibe River is replaced by Eastern Grey Bamboo Lemur.

In addition to habitat loss and fragmentation, Northern Grey Bamboo Lemur suffers direct persecution from hunting, retaliation for raiding crops and collection for the illegal pet trade (Borgerson 2015, Reuter *et al.* 2016, Brook *et al.* 2019, Eppley *et al.* 2020d). IUCN Vulnerable.

BEHAVIOUR Information mainly from short studies or anecdotal observations. Group sizes between one and four recorded in Manongarivo (Raxworthy & Rakotondraparany 1988, Goodman & Schütz 2000); elsewhere in Sambirano region groups of up to six seen (Mutschler 2000, Mutschler & Tan 2003). In eastern forests group sizes of 3–10 recorded (Patel *et al.* 2014).

Probably cathemeral, although activity patterns may vary seasonally. In secondary forest on Masoala Peninsula, 37% of time was spent feeding, with peak activity periods 05.00–08.00 and 17.00–21.00 (Martinez 2008). During wet season, in Bemaraha and Manongarivo active by day (Raxworthy & Rakotondraparany 1988, Garbutt 2007), during dry season (July–September) observations in both Bemaraha and Sambirano region suggest it is more

Northern Grey Bamboo Lemur.

cathemeral, when activity begins *c*.16.30 with vocalisations (not heard by day) and continues well into hours of darkness (Mutschler 2000).

Forages at all levels of forest, mainly canopy and understorey vegetation, but occasionally on ground (Garbutt 2007). Diet contains high proportion of bamboo: in Tsaratanana and Marojejy several species recorded in diet including *Ochlandra capitata*, *Phyllostachys aurea* and *Dendrocalamus giganteus*. (Schwitzer *et al.* 2013c); in Bemaraha also feeds on liana flowers (Mutschler & Tan 2003). In eastern regions, in semi-agricultural and degraded areas three bamboo species dominate, *Valiha diffusa*, *Bambusa vulgaris* and *Dendrocalamus giganteus*, and all feature in diet of *H. occidentalis*. In studies at Antanetiambo, a degraded forest reserve near Andapa, non-native Chinese Dwarf Bamboo *Phyllostachys aurea* formed *c*.80% of diet, with new stalks, shoots and leaf bases particularly targeted. Also soil consumption (geophagy) is important (Patel *et al.* 2014).

Other aspects of behaviour not known, but probably similar to Eastern Grey Bamboo Lemur.

WHERE TO SEE Most accessible sites are in east. Often seen in Marojejy National Park, especially around Camp I (Mantella), or Antanetiambo Nature

Reserve near Andapa. Also, Farankaraina Reserve near Maroantsetra or peripheral areas of Masoala National Park (where bamboo is common). In north-west, best looked for along Sambirano River valley, south-east of Ambanja. Also known from Ambato Massif, and Nosy Faly Peninsula, north of Ambanja.

SOUTHERN GREY BAMBOO LEMUR *Hapalemur meridionalis*

Warter, *et al.* 1987

Other names: Southern Grey Gentle Lemur, Rusty-grey Lesser Bamboo Lemur, Southern Lesser Bamboo Lemur
Malagasy name: Halo

MEASUREMENTS
Head-body: *c*.260–300mm.
Tail: *c*.320–370mm. Weight: *c*.900–1,300g. (Eppley *et al.* 2015, Eppley 2021)

DESCRIPTION & IDENTIFICATION Medium-sized species; smallest diurnal lemur within its range. Coloration broadly similar to Eastern Grey Bamboo Lemur, but slightly darker and more reddish. Upperparts grey to olive-grey with variable russet areas around shoulders and on back. Face and underparts paler grey. Tail darker grey and slightly short (versus other *Hapalemur* spp.) (*c*.100–110% head-body). Ears relatively large, but still hidden by fur (Groves 2001).

Throughout most of range, sympatric with only one other diurnal species, Red-collared Brown Lemur *Eulemur collaris*, with which it cannot be confused. In humid forest of Andringitra National Park (see below), potentially sympatric with two other bamboo lemurs: distinguished from Golden Bamboo Lemur by smaller size and grey, rather than golden-brown, coloration, whilst Greater Bamboo Lemur has broadly similar coloration, but is noticeably larger with prominent whitish ear-tufts.

HABITAT & DISTRIBUTION Littoral forest, sub-tropical lowland and montane rainforest, secondary forest and degraded forest, generally with, but not limited to, dense stands of bamboo, from sea level to *c*.1,600m. Originally described from Mandena littoral forest, north of Tolagnaro. Range extends from southern extreme of rainforest region, including

Mandena Conservation Area, Andohahela National Park (Anosyenne Mountains), Ambatotsirongorongo Protected Area, Tsitongambarika Protected Area (Vohimena Mountains), Manangotry corridor, and north to at least Mananara River (including Midongy-Sud National Park and Kalambatritra Special Reserve) and possibly as far north as Andringitra National Park. Specific identity of grey *Hapalemur* in Andringitra needs clarification, as boundary with *H. griseus ranomafanensis* is unclear. Andringitra may be a zone of hybridisation. Also, possible range of Southern Grey Bamboo Lemur extends as far north as Ranomafana National Park (Eppley *et al.* 2011, 2015a, 2020g, Nguyen *et al.* 2013, Rakotonirina *et al.* 2017, Donati *et al.* 2020a). In Tsitongambarika recorded density of *c.*18 individuals/km^2. Threatened by habitat loss and hunting: often targeted as it raids crops near forest edge (Campera *et al.* 2019b). Also occasionally kept illegally as pets (Reuter & Schaefer 2017, Reuter *et al.* 2019). IUCN Vulnerable.

BEHAVIOUR Studied intensively in Mandena littoral forest, where it lives in social groups of 2–9 individuals, which contain one or two breeding adult females and one or two adult males. Females are dominant and initiate groups movements. Home ranges variable, from *c.*6.5ha to 18ha (Eppley *et al.* 2015b, 2016c, 2017d,e).

Southern Grey Bamboo Lemur is primarily cathemeral throughout the year; peaks of activity occur after sunrise and prior to sunset. There is increased nocturnal activity during periods of strong moonlight (Eppley *et al.* 2015b). Spends roughly equivalent periods of time feeding and resting (*c.*42% each), and remainder travelling (Eppley *et al.* 2011). Cathemeral behaviour may help reduce predation or be linked to maintaining specific dietary needs (Eppley *et al.* 2017d).

Adaptable and can persist in degraded habitats (Eppley *et al.* 2017a). Consequently, diet exhibits greater breadth than other bamboo lemurs, with >70 species recorded (Eppley *et al.* 2016a). In Mandena littoral forest Southern Grey Bamboo Lemur grazes (Eppley & Donati 2009): terrestrial grasses (Poaceae) and reeds (Cyperaceae) (rather than bamboo) dominate the diet and comprise *c.*75%, with nearly half of foraging time focused on Old World tropical grass *Panicum parvifolium* (Poaceae). The entire upper part of the plant (culm, leaves, grains) is consumed (Eppley *et al.* 2011). Other foods include herbaceous plants, fungi, liana stems, leaves and significant amounts of fallen fruit

Southern Grey Bamboo Lemur, Mandena.

(*c.*34 species, exceptionally high for folivorous primate) (Eppley *et al.* 2016a).

Largely feeding on terrestrial grasses, Southern Grey Bamboo Lemur spends *c.*66% of time foraging on the ground in more open marsh/swamp areas where these species flourish (rather than surrounding forest), with this feeding pattern more prevalent in cooler, winter months (June–July). Terrestrial feeding (in Mandena) may be a response to habitat alteration and limited availability of more regular foods (Eppley *et al.* 2016a). In more intact forest areas (e.g., Andohahela National Park and Tsitongambarika Protected Area), Southern Grey Bamboo Lemur does feed more regularly on woody and liana bamboos (Feistner & Schmid 1999, Donati *et al.* 2020a).

Defecates at visually conspicuous communal latrine sites, which groups use in turn, with adults going first followed by juveniles. Precise function is not clear, but likely serves a communicatory and mate-guarding function, and possibly helps demarcate territory (Eppley *et al.* 2016c).

When resting, in degraded littoral forest, often selects larger trees and individuals may huddle together, which is thought to aid thermoregulation, especially during prolonged resting bouts (Eppley *et al.* 2017c). In addition, huddling behaviour increases during breeding season, which may allow males to gain favour with receptive females (Eppley *et al.* 2017e).

In common with Eastern Grey Bamboo Lemur, mating probably occurs in June and July, as births have been noted in November. Twins also recorded (Eppley *et al.* 2016b).

Predators include Fosa, Ring-tailed Vontsira, large raptors and large snakes such as Dumeril's Boa *Acrantophis dumerili* (Eppley & Ravelomanantoa 2015, Eppley *et al.* 2016a). In some areas feral dogs may also pose a threat (Eppley *et al.* 2016a).

In Mandena, there is an extraordinary record of a Southern Grey Bamboo Lemur family unit accepting a lone female Ring-tailed Lemur, with the two species showing significant levels of behaviour overlap and mutual understanding, including the female Ring-tailed Lemur attending and transporting an infant bamboo lemur (Eppley *et al.* 2015c).

WHERE TO SEE Mandena Conservation Zone (part of QMM Titanium Dioxide Development Project), 12km north of Tolagnaro. Alternatively, Nahampoana Private Reserve and *Domaine de la Cascade* of Manantantely both near Tolagnaro, provide relatively easy viewing. Also recorded (a group of four) in forest fragments around summit of Pic St. Louis outside Tolagnaro (Garbutt 2007). The more intrepid should visit Malio, the rainforest area of Andohahela National Park (parcel 1). This is beautiful forest; the area should become more accessible as ecotourism develops.

ALAOTRA REED LEMUR
Hapalemur alaotrensis   Rumpler, 1975

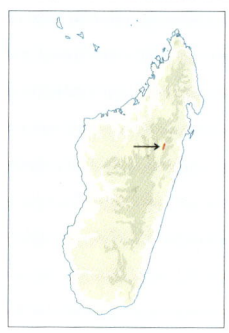

Other names: Lac Alaotra Reed Lemur,
Lac Alaotra Gentle Lemur
Malagasy name: Bandro

MEASUREMENTS
Head-body: *c.*380–400mm.
Tail: *c.*390–410mm. Weight: *c.*1.1–1.6kg (mean 1.24kg).
(Mutschler *et al.* 2000, Waeber & Hemelrijk 2003, Rabarivola *et al.* 2007)

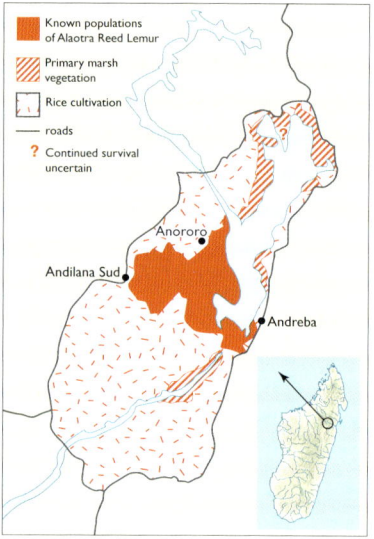

DESCRIPTION & IDENTIFICATION Largest of 'grey' bamboo lemurs, Golden Bamboo Lemur is only larger congener. No size difference between sexes (Mutschler *et al.* 2000, Waeber & Hemelrijk 2003). Pelage notably dense and woolly. Upperparts medium to darkish grey-brown, face and underparts paler grey. Crown and nape chestnut-brown, gradually fading over shoulders. Head rounded, ears small and unobtrusive. Medium-sized with rotund body and long tail. In very restricted range and unusual habitat, Alaotra Reed Lemur cannot be confused with any other species. Mouse lemurs, probably Simmons's Mouse Lemur, also occur in reedbeds.

HABITAT & DISTRIBUTION Unique among primates as exclusively inhabits reed and papyrus beds; the only primate permanently living in a wetland.

Alaotra Reed Lemur, reedbeds, Lac Alaotra.

Confined to marshes of Lac Alaotra (largest in Madagascar), dominated by *Cyperus madagascariensis* (Cyperaceae) and *Phragmites communis* (Poaceae), in centre-east up to 750m. Area around lake hugely important for agriculture: 120,000ha of rice fields, considered 'Rice Bowl' of Madagascar, as such exerts massive pressure on remaining marsh habitat (Ratsimbazafy *et al.* 2013, Waeber *et al.* 2019b). Alaotra Reed Lemur was probably once present around entire lake and also occurred near Andilamena, at least 35km to north (but disappeared in 1950s when marshes were converted to rice paddies) (Mutschler & Feistner 1995).

Range further restricted by lake drainage and destruction of habitat, especially by fire. Stronghold is a discontinuous area of primary marsh south-west of Lac Alaotra, largely within an area bounded by villages of Anororo, Andilana-Sud (in west), Ambodivoara and Andreba Gare (in east). Second smaller isolated population in fragment of marsh around Belempona Peninsula on north-west shore, but its continued existence is uncertain (Ralainasolo *et al.* 2006, Reibelt *et al.* 2019).

Survival severely threatened by marsh burning for conversion to rice paddies and access to fish ponds (Mutschler *et al.* 2001, Ralainasolo *et al.* 2006, Copsey *et al.* 2009a,b, Ratsimbazafy *et al.*

2013) and to lesser extent by hunting and capture for pets (Reuter & Schaefer 2017, Borgerson *et al.* 2018). Has one of most restricted ranges of any lemur (<220km^2); only around 2,000–2,500 animals remain, with numbers still declining (Reibelt 2017). IUCN Critically Endangered.

BEHAVIOUR Lives in groups of two to nine (mean *c.*3–5) comprising one or two breeding females, their offspring, and one or two breeding males (Nievergelt *et al.* 2002a). Occasionally groups of 13 or more recorded. Within a group, females are dominant and initiate aggressive encounters and group movements, although males lead channel crossings more often (Waeber & Hemelrijk 2003). Around 35–40% of groups contain two related breeding females and up to three adult males, although only one breeds. Solitary individuals of either sex encountered infrequently. Breeding females at core of group structure, whereas reproductive males transfer between groups relatively often. Mating system appears variable, from serial monogamy to polygyny within social unit (Mutschler *et al.* 2000, Nievergelt *et al.* 2002a, Mutschler & Tan 2003).

Alaotra Reed Lemur is territorial: groups occupy exclusive home ranges of *c.*1–8ha (normally *c.*2–5ha), a proportion of which may be water that

Alaotra Reed Lemur, reedbeds, Lac Alaotra.

varies seasonally in extent (Nievergelt *et al.* 1998). Group encounters are frequent; home ranges defended by vocalisations, scent-marking and display behaviour, with males tending to be more active in these roles. Group cohesiveness maintained by allo-grooming, where subordinates groom dominant individuals (generally females) and animals sit facing one another, divide partner's fur with their hands and groom with tooth comb (modified lower front dentition) (Mutschler 1999a,b, 2000, Waeber & Hemelrijk 2003).

Movement and locomotion unusual; quadrupedal, where animal walks up reed stem until it bends over, then walks along it to reach next stem. Technique allows them to cross narrow water channels: wider channels are traversed by jumping in conventional bamboo lemur fashion. Also, able to swim short distances across open water in 'dog-like' manner (Waeber *et al.* 2019b).

Cathemeral (active at any time in 2-hour cycle); principally active during daylight, with significant bouts of nocturnal activity (lasting 30 minutes or longer) each night. Main periods of activity concentrated around first and last three hours of daylight (Mutschler *et al.* 1998).

Diet strictly folivorous. Food diversity low, with 95% of feeding on four species: major elements are pith of papyrus stems *Cyperus madagascariensis* (Cyperaceae) and three species of Poaceae, young reed shoots *Phragmites communis* and surface grasses *Echinochloa crusgalli* and *Leersia hexandra* (Mutschler 1999a,b). Also plant part selective: February–April, they feed predominantly on pith of young papyrus and new shoots of reeds, later in year diet shifts towards older leaves. Diet of low quality, so lemurs remain active and feed throughout day.

Also spends considerable periods foraging at ground level, particularly during dry season (June–October) when more solid ground is exposed. During wet season, water levels rise but lemurs still descend onto floating vegetation (Mutschler *et al.* 1998).

Births occur between September and February; singletons are normal, but twins relatively common (*c.*35% of births) (Mutschler *et al.* 2000, Mutschler & Tan 2003). Gestation is *c.*135–145 days. Young relatively well developed and initially carried on mother's back. By 6–8 days start to move independently. From one to two weeks infants are 'parked' by mother in dense thickets of papyrus and left for up to 30 minutes. By three weeks this stops and young travel with mother. Females can swim across open channels while carrying offspring. Both sexes disperse from natal area: females leave before maturity and may breed for first time at two years, males delay departure until maturity and first breed when at least three years old (Mutschler *et al.* 1998, 2000).

No direct predation observed (other than by humans). Potential threats: large snakes, large raptors and possibly Ring-tailed Vontsira and Brown-tailed Vontsira *Salanoia concolor* (Waeber *et al.* 2019a).

WHERE TO SEE Not an easy species to see. Best looked for in rainy season (November–April) when water levels are higher allowing canoes to pass through reedbeds; necessary to be on the water before dawn to maximise chances when lemurs are active. Visit Park Bandro, a collaborative community association (VOI) project based near Andreba Gare on south-east shore of Lac Alaotra, where the 'Bandro' is a flagship species for local community conservation (Rendigs *et al.* 2015, Reibelt *et al.* 2017, Waeber *et al.* 2018). The modest-sized (85ha) park is home to *c.*150 individuals (Waeber *et al.* 2019b). Basic facilities at Camp Bandro, and local guides available for boat trips (Reibelt 2017).

GOLDEN BAMBOO LEMUR
Hapalemur aureus   Meier, *et al.*, 1987

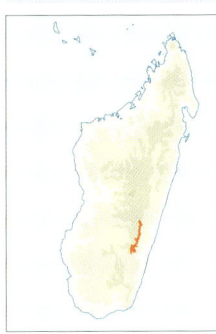

Malagasy names:
Bokombolomena,
Varibolomena

MEASUREMENTS
Head-body: *c.*340–380mm.
Tail: *c.*380–420mm.
Weight: *c.*1.25–1.65kg.
(Tan 1999, 2000)

DESCRIPTION & IDENTIFICATION Medium-sized, but largest *Hapalemur*. No size difference between sexes. Pelage dense and soft. Upperparts rich olive to russet/chestnut, with crown, nape, shoulders, dorsum and upper tail slightly darker. Tail darkens towards tip. Underparts, inner limbs and underside of tail golden-brown to ginger. Outer face, cheeks and chin golden-brown to ginger. Inner face around eyes dark brown, extending around muzzle. Paler areas above eyes. Nose may be more pinkish. Ears small without tufts and golden-brown.

Depending on location, sympatric with two other *Hapalemur*: larger than Eastern Grey Bamboo Lemur in Ranomafana, and Southern Grey Bamboo Lemur in Andringita, but smaller than Greater Bamboo Lemur in both localities. Golden-brown coloration distinctive and coupled with size differences should prevent confusion with these species.

HABITAT & DISTRIBUTION Primary mid- and higher-altitude rainforest with abundant bamboo, particularly Madagascar Giant Bamboo *Cathariostachys madagascariensis*, from *c.*600m (Ranomafana) to *c.*1,600m (Andringitra) (Arrigo-Nelson & Wright 2004). Known range extends from northern Ranomafana National Park, south through highly fragmented COFAV (Fandriana Vondrozo Midongy) forest corridor to north-east slopes of Andringitra Massif (Sterling & Ramaroson 1996, Lehman & Wright 2000, Goodman *et al.* 2001, Rakotondravony & Razafindramahatra 2004, Irwin *et al.* 2005). Indirect evidence alludes to range extending from about south of Onive River (Ambohimiadana) to the Befotaka-Midongy du Sud Protected Area (Marovato) (Goodman *et al.* 2018a). In range, distribution very patchy and occurs at very low densities, *c.*0.2 individuals/km² (Herrera *et al.* 2011, Wright *et al.*

Golden Bamboo Lemur eating bamboo, Ranomafana National Park.

Golden Bamboo Lemur, Ranomafana National Park.

2012). Extremely vulnerable to habitat fragmentation, in part due to heavy reliance on patchily distributed Madagascar Giant Bamboo (Eppley *et al.* 2020d). Total range less than *c.*2,500km², with perhaps fewer than 1,000 individuals remaining, the majority in Ranomafana National Park. IUCN Critically Endangered.

BEHAVIOUR Principally diurnal. Activity begins before dawn (*c.*05.00) continuing to mid-morning, with distinct rest period in middle of day, between 09.00 and 13.00, before foraging resumes until around dusk (*c.*19.00). Activity periods alternate between feeding and resting at *c.*30-minute intervals. Beds down just after dusk. Nocturnal activity occasional (Tan 1999, 2000, Mutschler & Tan

2003). Across annual cycle, overall proportion of time concerning major activities is 50–60% resting, 22–37% feeding and 8–14% travelling, depending on season.

Group size two to eight, three to four is the norm, consisting of monogamous pair, with slightly smaller subadults and juveniles. Home range 10–30ha (mean 26ha), with all areas used on weekly cycle (cf. Eastern Grey Bamboo Lemur that favours a core area). Mean daily travel distances *c.*500–600m (coverage *c.*275–950m) (Wright & Randriamanantena 1989, Tan 1999, 2006, Tan *et al.* in press).

Highly vocal; at least two distinct calls identified. First, a quiet, resonant *wuulp* ('sigh-miaow') with inquisitive inflection that is probably a contact call, as often issued when feeding in dense bamboo. Second, a loud, sharp, staccato, 'guttural honk' repeated on slowing, descending scale, decreasing in volume. Only one individual (a male) gives this call, sometimes heard in conjunction with *wuulp* call from others in group. Honk call generally heard around dusk or occasionally dawn, and is probably territorial.

Diet almost exclusively bamboo (*c*.90%), especially endemic Madagascar Giant Bamboo. Also eats bamboo creeper and bamboo grass. Highly plant part selective: bases of young leaves and new branch shoots preferred, which probably reduces competition, particularly with Greater Bamboo Lemur that prefers ground shoots and inner pith (Tan 1999, 2006). Branch shoots and young leaves are protein-rich but also contain high concentrations of cyanide toxins. Golden Bamboo Lemur may ingest quantities of cyanide 10–50 times the lethal dose for other mammals (Glander *et al.* 1989, Ballhorn 2016). They metabolise and detoxify the cyanide, with the help of high levels of protein in their diet. The proteins are broken down into amino acids to produce the enzyme rhodanese, which acts as catalyst in converting toxins to non-lethal thiocyanate, and is then excreted in urine (Yamashita *et al.* 2010, Eppley *et al.* 2017b).

Females may have two oestrous cycles per year. Mating takes place at night or dawn, in July–August. Births occur in November and December (start of wet season) after a gestation of *c*.138 days. Single infants are the norm (one twin birth witnessed). Young altricial: during first two weeks, infants shelter in dense tangles of vegetation; when moving carried by mother in her mouth. When slightly older infant is 'parked' by mother while she forages up to 250m away; bouts of 'parking' may last up to *c*.3 hours. Siblings may also care for infants. When strong enough, infant carried by mother as she forages. Infants first move away from mother at two weeks and begin to show interest in solid foods at *c*.6 weeks. Start ingesting bamboo and other plants at *c*.10 weeks. Weaning occurs before six months. Young remain in natal group for three years pre-independence (Tan 1999, 2006, Tan *et al*. in press).

WHERE TO SEE Ranomafana National Park is only accessible locality. The species' discovery brought Ranomafana to the world's attention and prompted the creation of the park in 1991. The best area is Talatakely (the main tourist centre), around stands of giant bamboo adjacent to main trail. Allow at least two or three days. It is essential to seek the assistance of a local guide.

GREATER BAMBOO LEMUR
GENUS *Prolemur*

Previously included in genus *Hapalemur*: based on distinctive chromosomal and cranio-dental features and behavioural aspects, the Greater Bamboo Lemur is now placed in monotypic genus *Prolemur* (Groves 2001). Like other bamboo lemurs, it has a broadened and short muzzle, but is significantly larger than *Hapalemur* species (Albrecht *et al*. 1990).

GREATER BAMBOO LEMUR
Prolemur simus   (J. E. Gray, 1871)

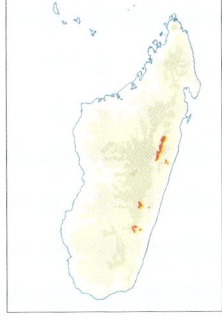

Other names: Greater Gentle Lemur, Broad-nosed Bamboo Lemur, Broad-nosed Gentle Lemur

Malagasy names: Godrogodroka, Godroka, Varibolo, Varikovoka, Reniben'ny Kotrika, Kotrikabe, Halogodro, Bokombolobe, Alakoto, Halokoto, Talembarika

MEASUREMENTS
Head-body: *c*.400–420mm. Tail: *c*.450–480mm. Weight: *c*.2.2–3.5kg, males heavier than females. (Meier *et al*. 1987, Kappeler 1991, Frasier *et al*. 2015)

DESCRIPTION & IDENTIFICATION Largest, most robust bamboo lemur. Males heavier than females, but size difference not obvious in field (Frasier *et al*. 2015). Pelage short and dense. Upperparts and tail sooty grey-brown, with russet tones. Head, neck, shoulders and upper arms more olive-brown. Noticeable chestnut-brown patch on pygal region. Tip of tail very dark grey. Underparts and throat paler creamy-brown. Inner face, nose-bridge and muzzle dark grey; areas just above eyes pale grey. Prominent pale grey to white ear-tufts distinctive. Some individuals on Andringitra Massif are predominantly deep golden-red to russet, with large ears and reduced ear-tufts, and may represent a distinct colour morph (Sterling & Ramaroson 1996). Medium-large, rotund lemur. At different locations sympatric with Northern Grey Bamboo Lemur, Eastern Grey Bamboo Lemur, Southern Grey Bamboo Lemur and Golden Bamboo Lemur. Appreciably larger size and distinctive ear-tufts of Greater Bamboo Lemur should prevent confusion.

Greater Bamboo Lemur eating Giant Bamboo, Ranomafana National Park.

HABITAT & DISTRIBUTION Lowland and montane primary and secondary humid forest associated with large-culmed bamboo species, from near sea level to c.1,600m (Wright *et al.* 2012, King *et al.* 2013b, Ravaloharimanitra *et al.* 2020). Prefers primary habitats, but can survive in degraded areas and plantations with reduced disturbance, provided large-culmed bamboo remains (Olson *et al.* 2013).

Historically (>60,000 years ago) widespread around island and probably very common. Two major population declines followed; one caused by climate change with significant warming shifting botanical diversity and causing localised disappearance of giant bamboos (Eronen *et al.* 2017), and the second caused primarily by human-induced factors, habitat destruction and hunting (Hawkins *et al.* 2018).

Most abundant lemur in subfossil record, with remains found at sites across west, north-west, north, centre and east (Godfrey & Vuillaume-Randriamananatena 1986, Godfrey & Jungers 2003); evidence found at Ampasambazimba east of Antananarivo, Montagne des Français and Ankarana Massif in north, Grottes d'Anjohibe north-east of Mahajanga, and Tsingy de Bemaraha National Park and Maintirana in west (Godfrey *et al.* 2004a, Burney *et al.* 2020). By late 19th century, range reduced to eastern rainforest and by late 20th century was thought to be restricted to handful of localities in south-east, and on verge of extinction (Mutschler & Tan 2003).

Extensive fieldwork over last 20 years has revealed a broader distribution. Now known from a number of forest fragments from approximate latitude of Toamasina, in Ankeniheny-Zahamena Forest Corridor in north, to region north of Vondrozo but south of Andringitra Massif in south. This includes several sites north of Andasibe-Mantadia National Park (e.g., Torotorofotsy) in Ankeniheny-Zahamena Forest Corridor (CAZ), lowland rainforests of Andriantantely, sites near Brickaville in lowland degraded landscapes, areas near confluence of Mangoro and Nosivolo Rivers, east of Ranomafana National Park around Kianjavato and Ifanadiana, and vicinity of Evendra and Karianga near Vondrozo (Dolch *et al.* 2004, 2008, 2010, Wright *et al.* 2008, Delmore *et al.* 2009, King & Chamberlan 2010, Rakotonirina *et al.* 2011, 2013, Ravaloharimanitra *et al.* 2011, 2020, Bonaventure *et al.* 2012, Lantovololona *et al.* 2012, Olson *et al.* 2012, Andrianandrasana *et al.* 2013, King *et al.* 2013a,b,c,d, Randriahaingo *et al.* 2014, Randrianarimanana *et al.* 2014, Ralison *et al.* 2015). In addition to these sites, also occurs in Ranomafana and Andringita National Parks. Range may extend further, as distinctive feeding signs (shattered and stripped giant bamboo) recorded in Zahamena National Park in north, and Midongy du Sud National Park and Kalambatritra Special Reserve in south (Rakotonirina *et al.* 2011, 2013, 2017c).

Continued threats include forest clearance and fragmentation, climate change and hunting. Since 2008, community-based conservation has resulted in known populations increasing from fewer than 100 individuals to >1,000 individuals. Across all fragmented populations, perhaps c.1,500 individuals remain: many subpopulations likely declining, whereas those associated with community conservation projects are thriving. IUCN Critically Endangered.

BEHAVIOUR Activity patterns summarised as cathemeral: more active by day, but regular bouts of nocturnal activity occur year-round. Group size and composition appears variable; smaller groups comprise a monogamous pair with offspring, whereas larger multi-male, multi-female groups, contain up to 30 and sometimes as many as *c.*50–90 individuals (Mihaminekena *et al.* 2020, Mihaminekena & King 2021) and are polygynous (Tan 2000, Frasier *et al.* 2015). Unusual amongst lemurs, males appear dominant (Roullet 2012).

In Ranomafana National Park occupies large home ranges, 40–60ha, although groups tend to remain in *c.*20% of range for up to two weeks before moving on, presumably in response to diminished resources (Tan 1999, 2000). Mean daily distance travelled *c.*650m (range *c.*160–1,560m) (Tan *et al.* in press).

At other sites, e.g., Torotorofotsy, home ranges of several hundred ha recorded (Dolch *et al.* 2008, Ravaloharimanitra *et al.* 2020). Home ranges size may vary seasonally, being larger in austral summer (December–March) (Mihaminekena *et al.* 2018) and can overlap with ranges of other bamboo lemurs (*Hapalemur* spp.), with mixed-species associations not uncommon (both *Hapalemur* and *Eulemur* spp.).

Greater Bamboo Lemur is very vocal, with wide repertoire that includes purrs, grunts, squeals and hums. These are thought to arrange into seven general types, with different ones associated with aggression, contact and group cohesion. Alarm calls are low-pitched roars (*grraaa*) that decrease in intensity and may morph into two-phased *ouik-grraaa* that resembles a duck quacking (Schwitzer *et al.* 2013d).

Diet almost exclusively large-culmed bamboo, particularly Madagascar Giant Bamboo at mid- and high-elevation sites (above 700m) and large woody bamboos, *Valiha diffusa* (endemic) and *Bambusa vulgaris* (pantropical) in lowland secondary habitats (Bonaventure *et al.* 2012, Lantovololona *et al.* 2012, Mihaminekena *et al.* 2012, King *et al.* 2013b). These constitute up to 95% of diet. Selects different parts of Madagascar Giant Bamboo to Golden Bamboo Lemur, concentrating on ground shoots and inner pith, so diet contains considerably fewer cyanide-toxins (Eppley *et al.* 2017b).

During rainy season (mid-November to March) often descends to ground when foraging, especially in January/February (>20% of foraging time), when new giant bamboo shoots erupt from ground and represent 98% of diet (Tan 2000, 2006, Frasier *et*

Greater Bamboo Lemur eating Giant Bamboo, Ranomafana National Park.

al. 2015, Mihaminekena *et al.* 2018). After March, when ground shoots decline, leaves and branch shoots eaten increasingly. At some sites, between July and November Greater Bamboo Lemur relies heavily on inner pith (Tan 1999, 2000). Other foods include flowers of Traveller's Tree *Ravenala madagascariensis*, palm fruits (*Dypsis* spp.), fruits of *Artocarpus*, *Ficus* spp. (Moraeceae) and leaves of *Pennisetum clandestinum*, *Aframomum angustifolium* and *Rubus alceifolius* (Meier & Rumpler 1987, King *et al.* in press). Also reported eating various agricultural crops, e.g., Mango *Mangifera indica*, lychee *Litchi sinensis*, custard apple *Annona squamosa*, jackfruit *Artocarpus heterophyllus*, breadfruit *A. altilis*, banana *Musa* sp. and sugar cane *Sacharum* sp., which causes conflict with human populations (King *et al.* 2013).

Shredded Giant Bamboo: tell-tale feeding evidence of Greater Bamboo Lemur.

Greater Bamboo Lemur has powerful jaws, adept at stripping outer layers from live stalks and breaking through bamboo poles to reach pith. This technique is destructive: ravaged, shattered stands of large-culmed bamboo are clear evidence of this species' presence.

Breeds annually. At mid-elevation locations (900–1,100m), e.g., Ranomafana and Ankeniheny-Zahamena Forest Corridor (CAZ), mating occurs in May–June, gestation is c.148–150 days and females give birth in October–November (Tan 1999, Randrianarimanana et al. 2012). At lower elevations (c.200m), e.g., Kianjavato, sequence around one

month earlier: mating in April–May, births in September–October (Frasier et al. 2015).

Females may remove themselves from group for up to six days prior to birth; single offspring is norm, twins reported rarely (Tan 2006, Randriahaingo et al. 2014, Frasier et al. 2015). Initially, infants completely dependent on mother and during first month they maintain almost constant body contact. During second month, contact reduces, infant begins to venture away and starts manipulating items (food and non-food) (Mihaminekena et al. 2020). Unlike *Hapalemur* spp., infants are never 'parked': instead, mothers continuously carry infants for around 3–4 months. From three months, manipulation of consumable items (young bamboo stalks and leaves) increases. Around four months begin to move independently, coinciding with period of increased terrestriality (January/February) that correlates with eruption of new bamboo shoots from ground (Tan 2000, Mihaminekena et al. 2020). Weaning occurs between eight and ten months, but solid bamboo first taken at c.2 months. Females may remain in natal area and begin breeding at three years old. Males disperse between three and four years, when tensions with father increase (Tan 2006).

Infant mortality can be high (c.50%), with majority of deaths coinciding with period of first independent movement, presumably due to increased vulnerability of predation. Recorded predators include large raptors, e.g., Henst's Goshawk and Fosa (Karpanty & Wright 2007, Wright et al. 2012, Frasier et al. 2015). No evidence of infanticide (Frasier et al. 2015).

WHERE TO SEE Ranomafana National Park is the most accessible locality. However, in recent years the population of Greater Bamboo Lemurs has declined and they are now more challenging to see around the main trail network in Talatakely. Enlisting the help of a park guide is essential. Alternatively, c.65km east of Ranomafana is Kianjavato Ahmanson Field Station, the site of an ongoing Greater Bamboo Lemur project. The population here has increased in recent years, with several groups now habituated. Visits must be by prior arrangement. Near Andasibe, can be seen at Torotorofotsy, adjacent to Mantadia National Park. Also easily seen, sometimes in large groups, at several community-managed sites in Brickaville District supported by the Aspinall Foundation. Access to most sites is challenging and visits should be arranged in advance.

True Lemurs Genera *Lemur*, *Eulemur* and *Varecia*

The informal grouping often referred to as 'true lemurs' or 'typical lemurs' includes some of the most familiar Malagasy primates. Three genera are recognised: *Lemur*, *Eulemur* and *Varecia* (Groves & Eaglen 1988, Simons & Rumpler 1988). They are medium to medium-large lemurs that adopt a horizontal body posture and move and leap quadrupedally. All are skilled climbers, but some also spend time on the ground to varying degrees, especially Ring-tailed Lemur. Previously, they have been regarded as principally diurnal, but it is now clear the majority can be active for discrete periods throughout the 24-hour cycle, i.e., they are cathemeral. This behaviour is particularly prevalent across *Eulemur* species, but is also known in genus *Lemur*. It has yet to be recorded to any appreciable extent in ruffed lemurs *Varecia* spp.

RING-TAILED LEMUR

GENUS *Lemur*

A genus containing a single species. Arguably Madagascar's most familiar inhabitant, the Ring-tailed Lemur is *the* flagship species, an icon synonymous with its island home. Paradoxically, the species is far from being typical: it is the most terrestrial of Madagascar's primates, lives in the largest social groups and tolerates a variety of extreme habitats that no other lemurs can (Jolly 2003). Alarmingly, there is increasing evidence that it is now amongst the most threatened of all lemurs (Gould & Sauther 2016, Sauther & Cuozzo in press).

At one time the genus *Lemur* also contained all 'true' lemur species now assigned to the genus *Eulemur*, largely because of morphological similarities (e.g., the skeletons of these genera are virtually indistinguishable). However, molecular DNA evidence and scent gland distribution suggest that closer evolutionary affinities may exist between Ring-tailed Lemur and bamboo lemurs (genus *Hapalemur*), consequently the genus *Eulemur* was proposed for other species to differentiate them from *Lemur* (Tattersall 1982, Groves & Eaglen 1988, Simons & Rumpler 1988).

Ring-tailed Lemurs are the most terrestrial of all Malagasy primates, Berenty Private Reserve.

RING-TAILED LEMUR
Lemur catta
Linnaeus, 1758

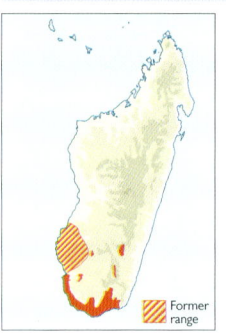

Former range

Malagasy names: Maki, Maky, Hira

MEASUREMENTS
Head-body: 385–455mm.
Tail: 560–625mm. Weight:
*c.*2.0–2.5kg. (Tattersall
1982, Sussman 1991,
Sauther *et al.* 2006, Fardi *et al.* 2018)

DESCRIPTION & IDENTIFICATION Medium-sized lemur, often seen on ground. Long tail, ringed alternately with 13–15 black and white bands ending in black tip, is most distinctive feature. Dorsal areas usually grey to grey-brown; rump, limbs and haunches grey. Underparts are off-white or cream. Neck and crown charcoal-grey, contrasting with white outer face, ears and throat. Large triangular dark eye patches just reach crown. Muzzle also dark. Ears prominent. Both sexes have apocrine and sebaceous glands in their genital regions. Males also have sebaceous glands in the armpit region (that secrete a brown paste) and apocrine glands on wrists, adjacent to a keratinised spur (traits also associated with *Hapalemur* spp.) that secrete clear fluid (Scordato *et al.* 2007).

Individuals from isolated subpopulations on Andringitra Massif vary markedly from typical form above, but are not taxonomically distinct (Yoder *et al.* 2000a). Their fur is denser and woollier in appearance. Dorsal region, rump and limbs noticeably darker and more rufous-brown than typical form. Tail has fewer dark rings (11–12) (Goodman & Langrand 1996a). All are plausible variations brought about by cold climate with high ultra-violet light levels that bleach fur.

Arguably the most distinctive and recognisable lemur; instantly identifiable by its ringed tail. Cannot be confused with any other species.

HABITAT & DISTRIBUTION Known from at least seven distinct habitat types; remarkably adaptable with broadest habitat tolerance of any lemur. Recorded in spiny forest, dry scrub, littoral forest, gallery forest, deciduous forest and degraded landscapes (Sauther *et al.* 1999, Goodman *et al.* 2006b, Gould 2006, Cameron & Gould 2013). Also, rocky outcrops and surrounding

moist fire-resistant forest in Isalo National Park and higher elevations (above treeline) on exposed rocks and subalpine ericoid bush in Andringitra National Park and adjacent areas. Recently discovered population inhabits humid rainforest on boundary of east–west watershed south of Ihosy and east of Betroka (western edges of Kalambatritra) (LaFleur & Gould 2020). Recorded from near sea level to over *c.*2,600m (Andringitra Massif) (Goodman & Langrand 1996a) and tolerates most extreme variation in climatic conditions, from hottest, driest areas, where temperatures reach 45°C (Tsimanampetsotse) to coldest zones, where nighttime temperature may drop to -10°C (Andringitra Massif).

Historic range probably covered much of south-west and south, from near Belo-sur-Mer on west coast through south and south-west regions to Tolagnaro, and north along western edge of east–west watershed ecotone to vicinity of Ambalavao (Goodman *et al.* 2006a). By turn of 21st century habitat increasingly fragmented and population densities very low (Sussman *et al.* 2003). Populations have crashed due to increased habitat loss, hunting and collection for pet trade. Today, range massively restricted and limited to fewer than 35 apparently isolated populations, which maintain reasonably high levels of genetic diversity (Chandrashekar *et al.* 2020), with only eight of these supporting more than 100 individuals and only three sites having >200 animals (Sauther & Cuozzo in press). Combined number remaining in studied populations estimated at fewer than *c.*2,500, with total numbers only a few thousand (Gould & Sauther 2016, LaFleur *et al.* 2017, 2018).

The eight locations are:
- Amoron'i Onilahy Protected Area
- Bezá-Mahafaly Special Reserve*
- Tsimanampesotse National Park
- Isalo National Park
- Anja Community Reserve*
- Andringitra National Park
- The 'Lost Forest', Ihorombe region south of Ihosy (western edges of Kalambatritra)
- Berenty Private Reserve*, including Ifotaka and surrounding forest fragments

* Sites known to support >200 individuals.

Population densities influenced by habitat type and degree of fragmentation, and vary considerably between *c.*6 individuals/ha and *c.*0.02 individuals/ha (Gould & Sauther 2016), e.g., in Anja Community Reserve near Ambalavao populations more crowded (Gould & Andrianomena 2015), in degraded dry

In the Andringitra region, where night-time temperatures are cold Ring-tailed Lemurs have luxuriant fur and bushy tails, Anja Community Reserve.

spiny thicket (Cap Sainte Marie) with no water and shade densities very low (Kelley 2013).

Populations at Anja Community Reserve, Andringitra National Park, Bezá-Mahafaly Special Reserve and Berenty Private Reserve are well protected and relatively secure. Those inhabiting Amoron'i Onilahy Protected Area and Tsimanampesotse National Park vulnerable to hunting and live capture for pet trade, whilst populations in Isalo National Park and 'Lost Forest' are susceptible to fire risk. Previously recorded in Kirindy-Mitea National Park and Zombitse-Vohibasia National Park, but these populations likely now extirpated (Gould & Sauther 2016).

Outside official protected areas, community conservation projects at culturally sacred sites, e.g., Anja and Tsaranoro Valley provide some sanctuary (Gould & Andrianomena 2015, Rafalinirina et al. 2020), but across all remaining populations, habitat loss and fragmentation, consequent genetic isolation, bushmeat hunting and local pet trade remain omnipresent threats (Gardner & Davies 2014, Clarke et al. 2015, Reuter et al. 2016, 2019, LaFleur et al. 2019), whilst increasingly extreme weather patterns and drought also have a negative impact (Fardi et al. 2017), e.g., in Tsimanampetsotse National Park the number of surviving infants drops markedly in drought years (Kasola et al. 2020). IUCN Endangered.

BEHAVIOUR Previously regarded as diurnal, but growing evidence suggests activity is cathemeral. Extent of nighttime activity varies between habitats and with lunar phase; nocturnal activity increases when intensity of moonlight is high (Donati et al. 2013, LaFleur et al. 2014). In extreme environments of south-west (Tsimanampetsotse National Park), nighttime activity occurs primarily in austral spring and summer (September–April) when days are very hot and nights cooler. More time devoted to feeding after dark than during daylight; feeding occupies c.50% of activity after dark and c.40% of during day. This is probably linked to avoiding thermal stress (LaFleur et al. 2014).

Ring-tailed Lemur is semi-terrestrial, spending more time on the ground than any other lemur, c.30% during daylight (Jolly 2003), probably dictated by semi-arid environments, where vegetation is not always continuous and resources may be sparse and unevenly distributed. During bouts of nocturnal activity, they spend only 5% of time on ground, probably to reduce potential for predation (LaFleur et al. 2014). In forest areas, large trees are favoured for sleeping. In Andringitra Massif and Tsaranoro Valley, where populations live in zones with huge rocky outcrops above treeline, Ring-tailed Lemur sleeps in caves (Goodman & Langrand 1996, Goodman et al. 2006, Cameron & Gould 2013). Similarly in extreme environments at edge of Mahafaly Plateau (Tsimanampetsotse National Park) in south-west, or along Mandrare River in south-east, small caves or cliff faces are favoured sleep sites, which may relate to thermoregulation, resource acquisition and predator avoidance (Sauther et al. 2013, Semel & Ferguson 2013).

Ring-tailed Lemurs live in larger groups than any other Malagasy primate: group size ranges from c.6–24 individuals (13–15 average), but >30 recorded (Sauther et al. 1999, Jolly 2003, Sauther & Cuozzo in press). Groups contain equal numbers of adult females and males, plus immatures and juveniles. Group core is matrilineal, mothers, sisters, daughters: males migrate from natal group around maturity (c.3 years), often brothers with brothers, and frequently (every 3–5 years) move from one group to another during adult life. Less often, females also transfer, when mothers and daughters or sisters with sisters move between groups (Jones 1983, Sussman 1992, Sauther & Sussman 1993, Parga et al. 2015, Sauther & Cuozzo in press).

Separate, well-defined and maintained, female and male hierarchies exist within group, where females are completely dominant and an alpha female forms focal point for group (Sauther 1993, Jolly 2003, 2004). Aggression between females establishing dominance in a group can result in serious wounds and ultimately expulsion of subordinates (Vick & Pereira 1989, Jolly et al. 2000). Male rank established through 'stink fights', where individuals anoint their tails with secretions from wrist glands and waft scent in direction of opposing males (Kappeler 1998b), and 'jump fights' in which rival males leap in the air, slashing downward with their upper canines (Sauther & Cuozzo in press).

The 'dominant' male interacts with females more often than other group males, but during breeding season male dominance relationships become blurred, so rank may not translate to increased mating success (Sussman 1989, 1991, Parga et al. 2016). Social and individual grooming is important; individuals groom themselves using 'toilet' claw on hindfoot.

Groups not strictly territorial but do have preferred home range, size of which determined by habitat and resources within. Seasonal changes mean home ranges vary from 6ha to >30ha. Scent-marking important in demarcating ranges. Females and males leave genital smears on saplings and branches (often adopting distinctive handstand position) (Palagi & Norscia 2009), and males use wrist gland armed with horny pad to gouge scent into bark of saplings. High degree of overlap between home ranges of neighbouring groups. When groups meet, dominant females are primarily responsible for defence, confronting members of other groups by staring, lunging approaches and occasional physical aggression. Following agonistic encounters, group retreats towards centre of their home range (Jolly et al. 1993, Sauther et al. 1999).

In addition to olfactory communication, Ring-tailed Lemurs have complex vocal repertoire: >20 different call types identified in adults, plus several others specific to infants (Macedonia 1993). A moderate repertoire compared to other primates, but highly elaborate in comparison with other lemurs (Kittler et al. 2015). Some calls function to maintain close-contact group cohesion (e.g., cat-like 'miaow' calls), whereas sharp high-pitched 'wails' serve to reunite separated group members. Male territorial call is a long plaintive howl (carries up to 1km). Other calls relate to maintenance of rank and spacing between nearest neighbours and companions ('moan' and 'hmm' calls) (Bolt 2017). Both males and females have anti-predator calls/responses that differentiate between aerial predators (large birds of prey) and terrestrial predators (Fosa, feral dogs and cats) (Sauther 1989, Bolt et al. 2015, Sauther & Cuozzo in press).

Diet very varied, consisting primarily of fruit, leaves, flowers, bark and sap. Occasionally (but avidly) eats large insects (e.g., locusts and cicadas) and even small chameleons (Oda 1996, Simmen et al. 2003, 2006). Fruits always preferred and proportion in diet varies with habitat and season, e.g., in gallery forests between October and November fruits and new leaves of Tamarind Tamarindus indica constitute majority of food intake (Sauther et al. 1999, Yamashita et al. 2015). In Bezá-Mahafaly

Ring-tailed Lemurs drinking in an underground cave, Tsimanampetsotse National Park.

after extreme climatic events (cyclones), when Tamarind fruits and leaves are unavailable, recorded feeding outside forests in cultivated areas on sweet potato *Ipomoea batatas* and non-native, invasive herb Mexican Prickly Poppy *Argemone mexicana* (LaFleur & Gould 2009). In dry rocky interior areas, e.g., Anja Reserve and Tsaranoro Valley leaves and fruit of *Melia azedarach* are important year-round and *Ficus* spp. (Moraeceae) in dry season (Gould & Gabriel 2015). Also licks soil rich in sodium and soil from termite mounds with high mineral content. In extreme arid environments of south-west (Bezá-Mahafaly) coprophagy recorded, with older individuals eating dried human, cattle and feral dog faeces in dry season (Fish *et al.* 2007).

Gestation and early lactation occur during dry season (July–October), when resources are scarce, consequently females prioritise plant foods low in fibre but high in protein and water (Gould *et al.* 2011, 2015). In gallery forests, Tamarind in particular provides majority of nutrition (protein and fibre) during peak period of lactation (September–March) (LaFleur & Sauther 2015). Also, in very dry habitats, Ring-tailed Lemur is able to gain water

from dew and succulent plants like endemic *Aloe* spp. and introduced prickly pears *Opuntia* sp. (Jolly 2003), and in extreme rocky spiny bush habitat of Mahafaly Plateau (Tsimanampetsotse) descends into limestone caves to drink from permanent ground-water pools (LaFleur 2012, Sauther *et al.* 2013).

Courtship and mating highly synchronous, occurring in 2–3-week period between mid-April and mid-May: females are in oestrous for only four to six hours during this period. Males compete violently to access receptive females. Bouts of aggression involve 'jump fights' where rivals leap in air slashing downwards with their upper canines, often inflicting nasty wounds (Jolly 2003). Male rank may influence mating success, in that they mate first and guard the female after mating. However, because females are dominant, they may choose multiple mates, including group resident and non-resident males. Even low-ranking males are able to copulate with some females. However, conception generally achieved by first mate, usually the highest-ranking male (Pereira & Weiss 1991). Mating regime appears variable between groups and flexible over time. In some groups a single male can monopolise mating in

Ring-tailed Lemur leaping in spiny forest, Anjampolo.

Infant Ring-tailed Lemur clinging to mother, Berenty Private Reserve.

one or more years, whereas in other groups two or more males may sire offspring, and up to c.30% of offspring may be sired by males from outside group (Parga *et al.* 2016).

Females are pregnant during austral winter, when resources are limited (Sauther 1991), although they do not rely on stored fat reserves (as do so-called 'capital breeders'), but instead maximise resource use during the reproductive period (i.e., 'income breeders) (Gould *et al.* 2003). Gestation around 135–145 days, young born mainly in September, a few in October (Koyama *et al.* 2001, Sauther & Cuozzo in press). A single offspring is the norm, with twins infrequent and triplets rare. Initially, infant clings to mother's underside, but moves to ride on her back (or that of another group member) after one to two weeks. All adult females participate in raising infants and they frequently 'allonurse', i.e., tolerate nursing by other females' offspring. Males rarely help in infant care. Weaning occurs in February, when food most abundant (Sauther 1991, Jolly 2003). Infant mortality high: c.50% die in first year and only c.30–40% reach adulthood. Females sexually mature at three years and most give birth annually.

Young predated by raptors, e.g., Madagascar Harrier-Hawk and Madagascar Buzzard. Adults and young predated by Fosa and very occasionally non-native Small Indian Civet *Viverricula indica* and feral cats (Goodman 2003, Goodman & Ganzhorn in press). Also, two exceptional records of infants being eaten by a hybrid *Eulemur* (Pitts 1995, Jolly 2004).

WHERE TO SEE It is ironic that one of the most range-restricted and endangered species, should be so easily and frequently seen. Berenty Reserve, west of Tolagnaro, is one of the island's best-known sites: the lemurs are completely habituated, so viewing is easy in the gallery forest. To see Ring-tailed Lemurs in spiny forest visit Anjampolo north of Berenty. Isalo National Park offers a very different habitat: Ring-tailed Lemurs may be seen clambering over impressive sandstone formations and in adjacent forests. Alternatively, visit Anjà Community Park or Tsaranoro Valley near Ambalavao at northern edge of Andringitra Massif, to see particularly beautiful animals with thick coats climbing on huge granitic boulders. The more adventurous wanting to see Ring-tailed Lemurs in spiny forest in the south-west, south of Toliara, should visit the Amoron'i Onilahy Community Protected Area, Bezá-Mahafaly Special Reserve or the western edge of Mahafaly Plateau in Tsimanampetsotse National Park.

'TRUE' LEMURS
GENUS *Eulemur*

The genus *Eulemur*, commonly referred to as the 'true' lemurs, encompasses a widespread and diverse group. Recent revisions have elevated a number of taxa in the genus to species level. Formerly, the Brown Lemur *Eulemur fulvus* was divided into six subspecies, but these taxa are all now considered species (Groves 2001), with distinction supported by detailed DNA, morphological and vocal analyses (Markolf *et al.* 2013). Furthermore, Red-fronted Brown Lemur *E. rufifrons* has been split from Rufous Brown Lemur *E. rufus* (Groves 2006). Black lemurs were previously split into two subspecies, both now elevated to species level, Black Lemur *E. macaco* and Blue-eyed Black Lemur *E. flavifrons*. Consequently, 12 species of 'true' or 'typical' lemur are now recognised.

The genus *Eulemur* are characteristically medium-sized, weighing 1–2.5kg, quadrupedal lemurs that move primarily by walking and running along branches, but are capable of significant leaps between branches and trees to cross canopy gaps (up to 6–8m). They can also hang by their rear feet while feeding.

Cathemerality is a characteristic (Johnson *et al.* in press), indeed the behaviour pattern was initially identified in respect of this genus (Tattersall 1987, 2006a): all *Eulemur* species show discrete patterns of activity throughout the 24-hour cycle, with the proportion of daytime and nighttime activity often varying seasonally or in line with the lunar cycle and intensity of nocturnal illumination (Curtis & Rasmussen 2002, 2006). In addition, the significant increase in human disturbance and habitat alteration may be causing increased levels of nocturnal activity in *Eulemur* species (Donati *et al.* 2016).

This genus exhibits a considerable degree of sexual dichromatism (males and females having different coloration/markings) (more than any other lemur group), although it varies from negligible in Brown Lemur to extreme in Black Lemur and Blue-eyed Black Lemur, where males are all black and females are ginger-blonde to orangey-brown. Beyond pelage coloration, there is no appreciable sexual dimorphism (Kappeler 1990).

Social structure is variable. Two species, Red-bellied Lemur *E. rubriventer* and Mongoose Lemur *E. mongoz* form monogamous pair-bonds, with groups consisting only of adult male and female plus dependent offspring, while the other ten species live in large multi-male/multi-female groups of up to *c.*20 individuals (Overdorff & Johnson 2003, Johnson *et al.* in press). Female dominance is seen in some species, but is not a prevalent trait throughout the genus.

Olfactory cues and scent-marking are particularly well developed as means of communicating physical state and location, and for individual recognition and territorial demarcation. Males tend to scent-mark using the ano-genital region, hand and head, whereas females mark with ano-genital smears and urine. Males often immediately over-mark a female's smear and will also scent-mark on the female directly. They are also highly vocal, exhibiting an array of calls, some of which are nasal grunts, others loud ascending barks, which serve different purposes such as within-group contact and cohesion, territorial defence and predator alarm; these vary between species.

Eulemur species are predominantly frugivorous, with fruits routinely comprising 60–90% of dietary intake, although this does vary seasonally, particularly in species inhabiting drier habitats where fruit is seasonally scarce (Sato *et al.* 2016). Flowers are also an important supplementary food source, especially in the dry season (Johnson *et al.* in press), hence 'true' lemurs can be important agents for pollination (Kress 1993). Further regional differences do exist, with species from wetter eastern regions generally having a broader more varied diet than congeners in drier western areas (Overdorff & Johnson 2003, Johnson *et al.* in press).

'True' lemurs are also important seed dispersers and therefore crucial in helping maintain forest health and quality. In humid forests at Ranomafana National Park, Red-bellied Lemur and Red-fronted Brown Lemur disperse seeds on average *c.*100–120m, which may result in clustered tree distributions (Razafindratsima *et al.* 2014). In the dry deciduous forests of Ankarafantsika National Park, Brown Lemur is responsible for dispersing the large seeds of >20 plant species that are dependent on the primate (Sato 2018). And in littoral forests of the south-east, Red-collared Brown Lemur *E. collaris* is the only disperser of large seeds (more than *c.*12mm in length) (Bollen *et al.* 2004).

As a genus *Eulemur* has a broad habitat tolerance and is found in most native forest regions, adjacent secondary or degraded forests, with the exception of spiny forest and spiny bush formations in the extreme south and south-west, including littoral forests near sea level and high-elevation areas up to *c.*2,500m. In many areas (particularly in drier western regions) only a single species is present. Major rivers appear to act as barriers and often delineate boundaries between species (Goodman & Ganzhorn 2004). Areas where two *Eulemur* species are sympatric include central and northern rainforests, where Red-bellied Lemur lives alongside one of the brown lemurs (the species depends on location), northern deciduous forests where Crowned Lemur *E. coronatus* and Sanford's Brown Lemur *E. sanfordi* occur together, and restricted areas in western forests where Mongoose Lemur is sympatric with Brown Lemur east of the Betsiboka River or Rufous Brown Lemur west of the river.

Predators of *Eulemur* include Fosa and large diurnal raptors like Madagascar Harrier-Hawk and Henst's Goshawk, which pose particular threats to infants. Also, there is a record of a non-native feral cat taking a Red-fronted Brown Lemur *E. rufifrons* (Merson *et al.* 2019).

It is increasingly apparent that the most severe anthropogenic threats are hunting and local bushmeat consumption, levels of which have increased significantly recently with expanding human populations (Golden *et al.* in press). Habitat destruction and fragmentation, particularly beyond the boundaries of protected areas continue to be major threats, as is collection of individuals from the wild, especially infants, for the local pet trade (Reuter *et al.* 2016, 2019).

BROWN LEMUR
Eulemur fulvus (É. Geoffroy Saint-Hilaire, 1796)

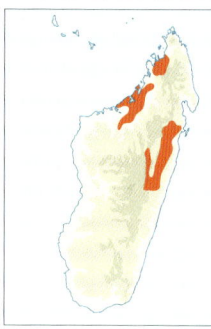

Other name: Common Brown Lemur
Malagasy names: Varika, Varikosy, Akomba, Gidro, Dedrika, Varikafasin, Varikamava

MEASUREMENTS
Head-body: 430–500mm. Tail: 415–510mm. Weight: 1.7–2.1kg. (Mittermeier *et al.* 2010).

DESCRIPTION & IDENTIFICATION Least dichromatic *Eulemur*: male and female similarly coloured, with females sometimes slightly paler. Pelage short but dense. Upperparts uniform brown to grey-brown, underparts paler and slightly greyer. Face, muzzle and crown dark grey to black (female face slightly paler

Cathemerality – periods of activity throughout the 24 hour cycle – is a feature of behaviour in the genus Eulemur.

than male), with paler faint eyebrow patches and paler brown-grey fur around moderately prominent ears, cheeks and underneath chin. Eyes rich orange-red. Tail long and slightly bushy towards tip.

A medium-sized lemur with long tail and horizontal posture which moves quadrupedally on the ground and in the canopy. In centre-east, might be confused with Red-bellied Lemur, which is considerably more reddish, males have distinctive white 'tear-smears' below eyes, whilst females have creamy-white throat, chest and belly. In the north-west, Brown Lemur is distinguished from Mongoose Lemur by its uniform brown, rather than grey, coloration.

HABITAT & DISTRIBUTION Occurs in variety of habitats including lowland and montane rainforests, moist evergreen forests (Sambirano) and dry deciduous forest, from sea level to c.1,880m. Once continuous range reduced to three isolated populations. In centre-east occurs in rainforest north of Onive and Mangoro Rivers probably to Anove River (vicinity of Ambatovaky Special Reserve). North of Anove River replaced by White-fronted Brown Lemur *E. albifrons*, but precise boundary unclear. At southern limit, south of Mangoro River isolated populations remain in Vohitrambo and Vohibe Forests (Andrianandrasana *et al.* 2013, Rajaonson & King 2013).

In north-west found in dry deciduous forest north of Betsiboka River north to vicinity of Baie of Loza and Maevaranao River. Further north-west another population is centred on Manongarivo in Sambirano region, extending from Andranomalaza River north to Mahavavy du Nord River and including western slopes of Tsaratanana Massif.

Between north-western and eastern populations, Brown Lemurs survive in isolated forest remnants in Central Highlands, e.g., Tampoketsa Analamaitso between Ankarafantsika and Marotandrano, and Ambohitantely north-west of Antananarivo (Stephenson *et al.* 1994). When forests were intact, these populations would have been continuous (Tattersall & Sussman 1998). An introduced population persists on Mayotte in Comoros Islands (Tattersall 1982, 1988a).

In intact primary forest (Ankeniheny-Zahamena Forest Corridor) population densities of c.40–60 individuals/km^2 recorded, but often much lower, especially at lower elevations (Seaman *et al.* 2018, Irwin & King 2020). In addition to deforestation (Morelli *et al.* 2020), threatened by hunting (traps,

Male Brown Lemur, Andasibe-Mantadia National Park.

blowpipes and firearms), when entire groups may be captured. Also trapped for pet trade. IUCN Vulnerable.

BEHAVIOUR Like other *Eulemur* species, Brown Lemur is cathemeral. Spends long periods in canopy feeding, with movement and foraging often continuing after dark. Extent of nocturnal activity influenced by lunar cycle: when moon full, nocturnal activity reaches its peak (Vololonirina *et al.* 2017). In western dry deciduous forests (Ankarafantsika National Park) nocturnal activity is highest in dry season, but during rainy season nighttime activity is minimal, so Brown Lemur is primarily diurnal (Ratsirarson & Ranaivonasy 2002, Rasmussen 2005).

Lives in groups consisting of multiple adult males and females, with subadults, juveniles and infants; group size 5–13, with 5–9 being average (Overdorff

Male Brown Lemur, Ankarafantsika National Park.

Home range size strongly influenced by habitat: in dry forests 7–8ha, in rainforests up to 20ha recorded. Groups scent-mark territory, but some overlap between ranges occurs; loud vocalisations help groups avoid one another. In dry deciduous forests (Ankarafantsika National Park) has been recorded occasionally leaving home range and travelling *c.*1–1.5km (habitat shifting) to feed on particular fruits (*Grewia triflora* and *Landolphia myrtifolia*) for up to two weeks, before returning to home range (Sato 2013a).

Diet varied, mainly fruits, seeds, leaves, buds, shoots and flowers, proportions of which vary seasonally (Ganzhorn 1988, Overdorff & Johnson 2003); occasionally supplemented with bark, sap, soil, and invertebrates (insects, centipedes, millipedes) and even vertebrates, like bird chicks, small chameleons and frogs (Irwin & King 2020). In far north-west recorded eating emerging cicadas (Birkinshaw 2003). In east has been observed feeding in plantations on flowers of introduced guava, pine and eucalyptus trees (Ganzhorn 1985), and also seen eating an orb-web spider (Dolch 2006). Unusual among lemurs (and other primates) in that individuals voluntarily share food with conspecifics (Schwitzer *et al.* 2013d). In dry deciduous forests of Ankarafantsika National Park, Brown Lemur is a crucial agent of seed dispersal (Sato 2012); seeds of >50 species are eaten and Brown Lemur is the only disperser of large seeds belonging to more than 20 species (Sato 2013b, 2018).

Mating between May and June, gestation *c.*120 days, births occur in September and October (Overdorff & Johnson 2003). A single offspring is usual, twins have been recorded. At 6–8 weeks young begin eating same solid food as mother. Gradually proportion of suckling time decreases and mother increasingly rejects attempts to suckle. During first 12 weeks, young rarely more than 1m from mother, but gradually separation increases. Weaning occurs at six to seven months, when young spend most of time playing with other juveniles. Sexual maturity reached at *c.*18 months.

WHERE TO SEE Most accessible location in eastern rainforests is Andasibe-Mantadia National Park and adjacent community forests. Readily seen close to main entrance or along ridge in Analamazaotra Special Reserve. Equally rewarding is Analamazaotra Forest Station, administered by local NGO Association Mitsinjo, or V.O.I.M.M.A. Community Reserve. An alternative eastern location is Anjozorobe Forest

& Johnson 2003, Irwin & King 2020). Larger groups recorded on Mayotte (Tattersall 1982). Agonistic interactions seem infrequent and there are no discernible dominance hierarchies.

Female Brown Lemur carrying infant, Anjajavy Private Reserve.

north-east of Antananarivo. In western Madagascar, best location is Ampijoroa, in Ankarafantsika National Park. Groups occupy territories adjacent to campsite and are encountered on walks throughout the system of forest trails. Alternatively, Brown Lemur is very easy to see at Anjajavy Hotel on north-west coast, where habituated groups live in gardens and adjacent forests.

SANFORD'S BROWN LEMUR
Eulemur sanfordi (Archbold, 1932)

Other name: Sanford's Lemur
Malagasy names: Varika, Ankomba, Beharavoaka

MEASUREMENTS
Head-body: 380–400mm.
Tail: 500–550mm. Weight: c.2kg.

Female Sanford's Brown Lemur, Montagne d'Amber National Park.

DESCRIPTION & IDENTIFICATION Sexually dichromatic. *Male:* Upperparts and tail medium-brown, slightly darker grey-brown on back and towards tip of tail. Underparts pale brown-grey. Pronounced

227

Male Sanford's Brown Lemur, Montagne d'Amber National Park.

Lemur also highly distinctive. Female Sanford's and female White-fronted Brown Lemurs are very similar; where ranges meet, differentiation can be problematic, but males are distinctive.

HABITAT & DISTRIBUTION Primary and secondary rainforest, gallery and dry deciduous forest and partially degraded forest up to 1,400m; elevations between 800m and 1,000m are preferred in Montagne d'Ambre (Freed 1996). Appears absent from very dry forest (e.g. Cap d'Ambre).

Restricted to far north. Range extends from south of Cap d'Ambre Peninsula, south to Manambato River in east and Mahavavy du Nord River in west, including Ankarana, Analamerana and Montagne d'Ambre. Populations in drier forests of Loky-Manambato Protected Area (Daraina) are disjunct. Remaining populations of *Eulemur* south of Manambato River and north of Bemarivo River (northern limit of White-fronted Brown Lemur) require investigation to determine taxonomic status.

Population densities very variable. Generally occurs at higher densities in more humid forests, e.g., Montagne d'Ambre National Park. Also very high densities in Ankarana National Park, with significant differences between primary forest (109/individual/km^2) and secondary forest (233 individuals/km^2) (Gudiel *et al.* 2017). Mean density across broader range *c.*25 individuals/km^2. In dry areas (Analamerana) densities drop to *c.*3–5 individuals/km^2 (Banks 2005), where sympatric Crowned Lemur more abundant and Sanford's Brown Lemur disappears from many very dry areas where Crowned Lemurs are able to survive.

Sanford's Brown Lemur particularly susceptible to habitat fragmentation (Eppley *et al.* 2020d) and does not survive in badly degraded habitats (Gilles & Reuter 2014). Also hunted and collected for pet trade (Reuter *et al.* 2016, Reuter & Schaefer 2017). IUCN Endangered.

BEHAVIOUR Cathemeral. Prefers to forage in mid-storey and forest canopy. Fruit constitutes up to 90% of diet, together with shoots, buds, flowers and some invertebrates, like millipedes and spiders. In wetter months, when fruit scarce, greater variety of plant species utilised than in dry season when favoured resources predominate. In Montagne d'Ambre National Park in wet season regularly forms mixed-species foraging groups with Crowned Lemurs for up to 30% of day, with very little interspecific aggression (Freed 1996). Similar

creamy-grey beard and prominent ear-tufts give males a 'maned' appearance. Crown also long and generally brown-grey. Nose-bridge and snout black. Between dark areas and mane are patches of short creamy-white hair on cheeks and above eyes. *Female*: Upperparts grey-brown to rufous brown, tail grey-brown becoming darker towards tip. Underparts pale grey. Face and head completely grey, which colour extends over crown, down neck and onto shoulders. Females lack ear-tufts and beard. Some have slightly paler areas above eyes.

Medium-sized lemur with long tail that adopts horizontal posture and moves quadrupedally. Shares range with similar-sized Crowned Lemur, but appearance is very different. Male Crowned Lemurs are orangey-brown, females are grey, and both sexes have distinctive orange-brown V-shape on crown. Cheek and ear-tufts of male Sanford's Brown behaviour also observed in drier forests of Ankarana during dry season (Gudiel *et al.* 2017).

Groups are multi-male/multi-female plus dependent offspring, with no evidence of female dominance. Group size varies with habitat: in rainforest (Montagne d'Ambre National Park) groups smaller, 3–9 (Freed 1996), whereas in dry forest up to 15 recorded (Wilson *et al.* 1989). Home ranges *c.*15ha, with considerable overlap. Agonistic interactions rare (Hawkins *et al.* 1990, Freed 1996).

Births occur in late September/early October, gestation 120 days. Infant clings to mother's belly for first month before transferring to ride on her back.

WHERE TO SEE Montagne d'Ambre National Park is most accessible location. Groups often encountered near *Station Roussettes* and *Petite Cascade* and along botanical trail. Ankarana National Park is another rewarding locality: fairly common around both *Campement Anilotra* (Camp des Anglais) and *Campement d'Andrafiabe* (Camp des Américans). Also possible to see in forests around Daraina, albeit less frequently than Crowned Lemur.

WHITE-FRONTED BROWN LEMUR
Eulemur albifrons (É. Geoffroy Saint-Hilaire, 1796)

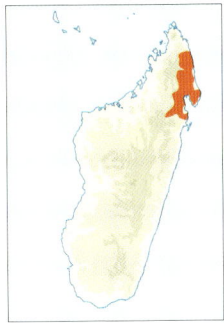

Other names: White-fronted Lemur, White-headed Lemur
Malagasy names: Varika, Varikosy, Varikosa, Alokosy

MEASUREMENTS
Head-body: 390–420mm. Tail: 500–540mm. Weight: *c.*2kg.

DESCRIPTION & IDENTIFICATION Sexually dichromatic: greatest pelage differences between sexes of 'brown' lemurs. *Male:* Upperparts and tail rich mid- to dark brown, becoming darker and more reddish towards rear. Underparts pale grey, becoming more creamy-white around chest and throat. Head creamy-white with bushy cheeks, muzzle and nose-bridge black. Eyes red-orange. *Female:* Upperparts

Female White-fronted Brown Lemur with infant, Masoala National Park

Young male White-fronted Brown Lemurs, Nosy Mangabe Reserve.

and tail mid- to dark brown, darkening towards rear and on tail, underparts grey. Head, face and muzzle dark grey, nose slightly darker. Unlike male, cheeks not bushy.

A medium-sized lemur with horizontal posture and long tail. In range, could be confused with Red-bellied Lemur, but highly distinctive white head of male White-fronted Brown Lemur makes this unlikely.

HABITAT & DISTRIBUTION Primary and secondary rainforests up to *c*.1,650m. Restricted to north-east humid forests from south of Bemarivo River (south of Iharana/Vohemar) to probably Anove River (south of Mananara), including Masoala Peninsula. At north-east limits of range, populations probably extend through COMATSA forest corridor onto Tsantanana Massif. Precise range limits south of Anove River and distributional relationship with Brown Lemur are unclear: appears to be significant zone of hybridisation (Mittermeier *et al.* 2010). Also an introduced population thrives on Nosy Mangabe (in Bay of Antolgil, off Maroantsetra) and an isolated population is found in Betampona

Reserve, considerably further south of main range. Anecdotal evidence suggests these animals may have been introduced.

Estimated population densities vary considerably with location, ranging between *c*.6.5 and 58 individuals/km^2 (Herrera *et al.* 2018, Brook *et al.* 2019). On Masoala Peninsula densities of *c*.58 individuals/km^2 in remote uninhabited areas and *c*.16 individuals/km^2 nearer villages, reflecting increased hunting pressure (Borgerson *et al.* 2019). Population on Nosy Mangabe may occur at abnormally high density, with detrimental effects due to increased competition: mean weight is 1.7kg, whereas *c*.2.0kg in mainland populations.

White-fronted Brown Lemur is extensively hunted throughout range, contributing to nutrition and health of human populations as well as local economies (Golden 2009, Golden *et al.* 2014, Borgerson 2015, 2016, Borgerson *et al.* 2016, 2017, 2019, Brook *et al.* 2019). Hunting most prevalent in cool austral winter when lemurs are caught in snares at seasonally fruiting trees (Borgerson 2015, Brook *et al.* in 2019). Majority of hunting (>95%) is at subsistence level in very poor rural areas, but hunting with guns is increasing near urban areas (Borgerson 2016, Borgerson et al. 2016, 2017). Live trapping for local pet trade also an issue (Reuter & Schaefer 2017). IUCN Vulnerable.

BEHAVIOUR White-fronted Brown Lemur is cathemeral throughout the year (Vasey 2004). Groups are multi-male/multi-female, ranging in size from 4–13, with 8–9 on average (Rakotondratsima & Kremen 2001). Almost exclusively arboreal, but primarily utilises understorey at 5–15m, where lianas more common, rather than canopy (Vasey 2000a, 2002). During midday resting periods, groups often retreat to dense lianas, where individuals huddle together and wrap tails around one another. Behaviours may help predator avoidance and have a thermoregulatory function when cold (Vasey 2004).

Diet dominated by ripe and unripe fruit (c.70%) and, to a lesser extent, leaves. Fruits of *Grewia* trees and various *Ficus* spp. (Moraeceae) are particularly relished. Flowers and nectar important in dry season, particularly to females. Fungi and invertebrates like millipedes, spiders and insects occasionally eaten (Vasey 2000b, 2002). Treefall gaps in canopy caused by cyclones often form localised food-rich areas (increased new growth and fruiting) that are exploited seasonally (Mogilewsky 2020).

Home range c.16ha (Vasey 1997a). Vocalisations heard around dawn and dusk, and also well into the night. Females give birth between mid-October and early December, often coinciding with flowering season. Other aspects of behaviour probably similar to other brown lemurs inhabiting rainforest areas (see Brown Lemur and Red-fronted Brown Lemur).

Fosa is probably a major predator. On Nosy Mangabe records of predation by Madagascar Ground Boa (Goodman et al. 1993c).

WHERE TO SEE Masoala National Park is a particular good location to see this species. All areas around the main trail network at Lohatrozona can be productive. Equally productive are the forests adjacent to Arol Lodge, Masoala Forest Lodge and Tampolo Lodge. Nearby Nosy Mangabe in the Bay of Antongil is very good as is Farankaraina Reserve, east of Maroantsetra. The more adventurous might wish to visit Makira, inland of Maroantsetra. Alternatively, Marojejy National Park is very rewarding with groups regularly encountered on the main trail between the park entrance and Camp Marojejia.

Adult male White-fronted Brown Lemur, Marojejy National Park.

RED-FRONTED BROWN LEMUR
Eulemur rufifrons   (Bennett, 1833)

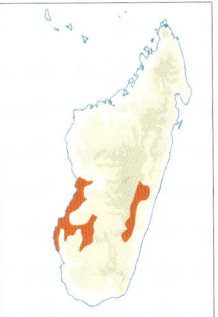

Other name: Southern Red-fronted Lemur

Malagasy names: Varikamavo (east), Gidro, Halo,

MEASUREMENTS
Head-body: 400–480mm. Tail: 450–550mm. Weight: 2–2.5kg.

DESCRIPTION & IDENTIFICATION Sexually dichro-matic, with noticeable variations between individuals and between eastern and western populations. *Male*: Upperparts and tail grizzled grey to grey-brown, underparts pale creamy-grey. Extremities of limbs sometimes tinged rufous-brown. Tail grey-brown, darkening towards tip. Face, muzzle and mid-forehead dark grey to black,

*Red-fronted Brown
Lemurs drinking at
waterhole, Kirindy Forest.*

with thin dark line dividing rich russet-brown crown. Prominent white eyebrow patches and distinctive bushy white cheeks and beard. Ears not prominent, eyes orange-red. *Female:* Considerable regional variation. Upperparts and tail grizzled grey-brown to rufous-brown, underparts pale grey. Face and muzzle dark grey to black, with dark line extending over forehead to grizzled grey crown. Large white eyebrow patches and creamy-coloured cheeks, although less bushy than in males. All infants born with adult male coloration: females gradually change to their adult coloration at 7–17 weeks (Schwitzer *et al.* 2013d).

Medium-sized lemur with long tail, adopts horizontal posture and moves quadrupedally. In rainforest areas may be confused with Red-bellied Lemur, which is more russet-red and lacks very pale cheeks and eyebrow patches. At south-eastern extreme of range hybrids with White-collared Brown Lemur *E. cinereiceps* occur. In western areas this species is unmistakable and is not sympatric with any other *Eulemur* species.

HABITAT & DISTRIBUTION Occurs in dry deciduous forest in western areas, and primary and secondary rainforests in eastern regions, from sea level to *c.*1,700m. Range once continuous, now found in two distinct populations in west and east. In west found from Fiherenana River north of Toliara, to Tsiribihina River. In east, northern limit is Onive and Mangoro Rivers, extending south to Andringitra Massif and further south to Manampatrana River. On south-east slopes of Andringitra Massif, ranges

of Red-fronted Brown Lemur and White-collared Brown Lemur meet and there is a 70km-wide zone of hybridisation around Iantara River (headwaters of Manampatrana River) (Irwin *et al.* 2005, Delmore *et al.* 2013).

In Ranomafana National Park population density 6.75 individuals/km^2 (Wright *et al* 2012), but lower densities of *c.*1.95–4.9 individuals/km^2 in Andringitra National Park (Rajaonson *et al.* 2014). In western deciduous forests estimates vary from 20–30 individuals/km^2 (Kelley *et al.* 2007, Johnson *et al.* 2020b). Continuing loss of habitat remains primary threat. It is also one of the most widely hunted lemur species across the island. IUCN Vulnerable.

BEHAVIOUR Eastern populations cathemeral throughout year (Overdorff & Rasmussen 1995), whereas western populations are more diurnal with increased nighttime activity during dry season (Donati *et al.* 1999, 2001, Kappeler & Erkert 2003). Overall behaviour appears very adaptable: almost totally arboreal, only occasionally descending to ground, sometimes to lick and eat soil. Groups contain several adult females and males plus younger animals at various stages of maturity, with closely related females forming core of group. Group size variable; western forests 4–17, with *c.*9 being average, in eastern forests 6–18, with *c.*8 average (Overdorff 1996b, Donati *et al.* 1999, Johnson & Overdorff 1999, Holmes *et al.* 2019). In western forests aggregations of 30+ reported feeding in single large fruiting fig tree *Ficus grevei*.

Male Red-fronted Brown Lemur, from dry forest, Kirindy.

Female Red-fronted Brown Lemur, from dry forest, Kirindy.

Group cohesiveness maintained via grunts, contact calls and other vocalisations, but no dominance hierarchies apparent, although females take the lead. Home range size heavily influenced by habitat and season. During wet season in western deciduous forests may be as small as 1ha and groups rarely move more than 125–150m per day. In dry, cold season, when food scarce and more dispersed, home ranges expand to up to 30ha, with a corresponding increase in daily distances travelled (Sussman 1975, Ostner & Kappeler 1999). In cold periods lemurs reduce activity levels and huddle together in small groups to help thermoregulation and conserve energy (Ostner 2002). In addition, ranging and pattern of movement also heavily influenced by water availability (Amoroso et al. 2020b).

In eastern rainforests home ranges much larger, up to 100ha, with daily travel distances between 150m and 2,000m, correlating with seasonal food availability and density of fruiting trees. Neighbouring territories overlap extensively, with peaceful movement of individuals between social units. However, aggressive inter-group encounters do occur, and loud vocalisations help avoid contacts

Male Red-fronted Brown Lemur, from rainforest, Ranomafana.

and maintain group spatial separation (Overdorff 1993a, 1996a, Johnson & Overdorff 1999).

Diet varies with habitat, but fruit dominates in both dry deciduous forests and rainforests. Via their faeces, Red-fronted Brown Lemurs are important agents for seed dispersal, although much is within less than 100m of original feeding tree and very rarely more than 500m distant (Ganzhorn & Kappeler 1996, Razafindratsima *et al.* 2014). In western areas, leaves also important, along with pods, stems, flowers and sap (Sussman 1975). In eastern areas fruit is dietary mainstay, including non-native species like Strawberry Guava *Psidium cattleianum*. Leaves less important, but fungi, insects and other invertebrates like millipedes also eaten. Millipedes first tossed from hand to hand to induce secretion of toxins: these then washed off with saliva before lemur wipes and dries millipede on its tail, before eating. Flowers and nectar consumed more frequently during warm wetter months (Overdorff 1992b, 1993b, 1996a).

In western populations during breeding season, one male monopolises all receptive group females. In eastern populations several males form reproductive pairings with different females. Mating occurs in May–June, young born September or early October, gestation *c.*120 days (Overdorff & Johnson 2003, Ostner *et al.* 2008). Breeding timed so peak fruit availability coincides with maximum lactation costs to mother (Wright *et al.* 2006). Birth weight *c.*75g. Initially infants carried on mother's belly, after one month or so begins to move about, transferring to ride on mother's back. Infants travel independently

at around three months. Daughters of dominant females may remain in natal group, otherwise both males and females transfer from natal group, females at 23–26 months, males at *c.*35+ months. Sexual maturity reached at two to three years.

In western regions during dry season, Red-fronted Brown Lemurs descend to ground-level waterholes to drink, but modify patterns of waterhole use to maximise avoidance of predators also using same resources (Amoroso *et al.* 2020a). Red-fronted Brown Lemur has specific alarm call for aerial predators and more general vocalisation for terrestrial predators (Fichtel & Kappeler 2002). Fosa is main predator in both rainforest and dry deciduous forest (Schnoell & Fichtel 2012). Large raptors, e.g., Henst's Goshawk and Madagascar Harrier-Hawk are also significant predators (Goodman 2003, Karpanty & Wright 2007). In some dry forest areas (e.g. Menabe) predation by feral cats *Felis sylvestris* recorded (Merson *et al.* 2019).

WHERE TO SEE In eastern rainforests, Ranomafana National Park is best place to see this lemur. April–May is a particularly good time, when Strawberry Guava *Psidium cattleianum* is in fruit, and lemurs may congregate: the *Belle Vue* area can be very good. Kirindy Forest north-east of Morondava is the best locality in dry deciduous forests. Alternative western locations are Kirindy-Mitea National Park, Zombitse-Vohibasia National Park and Isalo National Park.

RUFOUS BROWN LEMUR
Eulemur rufus   (Audebert, 1799)

Other names: Northern Red-fronted Lemur, Red Brown Lemur

Malagasy names: Varika, Halo

MEASUREMENTS
Head-body: 400–480mm. Tail: 450–550mm. Weight: *c.*2kg. (Gerson 1999)

DESCRIPTION & IDENTIFICATION Sexually dichromatic. *Male*: Upperparts mid brown-grey to dark olive-grey, with deep brown tones around lower back and tail. Tail rich brown, darkening towards tip. Underparts pale creamy-grey. Hands and feet tinged russet-brown. Crown rich brick-red, cheeks golden-

Female Rufous Brown Lemur, Tsingy de Bemaraha National Park.

Female Rufous Brown Lemur on limestone karst, Tsingy de Bemaraha National Park.

red. Dark mid-facial stripe extends from crown to nose-bridge, joining black muzzle. Areas either side of muzzle and eyebrow patches creamy-white. Ears not prominent and eyes orange-red. *Female*: Upperparts and tail rich gingery-red to orange-cinnamon-brown. Underparts orange-brown. Extremities of limbs dark russet-brown. Crown slate-grey with dark grey/black medial stripe extending from top of head to nose-bridge and becoming black on muzzle. Inner cheeks and patches above eyes pale grey-white.

Medium-sized quadrupedal lemur with a long tail and horizontal posture. In range should not be confused with any other species. Not sympatric with any congener, except very narrow zone west of Betsiboka River where hybridisation with Mongoose Lemur occurs (Zaramody & Pastorini 2001). Previously regarded as synonymous with *E. rufifrons*, when all populations were classified as *E. rufus* or *E. fulvus rufus*.

HABITAT & DISTRIBUTION Primary and secondary dry deciduous forest. Also recorded in coastal mangroves (Gauthier *et al.* 1999). Restricted to central-western regions and is patchily distributed: northern limit Betsiboka River, southern limit Tsiribihina River. Between Betsiboka and Mahavavy Rivers, occurs in most remaining forest patches from coast, to forests of Ambato-Boeny and Ankirihitra,

and in Maevatanana District in Mandrava Forests as far south as Ikay Forest (Rakotonirina *et al.* 2014). North of Manambolo River, occurs in Tsingy de Bemaraha National Park where able to move over razor-sharp limestone karst (*tsingy*) formations from one forest patch to another. In Ankirihitra Forest, densities of 44–53 individuals/km^2 recorded (Rakotondrabe & Razafindramanana 2019). Throughout range severely threatened by deforestation and habitat loss, and high levels of hunting and some collection for pet trade (Reuter & Schaefer 2017, Golden *et al.* in press). IUCN Vulnerable.

BEHAVIOUR Cathemeral year-round, with increased nocturnal activity in dry season. Previous studies of '*E. rufus*' correspond to populations now classified as *E. rufifrons*. Most aspects of Rufous Brown Lemur behavior are likely very similar to those of Red-fronted Brown Lemur in dry deciduous forests (e.g., in Kirindy Forest).

WHERE TO SEE Tsingy de Bemaraha National Park, north of Manambolo River is the best location. Access via village of Bekopaka. As well as viewing lemurs in forest, also possible to see them clambering over the jagged rock formations. Alternative sites are Tsiombikibo Forest near Mitsinjo or Baie de Baly National Park near Soalala (Mittermeier *et al.* 2010).

WHITE-COLLARED BROWN
LEMUR *Eulemur cinereiceps*
(A. Grandidier & Milne-Edwards, 1890)

Other names: White-collared Lemur, Grey-headed Brown Lemur, Grey-headed Lemur

Malagasy names: Varika Mena, Varika

MEASUREMENTS
Head-body: 390–400mm. Tail: 500–550mm. Weight: *c.*2-2.5kg. (Johnson *et al.* 2005)

DESCRIPTION & IDENTIFICATION Sexually dichromatic. *Male*: Upperparts grey-brown, tail and lower limbs darker. Darker stripe often on spine. Underparts paler grey. Head and face predominantly pale grey-brown, with dark grey crown grading to paler grey on neck and shoulders. Cheeks and beard white, bushy and pronounced. *Female*: Upperparts and tail reddish-brown (noticeably more rufous than males). Underparts similar but slightly paler. Feet darker. Head and face mid-grey with darker crown, reddish-brown cheeks and beard (match upperparts), cheeks less bushy than males.

Male White-collared Brown Lemur.

A medium-sized lemur with horizontal posture. Known range does not overlap with any congener, but range limits with Red-fronted Brown Lemur are imprecise and where they meet a zone of hybridisation occurs (see below) (Johnson & Wyder 2000). Female White-collared Brown Lemur and female Red-collared Brown Lemurs virtually indistinguishable but genetic analysis confirms species are distinct (Wyner *et al.* 1999). With extent of deforestation in region there is no overlap between these species.

Previously, White-collared Brown Lemur was referred to as *E. albocollaris*: historical investigation revealed *E. cinereiceps* describes the same taxon and is senior name (Johnson *et al.* 2008). Species name '*cinereiceps*' refers to 'grey head' as original description was of a female.

HABITAT & DISTRIBUTION Lowland and mid-altitude rainforest and coastal littoral forest from sea level to *c.*800m. Restricted to narrow strip of forest in south-east. Northern limit is just north of Manampatrana River (Andringitra Massif), extending south to Mananara River (Johnson & Wyner 2000). Vevembe is main block of remaining forest within range. Two isolated populations occur in Manombo Special Reserve (lowland humid forest) and Mahabo Forest (littoral forest) south of Farafangana. A small coastal population may also occur at Vohipaho, south of Mananara River: genetic analysis required to confirm whether this is White-collared Brown Lemur or Red-collared Brown Lemur (Johnson *et al.* 2020a).

South-east of Andringitra Massif, there is an up to 70-km wide hybridisation zone between Red-fronted Brown Lemur and White-collared Brown Lemur, around Iantara River and headwaters of the Manampatrana (Irwin *et al.* 2005, Delmore *et al.* 2013). Risk of hybridisation 'overrunning' populations of White-collared Brown Lemur appears low (Wyner *et al.* 2002, Delmore *et al.* 2011). Population densities of *c.*8.7–13.5 individuals/km², the lowest in genus *Eulemur* (Johnson *et al.* 2011, Brenneman *et al.* 2012).

Range is most restricted of any *Eulemur* species (Johnson *et al.* 2009), probably covering no more than 700km² (Irwin *et al.* 2005). Deforestation, extreme climatic events (e.g. cyclones), hunting and live trapping for pet trade remain constant threats. IUCN Critically Endangered.

BEHAVIOUR Studied in forests at Vevembe, Manombo and Mahabo. Cathemeral throughout year, with no obvious seasonal variation in activity.

Male (front) and female (behind) White-collared Brown Lemurs.

Groups are multi-male/multi-female containing up to 16 individuals when resources abundant, but split into smaller groups when food is scarce to avoid conflict. Frequent movement of both sexes between groups (Johnson 2002, 2006). Average activity patterns comprise 40% resting, 30% social interactions and grooming, 12% feeding and 12% travelling (Johnson *et al.* 2011).

Diet dominated by fruit, but leaves, flowers, nectar, fungi and invertebrates (centipedes, millipedes, spiders and insects) also eaten occasionally; flowers of *Pandanus* spp. particularly important during dry season (Johnson 2002) and their fruits also eaten, along with *Noronhia*, *Pyrostria* and *Uapaca* (Johnson *et al.* 2009). In Manombo, also observed eating spicy fruits of non-native species *Aframomum angustifolium* and shelf fungus growing on invasive *Cecropia* (Ralainasolo *et al.* 2008). Such flexibility may make White-collared Brown Lemur less susceptible to forest degradation and edge effects (Ingraldi *et al.* 2010).

WHERE TO SEE Seeing this lemur is a challenge, as limited range is time-consuming to reach. The best location is Manombo Special Reserve and associated classified forests, south of Farafangana. Access is difficult in wet season when local rivers swell and cut off portions of forest. An alternative is Agnalahaza Forest, 40km south of Farafangana (Mittermeier *et al.* 2010). Vevembe also offers a reasonable chance of success but is more challenging to reach.

RED-COLLARED BROWN LEMUR
Eulemur collaris (É. Geoffroy Saint-Hilaire, 1812)

Other names: Collared Lemur, Collared Brown Lemur, Red-collared Lemur

Malagasy name: Varika

MEASUREMENTS
Head-body: 390–400mm.
Tail: 500–550mm. Weight:
2.25–2.5kg.

DESCRIPTION & IDENTIFICATION Sexually dichromatic. *Male*: Upperparts brownish-grey, tail darker. Underparts pale brown-grey. Muzzle, forehead and crown dark slate-grey to black, extending and becoming paler on neck. A dark stripe runs down

237

Male Red-collared Brown Lemur.

spinal ridge. Eyebrow patches pale creamy-grey but vary individually. Cheeks and beard thick, bushy and orange to rufous-brown. *Female*: Upperparts browner than males, sometimes rufous. Underparts pale creamy-grey. Head and face grey, with faint grey stripe over crown. Cheeks rufous-brown, but less bushy than males. Eyes of both sexes orange-red.

Medium-sized lemur, adopts horizontal posture and moves quadrupedally. Range does not overlap with any congener so confusion is unlikely. While male Red-collared Brown Lemurs and male White-collared Brown Lemurs are noticeably different, females are virtually indistinguishable, but genetic analysis confirms species distinction (Djletati *et al.* 1997, Wyner *et al.* 1999).

HABITAT & DISTRIBUTION Lowland, mid-altitude and high-altitude primary and secondary rainforest and fragments of coastal littoral forest. Recorded from sea level to 1,875m in Andohahela (Feistner & Schmid 1999). Restricted to extreme south-east, from Mananara River south through Midongy du Sud to Tsitongambarika Protected Area, Andohahela National Park and Ambatotsirongorongo transitional forest, including isolated populations in littoral forests in Mandena and Sainte Luce Conservation Zones. Western limits appear to be forests of Kalambatritra on eastern side of east–west watershed and ecotone. South of Mananara River, in littoral forest

at Vohipaho, a small population requires genetic analysis to confirm species identity, Red-collared Brown Lemur or White-collared Brown Lemur.

Population densities vary with habitat and location. In littoral forest fragments in Sainte Luce and Mandena Conservation Zones, wide variation between 16 and 74 individuals/km^2 correlated to fragment size (Ganzhorn *et al.* 2007); in Midongy du Sud National Park, 14 individuals/km^2 (Irwin *et al.* 2005). In Andohahela National Park and southern Tsitongambarika Protected Area, 8–11 individuals/km^2 (Feistner & Schmid 1999, Donati *et al.* 2020c) and in primary forest in Ampasy Valley within Tsitongambarika Protected Area, 29 individuals/km^2 (Balestri 2018).

Habitat destruction and fragmentation remain major threats (Bollen & Donati 2006). Additionally, it is widely hunted and trapped for local pet trade (Campera *et al.* 2019b, Reuter & Schaefer 2017). Mineral mining also threatens populations in littoral forest fragments where isolation limits gene flow (Bertoncini *et al.* 2017). IUCN Endangered.

BEHAVIOUR In common with congeners, Red-collared Brown Lemur is cathemeral throughout year, with activity strongly influenced by lunar phase (nocturnal luminosity) and day length: nocturnal activity increases when moon is bright and diurnal activity increases with longer days. High humidity and

Male Red-collared Brown Lemur.

rainfall decrease nocturnal activity. Dusk is primary zeitgeber in coordinating onset of group activity (Donati & Borgognini Tarli 2006b). In addition, Red-collared Brown Lemurs in forests with high human disturbance tend to show increased levels of nighttime activity (Donati et al. 2016).

Red-collared Brown Lemur lives in multi-male/multi-female groups, with group size influenced by habitat type and quality. In littoral forests, group size varies from 2–17 depending on size of forest patches and levels of degradation, with groups no larger than six in degraded areas and larger groups only in better quality habitat (Donati et al. 2011, Campera et al. 2014). Groups up to 22 recorded in primary rainforest areas (Donati et al. 2020c). Home range size also varies considerably with habitat type and quality (c.20–110ha), with larger ranges associated with poorer quality, degraded areas, where food is less abundant (Donati et al. 2011, Campera et al. 2014). In degraded area small groups can disperse across semi-open habitat between forest fragments (Hyde Roberts et al. 2020).

In common with congeners, largely frugivorous (fruit >70% of diet), with smaller quantities of flowers, mature and young leaves, and invertebrates comprising rest of diet. Over course of year a wide variety of plant species eaten (>120 species), with some eaten only during day and some targeted only at night (Donati et al. 2007, 2011, Eppley et al. 2017a).

Red-collared Brown Lemur is only vertebrate within range able to swallow large seeds, so several tree species may rely on it for seed dispersal (Schwitzer et al. 2013d).

Females give birth in September/October after a gestation of c.120 days. Singletons normal, but twins not uncommon (Donati et al. 2020c).

WHERE TO SEE Most accessible localities are Mandena Conservation Zone and Sainte Luce Reserve, north of Tolagnaro. Also worth visiting Malio, the rainforest area of Andohahela National Park (parcel 1).

MONGOOSE LEMUR
Eulemur mongoz   (Linnaeus, 1766)

Malagasy names: Dedrika, Gidro

MEASUREMENTS
Head-body: 300–350mm. Tail: 450–480mm. Weight: 1.1–1.6kg. (Tattersall 1982, Pastorini et al. 1998)

Female (left) and male (right) Mongoose Lemurs, Tsiombikibo Forest.

DESCRIPTION & IDENTIFICATION Sexually dichromatic. *Male:* Upperparts slate grey-brown with darker tail tip, underparts paler creamy-grey. Some brownish tones on shoulders and a dark grey pygal patch. Face grey, muzzle pale grey with dark nose. Cheeks, beard, forehead and back of neck distinctively rufous-brown. Mature males sometime have triangular bald patch on crown caused by excessive head rubbing when scent-marking. *Female:* Upperparts grey (generally paler than males) with some faint brownish tones on rear flanks and rump, and a dark grey pygal patch. Tail tip darker, underparts creamy-grey. Face dark slate-grey fading to pale grey or white around muzzle. Cheeks and beard form creamy-grey to white ruff continuous with throat and underparts. Eyes in both sexes reddish-orange.

Medium-sized lemur with long tail and horizontal body posture that moves quadrupedally. Potential for confusion with Brown Lemur or Rufous Brown Lemur; Mongoose Lemur is slightly smaller with predominantly slate-grey rather than brown pelage.

HABITAT & DISTRIBUTION In Madagascar in dry deciduous forest and secondary forest from sea level to c.400m. Also recorded using mangroves (Donati *et al.* 2019). Introduced populations on Comoros Islands inhabit more humid forest. Natural range restricted to north-west Madagascar, from west of Mahavavy River in vicinity of Mitsinjo (Mahavavy-Kinkony Wetland Complex, including Tsiombikibo, Anjamena, Analabe, Analamanitra, Ankamangoa, Bemahazaka, Tsilaiza, Anoloky and Anoborengy forests) (Müller *et al.* 2000), to region of Boriziny, including coastal areas of Antrema (Katsepy), Bombetoka-Belemboka and Ankirihitra Forests (Ambato-Boeny area) on west side of Betsiboka River and Ankarafantsika National Park east of Betsiboka River, and Mariarano Classified Forest north-east of Mahajanga (Andriantompohavana *et al.* 2006b, Shrum 2008, Schwitzer *et al.* 2013a). Introduced on Grande Comoro, Anjouan and Moheli in the Comoros (Tattersall 1982, 1988a).

In Ankirihitra Forest densities of 5–7 individuals/km^2 recorded (Rakotondrabe *et al.* 2019). Hybridisation with Rufous Brown Lemur recorded west of Betsiboka River. Severely threatened by habitat loss and fragmentation; limited to largest remaining forest patches and does not move between fragments (Shrum 2008). Additionally, bushmeat hunting and illegal local pet trade are significant threats (Reuter & Schaefer 2017). IUCN Critically Endangered.

BEHAVIOUR Lives in small groups, comprising monogamous pair and 1–3 dependent offspring (Curtis & Zaramody 1998). Social bonds within family units strong; groups cohesive when feeding, travelling, resting and sleeping. Home ranges not large, with extensive overlap between neighbouring groups. Inter-group encounters infrequent but cause considerable agitation, vocalisation and scent-marking when they occur (Curtis & Zaramody 1999). Tolerates close proximity of Rufous Brown Lemur (sympatric in some areas west of Betsiboka River); the two species are sometimes seen in mixed groups, and hybridisation can occur (Zaramody & Pastorini 2001).

Cathemeral in both wet and dry seasons. During warm wet months (December–April) considerably more diurnal (or crepuscular) activity. With onset of dry season in May a shift occurs towards nocturnal behaviour, which becomes predominant (Curtis et al. 1999). At this time, groups travel and feed between dusk and 22:00 then rest for 2–4 hours before resuming foraging and continuing until just before dawn, when they return to sleep sites in dense foliage or tangled vines at tops of trees. Foraging at night, when temperatures are low, allows them to conserve energy. Also, given foliage cover is at its lowest at this time, nocturnal activity may reduce risk of predation from diurnal raptors.

At both seasons fruit dominates diet, with flowers, nectar, young leaves and shoots. In wet season, when active after dark, may be complemented by night-blooming flowers, particularly Kapok *Ceiba pentandra*, and nectar. During dry season, leaves form significant dietary elements, and beetles and insect grubs are also eaten (Curtis & Zaramody 1998, 1999).

Females give birth annually; young are born in October and November after a gestation of c.126–128 days. Infants able to move independently from mother at c.1 month, but initially do not stray far. Weaned at 6–7 months. Female primarily responsible for infant care, but males do contribute some of the time (Curtis & Zaramody 1999). Adult size and coloration attained between 14 and 16 months, sexual maturation takes longer.

WHERE TO SEE Ampijoroa (Ankarafantsika National Park) south-east of Mahajanga is most accessible site. They are less frequently seen than other lemurs and can require effort and patience to track down. Some more remote locations can offer a better chance of success, e.g., areas around Mitsinjo include Tsiombikibo Forest and forest patches adjacent to the town and around nearby Lac Kamonjo and Lac Kinkony; alternatives are forests around Anjamena on east (north) bank of Mahavavy River and Mariarano Classified Forest, north-east of Mahajanga.

CROWNED LEMUR
Eulemur coronatus   (J. E. Gray, 1842)

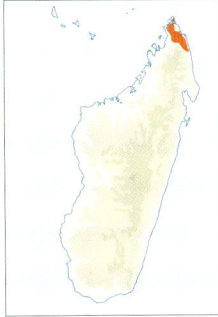

Malagasy names: Ankomba, Ankomba Fiaka, Gidro

MEASUREMENTS
Head-body: 340–360mm.
Tail: 410–490mm. Weight: 1.1–1.3kg. (Terranova & Coffman 1997)

DESCRIPTION & IDENTIFICATION Sexually dichromatic. *Male:* Upperparts grey-brown, browner to chestnut-brown on flanks and towards extremities of limbs. Underparts pale creamy-grey, with brown/chestnut wash. Tail grey-brown and slightly darker than body. Face and ears pale grey to white, tip of muzzle black. Conspicuous and distinctive chestnut-orange crown above eyebrow line and by side of ears, forming prominent V-shape with dark charcoal-grey patch on crown. Chestnut-orange coloration also extends around face forming bushy cheeks. *Female:* Upperparts, flanks and limbs mid-grey, underparts pale grey to creamy-white. Tail mid-grey darkening towards tip. Face and ears pale grey to white, tip of muzzle dark grey, nose black. Top of head and cheeks mid-grey, with distinctive chestnut-orange crown above eyebrow line, forming a V-shape with mid-grey area on top of head.

Medium-sized lemur with horizontal body posture. Can only be confused with Sanford's Brown Lemur, as both species are sympatric throughout majority of their ranges. Distinctive chestnut-orange crown and very marked sexual dichromatism of Crowned Lemur should prevent misidentification.

Female (left) and male (right) Crowned Lemurs, Ankarana Special Reserve.

HABITAT & DISTRIBUTION Prefers semi-deciduous dry lowland and mid-elevation forests, but also occurs in primary and secondary humid forests, from sea level to *c.*1,400m (Hawkins *et al.* 1990). Also in wooded savanna and agricultural areas, including plantations (Hending *et al.* 2018). In Ankarana also able to move between forest patches over razor-sharp limestone karst (*tsingy*) formations. Restricted to far north. Occurs from island's northern tip, Cap d'Ambre Peninsula, to region south of Ambilobe on west side (southern limit probably Mahavavy du Nord River) and in east to regions south of Daraina, and just south of Manambato River in Analamanara Forest. Also recorded just north of Bemarivo River. Population densities vary considerably with habitat type and location: overall mean across forest patches is *c.*20–30 individuals/km² (Banks 2005, Reuter *et al.* 2020b); in forested canyons of Ankarana

National Park, extraordinary densities of *c.*210–275 individuals/km² (Gudiel *et al.* 2017). Across much of range, sympatric with Sanford's Brown Lemur, which is more patchily distributed; Crowned Lemur tolerates drier and more degraded forests. Widely impacted by habitat loss and fragmentation. Hunted for subsistence consumption and local bushmeat trade (Gilles & Reuter 2014). Also suffers live collection for pet trade (Reuter *et al.* 2016, Reuter & Schaefer 2017). IUCN Endangered.

BEHAVIOUR Cathemeral: active both day and night throughout year (Freed 1996). Lives in groups of 5–15, with five or six the norm (Freed 1995), containing several adult females and males plus infants and juveniles. When feeding large groups may split into subunits of 2–4, where guttural grunts help maintain contact with adjacent subgroups.

Group size does not appear to vary with habitat (Gudiel *et al.* 2017). Occupies home range of 10–15ha, with considerable overlap between ranges of neighbouring groups. Travels between 800 and 1,000m per day while foraging (Freed 1996). One of only three *Eulemur* species with a degree of female dominance (Kappeler 1993).

Can occur in mixed-species associations with Sanford's Brown Lemur year-round, including the dry season (Gudiel *et al.* 2017), although more prevalent in wet season when resources are abundant and two species aggregate in foraging groups for 20–30% of day (Freed 1996). Crowned Lemur forages from ground level to canopy, with a preference for lower levels and understorey with lianas. This reduces competition with Sanford's Brown Lemur which prefers the upper levels; little interspecific aggression, but Sanford's Brown Lemur normally dominant.

Diet dominated by fruit, *c.*80–90%, during both wet and dry seasons (Reuter 2015). In dry season flowers, pollen and young leaves also taken, together with invertebrates and occasionally soil. In wet season greater diversity of plant species is utilised (Freed 1996). Also raids agricultural areas and plantations (Hending *et al.* 2018). Circumstantial evidence suggests may also take eggs from birds' nests: adult Hook-billed Vanga close to its nest seen violently mobbing female Crowned Lemur with small infant, causing infant to fall to ground (Garbutt 2021).

Mating occurs during late May and June, gestation *c.*125 days, births occur from mid-September through October with twins and singletons equally common (Kappeler 1987). Infants initially carried on mother's front, then move around to ride on back. Weaning occurs at six to seven months. Adult size and sexual maturity reached prior to two years.

Predated by Fosa and large raptors, and in Ankarana may also be eaten by crocodiles *Crocodylus niloticus* (Goodman 2003).

WHERE TO SEE A relatively easy species to see, especially in Ankarana National Park in *Canyon Forestière* and around *Campement Anilotra* (Camp des Anglais), *Campement d'Andrafiabe* (Camp des Américans) and *Lac Vert*. Also seen in Montagne d'Ambre National Park near main campsite, around *Station Roussettes* and at viewpoint overlooking *Grande Cascade*. Relatively easy to see in Andrafiamena-Andavakoera Protected Forest south of Analamerana (Black Lemur Camp) and near Daraina, especially forests around Camp Tattersalli. Latter location also very good for Golden-crowned Sifaka.

RED-BELLIED LEMUR
Eulemur rubriventer

(I. Geoffroy Saint-Hilaire, 1850)

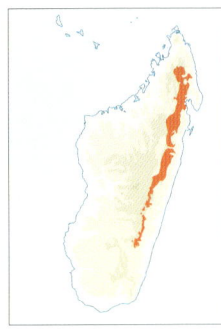

Malagasy names: Tongona, Barimaso, Tongo, Halimena, Soamiera, Kirioka, Varikamena

MEASUREMENTS
Head-body: 350–400mm. Tail: 430–530mm. Weight: 1.6–2.4kg. (Glander *et al.* 1992)

DESCRIPTION & IDENTIFICATION Sexually dichromatic. *Male:* Pelage long and dense. Upperparts and chest rich dark chestnut-brown to maroon, underparts slightly paler coppery-red. Tail noticeably

Male Red-bellied Lemur, Marojejy National Park.

Female Red-bellied Lemur, Ranomafana National Park.

darker, often almost black. Top of head, face and muzzle darker, often slate-grey. Conspicuous white patches under eyes extend to muzzle. No ear-tufts, but fur around ears dense and gives head a 'squarish' look. *Female*: Upperparts rich chestnut-brown, chest and underparts creamy-white. Tail dark grey to black. Head less 'squarish' than male and crown

not darker. Face and muzzle dark slate-grey, with white under eyes patches dramatically reduced and absent in some individuals. Lower cheeks and beard creamy-white.

A medium-sized lemur with horizontal posture and rich dense coat. In different parts of broad range occurs sympatrically with four other *Eulemur* species, all broadly similar in size and shape: White-fronted Brown Lemur in north, Brown Lemur in centre-east, Red-fronted Brown Lemur in south-central eastern areas and White-collared Brown Lemur at southern extreme of range. Distinguishing Red-bellied Lemur is straightforward: its coat is denser and richer russet-maroon, rather than brown. Also Red-bellied Lemur is encountered in small family groups, rather than larger multi-adult troops of congeners.

HABITAT & DISTRIBUTION Primary lowland, mid-altitude and high-altitude rainforest and occasionally second growth bordering primary habitat. Intolerant of degraded forest (Tecot 2008). Apparently sparsely distributed (prefers intact forest) and uncommon throughout eastern rainforests from Tsaratanana Massif in north to Manampatra River (south of Andringitra Massif) in south (Irwin *et al.* 2005) to, but not including, Masoala Peninsula. Previously ranged further south to Mananara River (Sterling & Ramaroson 1996). Recorded from 70m to 2,400m (Tsaratanana), although middle to high elevations apparently preferred.

Occurs at relatively low densities, e.g., in Ranomafana National Park *c.*5 individuals/km² (Irwin *et al.* 2005, Wright *et al.* 2012) and *c.*6.5 individuals/km² in Tsinjoarivo Forest (Rakotomalala *et al.* 2017a). Appears absent from some areas. Considerably rarer than sympatric *Eulemur* species.

Highly susceptible to habitat fragmentation and edge effect (Eppley *et al.* 2020d): in recent years has apparently disappeared from some relatively large forest fragments (Randriahaingo *et al.* 2014, King & Ravaloharimanitra 2017). Also suffers direct hunting and occasional collection for pet trade (Reuter & Schafer 2017). IUCN Vulnerable.

BEHAVIOUR Lives in family units of 2–6, comprising a monogamous adult pair and dependent offspring; some larger groups contain more than one adult of each sex (Overdorff 1996b, Overdorff & Tecot 2006, Tecot *et al.* 2016, Jacobs *et al.* 2018). Groups occupy home range of 12–15ha, which is defended, although some observations suggest neighbouring groups rarely show aggressive behaviour towards each other. Aggressive behaviour between two males in a group has been observed. Instigated and led by dominant female, groups travel 400–500m per day. Distance varies seasonally according to food availability; in times of shortage may travel more than 1km per day (Overdorff 1992a, 1996a).

Cathemeral, with activity patterns varying with seasons and food availability (Overdorff & Rasmussen 1995, Holmes *et al.* 2015). Fruits are mainstay of diet, including introduced species like Strawberry Guava *Psidium cattleianum*. They are important seed dispersers, either by dropping seeds when eating or via their faeces (Wright *et al.* 2012). When fruits unavailable, flowers and leaves also taken and feeding bouts often continue after dark. Also licks nectar from flowers, with 'brush-like' tongue in non-destructive way, so may transfer pollen between intact flowers and aid pollination. Utilises nearly 70 different plant species in a year. Invertebrates, especially millipedes, constitute a small, but important element of diet. Red-bellied Lemurs detoxify millipedes that produce noxious secretions, by repeatedly rolling them in their hands while salivating on them (Overdorff 1992b, 1993b, Tecot 2008, Fenosa & Razafindraibe 2020).

Single young or twins born in September–October after gestation of *c.*123–127 days (Tecot 2010). Breeding may occur annually or in alternate years; timed so peak lactation costs coincide with maximum fruit availability (Wright *et al.* 2006). Initially carried on mother's belly, later moves around to ride on to back. Infant also carried by male, who then may form a focus for other infants (Overdorff 1996b). After *c.*35 days female stops carrying offspring, although male may continue to do so until *c.*100 days of age. Weaning occurs between six and seven months. Infant mortality *c.*50% (Tecot 2010). Occasionally predated by Fosa and large raptors (Karpanty & Wright 2006).

WHERE TO SEE Ranomafana National Park is best location, where groups are habituated in both Talatakely (mid-elevation) and Vohiparara (high-elevation) zones. During late April–May, when Strawberry Guavas are fruiting, they can be particularly easy to find (together with Red-fronted Brown Lemur). Alternatively, can also be seen in Andasibe-Mantadia National Park or Marojejy National Park, generally at higher elevations, between Camp Mantella and Camp Marojejia or above Camp Marojejia.

BLACK LEMUR
Eulemur macaco
(Linnaeus, 1766)

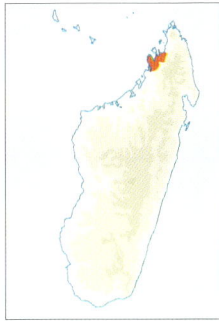

Malagasy name: Ankomba Joby

MEASUREMENTS
Head-body: 390–450mm. Tail: 510–650mm. Weight: 1.8–2.4kg. (Bayart & Simmen 2005)

DESCRIPTION & IDENTIFICATION Sexually dichromatic. *Male*: Entire pelage varies from dark chocolate-brown to almost black (in bright sunlight appears reddish-brown to dark chocolate-brown). Ears have lavishly long black tufts. Eyes yellow-orange to deep orange. *Female*: Upperparts golden-brown to mid-brown or rich chestnut-brown, becoming paler on flanks and limbs. Tail often darker rich chestnut-brown, particularly towards tip. Feet slate-grey to dark brownish-black. Throat and chin pale brown, gradually changing to pale grey-brown, even off-white on belly. Face and muzzle dark grey, top of head and temple pale grey. Ears extravagantly tufted with long white hair extending around cheeks and changing to orangey-chestnut beard. Eyes yellow-orange to deep orange.

In far south of range in zone of hybridisation (see below) coloration differs from typical form: ear-tufts and beard less well developed and female's throat, chin, top of head and temple all white (Schwitzer & Kaumanns 2005).

A medium-sized lemur with long tail and highly distinctive coloration; the most sexually dichromatic lemur. Confusion unlikely as allopatric with congeners over most of range. At southern and eastern limits of range (Golaka, Manongarivo and Tsaratanana Massifs) and in Ifasy and Ramena River valleys sympatric with Brown Lemur (Birkinshaw *et al.* 2000), and in western parts of Tsaratanana sympatric with Red-bellied Lemur; distinctive pelage and extreme sexual dichromatism of Black Lemur should prevent confusion.

HABITAT & DISTRIBUTION An adaptable species: occurs in moist Sambirano forests, similar humid forests on offshore islands and modified secondary habitats, including timber, coffee and cashew nut plantations.

Male (right) and female (left) Black Lemurs, Lokobe Reserve.

Recorded up to 1,600m (Goodman & Schütz 2000). Restricted to Sambirano region and adjacent offshore islands in north-west. Northern and north-east limit is Mahavavy du Nord River. Eastern limits unclear, but certainly includes some of Tsaratanana Massif. To west, range includes Ampasindava Peninsula, Nosy Faly Peninsula (Ambato Massif) and offshore islands of Nosy Be and Nosy Komba. Andranomalaza River previously thought to be southern limit, with hybridisation zone between Black Lemur and Blue-eyed Black Lemur *E. flavifrons* south and east of Manongarivo Massif and north of Andranomalaza River (Meyers *et al.* 1989, Rabarivola *et al.* 1991, Schwitzer *et al.* 2005, 2006). Recent extensive survey found no evidence of hybridisation zone and suggests southern boundary is Antsahakolana and Manongarivo Rivers (Tinsman *et al.* 2019).

Recorded at relatively high population densities in good quality habitat: in Ambato Massif Classified Forest, *c.*200 individuals/km², dropping to *c.*40–50 individuals/km² in poor-quality fragmented forest (Bayart & Simmen 2005). As many as *c.*400 individuals/km² in Manongarivo Special Reserve (Rakotoarinivo *et al.* 2011) and *c.*380 individuals/km² in Galoko-Kalobinono Reserve (Sahondrainjaka & Fanomezana 2018).

Habitat loss and forest fragmentation are ever-increasing threats (Tinsman *et al.* 2019) as is hunting for bushmeat and live trapping for local pet trade (Reuter *et al.* 2016, Reuter & Schaefer 2017). IUCN Endangered.

BEHAVIOUR Cathemeral: daytime activity concentrated around early morning and late afternoon, and extent of nocturnal activity varies seasonally or with lunar cycle. During middle of dry season (June–July) populations on Nosy Be completely inactive at night, and mornings are spent sunning at top of canopy. In contrast, at end of dry period (October–December) nocturnal activity predominates, coinciding with fruiting of favourite large trees, e.g., Ramy *Canarium madagascariensis* (Burseraceae) (Andrews & Birkinshaw 1998). On mainland, lunar phase is major influence on activity patterns: when moon is waxing, full nocturnal activity reaches its peak, probably related to increased light and lemur's improved nighttime visual acuity. Nocturnal activity also occurs year-round, but diurnal activity reaches its peak during early wet season (December–January) (Colquhoun 1996, 1998a).

Black Lemurs form groups of 2–15, with 7–10 the norm. Larger groups associated with anthropogenic and modified rather than primary forest habitats (Bayart & Simmen 2005). Groups comprise several adult males and females in roughly equal numbers, plus dependent offspring. Adult females form core of group, with activities dictated by dominant female. Adult males move between groups, especially during mating season (Bayart & Simmen 2005). Group cohesiveness maintained by regular guttural grunts and contact calls from all members. In good-quality forest (Lokobe Reserve, Ambato Massif), home ranges average 4–6ha, with considerable overlap between neighbouring groups. In more disturbed, human-altered forests (Ampasikely), mean home range size is *c.*18ha (Bayart & Simmen 2005). In some areas groups remain separate by day, but may come together after dark.

Foraging concentrated in middle and upper canopy. Diet varied, but dominated by ripe fruit, with flowers, leaves, fungi, bark, gums and occasionally invertebrates like millipedes and insects, e.g., cicadas (Birkinshaw 2003, Simmen *et al.* 2007). In degraded areas, non-native species, including cash crops (e.g., cashew, papaya, mango) are eaten (Simmen *et al.* 2007). In dry season nectar also eaten in significant quantities, making Black Lemur an important pollinator, especially of Traveller's Tree *Ravenala madagascariensis* (Strelitziaceae) and leguminous canopy tree, *Parkia madagascariensis* (Fabaceae). In Lokobe Reserve on Nosy Be, Black Lemurs may be the sole seed-disperser for a number of tree species (Birkinshaw 1999, Birkinshaw & Colquhoun 1998, 2003).

In degraded habitats they descend to ground and forage in leaf litter for fallen fruit and fungi, and even eat soil. By day they feed mainly in understorey, rather than canopy, and move from one resource patch to next with greater frequency. At night, foraging efforts are concentrated in canopy (when daytime avian predators are not active) and travel distances are reduced (Birkinshaw 1999).

Most mating takes place between late April and June, but can persist until August (Bayart & Simmen 2005). Gestation is *c.*125 days: most births occur between late August and mid-October, but can be as late as November to early December. A single infant is normal. Weaning occurs at 6–7 months, with sexual maturity reached around two years of age. Infanticide has been observed; one female killing infant of another and consequently assuming dominant status within group (Andrews 1998). Such behaviour thought to be extremely rare and observed in modified habitat where group survived under stressful circumstances.

WHERE TO SEE Easy to see in buffer zone of Lokobe Reserve on Nosy Be. From Ambatozavavy take a *pirogue* around the coast to Ampasipohy.

Female (left) and male (right) Black Lemur, Lokobe Reserve.

The modified forest area (second growth and some crops like vanilla) where Black Lemurs are easily seen is a short walk inland. Also a good place to see Nosy Be Sportive Lemur. Alternatively, the neighbouring island of Nosy Komba offers a chance to see totally habituated groups living in degraded forest on edge of village. Local people sell bananas to tourists to feed the lemurs. The contrived 'ecotourist experience' is unfulfilling and should not be encouraged.

BLUE-EYED BLACK LEMUR
Eulemur flavifrons (J. E. Gray, 1867)

Other names: Sclater's Black Lemur, Sclater's Lemur

Malagasy names: Akomba, Akomba sy Manga Maso

MEASUREMENTS
Head-body: 390–450mm. Tail: 510–650mm. Weight: 1.8-1.9kg. (Schwitzer *et al.* 2013d)

DESCRIPTION & IDENTIFICATION Sexually dichromatic. *Male*: Pelage shorter than *E. macaco* with softer appearance. Normally entirely black, sometimes dark brown tones apparent. Distinct ridge of hair on forehead forms short crest. No ear-tufts. Eyes characteristically blue-grey to mid-blue. *Female*: Upperparts and tail pale rufous-tan, sometimes rufous-grey. Underparts creamy-white to grey. In sunlight appears golden-orange. Hands and feet dark grey. Crown rufous-tan, face pale grey around eyes, darkening to slate-grey on muzzle and brown on nose. No ear-tufts. Eyes mid-blue to blue-grey. Females noticeably paler than Black Lemur.

Together with Black Lemur, most sexually dichromatic Malagasy primate. A medium-sized lemur with long tail and distinctive coloration, adopts horizontal posture and moves quadrupedally. Only similar species is Black Lemur, range boundary of two species is uncertain and confusion may be possible within this zone (see below). Allopatric with all other *Eulemur* species and cannot be confused with any other species. Other than humans, the only blue-eyed primate!

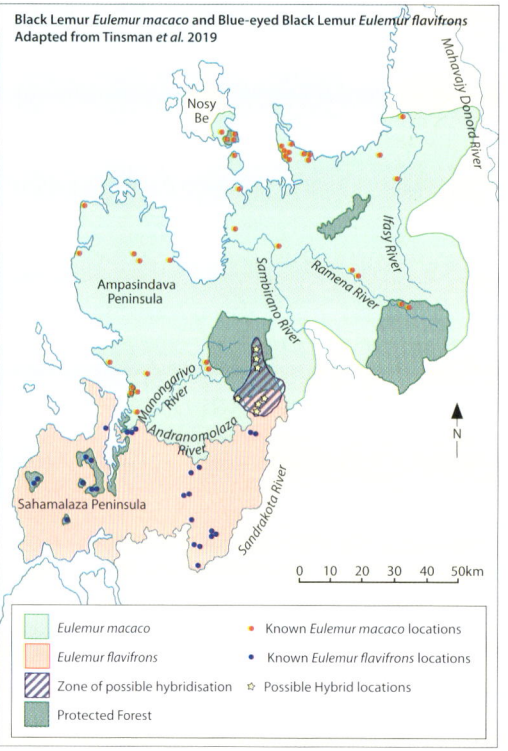

Black Lemur *Eulemur macaco* and Blue-eyed Black Lemur *Eulemur flavifrons*
Adapted from Tinsman *et al.* 2019

Nosy Be

Mahavoly Donord River

Ifasy River

Ampasindava Peninsula

Ramena River

Sambirano River

Manongarivo River

Andranomolaza River

Sahamalaza Peninsula

Sandrakota River

N

0 10 20 30 40 50km

Eulemur macaco	• Known *Eulemur macaco* locations
Eulemur flavifrons	• Known *Eulemur flavifrons* locations
Zone of possible hybridisation	☆ Possible Hybrid locations
Protected Forest	

Male Blue-eyed Black Lemur, Ankarafa Forest, Sahamalaza Peninsula.

Female Blue-eyed Black Lemur with male infant, Ankarafa Forest, Sahamalaza Peninsula.

HABITAT & DISTRIBUTION Inhabits more or less disturbed secondary sub-humid forest in southern Sambirano – a zone of transition between Sambirano to the north and typical western dry deciduous forest to the south. Also lives in adjacent coffee and citrus plantations. Recorded from sea level to 1,200m (Randriatahina & Rabarivola 2004). Extremely restricted range centred on Sahamalaza Peninsula and narrow strip of forest on adjacent mainland in north-west. Southern limit is combination of Maevarano and Sandrakota Rivers. To the east, range is bounded by Sandrakota River. Previously Andranomalaza River was thought to be northern limit, with a zone of hybridisation with *E. macaco* between Andranomalaza River and southern edge of Manongarivo Massif, with hybrids similar in appearance to pure-bred *E. flavifrons* but with pale brown eyes (Rabarivola

et al. 1991, Andrianjakarivelo 2004, Randriatahina & Rabarivola 2004). Recent extensive surveys of numerous forest patches found no evidence of such a hybridisation zone, suggesting the boundary between Blue-eyed Black Lemur and Black Lemur is more clearly defined and is defined by the Andranomalaza, Manongarivo and Antsahakolana Rivers (Tinsman *et al.* 2019).

In forest fragments in east of range mean population density of 24 indviduals/km^2 (Andrianjakarivelo 2004), with Ankarafa Forest on Sahamalaza Peninsula harbouring 60 individuals/km^2 (Schwitzer *et al.* 2007a) and 97 individuals/km^2 (Volampeno *et al.* 2011) recorded, although latter figure thought to be unusually high (Volampeno *et al.* 2020).

Principal threat is severe habitat destruction due to continuing slash-and-burn agriculture, selective logging, mining and forest fires (Seiler *et al.* 2010).

Blue-eyed Black Lemur is also hunted for food, especially by Tsimihety people in east of range. Also trapped for local pet trade (Reuter & Schaefer 2017). IUCN Critically Endangered.

BEHAVIOUR Blue-eyed Black Lemurs are cathemeral year-round with degree of nocturnal activity linked to lunar cycle, and maximum nighttime activity corresponding to high levels of lunar illumination (Schwitzer & Kaumanns 2005). Peak activity periods during early morning and evening twilight. Total daily activity and overall levels of nocturnal activity are higher in secondary forest than in primary forest, reflecting fewer or poorer quality resources (Schwitzer et al. 2007b).

An adaptable species able to cope with changing habitat and environment, and capable of crossing open ground between forest fragments (Prodger et al. 2018). Home range size and utilisation of resources differs between primary and secondary forests. In primary forest home ranges smaller (less than 6ha) and densities higher than in degraded forest where home ranges are larger (up to 20ha) and densities lower, reflecting diminished, more spatially separated, resources (Schwitzer et al. 2007a). Home range size also reduces in dry season and there is overlap with neighbouring groups.

Group size ranges from 2–15, with 7–10 the norm: groups are multi-male/multi-female, comprising one or two adult and subadult females which form the cohesive core. Males are more peripheral and switch groups (Andrianjakarivelo 2004, Randriatahina & Rabarivola 2004, Schwitzer et al. 2005, 2009, Volampeno et al. 2011).

Diet dominated by fruits and leaves. Over annual period, more than 70 different plant species are consumed, with c.52% fruit and c.48% leaves. Much smaller quantities of flowers, insects, insect exudate and fungi are eaten occasionally (Polowinsky & Schwitzer 2009). Seasonally, Blue-eyed Black Lemur may eat large quantities of cicadas.

Blue-eyed Black Lemur observed indulging in 'anting' behaviour. Both males and females allow carpenter ants to swarm over and attack their bodies, even allowing ants in their mouths while salivating. Bouts may last up to 30 minutes. The ants' formic acid is thought to protect the lemur against ticks and other parasites (Buckley et al. 2020).

Mating occurs late April to early June, infants born late August to late October after gestation of c.120–125 days. Infants greyish-black. Singletons the norm, twins rare. Initially infant carried on

mother's abdomen, by four weeks it rides on mother's back. After six weeks, it can leave mother for short bouts of independent activity. Play with other infants begins around seven weeks. Solid food is taken from ten weeks. Weaning at 6–7 months (Schwitzer et al. 2013d).

Because habitats on Sahamalaza Peninsula are fragmented and degraded, populations of Blue-eyed Black Lemur are under stress and consequently carry higher parasite loads than other lemurs. Parasite prevalence also higher for individuals inhabiting secondary forest compared to those in primary forest (Schwitzer et al. 2010).

WHERE TO SEE A challenging lemur to see, because of remote inaccessible range. Forests south of Maromandia (and south of Andranomalaza River), now part of Sahamalaza-Iles Radama National Park, provide good opportunties. Otherwise, larger forest fragments on Sahamalaza Peninsula (e.g. Ankarafa Forest) can be reached by boat from Analalava and a two-hour walk. Sahamalaza Peninsula can also be reached by boat from Maromandia.

RUFFED LEMURS
GENUS *Varecia*

Ruffed lemurs are among the most beautiful of Malagasy primates and are the largest extant members of the family Lemuridae. They are confined to eastern rainforests and are patchily distributed and uncommon to rare throughout their ranges.

They have characteristic long muzzles with 'dog-like' faces, luxuriant, patterned pelages and lavishly tufted or 'ruffed' ears. Their limbs are relatively short, but tails very long and bushy. Size and pelage pattern are the same for both sexes. Ruffed lemurs are extremely arboreal spending the majority of time in the canopy, 15–25m above the forest floor. They move quadrupedally and are capable of leaping considerable distances between branches.

Unlike other genera in the family Lemuridae, *Varecia* is primarily diurnal, exhibiting only limited nocturnal behaviour (Donati & Borgognini-Tarli 2006a, Bray et al. 2017), and is among the most frugivorous of lemurs, spending c.60–95% of foraging time feeding on fruits (Vasey et al. in press). Majority of time spent in middle to upper strata, relying heavily on large-crowned fruiting trees in primary forest (Vasey 2000a, Vasey et al. in press), something that makes them particularly vulnerable to habitat disturbance and fragmentation (Balko et al. 1995). They rarely descend to the ground and do not traverse large open areas (Vasey et al. in press). During feeding bouts, they often adopt a suspensory posture, hanging solely by their hindlimbs. In addition to fruits, flowers, nectar, seeds and leaves are also eaten seasonally and they are likely important seed dispersers, especially of Ramy Canarium madagascariensis (Burseraceae) and pollinators, especially of Traveller's Tree Ravenala madagascariensis (Strelitziaceae).

Ruffed lemurs are polygamous (both males and females may mate with multiple partners during the breeding season), have the shortest gestation period and largest litters, normally two or three infants, in the family Lemuridae. Varecia species are the only primates known to construct nests specifically for giving birth and initial infant rearing. Young remain in the nest for their first two weeks (Vasey 1997b, 2007, Schwitzer et al. 2013d). Rare in primates, allo-parenting is also common, with all members of the group participating in raising offspring. Non-parent group members often care for and guard infants while the mother is away foraging, and carers alarm call if potential danger is sensed. Males similarly provide care and guard infants for different mothers.

Initial infant growth is very rapid; they attain 75% of adult weight by four months (Pereira et al. 1987). Females have three pairs of mammae (one pectoral, two abdominal) and reach sexual maturity in less than two years, whereas males attain maturity after three to four years (Foerg 1982). There is no significant difference in size or pelage coloration between the sexes (Vasey 2003).

Ruffed lemurs are highly vocal, with an extensive repertoire of calls used in many contexts, including a raucous bark or roar that regularly develops into a simultaneous multi-individual group bout of calling that is highly distinctive, serves to denote territory and often gives an initial indication when trying to locate the lemurs in the wild.

Originally, ruffed lemurs were included in the genus Lemur. Subsequently, a single species, V. variegata, was recognised, divided into two subspecies, Black-and-white Ruffed Lemur V. v. variegata and Red Ruffed Lemur V. v. ruba (Tattersall 1982). More recent research concluded that each warrants species status, V. variegata and V. ruba (Vasey & Tattersall 2002). Further, there are numerous colour morphs, especially in V. variegata and these have been proposed as subspecies (Groves 2001) for which

genetic research provides partial support but the picture is far from clear, and these subspecific divisions are not conclusive (Vasey & Tattersall 2002).

Both species of ruffed lemur are severely threatened. They are particularly susceptible to habitat disturbance and fragmentation, and because of their large size are widely hunted for food, and are even considered a delicacy in some areas. In addition, they are commonly trapped and kept as illegal pets.

Ruffed lemurs' large size and distinctive coloration should prevent confusion with all other lemur taxa.

BLACK-AND-WHITE RUFFED LEMUR Varecia variegata (Kerr, 1792)

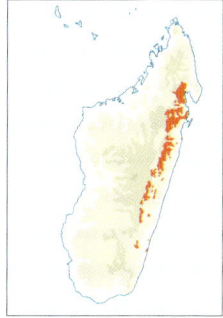

Other names: Variegated Black-and-white Ruffed Lemur, Pied Black-and-white Ruffed Lemur

Malagasy names: Vary, Varijatsy, Varikandra, Varikandana

MEASUREMENTS
Head-body: 430–570mm. Tail: 600–650mm. Weight: c.2.5–4.8kg. (Morland 1991a, Britt 1997, Balko 1998, Vasey 2003, Baden et al. 2008)

DESCRIPTION & IDENTIFICATION Pelage primarily black and white, with considerable variation in proportions across geographic range (Vasey & Tattersall 2002). Muzzle long, giving face a rather 'dog-like' appearance. Tail strikingly long (mean 60cm) and lavishly furred. Body fur also luxuriant, and predominantly consists of different-sized patches of black and white. In general, tail, hands and feet, inner surfaces of limbs, shoulders, face, muzzle and top of head are black, whilst back (or a portion of it), flanks, rump and majority of hindlimbs are white. Ears white and lavishly tufted; long hair forms a continuous white 'ruff' around cheeks and under chin.

Overall pattern varies throughout range; in very general terms from north to south, the proportion of black reduces as white areas increase. This pelage variation has prompted the description of different forms (Tattersall 1982), with apparent correlation to geographic range that has led to proposed subspecies (Groves 2001). Attempts to

Black-and-white Ruffed Lemur, Andasibe-Mantadia National Park.

confirm subspecies divisions in the field using genetic techniques provide only partial support, with some suggesting only *variegata* and *editorum* warrant recognition (Baden *et al.* 2014b). The continued acceptance of three subspecies based on pelage patterns must be treated with caution (particularly with respect to location): more likely *V. variegata* is a single taxon displaying multiple pelage patterns that cannot be differentiated into discrete geographic subspecies (Vasey *et al.* in press).

Nonetheless, as currently recognised by the IUCN (Louis *et al.* 2020p), the subspecies are:

Northern Black-and-white Ruffed Lemur

V. v. subcincta (A. Smith, 1833) (extreme north of range): black areas dominate dorsally, especially over lower back and outer thighs, white patches being restricted to lower forearms, a saddle over shoulders and rear hindquarters.

Variegated Black-and-white Ruffed Lemur

V. v. variegata (Kerr, 1792) (mainly central populations): black patches only on shoulders, upper arms, nape and head, on front of upper thighs and inner hindlimbs. Nape, upper and lower back and flanks are white.

Southern Black-and-white Ruffed Lemur

V. v. editorum (Hill, 1953) (mainly extreme south of range): large swathes of white over lower back and flanks, black areas restricted to shoulders, upper arms and inner thighs, and shoulder patches sometimes meet to form a mantle.

Horizontal posture, long tail and utterly distinctive black-and-white coloration mean this species should not be confused with any other lemur: Indri *Indri indri* are similarly black and white, but are larger, adopt vertical clinging posture and are (virtually) tail-less.

HABITAT & DISTRIBUTION Primary lowland and mid-altitude rainforest, from sea level to c.1,300m (Vasey 2003). Very patchily distributed throughout range, but can survive in isolated forest fragments, e.g., Kianjavato and Manombo (Ratsimbazafy 2002). Distribution and ranges of subspecies, *subcincta*, *editorum* and *variegata* uncertain and their recognition may not be valid (Vasey *et al.* in press). The ranges run from north to south (Vasey & Tattersall 2002).

Northern Black-and-white Ruffed Lemur *V. v. subcincta* (A. Smith, 1833) Northern limit is Antainambalana River, north-west of Maroantsetra (Red Ruffed Lemur *V. rubra* occurs on opposite side of river, although areas of hybridisation and intermediate coloration may occur north-east of Antainambalana River), and north to Antohaka-Lava Classified Forest, south-east of Andapa (Louis *et al.* 2020p). Range extends south to Anove River, south of Mananara-Nord. Known sites include Mananara-Nord, Marotandrano, Atialanankorendrina, Makira Forest, and introduced on Nosy Mangabe in 1930s.

Variegated Black-and-white Ruffed Lemur *V. v. variegata* (Kerr, 1792) Occurs south of Anove River, from approximately Ambatovaky Special Reserve, to region between Betampona and Zahamena National Park, although southern limit is ill-defined, but known to be north of Mantadia National Park and Torotorofotsy Ramsar Site (Louis *et al.* 2020w). Known localities include Zahamena National Park, Betampona and Ambatovaky Special Reserves.

Southern Black-and-white Ruffed Lemur *V. v. editorum* (Hill, 1953) Northern limits unclear, probably around Mantadia National Park; occurs south as far as Manombo Special Reserve, with additional populations further south at Ambalavera in Midongy-Vondrozo Corridor (Rakotonirina *et al.* 2013) and small lowland fragments east of Midongy-Vondrozo Corridor, where subspecific status yet to be confirmed (Rakotonirina *et al.* 2013, Louis *et al.* 2020p). Furthermore, form occurring in Mangerivola Special Reserve unknown (Louis *et al.* 2020p). Ranges of Southern Black-and-white Ruffed Lemur and Variegated Black-and-white Ruffed Lemur overlap, and intermediates may exist (Groves 2001). Known sites include Mantadia National Park, Maromahiza, Andriantantely, Ranomafana National Park, Kianjavato, Vatovavy, Lakia, Fandriana Protected Area, Manombo Special Reserve and Fandriana-Vondrozo Corridor (CO-FAV).

Population densities range from 0.4–2.5 individuals /km^2 in Manombo Special Reserve, to 10–15 individuals/km^2 in Antanamalaza and 24 individuals /km^2 in Mangevo (Ranomafana National Park) (Baden 2011). Populations in Manombo Special Reserve are extremely low after suffering severe impact from Cyclone Gretelle in 1997. On Nosy Mangabe densities are 29–43 individuals/km^2, artificially high due to isolation of 500ha island with no predators (Morland 1991a, Ratsimbazafy 2002, Louis *et al.* 2005). Such densities probably detrimental as individuals suffer over-competition for resources and are consequently smaller than mainland counterparts (average 2.5–2.95kg on Nosy Mangabe, versus 3.1–4.5kg on mainland) (Baden *et al.* 2008).

Captive-bred (in USA) individuals released into Betampona Reserve in 1997/98, with some integrating into wild groups and others predated by Fosa (Britt *et al.* 2001, 2003). Attempts in 2006 to re-introduce a small population from Mantadia National Park into nearby Analamazaotra Special Reserve (extirpated in early 1970s) were initially only partially successful, with losses due to hunting and other anthropogenic impacts (Rasoamanarivo *et al.* 2015, Louis *et al.* 2020p).

Deforestation and fragmentation caused by slash-and-burn agriculture remain primary direct threat. Fragmentation and other anthropogenic pressures negatively impact gene flow (Baden *et al.* 2019). Also due to large size, specifically targeted and one of most heavily hunted lemurs, particularly prized for flavour and quality of meat (Golden 2005, Borgerson *et al.* 2016). Increased ruffed lemur vocalisations, a consequence of seasonal rise in food availability, correlates to growing hunting pressure (Ratsimbazafy 2002, Vasey 2003). Very susceptible

Black-and-white Ruffed Lemur, Nosy Mangabe Reserve.

to disturbance and logging, and avoids forest edges; one of first species to disappear when humans encroach and impact on intact rainforest (Balko *et al.* 1995, White *et al.* 1995, Ratsimbazafy 2002, Balko & Underwood 2005, Seaman *et al.* 2018). Climate change probably also negatively impacting ruffed lemur populations (Morelli *et al.* 2020). IUCN Critically Endangered.

BEHAVIOUR Primarily diurnal, most active early in morning and late afternoon/evening. Principally a high-canopy dweller, preferring areas with tall tress (>90% of time spent in tree crowns (Vasey *et al.* in press). Year-round, *c.*30% of time dedicated to feeding, mainly in mid-canopy (11–20m above ground) (Beeby & Baden 2021). Very selective feeder, with diet dominated by fruit (*c.*70–90%, but can reach 100% at certain times of year), supplemented with nectar, flowers and leaves, the latter mainly during fruit-lean season (Morland 1991a, Balko 1996, 1998, Beeby & Baden 2021). Fruits of Ramy *Canarium madagascariensis* (Burseraceae) especially favoured, consequently Black-and-white Ruffed Lemur may be important seed disperser. Nectar available for only short periods each year but constitutes dominant food source when flowers in bloom. *Varecia* species use long snouts and tongues to reach deep inside flowers and lick nectar; they do not destroy the inflorescence, but collect pollen on their muzzles and fur, and transport this between flowers of different plants. Suspensory postures frequently adopted when feeding (Britt 2000). Traveller's Tree *Ravenala madagascariensis* (Strelitziaceae) is a favourite; size and structure of inflorescences, coupled with lemur's selectivity and method of feeding, strongly suggest ruffed lemurs are a major pollinator of Traveller's Trees and pollination system is co-evolved (Kress *et al.* 1994).

Group size and social system appear to vary considerably and be fluid (Vasey 2003, Baden *et al.* 2016). In mainland rainforests (e.g., Mangevo, Ranomafana National Park), groups are multi-male, multi-female units of 4–9 that may be cohesive and territorial (low fission-fusion dynamics) occupying home ranges of *c.*90–150ha (Balko 1998, Baden 2011, Baden *et al.* 2020a) or form communities with high levels of group fluidity (high fission-fusion dynamics) where members form small cohesive, typically mixed-sex, subgroups of unrelated individuals with regular movements between them, or spend time alone (community members never all in one place simultaneously), but collectively occupy a communally defended territory. Majority of subgroup movements are within a fraction of overall community range (Baden *et al.* 2020a,b). Subgroup dynamics alter with seasonal climatic variation, fruit abundance and availability, and female reproductive state (Baden 2011, Baden *et al.* 2014b, 2016). Overall, females use larger (*c.*20%) home ranges than males, but males and females have similar moderate home range overlap. May travel up to 2km per day (Baden *et al.* 2020a): travels less and rests more during fruiting season when food more abundant and locally clustered (Beeby & Baden 2021).

On Nosy Mangabe, groups are larger, consisting of 8–16 adults living in home ranges of *c.*30ha (Morland 1991a,b). Studies in Betampona Reserve on mainland suggest Black-and-white Ruffed Lemur forms monogamous groups of 2–5 individuals (adults with offspring) in home ranges of 16–43ha (Britt 1997). Overall variations may reflect seasonality and/or habitat. The strongest social bonds develop between females; those between males are much weaker. On Nosy Mangabe, group composition and ranging behaviour is seasonally variable: during warm wet season (November–April) females range widely either alone or in groups up to six; in cool dry season (May–October) smaller, more stable core groups occupy concentrated areas (Morland 1991a,b, 1993a).

Highly vocal: has an extensive vocal repertoire, broadly divided into three classes, high-amplitude, medium-amplitude and low-amplitude. High-amplitude include the raucous contagious 'roar/shriek' chorus participated in by all group members, including 3–4-month-old infants, which can carry up to 1km and allows groups to communicate and maintain spacing in forest (Vasey 2003). Such 'roar/shriek' choruses occur throughout day, but are concentrated at times of high activity and are more commonly heard during summer hot season. Similar high-amplitude 'abrupt roars' function as warning calls to aerial predators (large raptors), whereas 'pulsed squawks' signify terrestrial predators (e.g., Fosa). In response group members are agitated, wag their tails and quickly aggregate (Vasey 2003, Schwitzer *et al.* 2013d). Medium-amplitude calls include a 'growl' expressing mild alert or approach of another individual, or 'growl-snort' another general group alert call; 'chatter' and 'whine' calls are associated with subordinate status and submissive behaviour. Low-amplitude calls are 'grunts' signifying mild annoyance, 'huff' calls signal intense aggravation and a 'mew' call between mother and offspring or

Black-and-white Ruffed Lemur, Ranomafana National Park.

between adults when travelling. During mating periods, 'cough' calls signify aggression between males and females. Infants issue a 'squeak' call in distress (Schwitzer *et al.* 2013d).

Breeding is communal, but reproductive timing highly variable, with 'boom-bust' strategy, i.e. one synchronous reproductive event in several years (average 2–4 years, but one in six years recorded), which may be influenced by climatic factors (Morland 1993a, Baden 2011, Baden *et al.* 2013). In breeding years, mating occurs between May and July (female only receptive for one day per 30-day oestrous cycle), gestation *c*.102–109 days, most offspring born between September and early November (Rasmussen 1985, Morland 1990, 1993b, Baden 2011, 2019). During gestation, female builds several nests (3–15, mean *c*.8), location and density of which linked to distribution and availability of preferred food resources in area (Baden *et al.* 2013, Baden 2019). Breeding timed so peak lactation requirements coincide with maximum fruit availability (Wright *et al.* 2006).

Normal litter size two or three (Baden *et al.* 2013), but up to five recorded (in captivity). Young weigh just under 100g at birth (Brockman *et al.* 1987). Birth occurs in well-concealed nest – a platform or shallow bowl *c*.1m wide (Baden 2019) – constructed of twigs, leaves and vines, lined with mother's flank fur, generally 10–20m above ground. At birth, young have same fur pattern as adult, but pale blue eyes that change to adult's yellow-gold by end of second week. Young remain in natal nest for 3–22 days, with mother providing sole care (Baden 2019, Baden *et al.* 2013). Later mother carries young in her mouth, either transferring to alternate nest, or when older leaving infant concealed in canopy. When infants are three weeks or older, nests may be communal, with up to four females using one nest for their offspring, sometimes forming a crèche. Communal nesting correlates with increased infant survival (Baden *et al.* 2013). Infants may be 'parked' for several hours, allowing mother to forage more efficiently, over wide areas, during high-cost period of lactation (Morland 1990, 1993b).

Milk is rich (compared to other lemurs) (Tilden & Oftedal 1997) and young consequently develop more rapidly than other lemurs. At three weeks capable of following mother while regularly exchanging contact calls. Begin eating solid food from 40 days, weaned at 5–6 months, but may continue to suckle until seven to eight months (Morland 1989, 1990). By four months young are three-quarters grown and

as active and mobile as adults (Pereira *et al.* 1987, 1988). Infant mortality high, in some years 65% fail to reach three months; some die from accidental falls and related injuries. Sexual maturity reached at 18–20 months in females, 32–48 months in males (Foerg 1982, Morland 1990, 1991a).

Recorded predation by Fosa (Britt *et al.* 2001, 2003), but these animals were captive-bred and re-stocked into wild. Predation of truly wild animals occurs (Sefczek *et al.* 2018) but is probably rare, perhaps because Black-and-white Ruffed Lemurs are high-canopy specialists and difficult to catch.

WHERE TO SEE Northern Black-and-white Ruffed Lemur readily seen on island of Nosy Mangabe in Bay of Antongil, which is easily reached by boat from Maroantsetra. At reserves on mainland this lemur is more difficult to see. Two likely sites for Southern Black-and-white Ruffed Lemur are Ranomafana National Park, especially in April–May when non-native Strawberry Guava *Psidium cattleianum* is fruiting, and Mantadia National Park, especially around PK 15. Considerable patience, strenuous hiking and luck may be needed to ensure success.

Southern Black-and-white Ruffed Lemur also reintroduced to Analamazaotra Special Reserve (from where it was extirpated in early 1970s); in 2006 seven were translocated from nearby Mantadia National Park. Population grew to 25 individuals in five groups (in 2017), but now fewer than 15 individuals (Louis *et al.* 2020p), so consequently challenging to locate.

RED RUFFED LEMUR
Varecia rubra (É. Geoffroy Saint-Hilaire, 1812)

Malagasy names: Varimena, Varignena, Varinaina

MEASUREMENTS
Head-body: 500–550mm. Tail: 600–650mm. Weight: 2.1–3.6kg. (Vasey 2003, Dutton *et al.* 2008)

DESCRIPTION & IDENTIFICATION Pelage dense and luxuriant. Upperparts, legs and belly vary from deep chestnut-red, to red-orange, and honey-blonde. Distinctive ear-tufts, ruff around cheeks and throat also rich chestnut-red to honey-blonde. Inner limbs, belly, feet, long tail (mean 600mm), top of head, face and muzzle are black. Nape usually white. In some individuals further white, honey-blonde or pale red areas occur on wrists and ankles, rump and muzzle (Vasey 2003). Large size and distinctive coloration make confusion with any other species impossible.

HABITAT & DISTRIBUTION Primary lowland and mid-elevation rainforest from sea level to c.1,200m. Appears restricted to Masoala Peninsula and northern Makira region immediately north of Bay of Antongil. Western and northern limits ill-defined, but Antainambalana River probably forms boundary (Northern Black-and-white Ruffed Lemur occurs opposite side), with westernmost limits near confluence of Antainambalana and Sahantaha Rivers (Hekkala et al. 2007). Northernmost records c.30km south of Andapa in Besariaka Corridor and Antohakalava Forest (Vasey et al. in press). A zone of 'hybridisation' may once have occurred at confluence of Mahalevona, Vohimara and Antainambalana watersheds (Vasey & Tattersall 2002). Historical accounts suggest Red Ruffed Lemur once ranged north as far as Antalaha and Ankavanana River (Tattersall 1977).

Historical population density estimates vary from 31–53 individual/km^2 at Andranobe (Vasey 1997a) to 21–23 animals/km^2 at Ambatonakolahy (Rigamonti 1993, 1996). More recent surveys show densities near villages are lower: mean 2.9 individuals/km^2 near villages, compared to mean 8.3 individuals/km^2 for uninhabited interior of Masoala Peninsula (Borgerson et al. 2019).

Habitat loss and hunting are main threats. Very susceptible to disturbance: largest trees in forest often first to be removed, which are vital for feeding and provide crucial infant nest and 'parking' locations (Vasey et al. 2018). Widely (illegally) hunted throughout range, contributing significantly to local nutrition, subsistence economy, and health (Golden et al. 2014, Borgerson 2015, 2016, Borgerson et al. 2016, 2017, 2019). Red Ruffed Lemur loud calls convey their whereabouts to hunters. Hunting peaks during austral winter when lemurs loud call more frequently and are trapped when they visit seasonally fruiting trees (Vasey 2003, Borgerson 2015). Also sometimes trapped for illegal local pet trade (Reuter & Schaeffer 2017). IUCN Critically Endangered.

Red Ruffed Lemur, Masoala National Park.

Red Ruffed Lemur, Masoala National Park.

but range more widely through communal territory during hot rainy season (January–March) (Vasey 2005b). When habitats disturbed, Red Ruffed Lemur expands range and reduces size of community (Vasey 2000, 2005a).

Smell also important in territory defence; females scent-mark with ano-genital glands, males primarily use glands on muzzle, neck and chest by embracing branches and repeatedly rubbing themselves back and forth (Vasey 2003).

Red Ruffed Lemurs are highly vocal with a repertoire that varies seasonally; loud calls more frequent during hot summer when animals are more active, range widely and interact more regularly with dispersed members of their communities (Vasey 2003) (also see vocalisations of Black-and-white Ruffed Lemur).

Primarily frugivorous: c.75–95% of diet fruit (can reach 100% at certain times of year), remainder made up of flowers, nectar and leaves (Vasey 2002, 2004, Martinez 2010). Wide variety of plants eaten throughout year: >130 species from c.36 families recorded (Vasey 2000b, 2004, Martinez & Razafindratsimba 2014). Often hang upside-down by hindlimbs to reach fruit and flowers (suspension feeding) (Vasey 1999). High fruit diet means *V. rubra* is an important agent of dispersal, with seeds in faeces dropped up to c.500m from parent fruit tree: native tree species, *Calophyllum milvum* and *Garcinia verrucosa* (Clusiaceae) and multiple native *Ficus* species (Moraeceae) account for 70% of passed seeds (Martinez & Razafindratsimba 2014). Females feed consistently throughout seasons, but eat a greater proportion of leaves (protein) during pregnancy and lactation, reflecting increased energy costs of reproduction (Vasey 2000a, 2002, 2005b). Males feed more during the resource-rich, hot dry season (Vasey 2005b). Furthermore, treefall gaps in canopy caused by cyclone damage, often form localised food-rich areas that are seasonally exploited (Mogilewsky 2020).

Red Ruffed Lemurs are polygamous, with non-mating partner often helping with infant care (Vasey 2007). Reproduction occurs every one to two years, with longer inter-birth intervals or 'reproductive collapse' triggered by unfavourable resource availability/abundance or climatic events (Vasey & Borgerson 2009).

Mating season May–July, with births occurring from September through to early November after gestation of c.102–109 days (Vasey 2007, Vasey *et al.* in press). Constructs a nest of twigs and leaves in tree crowns concealed by dense foliage or

BEHAVIOUR Primarily diurnal, occupying very narrow niche: almost exclusively a high-canopy dweller, with >90% of time spent in crowns of tall trees (Vasey 2000a, Borgerson 2016). Lives in multi-male/multi-female communities where members form small cohesive, typically mixed-sex, subgroups with regular movements between groups (high fission-fusion dynamics) or spend time alone (community members never all in one place simultaneously), but collectively occupy a communally defended territory.

At Andranobe River watershed, Red Ruffed Lemur community contains 18–31 individuals with home range c.58ha (Vasey 2006). Structure of community varies seasonally. Males reside year-round in core areas, while females remain within core areas during cold wet season (June–August),

Red Ruffed Lemurs are primarily high canopy dwellers, Masoala National Park.

vines (Vasey *et al.* 2018) and gives birth to litters of two or three (up to five in captivity) (Vasey 2003). Young remain in nest and are tended solely by mother during early stages of development. After *c.*2 weeks infants may be moved (carried by mother in mouth) to different nests and when older 'parked' in concealed location (Vasey 1997b, 2000a). During first two months, nests and infant 'parking' sites are located within each mother's core area, and inhabitants of each core area within communal home range (Vasey 2007). When infants are 'parked', sites concealed and high in canopy, normally in largest trees in core area, often growing in valleys and festooned with lianas, providing maximum concealment and protection. As many as 40 different trees used per litter (Vasey *et al.* 2018). When 'parked' other adults (of both sexes and all age classes) provide care for young, allowing mothers to travel and forage distantly. Allo-parental behavior includes infant guarding, co-parking with other infants, infant transport, and allo-nursing (Vasey 2007). At 70 days infants can move around the canopy wholly independently. Females reach sexual maturity at just under two years, while males take three to four years to attain maturity.

There are no documented accounts of predation by Fosa, but unsuccessful hunting attempts have been observed, where Fosa have climbed into canopy in pursuit, with the lemurs responding with highly agitated alarm calls and fleeing through treetops (Garbutt 2021).

WHERE TO SEE Not a difficult lemur to see, but reaching suitable sites on Masoala Peninsula requires time and effort; take a boat from Maroatsetra (two hours). The best sites are Andranobe, and Masoala National Park trail network at Lohatrozona. Several territories are well known, but terrain is steep and tough. Equally productive are forests around Arol Lodge, near coastal village of Ambodiforaha.

Avahis, Sifakas and the Indri

Family Indriidae

The family Indriidae contains 19 species arranged in three genera. Two genera, *Indri* and *Propithecus*, are diurnal and include the largest extant Malagasy prosimians (*c.*3.5–9.5kg), whilst the genus *Avahi* is considerably smaller (*c.*1kg) and nocturnal. Indriidae are saltatory and characterised by long powerful hindlimbs (35% longer than forelimbs) and prefer to cling upright to, and leap between, vertical trunks and branches. Such posture and locomotion is unusual among primates (but mirrored by sportive lemurs *Lepilemur* spp.), but ideal for exploiting the lower branch and vertical trunk niche typical of the forest understorey. Leaps may be spectacular, covering up to 12m.

Using either arms or legs, they can also hang suspended from finer branches; postures often adopted when feeding or playing. Hands and feet are narrow and elongate (feet much larger than hands), with wide spans. The palms and soles are padded. Thumb and big toe both slightly opposable to the other digits and provide very good grasping ability, and also sufficiently dexterous to pull branches to mouth when feeding or hold food items, conveying them to the mouth directly. When climbing they employ a deliberate hand-over-hand technique and descend tail first (Mittermeier *et al.* 2013).

Indriidae are characterised by a broad, short snout and relatively small ears largely concealed by fur (except Indri). They have fewer teeth (30) than other lemurs, with large upper incisors and canines. The four lower front teeth project forwards and upwards, forming a tooth comb (as in Lemuridae) which is used in grooming and feeding (Mittermeier *et al.* 2013).

All Indriidae are primarily folivorous and have evolved digestive systems to cope with low-quality diets primarily composed of tough young leaves. Their salivary glands are substantially enlarged (to provide appropriate lubrication) and their intestines are proportionately long, in part due to an elongated cecum – a pouch containing gut bacteria that assist the breakdown of plant cellulose (Mittermeier *et al.* 2013).

Indriidae have slow rates of reproduction, particularly the larger diurnal species (sifakas, genus *Propithecus*, and Indri) compared with other lemurs and other primates: some sifakas have lower rates and longer inter-birth intervals than many Old World and New World monkeys.

Indriidae occur throughout the various major native forest types, although only the sifakas inhabit the arid spiny bush and forests of the south and south-west.

Interestingly, no extant species appears to be related to any known extinct lemurs: subfossil remains of extant Indriidae spp. from various sites date back *c.*2,000 years (Mittermeier *et al.* 2013).

WOOLLY LEMURS OR AVAHIS

GENUS *Avahi*

Avahi is the only nocturnal genus within Indriidae. The vernacular refers to the dense, curly 'woolly' fur that is characteristic. Like all Indriidae, woolly lemurs have proportionately long hindlimbs and are powerful leapers capable of covering gaps of *c.*2–3m between branches. Their tails are similarly proportionately long, generally exceeding combined head-body length. There is minimal sexual dimorphism, with females sometimes fractionally heavier than males.

Woolly lemurs exhibit some unusual traits. They are the only truly folivorous nocturnal primates (Ganzhorn 1988) and unique among nocturnal prosimians in moving and feeding as a cohesive family group (generally 3–5 individuals) containing an adult pair plus offspring up to two years of age (all other nocturnal prosimians tend to be solitary when foraging). During the day, family units often sleep huddled together in a tight ball (also rare in similar-sized primates), usually in clumps of dense foliage or vines, or the fork of a tree, in the lower or midstorey (Thalmann 2003). Groups are highly territorial,

aggressively defending home ranges of *c.*1–3ha. Moreover, they show parental care and females carry infants during nocturnal activity periods, another rarity amongst nocturnal prosimians (Thalmann, 2003, Kappeler, 2014).

Woolly lemurs may be secondarily nocturnal, meaning that in their evolutionary past ancestors were once diurnal (like other Indriidae) and that they subsequently became increasingly active at night (perhaps to avoid competition) but retained the gregarious lifestyle common to other family members (Ganzhorn *et al.* 1985, Müller & Thalmann 2000, Roos *et al.* 2004, Santini *et al.* 2015). Most activity and foraging bouts are concentrated around the first and last hour of the night, primarily in the middle and upper layers of the canopy (*c.*2–9m above ground). They occasionally may descend to ground level. When foraging they are generally solitary, but via cohesion calls, keep in close contact with other family unit members (Mittermeier *et al.* 2013).

The diets of woolly lemurs are principally immature leaves and buds (*c.*75%), with smaller quantities of flowers and unripe fruits (Donati *et al.* in press). Only the outer softer parts of the leaf blade are selected, with the petiole or midrib generally rejected. Targeted plant species have leaves high in

Female Eastern Woolly Lemur with infant, Andasibe-Mantadia National Park.

proteins and sugars, but low in alkaloids (Mittermeier *et al.* 2013). Such nutritionally low-value diets are possible due to adaptations that facilitate mid-gut fermentation such as a sacculated cecum and a looped colon (Chivers & Hladik 1980, Martin 1990). Woolly lemurs' leaping mode of locomotion requires high energy levels, and this coupled with low-quality diet means they rely on energy-saving strategies, such as the generally low levels of activity seen across the genus (Warren & Compton 1998, Thalmann 2003, Norscia *et al.* 2012).

Woolly lemurs breed annually, generally producing a single offspring in the austral winter (dry season). Gestations periods are *c.*120–150 days. Infants initially cling to the mother's underside, and later move around to ride on her back. Infants may occasionally be 'parked' in dense foliage, while the mother forages elsewhere. Offspring tend to stay with their parents for the first two years, before leaving the natal group (Mittermeier *et al.* 2013).

Formerly regarded as a single species, *A. laniger*, subdivided into eastern (*A. l. laniger*) and western subspecies (*A. l. occidentalis*) (Tattersall 1982, Harcourt & Thornback 1990). Subsequently these have been elevated to species. With more recent taxonomic revisions and descriptions of new forms, nine species are currently recognised: Eastern Woolly Lemur *Avahi laniger*, Peyrieras's Woolly Lemur *A. peyrierasi* (Zaramody *et al.* 2006), Masoala Woolly Lemur *A. mooreorum* (Lei *et al.* 2008), Betsileo Woolly Lemur *A. betsileo* (Andriantompohavana *et al.* 2007), Manombo Woolly Lemur *A. ramanantsoavanai* (Zaramody *et al.* 2006), Southern Woolly Lemur *A. meridionalis* (Zaramody *et al.* 2006), Western Woolly Lemur *A. occidentalis*, Bemaraha Woolly Lemur *A. cleesei* (Thalmann & Geissmann 2005) and Sambirano Woolly Lemur *A. unicolor* (Thalmann & Geissmann 2000). The recognition of some species remains contentious, with those inhabiting central-eastern and south-eastern rainforests – Peyrieras's Woolly Lemur, Betsileo Woolly Lemur, Manombo Woolly Lemur and Southern Woolly Lemur – especially requiring further clarification (Rumpler *et al.* 2011, Markolf *et al.* 2011). It is likely that the taxonomy of the genus *Avahi* will be re-evaluated and modified in the future.

The generic name *Avahi* is onomatopoeic and derives from the typical high-pitched alarm calls (*wo-he, va-hii, vou-hiiava-hee*) (Thalmann 2003), and is often also used as the vernacular. Otherwise, woolly lemurs do not have an extensive vocal repertoire (compared with larger diurnal Indriidae).

Due to their small size and nocturnal habits, woolly lemurs cannot be mistaken for any other member of Indriidae. When active at night they can be confused with sportive lemurs (*Lepilemur* spp). of similar size and posture. With clear views woolly lemurs are easily distinguished by their more rounded 'owl-like' face and prominent white patches on the rear of their thighs. The latter feature is so distinctive that it gives rise to one of the commonest Malagasy names for these lemurs, '*fotsife* or *fotsy-fe*' meaning 'white leg' (Thalmann 2003, Donati *et al.* in press).

Woolly lemurs occur in primary and secondary eastern humid forests at all elevations from sea level to higher montane areas (up to *c.*1,800m), and in lower-elevation western deciduous forests. They are absent from the arid spiny bush of the south and south-west. The ranges of the various species currently recognised and how they relate to one another are far from fully understood and must be treated as tentative; the picture will only begin to clarify after further fieldwork. As currently understood, the ranges of no two species overlap, so all are allopatric. All woolly lemurs are broadly very similar in size and coloration, and would be very difficult to tell apart in the field, so the most reliable means of species identification is geographic location.

EASTERN WOOLLY LEMUR
Avahi laniger (J. F. Gmelin, 1788)

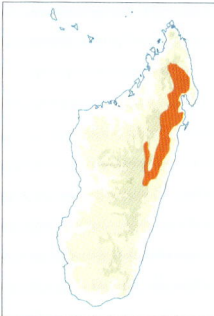

Other names: Eastern Avahi, Gmelin's Woolly Lemur, Gmelin's Avahi

Malagasy names: Avahy, Ampongy, Fotsifé, Fotsy-fe, Fotsifaka

MEASUREMENTS
Head-body: *c.*270–320mm. Tail: *c.*300–370mm. Weight: 1.0–1.4kg (males), 1.1–1.6kg (females). (Glander *et al.* 1992, Zaramody *et al.* 2006, Andriantompohavana *et al.* 2007, Lei *et al.* 2008)

DESCRIPTION & IDENTIFICATION Pelage thick, curled and woolly in texture. Upperparts greyish-brown with reddish tones, becoming paler towards rump. Tail more rufous than body and often rusty-red. Chest and underparts pale grey, insides of thighs have distinctive white patches. Head rounded with small hidden ears, muzzle short and dark. Facial disc brown with lighter cream eyebrow patches. Eyes large and brown-orange. A medium-small lemur with a vertical posture. Because of similar size and posture, confusion with sportive lemur species is possible; Eastern Woolly Lemur is generally larger with concealed, non-prominent ears, rufous tail and characteristic white thigh patches, white/grey cheek and eyebrow patches. Eastern Woolly Lemur also often seen in pairs/groups, whereas sportive lemurs are largely solitary.

HABITAT & DISTRIBUTION Primary and secondary lowland, mid-altitude and montane rainforest. Occurs throughout central-eastern and north-east rainforests, excluding Masoala Peninsula (see Masoala Woolly Lemur) from Bemarivo River in north (Marojejy, Anjanaharibe-Sud and COMATSA forest corridor) to Mangoro/Nesivolo Rivers in south (Mittermeier *et al.* 2010, Louden *et al.* 2017). An isolated population remains in forest fragment at Ambohitantely in central highlands. Formerly range extended to south-eastern extreme of rainforest: populations south of Mangoro/Nesivolo Rivers are now regarded as distinct species (Peyrieras's Woolly Lemur, Betsileo Woolly Lemur, Manombo Woolly Lemur and Southern Woolly Lemur).

May attain high population densities: in Analamazaotra Special Reserve (mid-altitude eastern rainforest) 72–100 individuals/km^2 estimated, but unlikely the species is so abundant throughout range (Ganzhorn 1988). May become more abundant in second growth and disturbed areas: in Manompana Forest (south of Mananara-Nord) densities in mature forest, 38 individuals/km^2 (Miller *et al.* 2018) to 41–55 individuals/km^2 (Sabin *et al.* 2013), whereas in regenerating secondary forest densities of 134 individuals/km^2 (Miller at al. 2018).

Habitat destruction and fragmentation due to slash-and-burn agriculture and logging are major threats. Hunting for food, particularly in Mananara-Nord and Makira Forest regions, also poses significant threat (Golden 2009, Brook *et al.* 2019, Golden *et al.* in press). IUCN Vulnerable.

BEHAVIOUR Nocturnal and lives in family groups of up to five individuals, an adult pair plus offspring from previous years up to *c.*2 years old. Day spent concealed in thick foliage or vines, 3–5m above ground, usually sitting on branch and lodged against vertical trunk (Harcourt 1991). Occasionally rests

Eastern Woolly Lemur, Association Mitsinjo Forest, Andasibe.

alone, but more usually pairs and offspring sleep tightly huddled together.

Activity begins after dusk. Families spend some time grooming themselves or each other before moving off to forage. Foraging largely solitary but family members remain in close proximity. Most activity and foraging takes place in first two hours of darkness and prior to dawn, but may continue for short periods in between. Intervening time spent resting and grooming when family re-unites. When apart, individuals keep in close contact with regular high-pitched whistles. They return to sleep sites just before first light (Harcourt 1991, Roth 1996).

Pairs occupy home range of *c.*1–2ha, aggressively defended by calling (distinctive *wo-he* or *va-hii*,

Eastern Woolly Lemur, Marojejy National Park.

vou-hii) call) followed by chasing if intruders persist (Harcourt 1991). Does not range widely, travelling c.300–500m per night. Diet mainly consists of young and to lesser extent mature leaves and buds from variety of tree species, plus small quantities of flowers and fruit (Ganzhorn *et al.* 1985, Roth 1996). Reduces competition with sympatric sportive lemur species by concentrating on better-quality vegetation, restricting sportive lemurs to leaves of lower nutritional value.

Breeds annually, mating in April or May, single infant born August or September (Ganzhorn *et al.* 1985, Ganzhorn 1988). Initially infant carried on mother's underside; when older rides on her back. Infants occasionally 'parked' in dense foliage when female is foraging (Thalmann 2003).

Predated by Fosa, Henst's Goshawk and Madagascar Harrier-Hawk (Goodman 2003, Karpanty & Wright 2007, Goodman *et al.* 2014b).

WHERE TO SEE Easily seen in various community forests around Andasibe-Mantadia National Park near Andasibe, c.4 hours east of Antananarivo. Easily seen on night walks along road near park entrance. Also Analamazaotra Forest Station, operated by Association Mitsinjo, V.O.I.M.M.A. Community Reserve near Andasibe village, or Vohimanana-Ankera, another community forest, c.15km from Andasibe. Local guides are adept at finding daytime sleep sites.

MASOALA WOOLLY LEMUR
Avahi mooreorum   Lei et al., 2008

Other names: Moore's Woolly Lemur, Masoala Avahi, Moore's Avahi

Malagasy names: Avahy, Ampongy, Fotsifé, Fotsy-fe, Fotsifaka

MEASUREMENTS
Head-body: c.280–330mm.
Tail: c.290–370mm.
Weight: c.900–1,000g.
(Andriantompohavana *et al.* 2007, Lei *et al.* 2008)

DESCRIPTION & IDENTIFICATION Fur dense and woolly. Overall slightly paler than *A. laniger*. Upperparts mixture of mid-brown and light brown, gradually becoming paler towards rump and tail base, which is pale cream/whitish. Very distinctive white patch on inner upper thigh. Tail pale rusty-red. Underparts pale grey. Head darker brown/grey. Facial mask distinct but not pronounced. No white eyebrow patches as in congeners. Small whitish areas under chin. Ears small and hidden. Adult eyes rich orange, paler yellow in juveniles (Andriantompohavana *et al.* 2007, Lei *et al.* 2008). A medium-small lemur with long tail and vertical posture. May be confused with similar-sized Masoala Sportive Lemur, which also adopts vertical clinging posture. Masoala Woolly Lemur slightly larger with very distinctive white thigh patches, reddish tail, and rounded face with short muzzle and concealed ears.

Female Masoala Woolly Lemur with infant, Masoala National Park.

Masoala Woolly Lemur, Masoala National Park.

HABITAT & DISTRIBUTION Primary and secondary lowland and mid-altitude rainforest. Range limited to Masoala Peninsula in north-east. Limits of range and boundary with Eastern Woolly Lemur, particularly west of Antainambalana and Sahantaha Rivers, and intervening forests between Masoala and Makira Forest to west and Anjanaharibe-Sud to north, need to be clarified. In touristic areas of Masoala National Park near Ambodiforaha density *c.*94 individuals/km², with large densities associated with high abundances of *Garcinia commersonii* and *Eugenia* spp. (Sawyer *et al.* 2017), but unlikely to be as numerous throughout Masoala Peninsula. Threatened by ongoing destruction and fragmentation of forests and by opportunistic localised hunting for food (Borgerson 2016). IUCN Endangered.

BEHAVIOUR Nocturnal. Limited specific studies. Observed in groups of up to four individuals. In disturbed forests singles more regularly seen, but pairs more often observed in primary habitat (Sawyer *et al.* 2017). Also utilises forest understorey more frequently in disturbed forests (Fenosa *et al.* 2018). Aspects of behaviour probably similar to Eastern Woolly Lemur.

WHERE TO SEE Relatively easy to see in Masoala National Park; take a boat from Maroatsetra (two hours). Night walks along the beach trail between Lohatrozona and Tampolo are productive. Equally good are the forests around Arol Lodge, near coastal village of Ambodiforaha.

PEYRIERAS'S WOOLLY LEMUR
Avahi peyrierasi   Zaramody *et al.*, 2006

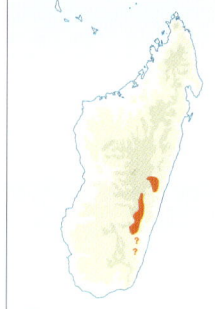

Other name: Peyrieras's Avahi

Malagasy names: Avahy, Ampongy, Fotsifé, Fotsy-fe, Fotsifaka

MEASUREMENTS
Head-body: *c.*260–310mm.
Tail: *c.*280–340mm.
Weight: *c.*900–1,200g.
(Zaramody *et al.* 2006, Mittermeier *et al.* 2010)

DESCRIPTION & IDENTIFICATION Slightly smaller than very similar Eastern Woolly Lemur. Pelage thick and woolly. Upperparts grey-brown with reddish tail. Underparts pale grey to whitish, rear and inner thighs distinctively white. Head rounded with small hidden ears; muzzle short and dark. Facial disc mid-brown with whitish border. Also whitish eyebrow patches and areas around cheeks and under chin. Eyes large and brown-orange (Zaramody *et al.* 2006). A medium-small lemur with vertical posture. Similar size and posture make confusion with sportive lemur species possible; Peyrieras's Woolly Lemur has concealed ears, a rufous tail, characteristic whitish thigh patches, whitish areas on face and whitish eyebrow patches.

HABITAT & DISTRIBUTION Primary and secondary lowland and mid-altitude rainforest up to *c.*1,600m (Andringitra) (Sterling & Ramaroson 1996). Occurs south of Mangoro/Nesivolo Rivers through south-

Male, female and infant Peyrieras's Woolly Lemurs, Ranomafana National Park.

central eastern and south-eastern rainforests to around Andringitra and Midongy du Sud National Parks, but precise southern limit and relationship with congeners, Southern Woolly Lemur and Manombo Woolly Lemur, yet to be determined. Formerly woolly lemurs from this range were classified as *A. laniger*. Some studies suggest densities may be greater in secondary forest, 40 individuals/km^2, than in primary forest, 19 individuals/km^2 (Lehman *et al.* 2006c), whereas other studies show encounter rates in undisturbed forests are higher than in secondary forests (Herrera *et al.* 2011). Habitat destruction and fragmentation due to slash-and-burn agriculture is a major threat. Hunting for food is increasing. IUCN Vulnerable.

BEHAVIOUR Little behavioural research to date. In Ranomafana (when classified as *A. laniger*), activity and feeding peaked around dusk and during early part of night, with a second smaller peak prior to dawn. Resting dominated between 20.00 and midnight and overall comprised *c.*60% of active period, with feeding occupying *c.*22%, travelling *c.*13% and grooming *c.*5%. Home ranges *c.*1ha. Diet consists almost exclusively of leaves (*c.*98%), along with very small quantities of flowers and fruits. Young leaves

Peyrieras's Woolly Lemur, Ranomafana National Park.

of *Harungana madagascariensis* and *Syzygium* spp. *c.*70% of diet (Faulkner & Lehman 2006). Predators include Fosa, Madagascar Harrier-Hawk and Henst's Goshawk (Karpanty & Wright 2007, Goodman & Ganzhorn in press). All other aspects of behaviour are likely similar to Eastern Woolly Lemur.

WHERE TO SEE Most easily seen in and around Ranomafana National Park, in areas where night walks are permitted.

BETSILEO WOOLLY LEMUR
Avahi betsileo   Andriantompohavana *et al.*, 2007

Other name: Betsileo Avahi

Malagasy names: Avahy, Ampongy, Fotsifé, Fotsy-fe, Fotsifaka

MEASUREMENTS
Head-body: *c.*260–310mm.
Tail: *c.*280–340mm.
Weight: 900–1,200g.
(Andriantompohavana *et al.* 2007, Lei *et al.* 2008)

DESCRIPTION & IDENTIFICATION Pelage thick and woolly. Upperparts and limbs reddish-brown. Central underparts dark grey, becoming pale grey on outer areas. Tail darker rufous-brown dorsally and paler ligher reddish-blonde ventrally. Chest and underparts pale grey, insides of thighs have distinctive white patches. Facial disc distinct, with lighter grey areas under chin and cheeks. Eyebrow patches cream. Fur particularly thick on head (more so than eastern congeners) giving head more 'rounded' appearance (Andriantompohavana *et al.* 2007). A medium-small lemur with vertical posture. Similar size and posture may cause confusion with sportive lemur species: Betsileo Woolly Lemur has concealed, not prominent ears, short muzzle and characteristic white thigh patches.

HABITAT & DISTRIBUTION Primary higher-elevation and montane rainforest. Very limited range: currently known only from Bemosary Classified Forest, near Fandriana, in central-eastern highlands. Range possibly extends from Mangoro River in north to Mananjary River in south (Andriantompohavana *et al.* 2007, Mittermeier *et al.* 2010). Habitat destruction and forest fragmentation are a major threat. IUCN Endangered.

BEHAVIOUR Nocturnal. No wild studies. Aspects of behaviour probably similar to Eastern Woolly Lemur.

WHERE TO SEE Only known site is Bemosary Classified Forest, a fragment *c.*1 hour's walk from small town of Fandriana, which is north-east of Ambositra and south-east of Antsirabe (Mittermeier *et al.* 2010).

MANOMBO WOOLLY LEMUR
Avahi ramanantsoavanai   Zaramody *et al.*, 2006

Other names: Manombo Avahi, Ramanantsoavana's Woolly Lemur, Ramanantsoavana's Avahi Malagasy names: Avahy, Ampongy, Fotsifé, Fotsy-fe, Fotsifaka

MEASUREMENTS
Head-body: *c.*240–310mm.
Tail: *c.*330–400mm. Weight: 900–1,100g. (Zaramody *et al.* 2006, Lei *et al.* 2008)

DESCRIPTION & IDENTIFICATION Pelage dense and woolly. Manombo Woolly Lemur appears slightly smaller than Eastern Woolly Lemur. Upperparts grey-brown. Underparts grey. Tail rufous-brown. Insides of thighs white. Facial disc may be lighter and more pronounced than in Eastern Woolly Lemur. White eyebrow patches distinct (Zaramody *et al.* 2006). A medium-small lemur with vertical posture. Similar size and posture of sympatric Manombo Sportive Lemur may cause confusion: Manombo Woolly Lemur has concealed ears, reddish tail and characteristic white thigh patches.

HABITAT & DISTRIBUTION Lowland rainforest. Very limited range: currently known only from Manombo Special Reserve and nearby Agnalahaza Forest in coastal lowlands of south-east, south of town of Farafangana (Zaramody *et al.* 2006, Andriantompohavana *et al.* 2007). Further field studies are required to establish range limits and relationships with Peyrieras's Woolly Lemur and Southern Woolly Lemur. Habitat destruction and forest fragmentation are major threats. IUCN Vulnerable.

BEHAVIOUR Nocturnal. No wild studies. Aspects of behaviour probably similar to Eastern Woolly Lemur.

WHERE TO SEE Only known sites are Manombo Special Reserve and nearby Agnalahaza Forest (formerly Mahabo Forest Reserve) c.30km south of Farafangana on south-east coast (Mittermeier *et al.* 2010).

SOUTHERN WOOLLY LEMUR
Avahi meridionalis Zaramody *et al.*, 2006

Other name: Southern Avahi

Malagasy names: Avahy, Ampongy, Fotsifé, Fotsy-fe, Fotsifaka

MEASUREMENTS
Head-body: c.230–290mm.
Tail: c.300–330mm.
Weight: c.950–1,400g.
(Zaramody *et al.* 2006, Lei *et al.* 2008, Balestri 2018)

DESCRIPTION & IDENTIFICATION Pelage dense, curled and woolly. Southern Woolly Lemur is similar size to Eastern Woolly Lemur. Upperparts grey-brown, grading more mid- to pale grey on lower back, with rufous-russet tones on shoulders and limb extremities. Underparts mid-grey. Tail rufous-brown with slightly darker tip. Insides of thighs white and visible. Facial disc pale and apparent. Whitish under chin and neck. Small whitish eyebrow visible (Zaramody *et al.* 2006). A medium-small lemur with vertical posture. Similar size and posture to sympatric Andohahela Sportive Lemur may cause confusion: Southern Woolly Lemur has concealed ears, reddish tail and characteristic white thigh patches.

HABITAT & DISTRIBUTION Lowland and mid-elevation rainforest and humid littoral forests. Also observed in *Eucalypus* plantations adjacent to littoral forest (Scobie *et al.* 2017). Range limited to humid areas of south-east, from southern extreme at Petriky (south-west of Tolagnaro), to Andohahela National Park, Anosyennes Mountains, Vohimena Mountains, Tsitongambarika Protected Area and littoral forest fragments at Sainte Luce and Mandena (Zaramody *et al.* 2006). Further field studies needed to establish range limits, especially north

Southern Woolly Lemur, Andohahela National Park.

of Tsitongambarika and towards Midongy du Sud National Park, and relationships with Peyrieras's Woolly Lemur *A. peyrierasi* and Manombo Woolly Lemur. Densities in littoral forests appear relatively high, 55–240 individuals/km^2 (Norscia 2008). In rainforests densities lower: c.17 individuals/km^2 in Andohahela National Park (Feistner & Schmid 1999), 19–55 individuals/km^2 in northern Tsitongambarika Protected Area (Balestri 2018) and 24–95 individuals/km^2 in southern Tsitongambarika Protected Area (Norscia *et al.* 2006, Donati *et al.*

2020d). Habitat destruction and forest fragmentation remain major threats, whilst opportunistic hunting is a less serious concern (Bollen & Donati 2006, Campera *et al.* 2019b). IUCN Endangered.

BEHAVIOUR Lives in pairs that sleep together and show some cohesiveness when active and foraging. Primarily nocturnal with occasional cathemeral behaviour: peak activity in twilight around dusk and dawn. Also shows sporadic diurnal activity in certain environmental conditions, when daily food intake needs to be maximised or to reduce competition with sympatric nocturnal Andohahela Sportive Lemur (Balestri 2018). Also tends to utilise lower strata in forest canopy and understorey than Andohahela Sportive Lemur (Balestri 2018).

Home range size varies from *c.*2.5–3.5ha at Sainte Luce (Norscia & Borgognini-Tarli 2008) and *c.*4.5–10.4ha at Ampasy (Balestri 2018). When foraging, nightly distance travelled longer (*c.*640m) in September–February (season of high food abundance, with increased new leaf growth) than during lean season (March–August), when distances are shorter (*c.*500m), necessitated by energy and resource conservation (Balestri 2018).

Like congeners, Southern Woolly Lemur is primarily folivorous, with young leaves preferred over mature leaves in September–February when new growth is abundant. Flowers, especially green sepals, occasionally eaten, when available September–December (Norscia *et al.* 2012, Balestri 2018). In littoral forests more than 40 different plant species eaten over seven-month period (Norscia & Borgognini-Tarli 2008); in lowland rainforest diet may be slightly broader with *c.*50 plant species utilised over ten-month period (Balestri 2018).

Births occur in August. Infants carried by mother for first four months. Independence attained after 5–6 months (December–January).

WHERE TO SEE Most accessible location is Mandena Conservation Zone (part of QMM Titanium Dioxide Development Project), 12km north of Tolagnaro. The more adventurous should try Malio, the rainforest parcel of Andohahela National Park or Tsitongambarika Protected Area further north.

WESTERN WOOLLY LEMUR
Avahi occidentalis   (Lorenz von Liburnau, 1898)

Other name: Western Avahi, Lorenz von Liburnau's Woolly Lemur

Malagasy names: Fotsifé, Fotsy-fe, Tsarafangitra

MEASUREMENTS
Head-body: *c.*270–300. Tail: *c.*310–370. Weight: *c.*800–1,100g. (Warren 1994, Thalmann & Geissmann 2000, Zaramody *et al.* 2006, Lei *et al.* 2008)

DESCRIPTION & IDENTIFICATION A small woolly lemur. Pelage dense, slightly curled and woolly. Upperparts light to mid-grey with sandy-brown or olive-brown tones that fade towards rear. Tail greyish with reddish-ochre tones near base, more reddish in some individuals. Underparts and throat sparsely furred, pale grey to light beige, tinged apricot. Pygal a distinctive triangle of cream or pale beige fur. Inner thighs clearly white. Face rounded and whitish-cream to pale grey, forming characteristic contrasting mask. Distinct darker triangle above black hairless nose. Ears small and unobtrusive. Eyes orange and encircled by black eye-rings.

A medium-small vertically clinging lemur. Smaller and paler than Eastern Woolly Lemur. Round face, concealed ears, reddish tail and visible white thigh patches distinguish from similar-sized Milne-Edwards's Sportive Lemur. Local name *Tsarafangitra* (Mahajanga area) means 'well-dressed hair' (Donati *et al.* in press).

HABITAT & DISTRIBUTION Dry deciduous and secondary forests. Restricted to north-west: precise range limits unclear. Known from area north and east of Betsiboka River, from Ankarafantsika National Park to region of Bay of Narinda (Thalmann & Geissmann 2000). Also recorded in Mariarano Classified Forest, north-east of Mahajanga (Andriantompohavana *et al.* 2006b). Further north, in Sambirano region and Ampasindava Peninsula, replaced by Sambirano Woolly Lemur *A. unicolor*, but boundary between them not precisely established. Towards far north, in Ankarana National Park, woolly lemurs previously regarded as *A. occidentalis*, but specific identity of this population requires clarification (Hawkins *et al.* 1990, Thalmann & Geissman 2005).

In Ankarafantsika National Park population density 67 individuals/km^2 (Ganzhorn 1988). In Mariarano Forest density estimates much higher, 130–155 individuals/km^2 (Ibouroi *et al.* 2013). Threatened by major habitat loss, often from seasonal burning to create cattle pasture (Eppley *et al.* 2020b). Species may be particularly susceptible to forest fragmentation: in smaller fragments (less than *c.*120ha) in western Ankarafantsika no woolly lemurs recorded (Steffens & Lehman 2018). Also probably hunted for food. IUCN Vulnerable.

BEHAVIOUR Almost exclusively nocturnal. Lives in small family groups of two to four, consisting of adult pair with dependent offspring up to two years old (groups of five have been seen, but composition uncertain). Females dominant. Family units occupy almost exclusive home ranges of 1–2ha, including a central core area where most time is spent. Modest overlap between ranges of adjacent social units, but territories defended with vocalisations, particularly loud *ava-hee* call (Warren & Crompton 1997, Ramanankirahina *et al.* 2011).

Vocalisations very important in maintaining family cohesion and group spacing. Three acoustically distinct types identified: onomatopoeic *ava-hee* or *va-hii*, whistle call and growling call. The loud *ava-hee* call is less frequently used and given when individuals are more distantly separated, the whistle call is loud and more frequently used when partners are visually separated, and the soft growling call is used more in close contact situations (Ramanankirahina *et al.* 2016).

Family group spends day resting, huddled close together in dense clumps of foliage or lianas, 3–13m above ground. Families use several favoured sleep sites and may move during day if initially chosen site is exposed to sun (Warren 1997). Sleep site choice may vary seasonally: in late dry season when foliage less, prefers central parts of trees to maximise cover and fewer sites are used; in wet season when foliage dense, a greater variety and number of sites used (Ramanankirahina *et al.* 2012).

Becomes active around dusk, but spends time grooming prior to vacating sleep site. Family unit largely remains together when foraging; individuals maintain contact with soft growling call. Foraging follows a regular itinerary, animals often re-visit same feeding sites night after night (Warren & Crompton 1997, Thalmann 2003).

Western Woolly Lemur, Ankarafantsika National Park.

Much of night spent travelling; average distance covered 750–1,200m (Warren & Crompton 1997, Thalmann 2003). Feeding mainly concentrated in first half of night. Returns to sleep sites before dawn, but during dry season (May–October) when foliage reduced, they sometimes remain active after sunrise (Warren & Crompton 1998a,b).

Diet consists of leaves (>70%); at certain times of year they are almost exclusively folivorous. Other food includes flowers, leaf buds and young leaf shoots, and they can be highly selective, choosing to feed predominantly on small number of species (Warren 1994, Thalmann 2001). Females dominant when feeding (Ramanankirahina *et al.* 2011). Mating occurs in April–May, a single offspring is born between September and October. Infants cling to mother's belly for first three weeks or so, then moves round to ride on her back. Offspring remain with parents until following breeding season, occasionally longer (Thalmann 2003).

There are predation records by Fosa (Dollar *et al.* 2007).

Female Western Woolly Lemur with infant, Ankarafantsika National Park.

WHERE TO SEE Relatively common and easy to find on night walks in Ankarafantsika National Park (Ampijoroa), south-east of Mahajanga. Trails either side of *Route Nationale 4* are productive, and often found on trails near entrance and visitor area. An alternative location is Mariarano Classified Forest, north-east of Mahajanga, but this site is more challenging to reach.

BEMARAHA WOOLLY LEMUR
Avahi cleesei   Thalmann & Geissmann, 2005

Other names: Cleese's Woolly Lemur, Cleese's Avahi

Malagasy name: Dadintsifaky

MEASUREMENTS
Head-body: *c.*230–310mm. Tail: *c.*320–360mm. Weight: *c.*750–1,300g. (Thalmann & Geissmann 2005, Zaramody *et al.* 2006, Lei *et al.* 2008)

DESCRIPTION & IDENTIFICATION Head and upperparts brown-grey, fur curled and woolly with flecked appearance. Underparts pale grey and sparsely furred. Tail beige to brown-grey and slightly reddish at base. Inner thighs distinctively white. Face slightly paler than forehead and crown, with upper line of face mask indented upwards towards crown, rather than downwards towards nose (as in Western Woolly Lemur and Sambirano Woolly Lemur). Forehead above face mask blackish and forms dark chevron. Eyes orange with narrow dark rings. Snout very dark, areas immediately around nose and mouth whitish (Thalmann & Geissmann 2005).

Taxonomic validity of Bemaraha Woolly Lemur has been questioned, with suggestion it is a junior synonym of *A. occidentalis*. Genetic studies confirmed species status is warranted (Andriantompohavana *et al.* 2007, Lei *et al.* 2008).

Within range can be confused only with similar-sized and nocturnal Bemaraha Sportive Lemur: Bemaraha Woolly Lemur has more rounded face, concealed ears and white inner thighs. Distinguished from allopatric congeners like Western Woolly Lemur by lack of white facial mask and broad dark eye-rings, and from both Western Woolly Lemur and Sambirano Woolly Lemur by dark chevron on forehead. Local name *Dadintsifaky* translates as 'grandparent or ancestor of the sifaka' (Donati *et al.* in press).

HABITAT & DISTRIBUTION Subhumid dry deciduous forest restricted to limestone karst (*tsingy*) areas. Known only from around Tsingy de Bemaraha National Park, north of Manambolo River in centre-west (Thalmann & Geissmann 2005). Originally recorded at just three sites: two within the park, Ankindrodro and type locality (forest 3–4 km further east, north-east of Ambalarano), and one outside park in disturbed forest near village of Ankinajao (Thalmann & Geissmann 2000). Forest at latter site now completely disappeared (Thalmann & Geissmann 2006). In known localities regarded as rare (Thalmann & Geissmann 2000). Recent surveys reveal major decline in forests near Ambalarano and searches of similar forests in region have failed to locate the species elsewhere (Louis *et al.* 2020l). Primarily threatened by continued habitat loss in very restricted range, often from seasonal, intentionally started bush fires. IUCN Critically Endangered.

BEHAVIOUR Studied only briefly. Nocturnal, forming groups (presumably family units similar to congeners) with home range *c.*2ha. Diet consists of buds, shoots and young leaves. Most active just after dark, for a period before midnight and again before dawn. Very vocal like other woolly lemur species. (Thalmann & Geissmann 2006).

WHERE TO SEE Can only be seen in forests in Tsingy de Bemaraha National Park near Bekopaka, north of Manambolo River. Sometimes challenging to find. Search forested areas surrounding *Grand Tsingy* and *Petit Tsingy*, e.g., Andadoany forest trail.

SAMBIRANO WOOLLY LEMUR
Avahi unicolor　　Thalmann & Geissmann, 2000

Other name: Sambirano Avahi, Unicolor Woolly Lemur, Unicolour Avahi

Malagasy name: Fotsifé, Fotsy-fe

MEASUREMENTS
Head-body: *c.*230–310mm.
Tail: *c.*260–300mm.
Weight: 700–1,000g.
(Zaonarivelo *et al.* 2007b, Mittermeier *et al.* 2010)

Bemaraha Woolly Lemur, Tsingy de Bemaraha National Park.

DESCRIPTION & IDENTIFICATION One of smallest woolly lemur species. Pelage curled and woolly. Head and body light sandy brown-grey. Face paler than head with indistinct facial mask result of differences

273

in fur texture (short and straight, opposed to curled). Tail darker grey-brown to reddish-brown; base paler grey-cream. Triangle of creamy light beige fur in pygal region. Underparts more thinly furred; pelage on chest, belly and inside of upper limbs pale grey, thin and downy. Inside lower limbs whitish. Eyes orange-yellow with obvious dark rings (Thalmann & Geissmann 2000).

In range can only be confused with similar-sized and nocturnal Mittermeier's Sportive Lemur: Sambirano Woolly Lemur has more rounded 'owl-like' face, concealed ears, greyish (rather than brownish coat) and white inner thighs. From allopatric Western Woolly Lemur by lacking white facial mask and broad dark eye-rings, and from Bemaraha Woolly Lemur in lacking chevron on forehead (Thalmann & Geissmann 2000).

HABITAT & DISTRIBUTION Subhumid transitional forest and moist lowland mid-elevation forest to *c*.1,600m and higher elevations in Sambirano forests (Louis *et al*. 2020c). Restricted to western Sambirano region, including Ampasindava Peninsula. Southern and northern range limits unclear; formerly possibly Andranomalaza (= Maetsamalaza) River, northern boundary perhaps Sambirano River (Thalmann & Geissmann 2000, Mittermeier *et al*. 2013). Recorded on western, but not north-eastern, slopes of Manongarivo Massif (Raxworthy & Rakotondraparany 1988, Goodman & Schütz 2000). Also recorded in isolated Anaborano Forest, east of Ankaramibe village and adjacent to Manongarivo Special Reserve (Zaonarivelo *et al*. 2007b). Woolly lemur populations further north in dry forests of Ankarana very rare (Hawkins *et al*. 1990) and taxonomic status requires verification. Sambirano Woolly Lemur appears rare throughout its range and is threatened by ongoing habitat destruction and forest fragmentation (Louis *et al*. 2020c). Mineral mining projects on Ampasindava Peninsula further threaten the region (Wilmet *et al*. 2017a). IUCN Critically Endangered.

BEHAVIOUR Nocturnal. Not yet studied in the wild. Behaviour probably similar to Western Woolly Lemur.

WHERE TO SEE Restricted range difficult to reach. Forests on Ampasindava Peninsula probably provide the best opportunity.

SIFAKAS OR SIMPONAS
GENUS *Propithecus*

Sifakas (locally called simponas or sadabe in eastern rainforest regions) are primarily diurnal large lemurs, ranging from the smallest (*c*.3kg), Verreaux's Sifaka *P. verreauxi*, Decken's Sifaka *P. deckeni*, Golden-crowned Sifaka *P. tattersalli*, to the largest (*c*.6.5kg), Diademed Sifaka *P. diadema*. Species inhabiting rainforests tend to be larger. There is no significant size difference between males and females (Lawler & Richard in press).

They are primarily arboreal with long, powerful legs and long tails. Like other Indriidae, sifakas are 'vertical clingers and leapers' and move through the trees in athletic bounding leaps between vertical trunks and branches. They are capable of adopting suspensory postures (hanging by their hindlimbs), particularly when feeding and also periodically descend to the ground to eat soil or occasionally terrestrial parasitic flowers. When on the ground they can hop on their hindlegs. Furthermore, dry forest species from northern, western and southern areas can be seen moving across open spaces on the ground by bounding and hopping bipedally, often with arms held out to the side or upwards in counter-balance. This behaviour is most notable in Verreaux's Sifaka (especially in locations like Berenty and Nahampoana Private Reserves). There is growing evidence that open savannas in northern, western and central regions are natural features that have contracted and expanded over time with changes in climate (Quéméré *et al*. 2012, Salmona *et al*. 2017a), hence sifakas may have developed this locomotive ability to move between forest patches. In semi-captive situations with open spaces, this behaviour is also seen in some rainforest species like Diademed Sifaka, so it seems likely all sifakas are capable of this type of locomotion if required.

Sifaka social structure is unusual among primates in that it appears flexible and variable, both within and between species: groups may be simple male/female pairs, one male/multi-female, multi-male/one female or multi-male/multi-female (Pochron & Wright 2003). Females are socially dominant over males, often displacing them at feeding sites and in other social contexts (Lawler & Richard in press). Species in rainforest regions (Diademed Sifaka, Milne-Edwards's Sifaka *P. edwardsi* and Silky Sifaka *P. candidus*) tend to live in smaller groups (*c*.3–9 individuals), occupy larger home ranges (up to *c.*50ha, exceptionally 250ha) and occur at lower densities than those in drier forest regions (Verreaux's Sifaka, Coquerel's Sifaka *P. coquereli*, Decken's Sifaka, Crowned Sifaka *P. coronatus*, Perrier's Sifaka *P. perrieri* and Golden-crowned Sifaka), where group size is *c.*2–14 individuals and home ranges are generally no more than 12ha (although can be *c.*30ha) (Richard 2003, Lawler & Richard in press).

Sifaka diets are dominated by leaves, seeds and fruits, with smaller quantities of flowers, and proportions varying seasonally (Sato *et al.* 2016). Their dentition is suited to slicing and grinding plant matter, and their gastrointestinal tracts are specialised for folivory, being particularly long to maximise fibre digestion and water extraction (Lawler & Richard in press). Sifakas are seed predators, destroying seeds during mastication and digestion, hence they are not agents of seed dispersal (cf. ruffed lemurs, genus *Varecia*). They also occasionally eat soil, presumably to aid digestion and/or help counteract potential toxins and tannins that might accumulate from their diet (Powzyk & Mowry 2003, Semel *et al.* 2019).

Olfactory communication is presumably well developed in sifakas, as evidenced by extensive scent-marking across all species. Males in all species have sebaceous chest glands, and both males and females have ano-genital glands, which they rub on branches and trunks regularly in a wide variety of contexts (Lawler & Richard in press). Males often over-mark where a female has recently marked. Males also gouge tree bark prior to scent-marking with chest glands.

Compared to other lemurs, e.g., Ring-tailed Lemur and ruffed lemurs, sifakas have a more limited vocal repertoire. Species in drier regions (e.g. Verreaux's Sifaka, Decken's Sifaka and Crowned Sifaka) are best known for their onomatopoeic *shi-fakh shi-fakh shi-fakh* or *tchi-fak-tchi-fak-tchi-fak* alarm calls that give rise to their common name 'sifaka'. Generally involving all members of the group, these loud calls are given rapidly while individuals characteristically jerk their heads backwards several times in succession. Potential aerial threats, e.g., large raptors, illicit a sequence of very load roars. Sifakas in rainforest regions (e.g., Diademed Sifaka, Silky Sifaka and Milne-Edwards's Sifaka) have a very loud forced sneeze-like *tzisk-tzisk-tzisk* or *zzuss-zzuss-zzuss* alarm call (described locally as *shim-poon*, hence the Malagasy name simponas). Other vocalisations for all species include low 'hums' and moans, and various contact and group cohesion calls (Irwin 2006, Mittermeier *et al.* 2013).

Suspensory feeding is common in the genus Propithecus: *Verreaux's Sifaka, Berenty Private Reserve.*

All sifaka are capable of bounding bipedally across open ground. This behaviour is well-known in Verreaux's Sifaka, Berenty Private Reserve.

Sifakas breed seasonally, although timing varies with species, location and habitat. Mating generally occurs between November and February, with females receptive only for a single day per annual oestrous cycle; receptivity is advertised by coiling the tail. Gestation lasts *c*.5 months. Single infants are the norm. Infants initially cling to their mothers underside, but when older (*c*.3–4 weeks) switch to riding on her back. From around two months they are able to move independently. Infant mortality over the first few months is high (*c*.40%+), Fosa being the main predator (Wright *et al.* 1997, Hawkins 2003, Patel 2005). Infanticide also occurs, with infants sometimes killed by unrelated adult male sifakas (Lewis *et al.* 2003, Littlefield 2010). Allo-parenting is common, with all members of the group interacting with infants, playing, grooming, carrying and nursing (Patel *et al.* 2003a, Wright *et al.* 2012). Independence is reached at 6–7 months, near-adult size is reached around 12 months (growth may continue up to four years) and sexual maturity in both sexes can be reached as early as two and a half years in some species, and up to five to six years in others. Mean longevity is 20–25 years, but can reach 30 years or more (up to 33 years recorded for Verreaux's Sifaka) (Irwin 2006, Irwin *et al.* 2019, Bronikowski *et al.* 2016).

In most areas, Fosa is the principal natural sifaka predator; they are often ambushed while sleeping at night (Karpanty & Wright 2007, Goodman & Ganzhorn in press). Large raptors like Madagascar Harrier-Hawk and Henst's Goshawk also pose a threat, especially to juveniles (Wright *et al.* 1997, Patel 2005, Irwin 2009). Sifakas are widely hunted by humans for food, both in eastern and western regions (Salmona *et al.* 2014, Golden *et al.* in press), although in some areas local taboos (*fady*) afford some protection to some species, e.g., Perrier's Sifaka (Anania *et al.* 2018), Coquerel's Sifaka (Salmona *et al.* 2014a), Verreaux's Sifaka (Louden *et al.* 2006) and Golden-crowned Sifaka (Vargas *et al.* 2002).

Formerly three species were recognised: *P. diadema* in eastern and north-eastern areas; *P. verreauxi* in drier southern and western regions; and *P. tattersalli* in a small enclave of forest in the far north-east. Both *P. diadema* and *P. verreauxi* were each further divided into four subspecies: *P. d. diadema*, *P. d. edwardsi*, *P. d. candidus* and *P. d. perrieri*, and *P. v. verreauxi*, *P. v. coquereli*, *P. v. deckeni* and *P. v. coronatus* (Mittermeier *et al.* 1994, Garbutt 1999). There has been much subsequent debate surrounding the taxonomy of the genus *Propithecus* (Pastorini *et al.* 2001a, Rumpler *et al.* 2004) which remains ongoing; however, the broad consensus is now that all subspecies warrant recognition at species level (Mayor *et al.* 2004, Mittermeier *et al.* 2010), such that the genus currently encompasses nine species, each with an apparently distinct range that together form a discontinuous ring around the periphery of the island. Major rivers are barriers and often delineate boundaries between species' ranges (Goodman & Ganzhorn 2004). Some species, e.g., Perrier's and Golden-crowned Sifakas, have among the most restricted known ranges of any lemur (or any primate).

DIADEMED SIFAKA
Propithecus diadema
Bennett, 1832

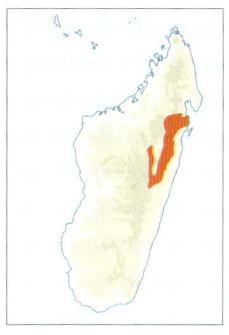

Malagasy names: Simpona, Simpony, Sadabe (Tsinjoarivo region)

MEASUREMENTS
Head-body: 500–550mm. Tail: 400–500mm. Weight: *c*.4.2–6.8kg. (Glander & Powzyk 1998, Irwin *et al.* 2019)

DESCRIPTION & IDENTIFICATION Largest sifaka species. Pelage moderately long and silky. Considerable variation in coloration across species' range. Typically,

Diademed Sifaka, reintroduced into Analamazaotra Special Reserve.

Diademed Sifaka: rare all black variant, Tsinjoarivo Forest.

head principally white with black crown extending on nape. Face bare and dark grey to black with rich red-brown eyes. Shoulders and upper back deep slate-grey, fading to silver-grey on lower back. Lower body and tail pale grey to white, often with a golden-yellow area around pygal region. Arms and legs rich orange to yellow-gold, hands and feet black. Chest and belly off-white to pale grey.

At southern extreme of range (vicinity of Tsinjoarivo), there is considerable variation: some animals are very much darker (with several melanistic all-black individuals), being black on head and shoulders, slate-grey on upper body and upper arms, with rich orange limbs, black hands and feet, and small white areas around face, whereas others are more typically white, slate-grey, sliver and orange. Despite previous suggestions that Tsinjoarivo population represented

a distinct taxon, molecular genetic studies indicate no clear separation, hence continued inclusion in *P. diadema* (Mittermeier *et al.* 2013).

A large lemur widely regarded as the most beautiful of Malagasy primates. Largest sifaka and second largest extant Strepsirrhine primate after Indri. Confusion between these two species extremely unlikely, with marked differences in coloration and morphology: Indri is black and white with virtually no tail, whereas all sifakas have long tails.

HABITAT & DISTRIBUTION Primary lowland and mid-altitude rainforest between *c.*200m and *c.*1,700m. Prefers intact forests but can survive in fragments as small as *c.*30ha (Irwin 2020). Restricted to rainforests of centre-east and north-east, precise limits unclear. Northern limit probably around Mananara River, west to Marotandrano Massif. Southern limit (vicinity of Tsinjoarivo) corresponds to west–east line formed by Onive and lower Mangoro Rivers. Most widely distributed rainforest sifaka.

Occurs at low densities throughout range. At Tsinjoarivo *c.*5–7 individuals/km^2 in larger tracts of intact forest, 14–20 individuals/km^2 in smaller isolated fragments (Irwin 2008a, Rakotomalala *et al.* 2017a). In areas where hunting occurs densities are reduced (Irwin 2020).

Diademed Sifakas successfully reintroduced to Analamazaotra Special Reserve (part of Andasibe-Mantadia National Park), from where it was extirpated in 1970s. Between 2006 and 2007, 27 were translocated from nearby threatened forests (Day *et al* 2009). Population has subsequently grown to *c.*60 individuals (Irwin 2020). Severely impacted by deforestation and forest fragmentation driven by slash-and-burn agriculture (Eppley *et al* 2020), which causes increased stress, lower body mass and negatively affects other health parameters (Irwin *et al.* 2010a, Irwin *et al.* 2019, Tecot *et al.* 2019). Deforestation resulting from cultivation of sugar cane for local rum production in some areas (Irwin & Ravelomanantsoa 2004). Hunting for food is also a major threat in many places and they are occasionally trapped for pet trade (Reuter & Schaefer 2017, Irwin 2020). IUCN Critically Endangered.

BEHAVIOUR Diurnal and arboreal. Group social structure highly variable: lives as male/female pairs, one male/multi-female, multi-male/one female or multi-male/multi-female groups of up to eight (at Tsinjoarivo 90% pairs), comprising 1–3 adult females and 1–2 males, plus offspring. One or two pairs

Female Diademed Sifaka with infant, Andasibe-Mantadia National Park.

breed per group: when two pairs breed, females are mother and daughter (Irwin 2021). Generally, both sexes disperse to non-natal group at maturity. Females sometimes remain when new unrelated male moves into group (Powzyk 1997, Irwin 2021). Females dominant: leading group movements, during bouts of feeding and social interaction (e.g., grooming), and over dominant males when breeding (Rasolonjatovo & Irwin 2019).

Groups occupy largely exclusive home ranges of 20–80ha (20–35ha in fragments, 50–80ha in continuous forest) (Irwin 2008a); defended primarily by scent-marking and occasional inter-group aggression. Compared to sympatric Indri, Diademed Sifaka spends more time patrolling and defending territory, travelling *c*.900–1,600m/day (*c*.700m/day for Indri) (Powzyk 1997, Irwin 2006a,b).

Groups spend time in all levels of canopy: *c*.55% of time resting, *c*.32% foraging and feeding, *c*.6% moving and *c*.5% in social behaviour. Groups may split into temporary subgroups to forage (up to *c*.700m apart), remaining separated for up to 48 hours before reuniting (Rijamanalina *et al.* 2020). Occasionally descends to ground to search for fallen fruits, subterranean fleshy flowers and soil, and indulge in bouts of play-wrestling (Powzyk 1997, Irwin *et al.* 2007). Group cohesion influenced by forest fragmentation: cohesion reduced in groups in smaller fragments (compared to groups in large continuous forests), which in turn is a consequence of greater reliance on mistletoes and other more dispersed and smaller food resources (Irwin 2007).

Vocalisations used primary to maintain group cohesion, signal alarm and convey aggression. Vocal repertoire relatively limited (compared to other lemurs, e.g. ruffed lemurs), but several different calls have been identified (Patel *et al.* 2005a). When alarmed by potential terrestrial threats, adults give intense sneeze-like *zzuss-zzuss-zzuss* call; when threatened from overhead (large raptors), call is loud resonant *honk-honk-honk* or roar (Garbutt 2007).

Diet primarily fruit (ripe and unripe), seeds, flowers and young leaves; respective proportions vary with seasonal abundance. More than 25 different plant species regularly utilised daily. In rainy season (November–April) ripe fruit and seeds constitute at least *c.*80% of diet. In dry season (May–October) fruit and seeds comprise only *c.*20%, with majority of diet young leaves and flowers (Powzyk 1997, Irwin 2008b, Irwin *et al.* 2014, 2015, Kerker Oliver 2017). At Tsinjoarivo relies heavily on hemiparasitic mistletoe (*Bakerella* sp.) for leaves, flowers and fruit year-round (Irwin 2008b). All foods contain important trace minerals, but leaves contribute higher concentrations than fruits or flowers (Irwin *et al.* 2017). Diademed Sifaka also uses sense of smell to detect/forage for flowers of underground parasitic plants (Balanophoraceae: *Langsdorffia* spp. and Cytinaceae: *Cytinis* spp.) hidden beneath leaf litter. This behaviour is likely learned from conspecifics (Irwin *et al.* 2007). Periodically descends to ground to eat soil; likely provides important trace nutrients and/or helps nullify toxins or tannins accumulated from diet (Powzyk & Mowry 2003). Minimal overlap in plant species eaten by Diademed Sifaka and broadly sympatric Indri: former eats higher energy foods, thus reducing food competition to minimum (Powzyk & Mowry 2003).

Females only receptive for one day per year, between November and January. Gestation *c.*170–180 days, single offspring born May–July. Offspring weighs *c.*140g at birth (Irwin *et al.* 2019): initially clings to mother's lower belly, progressing to riding on her back when 3–4 weeks old. Near-adult size is reached at *c.*12 months (growth may continue up to four years), sexual maturity in both sexes attained at *c.*3.5 years. Males migrate from natal group to neighbouring group at *c.*5 years. Females may remain or move to another group and first breed at five years. Probably lives 20–25 years or more in wild (Irwin 2006b, Irwin *et al.* 2019).

Infants and adults predated by Fosa (Garbutt 1994, Irwin 2009). Large raptors, e.g., Madagascar Harrier-Hawk and Henst's Goshawk also pose potential threat, especially to juveniles (Irwin 2020).

WHERE TO SEE Such is its beauty that all reasonable attempts should be made to see this lemur. Most accessible locality is Andasibe-Mantadia National Park, 140km east of Antananarivo. More remote Mantadia portion, 15km north of Andasibe, offers excellent opportunities as several sifaka groups are habituated. Enter the forest at PK15. Also easily seen in more central Analamazaotra Special Reserve portion, where it has been re-introduced. South of Andasibe (south side of *Route Nationale* 2) lies Maromizaha Reserve managed by lemur research group GERP. Forest similar to Mantadia, with steep trails, but Diademed Sifakas can be tracked down with patience. Alternatively, higher-elevation forests at Anjozorobe (part of Anjozorobe-Angavo Forest Corridor), *c.*90km north-east of Antananarivo are very rewarding. Tsinjoarivo, south-east of Ambatolampy, is another site for the more adventurous.

MILNE-EDWARDS'S SIFAKA
Propithecus edwardsi   A. Grandidier, 1871

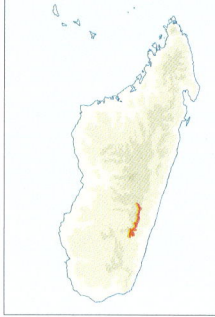

Malagasy names: Simpona, Simpony

MEASUREMENTS
Head-body: 420–520mm. Tail: 410–480mm. Weight: 5–6.5kg. (Glander *et al.* 1992, Lehman *et al.* 2005, King *et al.* 2011)

DESCRIPTION & IDENTIFICATION Large, distinctive sifaka. Pelage soft and dense; varies from dark chocolate-brown to almost black on head, upper body, limbs and tail. Chest and belly also dark, but more sparsely furred. Face bare, dark grey to black, eyes orange-red. Lower flanks and back form creamy-white saddle with reddish-brown at margins grading into main dark brown regions. A thin dark area runs down spine to base of tail. Large, dark, vertically clinging and leaping lemur with long tail. Should not be mistaken for any other species in range: Black-and-white Ruffed Lemur is large with vaguely similar coloration, but adopts a horizontal rather than vertical posture.

HABITAT & DISTRIBUTION Middle to high-elevation primary and slightly degraded rainforests, from *c.*600 to *c.*1,600m. Restricted to southern portion of eastern rainforest: northern limit is west–east line formed by Onive and lower Mangoro Rivers, and range extends south to eastern slopes of Andringitra Massif (southern limit may be Iantara and Rienana Rivers). Populations in north of range, i.e. north of Ranomafana National Park (now Marolambo-Fandriana National Park) are widely hunted and may have completely disappeared; no sightings since late 1990s (Irwin *et al.* 2005, Lehman *et al.* 2005b, 2006d, Wright *et al.* 2020b). Similarly, populations in COFAV Forest Corridor connecting Ranomafana National Park and Andringitra Massif have largely disappeared (Wright *et al.* 2020b).

Population densities vary reflecting habitat differences and effect of hunting pressure. Ranomafana National Park certainly constitutes the species' stronghold (perhaps 50% of remaining population): within the park densities vary from *c.*5–8 individuals/km²; in areas south of park (COFAV Forest Corridor) mean densities only *c.*3 individuals/km² (Wright 1987, Irwin *et al.* 2005, Wright *et al.* 2012).

Habitat destruction and forest fragmentation driven by slash-and-burn agriculture, logging, mining and illegal rum production are primary threats. Habitat disturbance also increases susceptibility to detrimental parasites and affects ranging behaviour (Wright *et al.* 2009, Gerber *et al.* 2012c). Hunting for food, using slingshots, blowpipes and firearms, is also a major threat (Dunham *et al.* 2008). IUCN Endangered.

BEHAVIOUR Lives in multi-male/multi-female groups of 3–9 individuals containing two or more mature members of each sex. Females dominant (Pochron *et al.* 2003). Groups largely stable; adults and offspring remain together throughout year. Adults may remain resident up to 12 years; breeding relationships stable over 6–10 years (Wright 1988, 1995). Periodically, both females and males migrate between groups: average during lifetime, males three times, females twice (Pochron & Wright 2003, Morelli *et al.* 2009, Wright *et al.* 2012). Dispersal distance between groups up to 12km (Wright *et al.* 2020b) but dispersal low across fragmented habitats (Pochron *et al.* 2004, Gerber *et al.* 2012).

Milne-Edwards's Sifaka feeding on unripe fruit, Ranomafana National Park.

Milne-Edwards's Sifaka, Ranomafana National Park.

Groups maintain large home ranges (c.45–55ha) with virtually no overlap. Home ranges boundaries remain consistent, even if group size fluctuates. Both sexes regularly scent-mark; serves as territorial defence and conveys information on status to other group members (Pochron et al. 2005a,b). Daily distances travelled considerable, c.650–1,250m. Prefers higher regions of canopy (above c.15m), especially females with infants, which helps predator detection (Minkus & Arrigo-Nelson 2015). At home range boundaries, neighbouring groups sometimes

approach within a few metres with no apparent aggression or vocalisations, but increased scent-marking activity; group intermingling has even been observed (Wright 1987, 1988, Pochron et al. 2005a,b). Sleep sites normally associated with ridgetops, preferring branches 8–10m above ground, probably to reduce predation (Wright 1998).

Diet primarily folivorous but varied: mainly equal amounts of leaves, fruits, seeds and flowers with >25 different species eaten per day. In spring (October–November) new leaves and shoots, rich in protein, are preferred (Wright 1987, Hemingway 1995, 1996, 1998, Wright et al. 2005). Individuals in disturbed forests eat higher proportion of leaves (rather than seeds) and spend proportionately longer feeding than counterparts in undisturbed forests (Arrigo-Nelson 2006, Matos 2017, Krauss 2018). During foraging they move and feed at all heights in forest canopy, occasionally descending to ground level to eat soil. This may provide vital trace nutrients or help counteract toxins and tannins accumulated from regular diet (Semel et al. 2019).

Infants born in June and July (peak June), gestation 180 days (Pochron et al. 2004). Even for predominately leaf-eating sifakas, fruit key to reproductive success. Breeding timed to synchronise with peak fruiting season: protein-rich resources vital for females during late lactation/early weaning period when energy costs are highest (Wright et al. 2012). Birth weight c.150g. For 3–4 weeks young hold onto mother's belly, at around one month transfer to ride on back. At two to three months begin taking solid food, by six months over half infant's nourishment is non-milk foods. Mothers sleep with offspring for up to two years. Females breed every other year, except if infant is lost when they breed the following year (Wright 1995, Pochron et al. 2004). Females continue to reproduce until death, even when 25 years old or more (Pochron & Wright 2003, Wright et al. 2008), although survival rates for young of older females are reduced (King et al. 2005).

Infant mortality very high: c.50% die in first year and c.75% fail to reach sexual maturity (Pochron et al. 2004, Morelli et al. 2009). Most infants (and some adults) predated, mainly by Fosa and large raptors (Wright et al. 1997, Wright 1998, Karpanty & Wright 2006). Some are victims of infanticide by immigrant adult males and females (Wright 1995, 1998, Morelli et al. 2009). At one year, individuals weigh c.50% of adult body weight. Sexual maturity reached at four years in females and five in males

Female Milne-Edwards's Sifaka with infant, Ranomafana National Park.

(King *et al.* 2011, 2012, Wright *et al.* 2012). From c.3.5–4.5 years, both sexes equally likely to disperse from natal group: if natal group contains unrelated opposite-sex adult, then male or female remains to breed; if only related adults are present, natal male or female disperses without breeding. Adult males and females entering new group kill group infants (Morelli *et al.* 2009). Life expectancy in wild is up to 32 years for females, and 19 years for males (Tecot *et al.* 2013).

Fosa is major predatory threat: both adults and juveniles are taken, often being ambushed when they are sleeping (Karpanty & Wright 2007, Irwin *et al.* 2009).

WHERE TO SEE Ranomafana National Park is best place to see this species. Groups are habituated in both mid-elevation (Talatakely) and high-elevation forests (Vohiparara) and it is often possible to approach individuals with ease. Some individuals have collars. Trails are steep, especially in Talatakely. Park guides are very good at finding these and other lemurs, and their services should always be sought.

SILKY SIFAKA
Propithecus candidus　　　A. Grandidier, 1871

Malagasy names: Simpona, Simpony, Simpona Fotsy

MEASUREMENTS
Head-body: 480–540mm.
Tail: 450–510mm. Weight:
c.5–6.5kg. (Lehman *et al.*
2005a, Garbutt 2007)

DESCRIPTION & IDENTIFICATION Large sifaka species. Pelage luxurious, silky and uniformly creamy-white, sometimes with tints of silver-grey around crown, back and limbs. Upper/inner arms and legs may have faint tinges of pale yellow/orange. Some have dark grey across upper back and shoulders. Pygal area (lower back and base of tail) sometimes darker and discoloured. Adult males readily

Female Silky Sifaka with infant, Marojejy National Park.

distinguished from females by large brown 'chest patch', consequence of scent-marking with sternal-gular gland: patch enlarges with increased scent-marking during breeding season (Patel 2006b). Face bare, slate-grey to black. Some (in Anjanaharibe-Sud and Marojejy) lack skin pigment on faces and other areas to varying extent: areas appear totally pink or mix of pink and slate-grey/black. Ears small and slate-grey, or blotched pink and protrude just beyond fur. Eyes deep orange-red. Only *Propithecus* species that exhibits extreme individual variation in partial skin pigmentation loss (leucism). All are born with black faces, but with age, some lose pigmentation to varying degrees and develop pink-blotched faces and hands and feet (Louden *et al.* 2017). A large, vertically clinging and leaping lemur with very distinctive all-white coloration. Cannot be mistaken for any other lemur within its range.

HABITAT & DISTRIBUTION Primary mid- and high-elevation rainforest at *c.*700–1,900m, and sometimes sclerophyllous forest and even low ericoid bush at highest elevations (in Marojejy and Anjanaharibe-Sud: Schmid & Smolker 1998, Sterling & McFadden 2000). At southern extreme of range (Adaparaty Forest, Makira) occurs in low-elevation (*c.*250–600m) forest fragments (Patel & Andianandrasana 2008). Restricted to northern extremity of eastern rainforest. Southern range limit just north of Antainambalana River (Adaparaty Forest, Makira), extending via north-east edge of Makira National Park (Antohaka Lava Forest) to Anjanaharibe-Sud Special Reserve, COMATSA-Sud/Betaolana Forest corridor and Marojejy National Park. At least one group in unprotected Maherivaratra Forest (Rasolofoson *et al.* 2007b, Patel, 2014, 2016).

Marojejy National Park and Anjanaharibe-Sud Special Reserve are species' relative strongholds. Occurs at extremely low densities. In Anjanaharibe-Sud Special Reserve *c.*0.3–2.6 individuals/km^2 (Patel 2016). In Marojejy National Park encounter rates are very low (line transects, 0.015 groups/km^2) (Loudon *et al.* 2016).

Forest destruction and fragmentation due to slash-and-burn agriculture, logging and hunting are principal threats. Hunted for food in many areas, e.g., Makira, peripheral areas of Marojejy and Anjanaharibe-Sud, and other areas around Andapa

Basin (Duckworth *et al.* 1995, Patel *et al.* 2005c, Golden 2009, Patel 2009, 2010). IUCN Critically Endangered.

BEHAVIOUR Diurnal and arboreal. Lives in multi-male/multi-female groups of 2–9 individuals; smaller groups the norm (Patel 2006a). Social structure appears variable. Smaller groups of 3–4 consist of an adult pair (pair bonded) plus offspring: larger groups are polygynous with single adult male and multiple adult females (rarely more than two) plus juveniles. Group movements led by females. Home range size varies with location and habitat quality: in undisturbed primary forest (Marojejy) *c.*55ha (group spends 95% of time in core *c.*40ha), in more disturbed forests (Makira) *c.*100ha (95% of time in core *c.*65ha). Groups travel *c.*700m per day, sometimes ascending 500m vertically up slopes (Patel 2006a, Santorelli *et al.* 2006, Rajaonarison 2015).

Olfactory communication well developed. Females scent-mark tree trunks by rubbing genital glands rhythmically up and down. Males scent-mark with both sebaceous chest and genital glands. Males scent-mark two or three times as frequently as females, but female scent-marks elicit swifter and more frequent responses, often over-marking by males (Patel 2006b, 2011). Males routinely gouge trees with toothcomb prior to chest marking and female over-marking, leaving lasting visible marks, often repeatedly on same key trees (known as 'totem trees') (Ritchie & Patel 2006, Patel & Girard-Buttoz 2008). Scent-marking generally occurs within core area of home range, rather than boundaries of territory. Both sexes regularly urinate while scent-marking (Patel 2006b).

Foraging activity begins around dawn; poor weather may delay this (Garbutt 2007). Approximately 45% of day spent resting in canopy, 16% in social behaviour, 22% feeding and rest travelling between foraging sites (Kelly & Mayor 2002, Patel 2006a, Santorelli *et al.* 2006). When feeding, females are dominant, although male submissive signals not always apparent (Patel 2003). During rest, regular social interactions such as grooming and play between preferred partners: play bouts on ground may last >30 minutes (Patel 2006a).

Silky Sifakas are folivorous seed predators. Diet varied, >100 species of mainly trees, with some

Female (centre) and male (bottom left) Silky Sifakas with infants around one month (on back) and five months, Marojejy National Park.

Female Silky Sifaka carrying infant on underbelly, Marojejy National Park.

lianas and epiphytes consumed; seeds and fruits comprise *c.*42%, mature or young leaves *c.*48% and flowers *c.*10%. Bark and soil also occasionally eaten. Most important foods (account for more than *c.*35% of feeding time) are fruit of *Pachytrophe dimepate* (Moraceae), seeds of *Senna* spp. (Fabaceae), immature leaves of *Plectaneia thouarsii* (Apocynaceae) and fruit of *Eugenia* spp. (Myrtaceae) (Kelly & Mayor 2002, Patel 2020).

Females receptive only one day per year: mating occurs between December and January. Gestation *c.*175–180 days. Young born in June or July. Infants initially grasp the mother's belly and later ride on her back, 'jockey style'. Offspring sleep with mothers until maturity approaches. All group members interact with infants: allo-parenting involves grooming, playing, carrying and nursing. Mothers will nurse their own and other offspring simultaneously (Patel *et al.* 2003b, Patel 2007). Females probably breed every two years, unless infant is lost, when offspring produced in consecutive years (Patel 2006a).

Moderately vocal: several distinct adult calls recognised. When disturbed, general response is repeated *zzuss- zzuss- zzuss* call: these are individually distinctive and differ between males and females, so convey identity and sex of caller to other group members, and also serve as a general group coordination signal (Patel *et al.* 2005a,b, Patel *et al.* 2006, Patel & Owren 2012). Also gives 'aerial disturbance' roars when large raptors fly overhead (Patel *et al.* 2003a).

Fosa is principal natural predator (Patel 2005). Silky Sifakas alarm call at potential 'aerial predators', e.g., Madagascar Buzzard, suggesting other large raptors, e.g., Henst's Goshawk and Madagascar Harrier-Hawk may pose potential threat, especially to juveniles (Patel *et al.* 2003a).

WHERE TO SEE A challenging lemur to see. Only accessible location Marojejy National Park: best looked for between Camp Marojejia (Camp 2) and Camp Simpona (Camp 3). Camp Marojejia is five to six hours walk from Manantenina (on *Route Nationale* 3b), Camp Simpona is a further three hours walk. Rustic huts with bunk beds at Camp Marojejia, tents at Camp Simpona. Terrain tough, hills very steep, undergrowth dense and often very wet and slippery. Experience is one of the best in Madagascar; worth staying at Camp Marojejia for view alone.

PERRIER'S SIFAKA
Propithecus perrieri   Lavauden, 1931

Malagasy names: Radjako, Ankomba Joby

MEASUREMENTS
Head-body: 430–470mm. Tail: 420–460mm. Weight: 3.7–5kg. (Lehman *et al.* 2005, Ranaivoarisoa *et al.* 2006)

Perrier's Sifaka (*Propithecus perrieri*)

Indian Ocean

Irodo River

Analamerana Special Reserve

Andrafiamena Mountains

Loky River

N

☐ Protected areas
▨ Forest where Sifakas occur

0   10   20   30km

Perrier's Sifaka, Analamerana Special Reserve.

DESCRIPTION & IDENTIFICATION Medium-sized sifaka. Dorsal pelage dense, silky and uniform lustrous black, while ventral fur slightly shorter and sometimes tinged russet-brown around chest and lower abdomen. Face bare and dark grey-black, eyes deep orange-red, ears small and largely concealed by fur. A highly distinctive, large, vertically clinging and leaping lemur with jet-black fur. Largest lemur in its limited range. Cannot be confused with any other species.

HABITAT & DISTRIBUTION Dry deciduous, dry evergreen and semi-humid forests, from near sea level to *c.*400m. Most restricted range of any sifaka species (Salmona *et al.* 2017b). Range extends from eastern edge of Analamerana limestone massif, along Indian Ocean coast to sandstone forests of Andrafiamena Mountains, and west to Ambohibe, north-east of Marivorahona, including Andrafiamena and Mahanoro Forests. Bound in north by Irodo River and in south by Andrafiamena Mountains (Ranaivoarisoa *et al.* 2006, Zaonarivelo *et al.* 2007c, Salmona *et al.* 2013, Banks *et al.* 2015). Formerly recorded further west in Ankarana Massif (Hawkins *et al.* 1990), where it may not have been resident (Heriniaina *et al.* 2020). No recent records there, but

287

Perrier's Sifaka, Andrafiamena Forest.

BEHAVIOUR Lives in relatively small multi-male/multi-female groups of 2–6: typically 1–3 females and 1–2 adult males plus dependent offspring. Exclusive home ranges c.30ha, with little aggression at boundaries (Meyers & Ratsirarson 1989, Mayor & Lehman 1999). Females dominant. Males emigrate from natal to neighbouring group at c.5 years, females may emigrate or remain in natal group. Within group only one adult pair reproduces per year (Meyers 1996). Infants born in June and July.

Diet mainly young and mature leaves, leaf petioles, unripe fruits, young shoots and flowers: variety of different trees and plants across nine families have been identified as primary food sources (Lehman & Mayor 2004). When feeding, group may be spread out in trees more than 50m apart, but individuals maintain contact with one another via regular quiet calls. At end of dry season (November) groups may move into ribbons of more humid forest bordering dry riverbeds and feed in trees of introduced Mango *Mangifera indica* (Anacardiaceae) (Garbutt 2007).

Regularly descends to ground to cross open savannas (sometimes >500m wide) between forest fragments and to drink from riverbeds (Mayor & Lehman 1999); makes them particularly vulnerable to predation, especially by Fosa (Mayor & Lehman 1999) and occasionally feral dogs (Banks 2013). *P. perrieri* populations show low levels of genetic diversity: ability to cross open areas between forest patches may mitigate negative affects of habitat fragmentation on genetic flow (Salmona et al. 2015, Bailey et al. 2016).

Distinctive alarm call, *zzuss-zzuss-zzuss*, is given to human intruders and potential predators. Reacts in highly agitated manner to Fosa: when predator on ground, all group members gather into adjacent trees, watch intently and alarm call, then quickly bound away through canopy (Garbutt 2007, 2021). Fosa densities high in region and may have negative impact on Perrier's Sifaka populations (Heriniaina et al. 2020).

seen in forest fragments between Analamerana and Ankarana (Banks et al. 2007, Salmona et al. 2013).

Overall densities are very low, but considerable variation between sites and forest fragments: densities in forest on sandstone soils more than double those on limestone soils (Banks 2013). Highest estimates, c.46 individuals/km² (Ambatovazaha) and c.35 individuals/km² (Andampibe), with lowest estimates c.3.5 individuals/km² (Ampasimaty) and c.1.5 individuals/km² (Ampondrabe). Overall mean density c.8.75 individuals/km², with perhaps c.500 individuals, including c.125 adults, in total (Heriniaina et al. 2020).

Forest destruction and fragmentation for slash-and-burn agriculture, charcoal production and fires set to increase cattle pasture are major threats (Salmona et al. 2017). Perrier's Sifaka is strongly protected by local Antankarana taboo (*fady*) (shared by c.95% local residents) prohibiting hunting and consumption (Anania et al. 2018), but hunting by immigrant miners is serious (Meyers & Ratsirarson 1989, Meyers 1996). IUCN Critically Endangered.

WHERE TO SEE A challenging lemur to see: in most areas shy and flees from humans (Salmona et al. 2015). Best locality is Andrafiamena Forest, part of Andrafiamena-Andavakoera Protected Area. Stay at Black Lemur Camp, near village of Anjahankely, an environmentally sustainable tourism and reforestation project operated by FANAMBY.

VERREAUX'S SIFAKA
Propithecus verreauxi   A. Grandidier, 1867

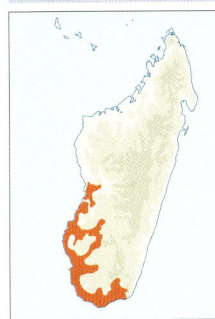

Malagasy names: Sifaka, Sifaka-Bilany (Isalo region)

MEASUREMENTS
Head-body: 400–475mm. Tail: 500–600mm. Weight: 3–3.5kg. (Tattersall 1982, Kappeler 1991, Richard *et al.* 2000, 2002)

DESCRIPTION & IDENTIFICATION Relatively small sifaka. Pelage longish, thick and soft. Overall colour, including tail, white, with dark brown crown extending to nape. Ears black and prominent surrounded by white tufted fur. Face and muzzle very dark grey to black, eyes vivid yellow. Fur on chest, belly and underarms sparse, allowing grey skin to show through; gives underparts a darker appearance. Upper chest in males sometimes tinged with reddish-brown from glandular secretions. Considerable variation exists: some in Isalo, Zombitse and Andohahela National Parks have dark brown areas on back, chest, upper arms, upper thighs and tail (previously referred to as 'subspecies' *P. v. 'majori'*) (Richard 2003). Others may be almost all white (e.g. in Berenty). Variants generally occur in mixed groups also containing 'normal' individuals (Garbutt 2007). A medium-large, vertically clinging and leaping lemur; impossible to confuse with any other species in its range.

HABITAT & DISTRIBUTION Broad habitat tolerance: dry deciduous forest in west, scrub thicket, spiny forest, spiny bush and gallery (riparian) forests in south-west and south, and transitional forest in extreme south-east of range, in Mangatsiaka (parcel 2) and Tsimelahy (parcel 3) Andohahela National Park. Exceptional record further east in humid forest zone (parcel 1) of Andohahela National Park at *c*.440m (Feistner & Schmid 1999, Rasoarimanana 2005). Found from sea level to (exceptionally) *c*.1,700m (Louis *et al.* 2020n): most records below *c*.400m.

Verreaux's Sifaka, Berenty Private Reserve.

Most widely distributed sifaka species. Range extends south from Tsiribihina River in centre-west, to Tsimanampetsotse National Park and Mahafaly Plateau Protected Area in south-west, and east to western edge of Anosy Mountains in extreme south-east, and inland to Zombitse-Vohibasia National Park and Isalo Massif.

Population densities vary considerably with habitat type and quality. In gallery forest (Berenty) densities can reach c.150–200 individuals/km^2 (O'Connor 1987), and at Antserananomby the equivalent of c.400–500 individuals/km^2 (Sussman 1975). Densities in deciduous forest (e.g. Kirindy Forest and Kirindy-Mitea National Park) are c.40–50 individuals/km^2, while in spiny forest c.37 individuals/km^2 (Kelly et al. 2007, Richard 2003).

Able to tolerate habitat fragmentation and some anthropogenic disturbance: groups can be found in small forest patches in highly fragmented environments. Severely threatened by massive habitat loss throughout range, primarily due to logging, charcoal production and slash-and-burn agriculture. Deliberate burning to improve cattle grazing further damages remaining forest areas. Additionally, introduced prickly pear (Opuntia sp.) causes detrimental alteration of spiny forest habitats. Hunting is a further serious threat:

although taboo (fady) for several tribal groups within species' range (e.g., Antandroy, Mahafaly) (Loudon et al. 2006), sifakas are hunted by other tribes (e.g., Sakalava) and regional immigrants. In Isalo region, Verreaux's Sifaka is known as sifaka-bilany meaning 'sifaka of the cooking pot'. IUCN Critically Endangered.

BEHAVIOUR Lives in multi-male/multi-female groups of 2–14, with 4–8 the norm. Larger groups contain more than one breeding female, but unusual for more than one infant to survive annually. Smaller groups (six or less) represent family units: larger groups comprise mutually familiar foraging parties. At least one and sometimes all females are socially dominant over males, and displace adult males at feeding sites (Richard 1974b, 1985, 1992, 2003, Richard & Nicoll 1987).

In spiny forest home ranges are 7–8ha; in gallery forest, with more abundant resources, c.2–3ha (Jolly et al. 1982). Furthermore, in dry habitats, home ranges in disturbed forests up to three times larger than in undisturbed areas (Wilson & Ferguson 2014). In dry deciduous forest, home ranges c.12–25ha; increase correlated with more intense competition, as dry deciduous forest supports greater density and diversity of other lemur species.

*Verreaux's Sifaka eating terrestrial parasitic flower (*Hydnora esculenta*), Berenty Private Reserve.*

Home ranges contract when females give birth. Ranges overlap, often considerably, but core territory remains exclusive to group and is defended (Benadi *et al.* 2008). Home ranges are defended by scent-marking, boundary disputes generally resolved with few or no vocalisations (Jolly *et al.* 1982, Richard 1985). Groups mutually avoid one another, but agonistic inter-group encounters can occur year-round. In deciduous forest (Kirindy Forest), rarely do all group members participate simultaneously, but males and females do so equally, except when females have infants (reducing risk of infanticide); dominant males participate more often than subordinate males (Koch *et al.* 2016).

Movement involves leaping between vertical trunks, even among the viciously thorned trees (e.g. *Alluaudia* spp.) of Didiereaceae spiny forest. May also descend to the ground and cross open spaces by skipping/bounding sideways on hindlegs with arms held aloft for balance.

Activity varies with habitat and season: in cool dry season often remains inactive for up to two hours after sunrise, then moves into canopy to sunbathe, and afterwards forages almost continuously until mid-afternoon before settling down for night (Norscia *et al.* 2006). Rarely travels more than *c.*500m per day. In warm wet season often active before sunrise and stops foraging by mid-morning, then rests during midday hours, and resumes foraging late afternoon, sometimes continuing after sunset, often travelling >1,000m in the day (Jolly *et al.* 1982). During midday heat of late dry season at Bezá-Mahafaly Special Reserve, sifakas thermoregulate by 'terrestrial tree hugging', resting on ground while holding tree trunk against thinly furred underparts. Ambient temperatures at base of trees cooler than body temperature (36°C) and cooler than air in canopy (Chen-Kraus 2019).

Diet seasonally variable in proportions eaten, principally leaves, fruits, seeds and flowers, and occasionally bark, dead wood, terrestrial parasitic flowers and termite soil, the latter may provide trace nutrients or help alleviate gastro-intestinal irritation (Richard 1978a, Norscia *et al.* 2005). Leaves predominate in dry season, with a shift to fruit, seeds and flowers during wet season. Most seeds are destroyed. Dietary diversity constricted significantly in disturbed forests (Wilson & Ferguson 2014). Can tolerate prolonged droughts: groups inhabiting spiny forest appear to gain all necessary moisture from leaves of Didiereaceae. In dry season, when heavy dews common, sifakas lick moisture from their own

Female Verreaux's Sifaka with infant feeding in spiny forest, Anjampolo.

coats and may also gain moisture by eating bark and cambium of Elephant Tree or Jabily *Operculicarya decaryi* (Anacardiaceae) (Richard 1974b, 1977, 1978a, Erkert & Kappeler 2003). Both sexes lose weight and condition during dry season (Lewis & Kappeler 2005). During pregnancy and lactation, females (in deciduous forests) significantly increase proportion of tannin-rich leaves in diet; tannins may help in milk secretion and also have other beneficial

Female Verreaux's Sifaka with 3–4 month old infant, Berenty Private Reserve.

medicinal properties during periparturient period (Carrai *et al.* 2003).

Mating season brief and highly synchronous: females in oestrous for single day between late January and early March (Lawler 2021). Around 45% of females breed each year, rest every other year. Male competition for access to females is intense and may involve conflicts (biting, grabbing, cuffing, occasionally resulting in serious injury) and prolonged canopy chases and lunges (Lawler 2009). In spiny forest (Bezá-Mahafaly Special Reserve) successful males move from female to female within group to find receptive mate, and sometimes roam to neighbouring groups to father offspring, which helps increase their lifetime reproductive success (Richard 1974a, Richard *et al.* 1991, Lawler *et al.* 2003, 2005, Lawler 2007). In deciduous forest (Kirindy Forest) single dominant male monopolises mating and paternity within his group (Kappeler & Schäffler 2008).

Single offspring born in dry season, usually in July–August, gestation *c.*162–170 days. Females observed giving birth in canopy in late afternoon (as other group members settled to sleep). Mother pulls infant from birth canal and immediately cleans it. Infant then moves to her breast to nurse and mother eats placenta and umbilical cord (Chen-Kraus & Raharinoro 2020). Initially, infants carried on mother's belly, after 10–12 weeks shift to ride on her back. Period of high lactation requirements coincides with increasing food availability (onset of rainy season) (Lewis & Kappeler 2005). Allo-parenting has been observed, with all group members participating in grooming, nursing, carrying and playing. Infanticide by non-paternal males also occurs (Lewis *et al.* 2003, Littlefield 2010). Independence reached at six months (Jolly 1966, Richard 1976, 2003). Adults particularly vigilant when group has young, always keeping eye out for predators.

Alarm call for ground predators is characteristic nasal bark, *shi-fakh, shi-fakh, shi-fakh,* and simultaneously individuals throw back their heads in a jerky motion. Potential aerial predators induce a loud bellow or roar (Fichtel & Kappeler 2002, Fichtel & van Schaik 2006). Other lemur species, e.g., Red-fronted Brown Lemur also recognise and react to sifaka alarm calls (Fichtel 2004).

Infant predation is high: in dry deciduous forests *c.*30% are eaten by Fosa within first year (Rasoloarison *et al.* 1995, Goodman 2003), with smaller numbers predated by large raptors, e.g., Madagascar Harrier-Hawk (Karpanty & Goodman 1999, Brockman 2003). Occasional records of predation by large snakes, e.g., ground boas (*Acrantophis madagascariensis* and *A. dumerili*) (Mittermeier *et al.* 2010). Domestic dogs and feral cats also likely predators, with evidence of feral cat predation at Bezá-Mahafaly (Brockman *et al.* 2008).

Age of sexual maturity appears to vary with habitat. In spiny forest (Bezá-Mahafaly Special Reserve) where conditions are harsh, majority of breeding females are >6 years old (Brockman 1999, Richard *et al.* 2002). In more resource-rich gallery forests (Berenty) sexual maturity reached earlier, *c.*3–5 years. Adult females remain within natal group, whereas males are encouraged by established group members to leave prior to maturity and transfer to neighbouring group. New groups often slow to accept incoming males, forcing them to spend time on periphery. Resident males also initially hostile towards newcomers. In some cases, males move between groups several times within their lifetime (Richard *et al.* 1991, 2002, Richard 1992, 1993, Brockman *et al.* 1996, 1998). Life expectancy *c.*25 years (Richard 2003).

WHERE TO SEE Easily seen at several locations. At Berenty Private Reserve, on banks of Mandrare River, groups are completely habituated with close encounters guaranteed. Most groups live in gallery forest with some in small spiny forest parcel. Berenty is best place to see sifakas 'skipping' across open spaces. To see them in more extensive spiny forest visit Anjampolo, north of Berenty, or Mangatsiaka or Ihazofotsy, part of Andohahela National Park. An alternative location with both spiny and gallery forest is Ifotaka Community Forest, north of Berenty, which straddles the Mandrare River. Verreaux's Sifaka is also easy to see in several deciduous forest locations. Kirindy Forest, north of Morondava, is the best. Further south, Kirindy-Mitea National Park offers opportunity for the more adventurous. Also possible to see in Zombitse-Vohibasia National Park. In nearby Isalo National Park they can be encountered in some ribbons of riparian forest that emerge from canyons at eastern edge of the park. The extremely adventurous could consider Makay Massif, c.100km north of Isalo, an incredibly remote area of canyons and forest patches that is home to Verreaux's Sifaka.

COQUEREL'S SIFAKA
Propithecus coquereli　　　(A. Grandidier, 1867)

Malagasy names: Sifaka, Tsibahaka

MEASUREMENTS
Head-body: 425–500mm. Tail: 500–600mm. Weight: 3.5–4.3kg. (Kappeler 1991)

DESCRIPTION & IDENTIFICATION Medium-sized sifaka. Pelage soft and dense; mostly white on head, body and tail, with distinctive deep maroon patches on thighs and arms, extending across chest. Some have brown to silvery-grey area at base of back. Face bare and black, surrounded by white fur around cheeks and crown, extending from forehead down nose to muzzle. Ears black and prominent. Eyes vivid

Coquerel's Sifaka, Anjajavy Private Reserve.

yellow to pale orange. Some individuals in aberrant populations deviate from 'typical' form in having grey-black coloration replacing maroon on thighs and arms (Mittermeier *et al.* 2013). A medium-large, vertically clinging and leaping lemur with highly distinctive coloration. Cannot be confused with any other species in its range.

HABITAT & DISTRIBUTION Dry deciduous and semi-evergreen forest up to *c.*380m, and bush, scrub and secondary forest fragments adjacent to primary forests. Also uses coastal mangroves (Anjajavy, Baie de Mahajamba and Mariarano Forest) for sleeping and feeding (Gardner 2016, Chell *et al.* 2020). Restricted to north-west: range extends from north and east of Betsiboka River, most southerly locality Betonendry/Maroakanga, south-east of Maevatanana on east bank of Betsiboka River (Rakotonirina *et al.* 2014), to western coastal forests (Anjajavy, Mariarano and Baie de Mahajamba), to northern limit around Bealanana, with eastern boundary near Antetemasy (west of Befandriana Nord) (Salmona *et al.* 2014a, Louis *et al.* 2020b, Ramilison *et al.* 2021).

Coquerel's Sifaka appears tolerant of some habitat fragmentation and degradation, and anthropogenic disturbance: able to survive in mosaic of small and large forest patches, scrub and grassland, and relatively close to human habitation (Salmona *et al.* 2014a, Chell 2018, Ramilison *et al.* 2021). Population densities vary with fragment size and habitat quality: in Ankarafantsika National Park densities range from 5–100 individuals/km^2 depending on site (Kun-Rodrigues *et al.* 2014).

Severely threatened by habitat destruction due to slash-and-burn agriculture, charcoal production and annual burning to generate new cattle pasture. Despite some local taboos (*fady*), hunting also increasingly prevalent including in and around Ankarafantsika National Park (Garcia & Goodman 2003, Salmona *et al.* 2014a). IUCN Critically Endangered.

BEHAVIOUR Lives in groups of 3–10, with 4–5 most common (Richard 1974b). Smaller groups are family units rarely containing more than one infant; larger groups have variable age and sex composition and are probably mutually familiar foraging units.

Female Coquerel's Sifaka feeding with infant, Ankarafantsika National Park.

Home range 4–9ha, but groups spend >60% of time in exclusive core area of 2–3ha. Considerable overlap of home ranges around periphery. Boundary encounters with other groups rarely aggressive, but rather mutual avoidance (Richard 1974b). In forest/scrub/grassland mosaic habitats often descends to ground to cross open areas between forest patches. Unlike other western sifakas, e.g., Verreaux's Sifaka that bound sideways, Coquerel's Sifaka leaps facing forwards in a kangaroo hop-like fashion.

Foraging occupies 30–40% of day, with seasonal changes in times of principal activity (Richards 1978a). During wet season feeding regularly begins before sunrise and peaks before mid-morning. After rest period, activity begins again in afternoon and continues until early evening. Average daily distance travelled c.1km. In dry season feeding begins later and ends earlier, with shorter rest period around midday. Daily distance covered c.750m.

Feeds primarily in forest canopy. Diet mainly mature leaves, buds and occasionally bark in dry season; in wet season proportion of young leaves (in particular), flowers and fruits increases. Known to eat nearly 100 different plant species, but 12 of these comprise 65% of diet (Richard 1978a,b). Groups also known to sleep and feed in coastal mangroves: at high tide seen feeding on mature leaves of red mangrove *Rhizophora mucronata* (Chell et al. 2020). Occasionally also feeds on introduced Mango *Mangifera indica* (Anacardiaceae) and Tamarind *Tamarindus indica* (Fabaceae), especially near villages (Salmona et al. 2014a).

Single offspring born in June/July, after gestation of c.160–165 days. Infant initially clings to mother's front. After 3–4 weeks, transfers to ride on back and may continue doing so to six months. Allo-parenting common: all group members interact with infants, playing, grooming, nursing and carrying. Adult size reached at c.1 year. Sexual maturity in both sexes at c.2.5 years (Richard 1976).

Fosa is principal predator (Dollar et al. 2007). Also a record of Madagascar Ground Boa attempting to take an adult Coquerel's Sifaka, with other group members then attacking the snake until sifaka was released (and survived). Snake later died from injuries inflicted by sifakas (Gardner et al. 2015).

WHERE TO SEE Ankarafantsika National Park (Ampijoroa Forest Station) is best place to see this stunningly beautiful lemur. Several groups live within close proximity of main campsite: trail network is well maintained and good views are almost guaranteed.

For those wishing to watch sifakas and enjoy beautiful coastal surroundings a trip to Anjajavy is highly recommended. Again several groups are habituated and are seen daily in hotel grounds. Alternatively, the more adventurous might visit Mariarano Classified Forest, north-east of Mahajanga.

DECKEN'S SIFAKA
Propithecus deckeni   Peters, 1870

Malagasy names: Sifaka, Tsibahaka

MEASUREMENTS
Head-body: c.425–475mm. Tail: c.500–600mm. Weight: 2.6–4.0kg. (Tattersall 1982, Mittermeier et al. 2013)

DESCRIPTION & IDENTIFICATION Medium-sized sifaka. Typical pelage creamy-white above and below, sometimes washed with tones of yellow-gold, silver-grey or pale brown on neck, shoulders, back and limbs. Face entirely black. Dark ears protrude from fur. Appears slightly more rounded and blunt-muzzled than similar species like Verreaux's Sifaka, but not as extreme as Crowned Sifaka.

Melanistic variants occur in several populations close to Mahavavy River (King et al. 2014). These range from lighter forms with pale heads and limited patches of dark fur, to intermediate forms with darker heads and overall appearance similar to typical Crowned Sifaka, to very dark brown forms, with extensive areas of dark fur on head, back, arms and legs that are unlike 'typical' forms of any sifaka species (Rakotonirina et al. 2014). In south-east of range, near source of Mahavavy River (vicinity of Ambohijanahary Special Reserve) c.20% of population intermediate to dark melanistic forms (Rakotonirina et al. 2014). In north-east of range (Analabe and Anktotrakotraka Forests), in Mahavavy-Kinkony Wetland Complex, c.25% of individuals melanistic, whereas at other forests in Mahavavy-Kinkony Wetland Complex all individuals conform to 'typical' white form (Curtis et al. 1998, Thalmann et al. 2002). Smaller proportions of melanistic individuals also in Kasijy Special Reserve (Randrianarisoa et al. 2001).

Medium-large, vertically clinging and leaping lemur. Confusion with any other lemur highly improbable. At southern and eastern extremities of range (Bongolava) distinction from Crowned Sifaka may be required; 'typical' Decken's Sifaka is all white, whereas 'typical' Crowned Sifaka has dark brown to black head. However, in areas where melanistic forms occur (of both species) identification may require more care.

HABITAT & DISTRIBUTION Dry deciduous forest from sea level and lower elevations on western coastal plain (Wilmé *et al.* 2006) to *c.*1,250m in Bongolova Massif (Rakotonirina *et al.* 2014). Restricted to central-western regions: in north of range found west of Mahavary River and in south of range occurs north of Manambolo River, including western edge of Bongolava Massif (Thalmann *et al.* 2002, King *et al.* 2014, Rakotonirina *et al.* 2014). In some localities, e.g., Tsingy de Bemaraha National Park and Beanka, it lives in forest surrounding limestone karst (*tsingy*) formations and is known to climb on and move over the rocks between forest patches.

Previous reports of Decken's Sifaka outside above range, i.e. east of Mahavavy River or south of Manambolo River, where populations of Crowned Sifaka occur (Tattersall 1982, 1988b, Thalmann & Rakotoarison 1994) probably refer to rare, localised events, where Decken's Sifaka individuals cross rivers or other biogeographical barriers by natural or human-induced methods, e.g. releases or escaped pet sifakas captured elsewhere (Rakotonirina *et al.* 2014). They do not represent existence of established Decken's Sifaka populations outside range outlined above (King *et al.* 2014).

Decken's Sifaka is regarded as reasonably common in areas of suitable habitat, e.g. Tsingy de Bemaraha National Park, but local populations are highly variable across forest fragments. No density estimates are available, but probably similar to those of Crowned Sifaka (*c.*50–300 individuals/km^2 (King *et al.* 2014).

Continued habitat loss and forest fragmentation, caused by logging, charcoal production and pasture burning is major threat (Andriamasimanana & Cameron 2014). Despite traditional taboos (*fady*), hunting for food widespread and increasing, and recorded at *c.*65–80% of localities surveyed (King *et al.* 2014, Rakotonirina *et al.* 2014). IUCN Critically Endangered.

BEHAVIOUR Group size varies from 2–10 individuals, with groups of 3–6 most common (Curtis *et al.* 1998, Müller *et al.* 2000, Dammhahn *et al.* 2013, King *et al.*

Decken's Sifakas, 'typical' white form, Tsiombikibo Forest.

Decken's Sifaka, melanistic form, Analabe Forest.

2014, Rakotonirina *et al.* 2014). Groups containing two adult females, both with young have been observed, but most groups contain just one reproductive female (King *et al.* 2014). Able to survive in quite degraded habitat, e.g. around Mitsinjo and Tsiombikibo Forest (Garbutt 2007). Appears to be primarily folivorous: diet likely similar to Crowned Sifaka. Also tolerant of encounters with other lemurs like Rufous Brown Lemur (Ramanamisata *et al.* 2014). Moderately vocal. Recognisable *tchi-fak* call is individually distinguishable between group members and noticeably longer in duration than Crowned Sifaka (Fichtel 2014). Other aspects of behaviour likely similar to Crowned Sifaka or Coquerel's Sifaka inhabiting dry deciduous forest.

WHERE TO SEE Readily seen in Tsingy de Bemaraha National Park, where locally common and relatively easily located. Forests near Bekopaka are worth exploring. Locally common further north, in region of Antsalova. Often seen in forests along west bank of Mahavavy River and around Mitsinjo, all part of Mahavavy-Kinkony Wetland Complex; these include Tsiombikibo Forest, forests adjacent to Lac Kamonjo and forests around Lac Kinkony.

CROWNED SIFAKA
Propithecus coronatus   Milne-Edwards, 1871

Malagasy names: Sifaka, Tsibahaka

MEASUREMENTS
Head-body: c.395–455mm. Tail: c.475–565m. Weight: c.3.2–4.3kg.

DESCRIPTION & IDENTIFICATION Medium-sized *Propthecus*. Overall colour creamy-white, with head, neck and throat dark chocolate-brown to black. Upper chest, shoulders, upper forelimbs and upper back variably tinged golden-brown, which lightens to golden-yellow lower on torso and fades to creamy-white by abdomen: discoloration due to secretions from glands on chest, more noticeable in males. Hindlimbs and tail creamy-white. Face bare, mainly dark grey to black; sometimes a paler grey to white patch across bridge of nose and whitish ear-tufts. Muzzle blunter and more bulbous, and face more squarish than other similar western sifaka species.

297

Crowned Sifaka, Mahajeby Forest.

Melanistic individuals occur in some populations in extreme south of range (Razafindramanana & Rasamimanana 2010, King *et al.* 2014), possibly resulting from historic gene flow with Decken's Sifaka (populations between Manombolo and Tsibirhina Rivers), or with Verreaux's Sifaka at southern range extremity between Mahajilo and Mania Rivers (King *et al.* 2012, 2014).

A medium-large, predominantly white, vertically clinging and leaping lemur. 'Typical' form (as above) cannot be confused with any other lemur in its range. Some colour variation has been observed; relationships and distinctions between this species and Decken's Sifaka yet to be fully elucidated (Curtis *et al.* 1998, Thalmann *et al.* 2002).

HABITAT & DISTRIBUTION Dry deciduous forest from sea level to *c.*1,100m (Rakotonirina *et al.* 2014). Occasionally seen resting in mangroves, but does not use habitat permanently (Gauthier *et al.* 1999, 2000). At northern extreme, range lies between Mahavavy and Betsiboka Rivers, in centre of range between Mahavavy and Ikopa Rivers, and at southern extreme between Manambola, Tsibirhina (east of 45°E), Mania and Sakay Rivers (Thalmann *et al.* 2002, Razafindramanana & Rasamimanana 2010, King *et al.* 2014, Rakotonirina *et al.* 2014). In north, Mahavavy River is boundary between Crowned Sifaka and Decken's Sifaka, whilst Betsiboka River separates Crowned Sifaka from Coquerel's Sifaka.

In north of range population densities vary considerably with locality, from *c.*50–300 individuals /km² and in well-protected Badrala Forest (within Andrema Protected Area) reach *c.*350 individuals /km² (Salmona *et al.* 2014). In south of range at sites including Mahajeby, Mandrava and Ankirihitra estimates vary from *c.*200–290 individuals/km² (Rakotondrabe *et al.* 2019).

Forest extremely fragmented and contracting throughout range, and destruction continues at all locations where Crowned Sifaka has been recorded, resulting in some local extinctions (Andriamasimanana & Cameron 2014, King *et al.* 2014, Rakotonirina *et al.* 2014, Salmona *et al.* 2014). Traditional taboos (*fady*) exist, protecting Crowned Sifaka at some localities, but hunting for food is widespread and occurs at *c.*80% of sites surveyed (King *et al.* 2012, 2014, Rakotonirina *et al.* 2014). IUCN Critically Endangered.

Crowned Sifaka, Antrema Forest.

BEHAVIOUR Social structure likely multi-male/multi-female with group size variable. At Anjahamena groups of 2–8 individuals recorded with variable sex and age compostion (Curtis *et al.* 1998). At other localities mean group size *c.*3–7 individuals (Pichon *et al.* 2010, King *et al.* 2012, 2014, Rakotonirina *et al.* 2014, Salmona *et al.* 2014, Sifaka Conservation 2018). Within groups, 'alpha' female appears dominant over males, winning majority of aggressive encounters and receiving more allo-grooming attention (Razanaparany *et al.* 2014).

Home ranges appear very small, *c.*1.2–2.3ha, with group members spending *c.*75% of time in core *c.*0.3–0.4ha (Curtis *et al.* 1998, Sifaka Conservation 2018). Within group, females dominant over males (Ramanamisata *et al.* 2014). Inter-group encounters sometimes peaceful, but can defend home ranges aggressively at times (Curtis *et al.* 1998, Müller *et al.* 2000, Ramanamisata *et al.* 2014). Also tolerant of encounters with other lemurs like Rufous Brown Lemur (Ramanamisata *et al.* 2014).

Crowned Sifaka visits many food patches per day, principally foraging in upper canopy. Rarely seen on forest floor, but occasionally eats soil. During dry season, diet primarily mature leaves with high fibre content and available immature leaves, buds and small quantities of unripe fruits. In wet season young leaves and flowers dominate with a greater proportion of fruit (up to *c.*25%) (Pichon *et al.* 2010, Rakotondrabe *et al.* 2017, 2019).

In dry season, groups active for >9 hours per day, foraging *c.*30–40% of day, covering average distance *c.*600m. Rest of day spent resting, grooming, playing, scent-marking, and in agonistic behaviours (Ramanamisata *et al.* 2014). At Anjahamena, groups choose tall trees (>20m), optimally exposed to sunlight and near major river (Mahavavy) to sleep in (Müller 1997, Curtis *et al.* 1998, Müller *et al.* 2000).

In common with congeners, moderately vocal. Highly recognisable *tchi-fak* call is individually distinguishable between group members and noticeably different in acoustic structure (e.g. in duration) to Decken's Sifaka (Fichtel 2014).

WHERE TO SEE Most accessible site is Antrema, the forest below the lighthouse north of Katsepy (opposite side of Betsiboka estuary from Mahajanga). There are few trails but sifakas are relatively easy to see. An alternative location is Belamboka-Bombetoka Forest, where a reasonable amount of walking is required. There are basic tourist facilities in Mataitromby (Mittermeier *et al* 2010). Another site is Anjamena, on east bank of Mahavavy River. This can be accessed from Namakia, but the logistics are challenging and arrangements ideally should be made through a tour operator.

GOLDEN-CROWNED SIFAKA
Propithecus tattersalli　　　　Simons, 1988

Other name: Tattersall's Sifaka

Malagasy names: Ankomba Malandy, Simpona

MEASUREMENTS
Head-body: 450–470mm.
Tail: 420–470mm. Weight:
3.4–3.6kg. (Meyers 1993a,
Lehman *et al.* 2005)

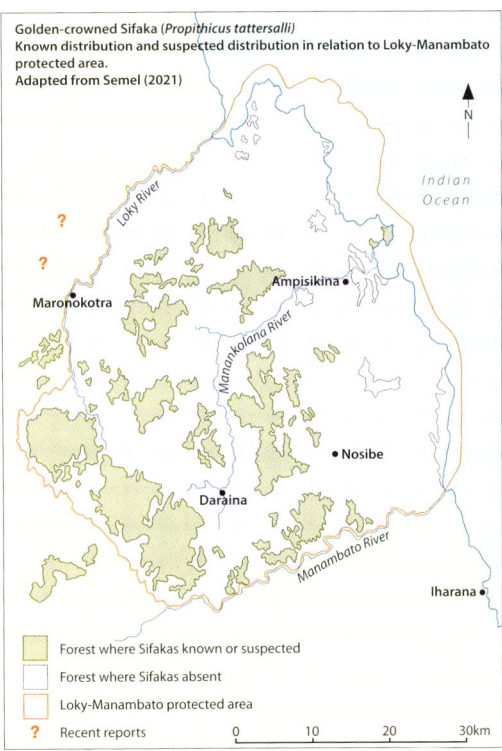

Golden-crowned Sifaka (*Propithicus tattersalli*)
Known distribution and suspected distribution in relation to Loky-Manambato protected area.
Adapted from Semel (2021)

Forest where Sifakas known or suspected
Forest where Sifakas absent
Loky-Manambato protected area
? Recent reports

DESCRIPTION & IDENTIFICATION Relatively small sifaka species. Pelage moderately long. Head and body predominantly creamy-white. Crown rich yellow-orange with similar tones on shoulders, upper arms, across chest and rump. Ears large and

Female Golden-crowned Sifaka with infant, forests near Daraina.

tufted, giving head a triangular shape. Face mainly bare, dark grey-black with white hairs extending below eyes onto cheeks. Eyes rich yellow-orange. On first discovery, in 1974, provisionally identified as variant of Silky Sifaka (then *P. diadema candidus*) (Tattersall 1982): later described as completely distinct (Simons 1988). A medium-large, vertically clinging and leaping lemur. Cannot be confused with any other species in its very limited range.

HABITAT & DISTRIBUTION Dry deciduous, gallery, subhumid semi-evergreen and littoral forests from near sea level to *c.*700m, with marked preference for elevations below *c.*500m (Meyers & Ratsirarson 1989, Vargas *et al.* 2002). Confined to small area primarily between Manambato River (southern limit) and Loky River in north-east. At north-west limit of range recorded in three forest fragments on west side of Loky River around Maromokotra. Recent reports of groups in Andrafiamena-Andavokoera Protected Area potentially extend known range by 15km west of Loky River (Semel 2021). Easternmost locations are forest patches in Amporaha Mountains and coastal forests near Ampisikinana (west of Bay of Andravina). Records at coastal localities further south unconfirmed: record from Analabe coastal forest probably an individual in transit (Meyers

1993b, 1996, Vargas *et al.* 2002, Semel *et al.* 2020). Across limited range many forest fragments now discrete, separated by matrix of anthropogenic grasslands, dry scrub, pasture and agricultural land (Quémére *et al.* 2013).

Within some remaining forests, including small fragments, Golden-crowned Sifaka may be locally common (Vargas *et al.* 2002). Able to live adjacent to human habitation, but avoids areas directly proximate to villages (Semel *et al.* 2020). Densities vary widely, from *c.*78 individuals/km^2 in Bekaraoka (near village of Andranotsimaty) to *c.*7 individuals/km^2 in Antsahabe. Mean density across range *c.*20 individuals/km^2, with total number of individuals estimated at *c.*10,200–12,600 (Quémére *et al.* 2010a, Semel 2021).

Major threats are ongoing forest loss due to slash-and-burn agriculture, uncontrolled grassland fires to create new cattle pasture, firewood and charcoal production, gold and gem mining, and harvesting precious hardwoods. Regional taboos (*fady*) largely deter local hunting, but non-local miners do hunt sifakas (Barrett & Ratsimbazafy 2009, Quémére *et al.* 2010b). IUCN Critically Endangered.

BEHAVIOUR Primarily diurnal, but seen moving before dawn and after dusk during rainy season

Golden-crowned Sifaka leaping sequence, Bekaraoka Forest.

(December–March). At night sleeps in taller trees. Group size varies from 3–10, most often 5–6, consisting of two or more mature members of each sex. Only one female in a group breeds successfully each year, with dominant male monopolising mating success (Parreira *et al.* 2020). Males move between neighbouring groups during mating season (Meyers 1993a).

Home range size varies with habitat, from *c*.3ha in subhumid semi-evergreen to *c*.50ha in very dry forests (Semel *et al.* 2020). Also varies seasonally: at one site between 6ha (wet season) and 12ha (dry season). Groups travel *c*.400–1,200m daily, with longer distances during drier months when food is less abundant (Meyers 1993a,b). Forages at all levels in canopy. Can move between forest patches via narrow ribbons of interconnecting riparian forest. Also moves across open spaces between patches.

Diet flexible and highly diverse (*c*.130–150 plant species, with genera *Poupartia*, *Olax*, *Landolphia*, *Marsdenia* and *Dalbergia* important), primarily a variety of young leaves, shoots, seeds, fruits and flowers, with seasonal variation. Ripe fruits, flowers and young leaves dominate in wet season (December–March) when availability peaks. In dry season mature leaves, unripe fruit and even bark are

Golden-crowned Sifaka, forests near Daraina.

eaten. Not known to eat soil. Also eats cultivated and non-native species, including Mango *Mangifera indica* (Anacardiaceae) and banana (*Musa* sp.) (Musaceae), particularly at forest edges (Meyers & Wright 1993, Quéméré *et al.* 2013).

Mating takes place in January, female in oestrous for only 24 hours with receptivity indicated by visible pink genital swelling. Male testes volume increases prior to breeding season. Copulation is brief, lasting only 30–90 seconds. Births occur mainly in July, after gestation *c*.165–175 days. Newborn infants sparsely covered in fur and initially carried by mother on belly, before later moving to ride on her back. Allo-parenting is common, with all group members interacting, grooming, carrying, playing and nursing infants. Weaning around five months (November–December), coinciding with increased abundance of high-quality immature leaves. Following weaning, mother repeatedly refuses all attempts by infant to suckle and only rarely tolerates dorsal riding for brief periods, for instance during predator scares. By one year, young animals attain *c*.70% of normal adult body weight. Both sexes sexually mature at *c*.2.5 years. Females breed every two years (Meyers & Wright 1993). Adult dispersal mainly among neighbouring groups within forest patch: despite high levels of fragmentation potentially limiting dispersal, there is little evidence of inbreeding depression (Quéméré *et al.* 2010c). Manankolana River is a known barrier to dispersal (despite being dry for several months per year), suggesting dispersal primarily occurs in wet season (January–April) when river is flowing (Quéméré *et al.* 2010b, Semel 2021).

Responds to large raptors, e.g. Madagascar Harrier-Hawk, with characteristic 'mobbing' alarm call; more familiar *tchi-fak* alarm call directed at terrestrial predators and threats, like Fosa or feral dogs (Garbutt 2007). Predation by Fosa is likely but not documented (Goodman & Ganzhorn in press).

WHERE TO SEE Despite its rarity, Golden-crowned Sifaka is easy to see. Bekaraoka Forest near Andranotsimaty, 5km north-east of Daraina (on *Route Nationale* 5a) offers excellent opportunities. Areas adjacent to Camp Tattersalli are particularly good with several habituated groups of sifakas (some come into camp). This site is also very good for Crowned Lemur and several nocturnal species including Daraina Sportive Lemur, Northern Fork-marked Lemur and even Aye-aye.

INDRI
GENUS *Indri*

A monotypic genus containing one of the most familiar and charismatic lemurs. The Indri is the largest surviving 'prosimian' and instantly distinguishable from all other lemurs by its size, distinctive black to black-and-white coloration and unique vestigial tail (the only living lemur possessing this trait). Its characteristic eerie wailing territorial song is unforgettable and provides an abiding memory from any visit to this species' rainforest home.

The territorial call is only one of a number of vocalisations that make up their repertoire, others being barks, roars and honks (alarm calls) and grunts, 'kisses' and wheezes (social cohesion and emotion conveying calls). Olfactory communication (scent-marking) is far less prevalent.

Indri are amongst the most arboreal of lemurs. In common with other Indriidae, they are 'vertical clingers and leapers', having evolved to exploit the lower branch and vertical trunk niche of the forest understorey. Their legs are extremely powerful, capable of propelling them on leaps spanning up to 12m. They are also able to bound vertically up tree trunks in a series of rapid hops.

A possible origin of the name 'indri' comes from the 18th century French naturalist Sonnerat. His Malagasy guide pointed the animal out in the field saying 'there' in Malagasy, which is '*iry*' or '*ery*', with 'indri' being an inadvertent corruption (Mittermeier *et al.* 2010). An alternative possibility is that the Malagasy also use the interjection '*ndri*' in admiration of something cute or beautiful (Dolch 2021). The range of the Indri largely coincides with that of the Betsimisaraka tribe, whose name for the animal, '*Babakoto*' literally means 'Father of Koto', 'Father of Man' or simply 'old man', which translates to the legend that they are the ancestors of humans and hence it is taboo (*fady*) to hunt them (Randimbiharinirina *et al.* 2021). Further, the Indri's stumpy tail, leads some Malagasy to believe it is related to a sacred ancestor (Powzyk & Thalmann 2003).

INDRI
Indri indri   (J. F. Gmelin, 1788)

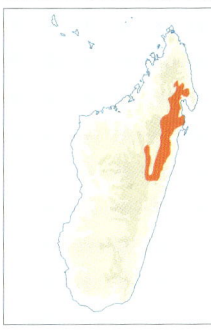

Malagasy names: Babakoto, Indry, Endrina, Amboanala

MEASUREMENTS
Head-body: 640–720mm. Tail: *c.*40–50mm. Weight: 6-9.5kg. (Glander & Powzyk 1998, Zaonarivelo *et al.* 2007a, Mittermeier *et al.* 2010)

DESCRIPTION & IDENTIFICATION Characterised by very long hindlimbs (equivalent to head-body length) and vestigial tail (<5cm). Only virtually tail-less lemur. Dorsal pelage long and very dense, underparts slightly shorter and less densely furred. Underlying skin black. Head rounded with prominent naked muzzle and large round ears. Eyes yellowish-golden, occasionally pale greenish blue.

Pelage coloration a mixture of black, pale grey and white, with considerable variation over range. Base colour black. At its most extreme, majority of body black, only white and pale grey areas being over eyes, around cheeks and muzzle, and small areas on extremity of lower limbs. In contrast others are piebald black-and-white, with distinctive creamy-white patches on crown, nape and throat, base of back, forearms, thighs and lower legs, and these areas may be tinged silver-grey or pale creamy-yellow. Intermediate individuals are primarily black with white facial disc and white extending below chin and areas on sides of abdomen to armpits. A white pygal triangle at base of back continues to rump and includes vestigial tail. Heels also pale grey or yellowish-white. White areas completely absent on forearms and upper hindlimbs. Slight sexual dimorphism: male chest generally much darker, female's paler. Male has white triangle at base of back, absent in female. Unclear whether these differences are consistent across entire range (Mittermeier *et al.* 2013).

Populations at northern limits of range generally predominantly black, whilst those at southern limits are black and white, with populations in centre of range exhibiting both variants. This is an over-simplification: description of black-and-white forms based largely on familiar individuals in Analamazaotra Special Reserve (Andasibe-Mantadia National Park). but almost completely black individuals (in mixed groups with black-

Indri feeding on fresh leaves, Andasibe-Mantadia National Park.

Adult Indri, Association Mitsinjo Forest, Andasibe.

and-white individuals) are known from Mantadia section of park, only 15km to north and Zahamena National Park further north.

The northern 'black' form and southern 'black-and-white' form have been proposed as subspecies (Groves 2001), with a transition in region of Mananara-Nord (Thalmann *et al.* 1993). There is neither morphological nor genetic evidence to support this, and the Indri simply exhibits clinal variation across its range (Zaonarivelo *et al.* 2007a, Brenneman *et al.* 2016).

A large, highly distinctive, vertically clinging and leaping lemur. In its range, only Diademed Sifaka is comparable in size; it is much paler with a long tail. Black-and-white Ruffed Lemur is similarly coloured, but smaller, has a long tail and a horizontal rather than vertical posture.

HABITAT & DISTRIBUTION Primary and secondary lowland and mid-altitude rainforest from near sea level to *c.*1,800m (exceptionally) (Goodman & Ganzhorn 2004); lower elevations, below 1,000m, are preferred. Confined to centre-east and north-east rainforests. Range extends from Anosibe

An'ala region in south to Antohaka Lava south-west of Andapa in north (near northern limit of Anjanaharibe-Sud Special Reserve). Does not reach further north into Marojejy National Park, nor does range extend to Masoala Peninsula. Subfossil evidence from Late Pleistocene and Holocene indicates range once more extensive: remains found in caves on Ankarana Massif in north, plus several locations in Central Highlands (Jungers *et al.* 1995, Godfrey *et al.* 1999).

Population densities low in pristine continuous forest: *c.*5 individuals/km² in Mantadia National Park (Powzyk 1997), *c.*7–13 individuals/km² in Betampona Reserve (Glessner & Britt 2005) and *c.*4 individuals/km² in Maromizaha Forest (Bonadonna *et al.* 2017). In isolated forest fragments densities are higher: *c.*23–27 individuals/km² in Analamazaotra Special Reserve and *c.*24 individuals/km² at Analamazaotra Forest Station (Mitsinjo Forest) (Pollack 1977, Bonadonna *et al.* 2017).

Habitat destruction for slash-and-burn agriculture, logging and fuelwood remains primary threat. Fragmentation leads to genetic isolation in some Indri populations (Nunziata *et al.* 2016). Does appear tolerant of forest edge, which likely increases vulnerability to human impacts (Seaman *et al.* 2018). Indri inhabiting fragmented forests, with increased human disturbance (including ecotourism), are more susceptible to parasitic infections and suffer other deleterious health issues (Junge *et al.* 2011). Protected by local taboos (*fady*) in some areas, but now widely hunted for food throughout range. In Makira Forest also hunted for pelts, which are worn as clothing (Golden 2005, 2009, Jenkins *et al.* 2011, Golden *et al.* in press). IUCN Critically Endangered.

BEHAVIOUR The most strictly diurnal lemur: moves at night only during bad weather or if flushed by a predator (in breeding season often calls at night). Lives in small family groups of 2–6 animals, comprising a (primarily) monogamous pair, confirmed via genetic paternity tests (Torti *et al.* 2013, Gamba *et al.* 2016, Bonadonna *et al.* 2019), with maturing offspring of varying ages. However, females have been observed leaving residential group to mate with male in neighbouring group (Bonadonna *et al.* 2014). Female dominant and has priority at food resources: at feeding sites males generally defer to approaching females and when aggression erupts females usually prevail (Pollock 1979b). Fights between females also occur (Dolch 2021). In continuous pristine forest, group size small; adult pair and single infant (Powzyk 1997). In fragmented forests, groups may contain several generations of offspring (Pollock 1975a,b, 1979a), probably a consequence of lack of uninhabited areas into which young animals can disperse (Powzyk & Thalmann 2003). Generally thought that individuals seek new partner only after death of mate (Powzyk & Thalmann 2003), but in larger groups in fragmented habitat group composition changes more frequently (Mittermeier et al. 2013).

Occupy exclusive territories defended by wailing calls. Little or no overlap with neighbouring territories (Bonadonna et al. 2017). In isolated forest fragments, like Analamazaotra Special Reserve, territories c.5–26ha (mean c.13ha) (Pollock 1979a, Bonadonna et al. 2017). Within expansive undisturbed forests, e.g. Mantadia National Park, territories can be considerably larger, c.40ha (Powzyk 1997). Inter-group encounters rare. Most (c.85%) resolved by territorial calls (see below), only rarely do disputes at periphery of territories result in physical fights between males (Bonadonna et al. 2017).

The Indri's song is its hallmark: the third loudest of any primate, after howler monkeys (Alouatta spp.) and the Siamang Symphalangus syndactylus. They generally move higher in canopy to call, which carries up to c.3–4km (Giacoma et al. 2010) and is generally answered sequentially by neighbouring groups (stimulating chain reaction that spreads around forest), so maintaining group spatial distribution in forest and preventing home range overlap (Pollock 1986, Geissman & Mutschler 2006). Most calling bouts between 06.00 and 13.00, peak period from 07.00 to 11.00, with occasional bouts between 14.30 and 16.30 (Powzyk 1997, Geissman & Mutschler 2006). Daily call frequency increases around September and peaks in October and November, when groups may also call at night or very early in morning (c.04.30–05.30) (Powzyk 1997).

Call typically consists of three distinct phases. First the 'roar' or waa notes (from adult male) lasting several seconds, then call proper, and finally descending sequence. Male call volume greatly increased by throat sac that acts as resonator. All group members contribute, except very young offspring: younger individuals participate in initial roar and first few seconds of song, subadults (3–6 years) participate in first half of song. Adult males and females call in coordinated manner ('duet') (Gamba et al. 2016), but there are sex differences, in presence or absence of notes, acoustic features of notes (male notes last longer) and number of notes produced (females have eight note types, males six) (Giacoma et al. 2010, Sorrentino et al. 2013). Acoustic structural differences between the sexes also convey information potentially meaningful to others, e.g. group size and sex composition and reproductive status (Giacoma et al. 2010). Calling bouts last 1–2 minutes but may continue up to 20 minutes. Call duration shows some correlation to number of group members singing (Torti et al. 2018). Group calls occur at least once per 24-hour cycle, but total number of bouts varies daily. Bouts may be spontaneous or in response to neighbouring group. Neighbouring groups may sing simultaneously, but group songs are typically contagious and sung sequentially (Gamba et al. in press). Elements of song closely resemble those of White-handed Gibbon Hylobates lar (Thalmann et al. 1993).

In addition to the 'song', Indris have eight other distinct call types (Maretti et al. 2010). Repeated 'roar' call warns of potential aerial predators (large raptors), while klaxon-like 'honk' (sometimes in conjunction with 'roar') may alert to ground

The territorial 'song' of the Indri is the species' hallmark, Andasibe-Mantadia National Park.

Indri are capable of prodigious leaps through the canopy, Andasibe-Mantadia National Park.

predators, like Fosa or feral dogs (Powzyk & Thalmann 2003, Dolch 2021). Gentle 'hums' often pre-empt group movement. Short and long tonal calls given during conflicts within group, while other 'softer' vocalisations include 'grunts', 'kisses' and 'wheezes' indicating increased anxiety and fear. Infants have a modest repertoire of 'hum' and 'grunt' contact calls for communicating with mother (Maretti et al. 2010).

Active for c.5–11 daylight hours, depending on season and weather conditions; on average

Indri leaping, Andasibe-Mantadia National Park.

c.8 hours spent resting (Pollock 1979a, Powzyk 1997). Groups capable of traversing entire territory in short time, but generally fairly sedentary, moving a few times per day (Bonadonna et al. 2017); normal daily movement 300–800m, mean c.700m (in continuous forest tracts) (Powzyk 1997). Group members normally remain within 100m of one another. Groups patrol majority of territory within 8–14 days (Powzyk 1997). Can move through canopy with spectacular bounds up to 12m between vertical branches and trunks. Occasionally moves on ground in bipedal manner similar to sifakas, but will also walk quadrupedally to traverse large forest gaps (Mittermeier et al. 2013). Sometimes scent-marks branches and trunks with ano-genital region or cheek glands, or both (Powzyk & Thalmann 2003). Prior to dusk, groups settle in a sleep tree, 10–30m off ground. Females sleep in contact with infants or subadult offspring, males typically 2–50m away. If males approach too close to sleeping female, they often get cuffed (Powzyk 1997).

Indri is primarily folivorous and feeds at all levels in forest, from near ground to canopy (c.2–40m). Leaves may be picked or brought to mouth while attached; fruits are generally taken initially into mouth and then grasped. Toothcomb used to scoop out flesh of unripe fruit (Powzyk & Mowry 2006). Diet dominated by young leaves and leaf buds (72%), also seeds and whole fruits (16%), some flowers (7%) and occasionally bark, with proportions varying seasonally (Powzyk 1997). Tree foliage dominates diet (>90% in a study in Maromizaha), with leaves of more than 80 species eaten (Randrianarison 2019).

Also occasionally (but year-round) descends to ground to eat soil (Gamba *et al.* in press); this provides micronutrients, helps neutralise toxins and tannins that accumulate from leaves and seeds, and also assimilates fungal communities from soil into the gut that aid digestion of plant matter (Britt *et al.* 2002, Borruso *et al.* 2021). Feeding bouts punctuated by periods of rest, before group moves to next feeding site. Preferred leaves and seeds different from those eaten by broadly sympatric Diademed Sifaka, thus reducing food competition (Powzyk & Mowry 2003).

Gut morphology typical of folivores: specialisations include enlarged salivary glands, proportionately large stomachs, sacculated caeca and looped colons, which aid efficient fermentation of leaf matter. Molars have high crowns, large crushing surfaces, and long shearing blades, which allows the mastication of leaf blades and seeds alike. All seeds are broken/crushed, none pass through intact, hence Indri are seed predators, not seed dispersers (Gamba *et al.* in press).

Females able to give birth every two to three years, so capacity for population growth is very slow. Mating between December and March, single offspring born between May and June, occasionally as late as August. Gestation *c.*150 days. Initially, infant carried on mother's lower stomach, but transfers to her back after four months. Infant practices leaping from five months, successful at landing and moving independently at eight months, but remains in close proximity to mother until well past its second year. Mother and infant always sleep together for first year, but afterwards do so only sporadically (Pollock 1977). Sexual maturity reached at 7–9 years (Pollock

1977), but individuals may disperse from natal group earlier (Torti *et al.* 2018, Gamba *et al.* in press); both males and females disperse and sex ratio at birth is approximately 1:1 (Kappeler 1997).

WHERE TO SEE No visit to Madagascar is complete without seeing this spectacular creature. Analamazaotra, part of Andasibe-Mantadia National Park, is the best location. In Analamazaotra Special Reserve, several family groups are habituated and good sightings are virtually guaranteed (although reserve is home to >30 groups and total of *c.*130 individuals). Adjacent is Analamazaotra Forest Station (Association Mitsinjo), which also has habituated groups. Be in the forest between 07.00 and 11.00 to hear the Indri call. Also, relatively common in Mantadia portion of park, but generally more time and effort required. Groups in forest around PK15 are reasonably habituated and can be approached once found. Some of the Indris in Mantadia are predominantly black in coloration. Nearby Maromizaha Reserve (south of *Route Nationale* 2), managed by lemur research group GERP, is similar forest to Mantadia: Indri are being studied and are habituated, but terrain can be steep and challenging. Alternatively, higher-elevation forests at Anjozorobe (part of Anjozorobe-Angavo Forest Corridor), *c.*90km north-east of Antananarivo can be very rewarding, with some almost black individuals often seen. It is also worth trying to see black variants in Anjanaharibe-Sud Special Reserve at the species' northern extreme. Indris here are not habituated so are much more difficult to find.

Aye-ayes

Family Daubentoniidae

A family containing a single extant species, the Aye-aye *Daubentonia madagascariensis*. Without doubt, the Aye-aye is the most anatomically and behaviourally diverse Strepsirrhine primate and as such is placed in its own infraorder Chiromyiformes. It is the only extant primate belonging to a monotypic infraorder, family, genus and species. Such is the bizarre appearance of the Aye-aye that, when first discovered, it was classified as a squirrel-like rodent with the scientific name *Sciurus madagascariensis* (J.F. Gmelin, 1788); not until *c*.1850 was the species widely accepted as a primate.

Unsurprisingly, the evolutionary affinities of the genus *Daubentonia* are regarded as contentious. Molecular evidence supports the theory that all lemurs, including Aye-ayes, are derived from a single primate colonisation event on Madagascar that occurred 50–60 mya, followed by a spectacular adaptive radiation, with Aye-ayes probably forming the first significant offshoot not long after initial colonisation (Yoder *et al*. 1996, Yoder 1997, 2003).

Fossil evidence from mainland Africa offers an alternative hypothesis. *Propotto* from Kenya and *Plesiopithecus* from Egypt are seen as being early African Chiromyiform lemurs that are related to Aye-ayes (based largely on shared dental morphology and molecular analysis). The last common ancestor of *Propotto* and *Daubentonia* is placed on the African mainland *c*.28 mya, with dispersal to Madagascar that ultimately gave rise to Aye-ayes occurring sometime after this (early Oligocene). Divergence of Chiromyiform and Lemuriform lineages on continental Africa is suggested at *c*.41 mya, which invokes separate colonisation events for the ancestor of Aye-ayes (Chiromyiformes) and the ancestor of all other lemurs (Lemuriformes) (Gunnell *et al*. 2018).

AYE-AYES
GENUS *Daubentonia*

Aye-ayes possess a variety of peculiar morphological traits setting them apart from all other primates: an unusual skeletal structure and dentition (18 teeth, instead of 30–36 like other lemurs, including incisors that grow continually, as in rodents, and no tooth comb), very large ears, piercing far-apart eyes, clawed digits, a skeletal middle finger, a third eyelid (nictitating membrane) and mammary glands low on the torso (Sterling 2003, Sterling & McCreless 2006).

Much of its anatomical uniqueness is derived from its foraging and feeding adaptations. The complex, hand, eye and auditory coordination necessitated by its 'percussive foraging' technique, has led to the evolution of a high degree of neurological complexity compared to body size, resulting in the largest brain of any Strepsirrhine primate (Simons & Meyers 2001). The huge, continuously growing, rodent-like incisors have roots extending well into the skull. The size of the upper incisors in particular has resulted in the eyes being spaced further apart (largest inter-orbital spacing of any lemur).

The Aye-aye's hands are proportionately longer than those of any other primate (only tarsiers, family Tarsidae, are equal) and its digits are perhaps its hallmark anatomical feature. The fourth finger is longest, while the third finger is exceptional, being both very long and thin (effectively bone and tendon covered in skin with no flesh). It is able to articulate completely independently of other digits, can rotate 360° around a 'ball and socket' metacarpophalangeal joint, and is used to extract wood-boring beetle grubs from decaying tree branches and trunks, as

Aye-aye, forests near Daraina.

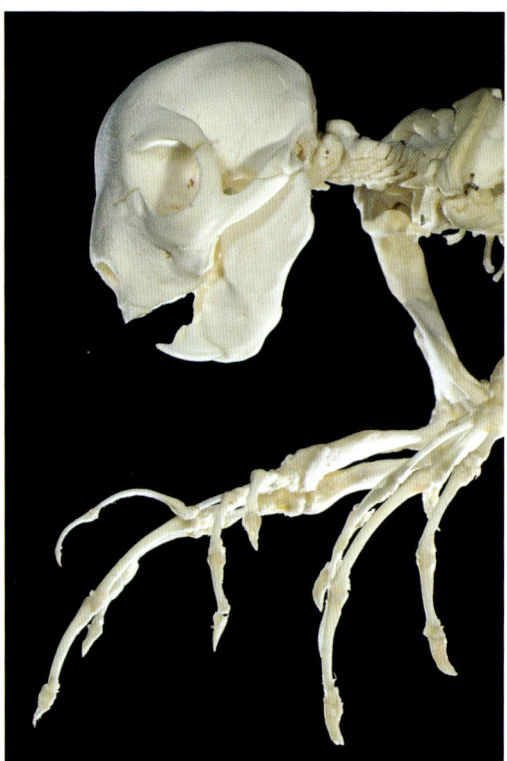

Aye-aye skeleton, showing its extraordinary hand morphology.

well as acquire most other foods, and for scratching and grooming (Soligo 2005, Sterling & McCreless 2006). The only other mammal to have digits resembling these is the Long-fingered Striped Possum *Dactylopsila palpator* (Petauridae) of New Guinea. Aye-ayes have claws on all digits, except their thumbs and big toes (which have flattened nails). The second toes have 'toilet' claws specialised for grooming. They also posses a 'pseudothumb' – a trait shared with the Giant Panda *Ailuropoda melanoleuca* – a bony/cartilaginous protuberance with specialised muscular attachment within the palm pad, which helps gripping when climbing (some of this capability has been reduced by the other extreme hand/finger adaptations) (Hartstone-Rose *et al.* 2020).

The Aye-aye's morphology is so unusual that determining its closest relatives amongst other lemurs and primates has proved extremely difficult (Sterling 1994a). It has previously been placed within the Indriidae (Schwartz & Tattersall 1985, Schwartz 1986) as an ancient sister taxon to all other lemurs (Yoder *et al.* 1996, Pastorini *et al.* 2002b, 2003, Yoder 2003); as the most ancient branch of the

Strepsirrhine primates (Delpero *et al.* 2001); and most recently linked with Strepsirrhine fossils in continental Africa (see above) (Gunnell *et al.* 2018). This species must be regarded as one of the most unusual and spectacular mammals on Earth.

A second species, the Giant Aye-aye *D. robusta*, is known only from subfossil remains in spiny forest areas in south and south-west Madagascar (the only region not inhabited by *D. madagascariensis*). It was 2.5–5 times heavier (up to *c.*14kg) than *D. madagascariensis* and yet still primarily arboreal, feeding in a similar way to its extant relative (Simons 1994, Godfrey & Jungers 2003). The Giant Aye-aye is known to have survived until at least *c.*900–1,000 CE, and may even have become extinct fewer than 300 years ago, due to human hunting (MacPhee & Raholimavo 1988, Schwitzer *et al* 2013e).

The origins of the name 'aye-aye' are uncertain. One theory suggests the term refers to '*aiee! aiee!*', a cry of 'exclamation and astonishment' uttered by a native Malagasy when shown specimens by the French explorer, Sonnerat, in 1782, but this seems unlikely as the name is common across the island, and therefore could not have originated with Sonnerat (Simons & Meyers 2001, Dunkel *et al.* 2012). Alternatively, the name may derive from '*heh heh*', which in Malagasy means, 'I don't know' and perhaps reflects their unwillingness to utter the name of an animal so intertwined with local beliefs and fears (Simons & Meyers 2001).

Aye-ayes can evoke powerful feelings and reactions among many of the Malagasy people: there is much local folklore and contradictory superstition surrounding the animal. Local taboos (*fady*) differ from region to region, and even from village to village within the same region. Most often reported are extreme negative attitudes towards Aye-ayes. For example, in some far north-western (Ambanja) and north-eastern regions (Andapa, Marojejy, Sambava) they are thought to bring bad luck and are killed, often with the whole body or tail then displayed on a roadside or at a crossroads outside the village. It is thought when passing strangers happen upon the corpse, its curse and ill-fortune will pass to them and be carried away (Simons 1993, König & Zavasoa 2006, 2008, Glaw *et al.* 2008). Also in north-eastern and some central-eastern areas, Aye-ayes are regarded as harbingers of evil, sickness and death and are killed, or if an Aye-aye is seen the village must be abandoned to avoid impending bad luck or doom (Simons & Meyer 2001). At its most extreme, in the Manongarivo region in the north-west,

Aye-ayes are thought by some to kill chickens and even humans, and are killed, with the corpse displayed on a roadside (Goodman & Schütz 2000). The Sakalava people from western Madagascar believe Aye-ayes are able to enter houses at night and murder the sleeping occupants by cutting their aortic arteries with their thin middle fingers! The Sakalava pre-empt this by killing any Aye-ayes they come across. In other western regions, beliefs are different and the animal is feared far less, but is still linked to bad luck (Simons & Meyer 2001).

In contrast there are places where Aye-ayes are regarded positively. In south-east Madagascar they are considered omens of good luck because they are thought to be linked to human origins. According to regional folklore, anyone who kills an Aye-aye, will themselves die within a year. In other areas the Aye-aye is believed to be the embodiment of ancestral spirits and is bestowed the same rites as a grand chief after death (Sterling 1993b, Simons & Meyers 2001).

However, studies in the Makira region of the north-east reveal a more complex, sometimes contradictory picture. In a wide-ranging survey of rural villages, fewer than half the people (47%) perceived Aye-ayes negatively, with c.35% holding neutral views and just under 20% actually regarding Aye-ayes positively, mainly because of the belief that they help control pests of crops (eating insect larvae on clove trees). Within any single village perceptions were largely consistent, but often differed significantly between neighbouring villages. In many villages, occupants may hold strongly negative views, although they have not actually seen an Aye-aye, nor indeed do they recognise one when shown a photograph. So much is simply transmitted by hearsay (Randimbiharinirina et al. 2021). It is easy to imagine how many negative beliefs evolved; the Aye-aye's 'sinister' appearance sets it apart. In other parts of the world, odd-looking animals are dealt a similar fate. In Borneo, the Puana people regard the goblin-like Western Tarsier *Cephalopachus bancanus* as the 'evil one'. Indeed, since Madagascar's original human colonists came from South-east Asia, the Malagasy may have inherited beliefs about weird-looking forest animals from their ancestors. Compounding the issue, the Aye-aye regularly feeds and nests in the Ramy tree *Canarium madagascariensis* (Burseraceae), which often grows adjacent to village tombs and sacred forest burial grounds, inextricably linking and associating the Aye-aye with death in local psyche (Sterling & Feistner 2000, Simons & Meyers 2001).

AYE-AYE *Daubentonia madagascariensis*
(J. F. Gmelin, 1788)

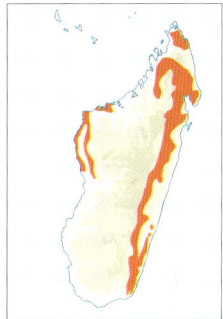

MEASUREMENTS
Head-body: 300–370mm. Tail: 440–530mm. Weight: 2.4–2.6kg. (Glander 1994, Feistner & Sterling 1995, Mittermeier et al. 2010)

DESCRIPTION & IDENTIFICATION Sexes similar. Head, upperparts, legs and tail brindled dark slate grey-brown with white flecks. Effect produced by long coarse dark grey-black guard hairs with white tips, overlaying dense but short pale grey to creamy-white undercoat. Ears black, very large, leathery in appearance and mobile. Head large, triangular in shape (in conjunction with ears). Face and muzzle thinly furred. Snout very short and blunt, nose

Aye-aye descending from canopy, forests near Daraina.

Aye-ayes forage widely and cover large distances each night, Bekaraoka Forest.

pinkish. Striking yellow/orange eyes surrounded by dark rings; whitish-grey patches above and below eyes, rest of face and throat pale grey. Tail long and very bushy. Hands highly distinctive: all digits elongated with curved claw-like nails (only first toe has flat nail). Middle finger extraordinarily thin – simply skin, tendons and bone. Incisors particularly long and rodent-like. Females have two inguinal mammary glands located abdominally near groin. Males have bacalum (penis bone). Largest fully nocturnal prosimian. Suite of peculiar features – huge ears, bushy tail, long shaggy coat, skeletal 'probe-like' middle finger and slightly awkward gait – make this species absolutely unmistakable.

HABITAT & DISTRIBUTION Adaptable: found in variety of habitats – low and mid-altitude rainforests, subhumid semi-evergreen forest, dry deciduous forest, littoral forest, and some degraded and cultivated areas like coconut, lychee and clove plantations (Ganzhorn 1986, Ganzhorn & Rabesoa 1986, Iwano et al. 1991, Andriamasimanana 1994, Farris et al. 2011). Appears dependent on tall forest trees (Schwitzer et al 2013e). Found from near sea level to c.1,800m.

In late 1960s thought to be on brink of extinction with perhaps no more than 50 individuals remaining, and thought to be confined to lowland rainforests of north-east, in vicinity of Mananara. Now known to be most widely distributed lemur species. Recorded localities cover entire extent of eastern rainforests, from Ampanefana in north to Andohahela National Park (humid parcel only) in south. Also found in moist forests of Sambirano region in north-west, drier forests further south in vicinity of Manasamody Hills, in far north deciduous forests of Ankarana, Analamerana and Loky-Manambato region, and humid forests of Montagne d'Ambre. Recorded at isolated localities in west, most notably Tsiombikibo Forest near Mitsinjo, the region south-east of Soalala, Tsingy de Bemaraha National Park, and forests south of Manambolo River (Simons 1993, Sterling 1998, 2003, Rahajanirina & Dollar 2004). No records south of Tsiribihina River or in south-western spiny forests.

Also introduced (in 1966–67) to island of Nosy Mangabe in Bay of Antongil (off Maroantsetra): at the time thought to be a last-ditch attempt to save

Female Aye-aye with young, emerges from nest, forests near Daraina.

species. Nine individuals (four males, five females) were originally released and a population still survives (Sterling *et al.* in press).

Due to nocturnal, secretive habits, there is little understanding of population densities and dynamics (Louis *et al.* 2020q). Sightings are rare, presence generally inferred from tree hole excavations and other feeding signs. Abundance difficult to estimate as one individual can leave multiple traces of feeding (Thompson *et al.* 2016, Aylward *et al.* 2018). Throughout range, irrespective of habitat, always appears to occur at very low densities (Sterling *et al.* in press). Has lowest levels of genetic diversity of any lemur, consequently no subspecific distinctions are justified (Perry *et al.* 2012, Kistler *et al.* 2015), with centre of endemism in northern Madagascar (Perry *et al.* 2013).

Habitat destruction and forest fragmentation are primary threats, especially as Aye-aye has enormous home range. Trees such as *Canarium* spp. and *Intsia bijuga* are dietary staples but are felled preferentially for construction of boats, houses and coffins (Iwano & Iwakawa 1988). Also killed in some areas as harbinger of evil/bad luck and as crop pest (e.g.

coconuts). Hunted for food in certain areas (Makira) (Golden 2009). IUCN Endangered.

BEHAVIOUR Strictly nocturnal and arboreal; spends majority of time high in canopy. Mainly solitary but does sometimes travel and forage in pairs or loosely associated groups of up to four (Sterling & Richards 1995). Day spent tucked away in messy oval-shaped nest constructed of interwoven twigs and dead leaves, located towards the canopy (*c.*7–20m above ground) in dense tangles of vines or branches. Roof of nest generally twigs laid flat, base lined with thick layer of shredded leaves. Nest material refreshed frequently. Nest entrance *c.*15cm in diameter, on side at one end. Nest trees often close to important food sources. Nests reusable and multiple individuals may use same nest over period of time. Sometimes individuals use new nest each night, not returning for considerable period; at other times nest may be occupied for several days. After giving birth females may occupy same nest for >1 month (Sterling *et al.* in press). Single individual may have up to five nests in use at one time. Typically, one nest used by one individual, but sharing occasionally occurs. High nest

Aye-aye nest in forest canopy, Masoala National Park.

turnover rate: on Nosy Mangabe, eight Aye-ayes used >100 nests in two years, with different Aye-ayes using same nest at different times. Large trees contain up to six nests (Sterling 1993a, Sterling & Richard 1995).

Primarily quadrupedal: can move quite nimbly around canopy, and leaps and climbs both vertically and horizontally, although horizontal movements sometimes more deliberate. Able to adopt suspensory postures, grasping with all four limbs or hindlimbs alone. Can descend both head-first and tail-first (Sterling *et al.* in press). Also descends to ground, where sometimes covers large distances. The diversity of positional behaviours likely evolved to maximise exploratory feeding opportunities and encounter rates with primary food source, wood-boring larvae (Teichroeb & Vining 2019, Sterling *et al.* in press).

Majority of initial wild studies carried out with artificially established population on 500ha island of Nosy Mangabe where male home ranges are 125–215ha, with considerable overlap; common area may be occupied by both males simultaneously and interactions occur, sometimes agonistic. Males capable of travelling *c.*2–4.5km per night. Female home ranges much smaller, *c.*30–50ha, and do not overlap with one another, but do overlap with home range of at least one male (Sterling 1993b, 1994b). Females interact rarely and are invariably aggressive towards one another. Males and females interact for brief periods, and sometime forage in tandem, suggesting a less solitary, more social aspect to behaviour (Sterling 1993b, Sterling & Richard 1995).

On Nosy Mangabe, activity begins up to 30 minutes before sunset, but may not start until three hours after. Males often active before females. Vocalisations frequent at this time. Up to 80% of night spent travelling and foraging in upper canopy. Foraging bouts punctuated by rest periods lasting up to two hours. Outside breeding periods, three to four individuals recorded communicating and travelling together and feeding at favoured sites (Sterling & Richard 1995).

In mainland disturbed humid forest (Kianjavato), male home ranges may be huge, up to *c.*900ha, and overlap considerably, whereas female home ranges still rather small, *c.*100ha (Randimbiharinirina *et al.* 2018). In more intact humid forest (Torotorofotsy) female home range *c.*750ha, with male home range over three times larger, *c.*2,500ha (Sefczek *et al.* 2020b). During dry season, activity starts around 18.00, but in wet season it may be *c.*30 minutes either side of 18.00. Prolonged bouts of self-grooming dominate first two hours), followed by moving and foraging. During nighttime activity *c.*40% time spent travelling, *c.*40% feeding, *c.*12% resting and *c.*8% self-grooming (Randimbiharinirina *et al.* 2018).

Aye-ayes communicate via vocal and olfactory signals. Vocal repertoire relatively limited. Most characteristic is loud metallic *creee-e-e-e* contact call, also used for cohesion between mother and infant. Other contact calls include long and short *eeep*. Alarm calls may be grunts or sneeze-like *rhon-tsit* indicating severe unease. When one individual follows another, *ggnnoff* call used and Aye-ayes may subsequently feed or groom together. When aggressive Aye-ayes may hiss like a cat (Stanger & Macedonia 1994, Schwitzer *et al.* 2013e).

Scent-marking also important. On Nosy Mangabe three types identified: ano-genital rubbing, head and chest rubbing, and over-marking, where individuals urinate or drag their genital region over area previously marked by another Aye-aye (Sterling & Richard 1995).

Diet highly specialised but adaptable, with variation between habitats and seasons; omnivorous or predominantly insectivorous. On Nosy Mangabe and at Ranomafana National Park, Ramy nuts *Canarium madagascariensis* especially important, and distribution of Ramy strongly influences Aye-aye distribution and foraging behaviour (Iwano *et al.* 1991, Sterling 1998, Sefczek *et al.* 2012). At other humid forest locations insect larvae dominate: Aye-ayes spend majority of foraging time searching, in both live and rotten wood, indicating they are particularly nutritious and energy-rich (Sterling *et al.* 1994). In studies at Kianjavato and Torotorofotsy insect larvae and other invertebrates comprised *c.*50–85% of diet, with smaller quantities of Ramy nuts and nectar from Traveller's Tree *Ravenala madagascariensis* (Strelitziaceae) (Sefczek *et al.* 2017, 2020a, Randimbiharinirina *et al.* 2018). Other dietary elements include cankerous growths on *Intsia bijuga*, seeds of palm *Orania trispatha* and tropical almond *Terminalia catappa*, some fungi, and occasionally raids coconut, lychee and mango plantations (Iwano & Iwakawa 1988, Iwano *et al.* 1991, Sterling 1993a, 1994c, Sterling *et al.* 1994, Sterling & McCreless 2006). Also extraordinary account of Aye-aye entering human habitation to raid and eat chicken eggs (Sefczek *et al.* 2018b).

Hard exteriors of Ramy nuts, other seeds and coconuts gnawed through with chisel-like incisors, before thin middle digit is used to scoop out pulp (Iwano & Iwakawa 1988, Iwano *et al.* 1991, Goodman & Sterling 1996, Sterling *et al.* 1994). Cankerous growths are removed from tree, then waxy substance from underlying cambium is scraped away with anterior teeth; eaten more often during cold season (May–October) when Ramy nuts less abundant (Sterling & McCreless 2006). Wood-boring insect grubs, especially longhorn beetles (Cerambycidae) and click beetles (Elateridae), are winkled out of bark and rotting wood, their cavities having first been located by Aye-aye tapping on wood with middle finger and listening for movement (percussive foraging) (Erickson 1995, 1998, Erickson *et al.* 1998)).

Percussive or tap foraging is an adaptation unique to the genus *Daubentonia*. Third digit used to tap repeatedly along branch or trunk (tapping finger moving substantially faster than other digits) to locate subsurface cavity. Not clear whether cavity pinpointed by touch or auditory cues, or combination. Once cavity located, large mobile ears listen for insect grub movement and Aye-aye chews opening to cavity with large incisors. Third finger then probes, ball-and-socket joint allowing movement in any direction. Third finger capable of bending up to 30° upwards, allowing curved claw to follow wall of cavity. Consequently, claw can move past grub, instead of pushing it deeper to irretrievable position.

Aye-aye probing for insect larvae.

Signs of Aye-aye excavation and feeding in rotten trunk, forests near Daraina.

When Aye-aye touches grub, fingertip curves over to encircle it for retrieval. Grub not pulped or impaled, but hooked and lifted out whole. Aye-aye's tactile sense is acute as they typically only lift finger from cavity when grub is on the claw (Sterling & McCreless 2006). Third finger also used to probe

Aye-ayes often groom for long periods after first emerging at dusk, Farankaraina Reserve.

for nectar and inserted into mouth (perhaps to clean teeth). More robust fourth finger used for scooping fruit and at times, e.g when eating coconuts, both fingers used simultaneously but performing different tasks (Lhota *et al.* 2008).

Aye-ayes have no fixed breeding season. Female reproductive cycle regular (in captivity *c.*47 days). Behavioural signs of female reproductive state apparent up to ten days before oestrous begins: female increases scent-marking activity and visits male nests (Sterling 1993b). Oestrous externally visible with genitals swelling and changing colour (Winn 1994). Males also increase scent-marking activity and testicular volume swells. At onset of oestrous females move rapidly around home range advertising with distinctive calls, which attracts attention of several males simultaneously (up to six recorded), who gather around female and fight one another for access. In the wild, copulation lasts around an hour (only minutes in captivity); extended period may have mate-guarding function. Afterwards, female moves to another location (500–1,000m away) and repeats her advertisement call. Males and females may mate with several partners. Both sexes increase scent-marking activity during

mating period (Sterling 1992, 1994b, Sterling & Richard 1995, Sterling & McCreless 2006).

A single is offspring born, gestation *c.*164–172 days. At birth infant weighs *c.*90–140g (captive animals). Mother carries infant in mouth and will often 'park' it in nest while she forages. Infants begin independent climbing at 9–12 weeks and leaping at 12–16 weeks, but full adult locomotive patterns may take nine months to acquire (Schwitzer *et al.* 2013e). Infant remains with mother 18–24 months, allowing them to learn the complex foraging techniques needed for survival. Inter-birth interval two to three years. Females begin breeding at three or four years. (Sterling 1993a, Feistner & Sterling 1994). Possible infanticide – inferred by physical injuries to infant, maternal vocalisation and mother chasing adult male – has been observed (Rakotondrazandry *et al.* 2021).

Other than humans, only known predator is Fosa (Sefczek *et al.* 2018a), although given rarity of both species, predation events are probably infrequent.

WHERE TO SEE Encountered extremely infrequently as it occurs at very low densities and is generally high in canopy. One of the most difficult lemurs to see. However, a handful of sites offer a chance. Bekaraoka Forest near Andranotsimaty, 5km north-east of Daraina (on *Route Nationale* 5a) is perhaps the best. Engage local guides to help find a nest and stake it out (sitting in silence on the forest floor) from before dusk. This site is also very good for Golden-crowned Sifaka and Crowned Lemur, and several nocturnal species including Daraina Sportive Lemur and Northern Fork-marked Lemur.

Aye-ayes are also seen at Farankaraina Reserve, 5km east of Maroantsetra. There are many Ramy trees *Canarium madagascariensis* (Burseraceae) and Aye-aye feeding signs are often evident. The canopy is very high in places, so considerable persistence is needed. Camp for at least two nights and use local guides (available in Maroantsetra). An alternative is Kianjavato Ahmanson Field Station, *c.*65km east of Ranomafana. Visits must be arranged in advance. Also occasionally seen at Domaine de Fontenay, near Montagne d'Ambre National Park.

Aye-ayes are seen rarely at other sites, like Andasibe-Mantadia, Ranomafana, Masoala and Tsingy de Bamaraha National Parks. In all locations, evidence of Aye-aye presence – gnawed holes in tree trunks, Ramy nuts with telltale teeth marks or nests in canopy – are far more likely to be seen than the animal itself.

Carnivorans

Order Carnivora

Madagascar is home to a remarkable and peculiar assemblage of carnivorans, the origins of which have prompted much debate. Although there are relatively few species, they exhibit considerable morphological and ecological diversity. Consequently, their taxonomic history has been contentious and confusing, with past suggestions variously placing some species with cats (Felidae) (genus *Cryptoprocta*), civets (Viverridae) (genera *Cryptoprocta*, *Fossa* and *Eupleres*) and mongooses (Herpestidae) (genera *Galidia*, *Galidictis*, *Salanoia* and *Mungotictis*) (Milne-Edwards & Grandidier 1867, Schreiber *et al*. 1989, Wozencraft 1989, 1993, Veron 1995). For detailed taxonomic history see Veron *et al*. (in press). Implicit in such thinking is the requirement for multiple and separate ancestral lineages and colonisations of the island in the distant past.

However, molecular research now confirms all Malagasy carnivorans evolved from a single common ancestor (Yoder *et al*. 2003). From the fossil record, we know the earliest placental carnivorans began evolving around 64 mya (after the Cretaceous-Paleogene, K-Pg boundary mass extinction event), by which time Madagascar was already an island, had long been isolated and established in its approximate current geographical position after the breakup of Gondwana. Therefore, the ancestral Malagasy carnivoran can only have arrived over water by swimming or more probably rafting (Ali & Huber 2010, Krause 2010) or possibly via a short-lived landbridge between Africa and the island (Masters *et al*. 2021) (see Biogeography).

It seems likely this ancestral carnivoran was 'mongoose-like' and crossed to Madagascar from Africa between 18 and 24 mya (Yoder *et al*. 2003, Poux *et al*. 2005, Veron *et al*. 2021) and subsequently underwent a spectacular adaptive radiation resulting in the evolution of diverse forms that superficially resemble several other carnivoran groups (cat-like, civet-like and mongoose-like) now found elsewhere in the world. This diversification was made possible by the prior absence of other carnivorans on Madagascar and the effective 'empty niche space' waiting for new colonists to exploit (Yoder *et al*. 2010, Veron *et al*. in press).

Malagasy Carnivorans Family Eupleridae

Because the origins of the Malagasy carnivorans can be traced to a single ancestor, all species are now placed in the endemic family Eupleridae: one of 16 families in the Order Carnivora (Do Linh San *et al*. 2021). Eupleridae has its closest evolutionary affinities outside Madagascar to 'true' mongooses (Herpestidae) and hyaenids (Hyaenidae) (Yoder *et al*. 2003, Yoder & Flynn 2003). Despite the molecular evidence binding this group together, there are few obvious anatomical traits that unite the family (Goodman 2009). In most instances, the natural history and distribution of the Eupleridae remains poorly understood; they are perhaps the least studied and most threatened of all carnivoran families (Brooke *et al*. 2014, Wampole *et al*. 2021), although the more recent advent of camera-trapping techniques has helped advance knowledge considerably (Farris *et al*. in press).

With no clear morphological characteristics defining the family, precise relationships within this group are still being investigated. Recent phylogenetic analyses suggest the ancestral 'founder' diverged shortly after arriving on the island, giving rise to three evolutionary lineages that now warrant recognition at subfamily level: the Fosa (*Cryptoprocta*) in Cryptoproctinae, the 'civet-like' species (*Fossa* and *Eupleres*) in Euplerinae, and the smaller 'mongoose-like' species (*Galidia*, *Galidictis*, *Mungotictis* and *Salanoia*) forming the Galidiinae (Veron *et al.* in press).

The number of species recognised within Eupleridae is also contentious. Within the recent past, three new species have been proposed: Giant-striped Vontsira *Galidictis gradidieri* (Wozencraft 1986), Western Falanouc *Eupleres majori* (Goodman & Helgen 2010) and Durrell's Vontsira *Salanoia durrelli* (Durbin *et al.* 2010). However, subsequent molecular genetic research has questioned the validity of these taxa (Veron *et al.* 2017, Veron & Goodman 2018) such that between seven and ten species may be recognised depending on the authority consulted (Goodman 2009, Hunter & Barrett 2018). Here eight species are recognised.

The vernacular (English) naming of Eupleridae species has also been problematic and confusing, largely because previously suggested evolutionary relationships are now known to be incorrect, e.g. Malagasy or Striped 'Civet' for *Fossa fossana* and Ring-tailed 'Mongoose' for *Galidia elegans*. To counter this, a standardised system based on Malagasy names has been proposed and is followed here (Duckworth *et al.* 2014).

Compelling and considerable recent evidence shows that all native carnivoran communities are under ever-increasing threats not only from habitat loss and fragmentation, but also directly from human hunting for food and retaliatory killing for poultry predation (Golden 2009, Jenkins *et al.* 2011, Gerber *et al.* 2012b, Borgerson 2016, Merson 2018, Merson *et al.* 2019a, Wampole *et al.* 2021) and predation and competition from non-native carnivorans (Farris *et al.* 2015b, 2016b, 2017a,b, Merson 2018).

Species particularly affected by hunting for food are Fosa, Narrow-striped Boky, Ring-tailed Vontsira and Spotted Fanaloka, although Falanouc and Broad-striped Vontsira are also eaten occasionally (Golden 2009, Golden *et al.* 2014, Farris *et al.* 2015b, Borgerson 2016, Campera *et al.* 2017, Merson 2018). In some areas the body parts of native carnivorans are used in traditional medicine and ceremonial potions (Goodman 2009).

Habitat fragmentation disproportionately diminishes the functional forest area suitable for native carnivorans as it significantly increases edges (borders of native forest and deforested zones). Native carnivorans (particularly smaller nocturnal species like Spotted Fanaloka, Falanouc and Broad-striped Vontsira) tend to avoid spaces within 500m of forest edge, often because these areas have higher densities of free-ranging domestic dogs *Canis familaris*, feral cats (*Felis silvestris*) and the non-native Small Indian Civet *Viverricula indica* that compete directly with them (Farris *et al.* 2015b, 2016b, 2017a,b, 2020, Merson *et al.* 2019b, Ross 2020), and also cause some native carnivoran species, e.g Fosa, to alter their normal daily activity patterns (Gerber *et al.* 2012a, Farris *et al.* 2015a, Merson 2018, Merson *et al.* 2018).

Further negative impacts result from viral and parasitic infection transmission from non-native to native species (Pomerantz *et al.* 2016, Rasambainarivo *et al.* 2017, 2018). Consequently, the populations of all native carnivorans are declining rapidly (Farris *et al.* 2017a).

Falanouc, Makira Natural Park.

Subfamily Cryptoproctinae

The subfamily Cryptoproctinae contains a single genus *Cryptoprocta*. Such are the anatomical, morphological and behavioural peculiarities of *Cryptoprocta* that it has been previously placed in Viverridae (civets and genets), Felidae (cats) and its own family Cryptoproctidae (Veron *et al.* in press). The extant species, *Cryptoprocta ferox*, is the island's largest native carnivoran (up to 11kg).

FOSA
GENUS *Cryptoprocta*

The genus *Cryptoprocta* contains one extant species, the Fosa and a second now extinct species, the Giant Fosa *C. spelea*, which is known from Holocene subfossil remains and disappeared as recently as 1,500–2,000 years ago (Goodman & Jungers 2014, Meador *et al.* 2019).

The generic name *Cryptoprocta* is from Greek and refers to the anus being surrounded and hidden within a pouch; '*krypto*' meaning to hide or conceal, and '*proktos*' meaning anus. The specific name *ferox* is Latin and means wild, savage, spirited, fierce or similar.

Female Fosa, Kirindy Forest.

FOSA	
Cryptoprocta ferox	Bennett, 1833

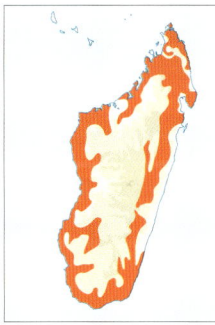

Other name: Fossa (in error)

Malagasy name: Fosa

MEASUREMENTS
Head-body: 75–80cm (males), 65–70cm (females). Tail: 70–90cm. Weight: 6–11kg (males), 5–8kg (females). (Hawkins 1998, 2003, Lührs *et al.* 2013)

DESCRIPTION & IDENTIFICATION Sexually dimorphic: some (but not all) males noticeably larger than females and other males (see below). Sleek, elongate and muscular with short powerful legs; shoulder height up to 35cm. Head relatively small, with shortish muzzle, prominent round ears and large eyes with vertical pupils. Tail slender and very long, almost equal to head-body length. Pelage short, smooth and dense. Upperparts sepia-brown blending into creamy underparts. In male underparts, especially between forelimbs and hindlimbs, stained with orange secretion (Hawkins 2003), produced by sternal glands, probably as a means of scent-marking (Gerber & Hawkins in press).

Male: testes relatively large and prominent (proportionately larger than many other carnivorans) (Lührs & Kappeler 2014). Penis supported by a bone (baculum) (mean length *c.*73mm) (Hawkins *et al.* 2002), protected by a sheath covering hard spines running along basal 65% (Gerber & Hawkins in press). *Female*: uterus paired, each with own cervix (Köhncke & Leonhardt 1986). Three pairs of functional nipples.

In several parts of the island indigenous people describe two contemporary forms of *Cryptoprocta*: *fosa mena*, or 'reddish fosa', which is considered smaller and corresponds to all known modern specimens, and the larger *fosa mainty*, or 'black (melanistic) fosa' reported from some rainforests (Goodman *et al.* 2003b). These may correspond to a black variant of feral cat, known as '*fitoaty*' in the

Female Fosa together with adult daughter, Kirindy Forest.

north-east, and sometimes referred to by locals as 'black fosa' (Farris *et al.* 2016a).

Shows several adaptations to tree climbing and arboreal hunting: long tail acts as counterbalance when climbing (but is not prehensile) (Laborde 1986a,b), large paw pads extend almost to heel, claws decurved and semi-retractile and ankles 'reversible' (Taylor 1989, Veron 1999, Gerber & Hawkins in press). When walking, gait is distinctive and can be either plantigrade or digitigrade (Hawkins 2003). Close morphological similarities (e.g. retractile claws, dental traits, facial resemblance) have led to past comparisons with cat family (Felidae) (Veron 1995). Fosa has attained such cat-like traits via convergent evolution.

Pugmark consists of five evenly spaced pads around large central pad, with no visible claw marks (cf. dog's paw mark has only four outer pads, with middle two closer together and often visible claw marks). Fosa scats also characteristic, grey cylinders with twisted ends, *c.*10–14cm in length, 1.5–2.5cm wide. Often clearly contain mammal hair and bone fragments; tenrec spines sometimes visible. When fresh, scats have strong, musky odour. In comparison, domestic dog scats generally broader, have blunt or round ends, contain little visible hair, and more pungent odour when fresh. In dry forest environments, e.g. Kirindy Forest, Ankarafantsika National Park, scats often visible; in humid, wet habitats rarely seen as broken down quickly by rain and invertebrates (Hawkins 2003, Gerber & Hawkins in press).

Largish, slender but powerful, low-slung carnivoran with very long narrow tail. Madagascar's largest extant carnivoran. Cannot be confused with any other species.

HABITAT & DISTRIBUTION Most widely distributed Malagasy carnivoran; found sparsely throughout native forest regions from sea level to *c.*2,600m (Andringitra Massif), including isolated native forest patches in Central Highlands (Goodman *et al.* 1997, Goodman 2009). Broad habitat tolerance from wettest rainforests to driest subdesert regions, and also high-altitude areas above treeline where nighttime temperatures fall below freezing. Also in degraded and human-altered habitats, including forest edges and corridors (Gerber *et al.* 2012b, Lührs & Kappeler 2013, Farris *et al.* 2015a, Wyza *et al.* 2020).

Occurs at low densities throughout its range: one individual per *c.*4–5km² in primary deciduous forest (Hawkins 1998, Hawkins & Racey 2005), whilst in eastern rainforests some studies have shown similar densities (Murphy *et al.* 2018), while others have found much lower densities (Dollar 1999a, Dollar *et al.* 1997, Gerber *et al.* 2010, Gerber *et al.* 2012a).

Fosa enter villages to kill chickens and consequently suffer direct persecution both inside and outside protected areas (Borgerson 2016, Merson 2018, Merson *et al.* 2019c). They are also hunted for food (Golden 2009, Farris *et al.* 2015a, Borgerson 2016, Merson 2018). Forest intrusion by feral cats and domestic dogs negatively impacts Fosa populations via direct competition (for space and food) and altering behaviour by affecting normal daily activity patterns (Barcala 2009, Gerber *et al.* 2012a, Farris *et al.* 2015a, Merson 2018, Merson *et al.* 2018b, 2019b). In addition, feral domestic carnivorans, especially cats, are responsible for transmission of viral and parasitic pathogens: at one site (Kirindy-Mitea) *c.*90% of captured adult Fosas had antibodies for *Toxoplasmosis* (Pomerantz *et al.* 2016). Road fatalities of Fosa have also been documented in Ankarafantsika (Wyza *et al.* 2018).

At best, total Fosa population *c.*8,500, of which 4,500 in rainforest areas and some 4,000 in dry forests zones; these totals are likely to be optimistic, with

The stocky, powerful profile of a male Fosa, Kirindy Forest.

lower estimates less than 40% of these values. Island-wide *c*.2,600 adults are estimated to live in protected areas (*c*.780 in dry forest and *c*.1,850 in rainforest) and there may only be four intact forest blocks, two in each region, large enough to support more than 300 adults (Gerber *et al.* 2012b). IUCN Vulnerable.

BEHAVIOUR Overall, activity patterns best summarised as cathemeral. Probably more frequently active at night, especially in rainforests, but also crepuscular and diurnal for up to 25% of time (Gerber 2010, Gerber *et al.* 2012a, Farris *et al.* 2015a); peak activity usually under cover of darkness, although this varies seasonally (Hawkins 2003) with increased daytime/crepuscular activity when breeding (Lührs 2012, Farris *et al.* 2015a). Individuals do not have regular sleep sites, instead use a variety of locations, caves or hollowed-out termite mounds, but most frequently prefer trees, usually resting on large branches. Females with young do return to same den. Also presence of non-native carnivorans, domestic dogs *Canis familaris* and feral cats *Felis silvestris*, tends to alter 'regular' activity patterns; Fosa are less likely to be active by daylight if these introduced species are present (Farris *et al.* 2017a, Merson *et al.* 2018, Farris *et al.* 2020).

Extremely agile climber, equally at home hunting in trees or on ground (Hawkins 2003, Lührs 2012), but large distances rarely covered in canopy. 'Reversible' ankles enable it to grasp both sides of slender tree trunk with hindfeet when ascending or descending (head-first) or leaping to an adjacent trunk (Hawkins 2003).

Previously regarded as solitary, but this is now known to be an over-simplification. Females appear strictly solitary, except when with offspring or for brief periods during breeding season, when they are not aggressive towards other females and may mutually groom, sleep together and cooperatively chase unwanted males (Garbutt & Lührs 2011). Males can be either solitary or social with related males. There are anecdotal and detailed observations of male-male associations, including cooperative hunting (Garbutt 1994, Lührs & Dammhahn 2010, Wyza *et al.* 2018, 2020). These comprise same-litter coalitions of two or three brothers, which facilitate increased hunting success of larger, more agile lemur prey (Lührs *et al.* 2013). Approximately half of male population forms such coalitions, while single offspring litter males remain solitary. Coalition males grow larger than solitary males: females and solitary

Same litter brother coalition, Andasibe-Mantadia National Park.

Fosas are extremely agile climbers, Andasibe-Mantadia National Park.

males are approximately equal size (6–8kg), whilst coalition males reach 9–11kg; this somatic difference is correlated to nutritional benefits of cooperative hunting (Lührs 2012). Coalition males also range over much larger territories, up to *c.*65km² in dry forests, versus 26–33km² for solitary males (Hawkins 1998, Lührs *et al.* 2013). Female territories smaller, up to 25km² (Lührs 2012, Hawkins 1998).

Female ranges are exclusive, but solitary male and male coalition ranges overlap with one another and those of two or more females. Male ranges significantly larger during breeding season (October–November) (Lührs 2012). Territorial boundaries marked using pungent scent from glands in ano-genital region. Fosa may cover large distances, 5–7km in a day (Hawkins 2003) and in mountainous regions, like Andringitra, daily climbs of at least 600m in elevation are known (Goodman *et al.* 1997).

Front paws used to pin larger prey down before a fatal bite is administered to throat or back of head (Britt *et al.* 2001). Victim often eviscerated and vital organs eaten first (Wright *et al.* 1997). Diet highly variable. Mammals are mainstay, although relative abundance and size of prey taken reflects habitat and location. In high mountains, like Andringitra, small prey such as shrew tenrecs (*Microgale* spp.) predominate (Goodman *et al.* 1997b), whereas in low-elevation native forests larger prey, particularly lemurs are preferred and may comprise more than

half the diet (Rasoloarison *et al.* 1995, Hawkins 1998, Dollar *et al.* 2007).

In both rainforests and dry forests, larger lemurs, like sifakas, are regularly taken (Wright *et al.* 1997, Patel 2005, Hawkins & Racey 2008, Irwin *et al.* 2009) especially by cooperatively hunting coalition males (Lührs & Dammhahn 2010, Lührs *et al.* 2013), and these may be ambushed at night while sleeping (Wright *et al.* 1997).

Other lemurs also eaten, including ruffed lemurs, 'true' lemurs (genus *Eulemur*), bamboo lemurs and sportive lemurs, and most Cheirogaleidae (Britt *et al.* 2001, Goodman 2003a, Karpantry & Wright 2007, Goodman 2009). Also known to predate smaller native carnivorans, including Narrow-striped Boky and Spotted Fanaloka. In deciduous forests near Morondava, Giant Jumping Rat *Hypogeomys antimena* is also taken (Rasoloarison *et al.* 1995). Various tenrecs (genera *Tenrec*, *Setifer*, *Echinops* and shrew tenrecs spp.) are eaten year-round, but especially in wet season when active; during dry season when aestivating, these animals are excavated from burrows or hideaways, as are dwarf lemurs (*Cheirogaleus* spp.) (Hawkins 2003). To avoid spines when tackling spiny tenrecs, Fosa dispatches victim with strike of powerful front paw or neck bite, then opens carcass from underside, where spines are reduced or absent (Goodman *et al.* in press a). Birds, reptiles, amphibians and invertebrates also predated occasionally, but constitute a smaller proportion of overall diet (Goodman 1996b). In high mountains (Andringitra National Park), shrew tenrecs, crustaceans and insects form part of diet (Goodman *et al.* 1997). Remains of introduced bush pigs (*Potamochoerus* sp.) and even cattle remnants have been found in scats, but were presumably scavenged (Goodman 2009).

Anecdotal observations suggest successful lemur hunts occur when stealth and complete surprise is maintained. If element of surprise is lost and lemurs begin alarm-calling hunts are generally unsuccessful. If a hunt is successful, other lemurs in the group do not alarm-call (Schnoell & Fichtel 2012). Also Fosa may be opportunistically attracted by prey species alarm calls (e.g. 'true' lemurs, mouse lemurs, dwarf lemurs) directed at alternative predator species (Fichtel 2009).

Mating system described as 'super-polyandrous'. Courtship is between October and December. In oestrous, females scent-mark and emit high-pitched, cat-like calls which attract attention of several males (up to eight in a day) (Hawkins 2003, Lührs 2012) and stay in close proximity, although female remains in same place, often a favoured tree branch (used year after year), for up to a week (Hawkins 1998, 2003, Hawkins & Racey 2009). She may mate repeatedly with the same male (up to ten times) and with several different males during this time (up to ten males) (Garbutt & Lührs 2011, Lührs & Kappeler 2014). Single copulations average *c.*40 minutes, total copulation events average 41, and total time mating averages 27.5 hours (Lührs & Kappeler 2014).

Somatic advantage makes coalition males more successful than solitary males in gaining access to females during peak mating period with correspondingly improved reproductive success, but after peak mating, larger males are discriminated against by females (Lührs *et al.* 2013, Lührs & Kappeler 2014). This extreme male sexual competition has led to males evolving disproportionately large penises and producing copious quantities of semen. Males also have sharp penile spines that help maintain coital union or induce ovulation.

During courtship females are surprisingly dominant and control coital encounters, which take place on a tree branch. Mating is noisy, with both sexes purring, snorting and shrieking and, if uninterrupted by other males, can last from several minutes up to six hours (Hawkins 2003). Male grasps female with forelimbs and often gently bites and licks her neck. Male and male/female together can fall to the ground during copulation.

Gestation period six to seven weeks (Hawkins 2003). Litter between two and four (six recorded) and females raise offspring alone (Albignac 1984). Den is usually a hollow tree, hollowed termite mound or similar. At birth, young are very pale grey, fully furred but toothless and blind, and weigh *c.*100–150g. Eyes open after 15 to 16 days; development is relatively slow. First solid food taken around three months and weaning occurs between four and five months when the young leave den for first time (Goodman 2009). Independence reached after one year (Albignac 1973) (less than 50% survival to this stage), with physical and sexual maturity not reached until three or four years (Hawkins 2003). Infants stay with mother for 12 months or more, so females breed only every other year.

Between one and two years old, juvenile females have a period of transient masculinisation, sharing a number of hormonal and morphological features with males that are not seen in adult females (Hawkins *et al.* 2002). The clitoris of juvenile females is partially covered with spines and supported by a small bone; proportionately this is much larger than in the adult

Fosas mating is often protracted and noisy, Zombitse National Park.

female. Juvenile female masculinisation does not appear dependent on androgen hormones (Hawkins *et al.* 2002). Rather it appears to correspond to age at independence (*c.*1 year) from their mother and is thought to reduce adult aggression by males and/or females (Gerber & Hawkins in press).

Much local folklore surrounds the Fosa. In some regions, e.g. Ranomafana National Park area, they are thought to scavenge the bodies of dead ancestors buried in the forest, so eating them is taboo (*fady*) (Jones *et al.* 2008). There are suggestions they pose a threat to humans, but there is no evidence for this and such stories become exaggerated over time. However, Fosas certainly behave in an irregular and erratic manner occasionally. Even in remote forest areas they have entered field camps, ransacked tent contents, chewed metal objects, leather boots, rucksacks and even soap (Duckworth & Rakotondraparany 1990, Garbutt & Lührs 2011).

WHERE TO SEE Nowhere common and consequently difficult to see in most localities. Kirindy Forest is the best location: Fosa often seen around field camp where some individuals scavenge. The more adventurous might try Kirindy-Mitea National Park, *c.*100km south of Morondava. Otherwise most likely to be seen in other western dry forests (Ankarafantsika, Ankarana), particularly in dry season when they visit pools to drink. Seems less wary (at times bold) in breeding season (October– November). Daytime encounters in rainforests areas like Mantadia, Ranomafana and Masoala National Parks are generally rare and fleeting.

Subfamily Euplerinae

The two endemic 'civet-like' carnivorans, Spotted Fanaloka *Fossa fossana* and Falanouc *Eupleres goudotii* form the subfamily Euplerinae (Veron *et al*. 2021) with each belonging to a monotypic genus. They are primarily nocturnal, native forest-dwellers and show degrees of morphological and ecological specialisation. They are the second and third largest of the island's endemic carnivorans, ranging from *c*.1.5kg (Spotted Fanaloka) to around 4.6kg (Falanouc).

SPOTTED FANALOKA
GENUS *Fossa*

Fossa is a monotypic genus, the sole species being *Fossa fossana*. In the past it has variously been named, *F. fossa*, *F. daubentoni* and *F. majori*, but these are all junior synonyms of the accepted current binomial (Kerridge *et al*. in press).

SPOTTED FANALOKA
Fossa fossana   (Müller, 1776)

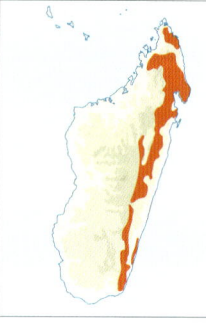

Other names: Malagasy Spotted Fanaloka, Malagasy 'Civet', Striped 'Civet'

Malagasy names: Fanaloka, Teza (Andasibe) (Dolch 2012)

MEASUREMENTS
Head-body: 400–450mm. Tail: 215–265mm. Weight: males *c*.1.5–2.1kg, females up to 1.3–1.8kg. (Farris *et al*. 2015b, Kerridge *et al*. in press)

DESCRIPTION & IDENTIFICATION A small 'civet-like' carnivoran approximately the size of a domestic cat with conspicuous body markings. Pelage dense and light brown with greyish areas around head and on back. Two blackish, largely complete, mid-dorsal lines and two irregular rows of black spots which may merge slightly, forming vague broken stripes, on flanks. Also irregular scattered spots on thighs. Underparts, including under neck and chin much paler and have few markings; often light grey or cream but can tend towards pale orange (Kerridge *et al*. 2003). Snout pointed; ears triangular with rounded tips. Tail cylindrical and similar colour to body, with faint darker bands and diffuse spots sometimes extending from back. Tail and dorsal markings more distinct in young than adults. Legs shortish and delicate with small paws and medium-sized claws. Gait obviously digitigrade (walks on toes) (Kerridge *et al*. in press).

Confusion possible with similarly marked non-native Small Indian Civet *Viverricula indica*, but Spotted Fanaloka is smaller, with thinner legs, a chunky body and lightly marked tail; Small Indian Civet larger, longer-bodied with smaller, more numerous spots on flanks often forming stripes, and a clearly ringed tail. Also, Spotted Fanaloka has ears more pointed and wider apart, whilst ears of Small Indian Civet are obviously rounded and close together. Spotted Fanaloka distinguished from Falanouc by less stocky build, spotted/striped flanks and blunter snout.

HABITAT & DISTRIBUTION Throughout lowland and mid-altitude rainforest areas of east and north, from sea level to 1,600m, including Sambirano region in north-west (Albignac, 1973, Goodman 1996b, Goodman & Pidgeon 1999, Goodman 2009, 2012) and littoral forests near Tolagnaro in south (Kerridge *et al*. 2003). Also occurs in isolated humid forests of Montagne d'Ambre and less humid deciduous forests of Ankarana and Daraina in far north (Ross *et al*. in press). Primarily in native forest, but can survive in adjacent degraded forests (Farris *et al*. 2015b), although numbers markedly reduced at forest edges and degraded areas (Gerber 2010, Farris *et al*. 2020): density estimates for intact forest 3.19 individuals/km², versus 1.38 individuals/km² for logged forest (Gerber *et al*. 2012b).

Research in Ranomafana National Park suggests, over short time period (six years), Spotted Fanaloka has shifted to higher-elevation forest, from lower areas that have become degraded and more disturbed (Beaudrot *et al*. 2018). Suffers negative impact from non-native carnivorans, especially in degraded areas, with feral cats *Felis* spp. and Small Indian Civet being particular culprits (Farris *et al.* 2015b): Spotted Fanaloka occupancy can decrease by 40% at sites where Small Indian Civet is present (Farris *et al.* 2021). Also hunted by rural populations for food throughout its range (Golden 2005, 2009). IUCN Vulnerable.

BEHAVIOUR Shy and primarily nocturnal, although some daytime activity occurs in undisturbed forests. In forest edge zones, Spotted Fanaloka alters its activity to avoid interaction with diurnal free-ranging dogs (Farris *et al.* 2015). Forages on forest floor and often along watercourses (Gerber *et al.* 2012a, Farris *et al.* 2015a). The diet is varied, but it favours aquatic organisms like crustaceans including freshwater crabs, amphibians and invertebrate larvae. It is also known to eat rodents, small tenrecs, reptiles and terrestrial invertebrates (Albignac 1972, 1984, Goodman *et al.* 2003c, Kerridge *et al.* 2003). There is seasonal dietary variation, with insects, reptiles and amphibians dominating (up to 96%) in the wet season, and insects, crabs and mammals (94%) in the dry season (Goodman 2009).

During wet season, Spotted Fanaloka lays down fat reserves, especially in its tail, in preparation for austral winter (June–August) when food is scarce and foraging difficult. These may constitute 25% of animal's body weight (Albignac 1984).

Male and female live as a pair and share a territory of up to 0.5km² (Kerridge *et al.* 2003). Boundaries marked with scent from glands around anus and on cheeks and neck. Areas around streams and marshes may be preferred (Kerridge *et al.* 2003). Sleeps by day in tree holes, under fallen logs or in rocky crevices. Vocalisations are limited: during antagonistic encounters a subdued growl is audible with louder calls between adult and young (Albignac 1984). Scent-marking is important for delimiting territories and during breeding, when glands on neck and cheeks and anal gland are well developed.

Courtship and copulation occur in August and September with single young born in austral summer after gestation of 82–89 days (Albignac 1984). Births occur in a secluded den. Young are very well developed at birth (unusual amongst carnivorans, but similar to Falanouc), weigh up to 100g (Goodman 2009), have eyes open, and are fully furred (Albignac 1984). Can walk after three days, but progress relatively slow thereafter. Females may carry very young infants when moving den sites (Kerridge *et al.*

Spotted Fanaloka, Ranomafana National Park.

in press). Solid food eaten after a month, but not fully weaned until two to three months. Probably independent at one year of age and begin breeding at two. Occasionally predated by Fosa (Kerridge *et al.* in press).

WHERE TO SEE Chance encounters during night walks are rare. Most regularly seen at Ranomafana National Park and also in Marojejy National Park (Camp Mantella). Also occasionally seen in Ankarana Special Reserve around *Campement Anilotra* (Camp des Anglais), but they are very shy.

FALANOUC

GENUS *Eupleres*

The genus *Eupleres* contains a single monotypic species, *E. goudotii*, although in the past two subspecies have been proposed: *E. g. goudotii* (from eastern regions) and *E. g. major* (in western areas), with some authors elevating these taxa to species (Goodman & Helgen 2010). More recent research does not support species or subspecies divisions, and the genus is now considered monotypic (Veron & Goodman 2018).

FALANOUC
Eupleres goudotii   Doyère, 1835

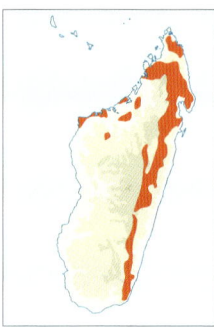

Other names: Malagasy Small-toothed 'Civet', Malagasy Small-toothed 'Mongoose'

Malagasy names: Falanouc, Ridarida, Amboa Laola (Rivera *et al.* in press)

MEASUREMENTS
Head-body: 450–650mm. Tail: 220–250mm. Weight: 1.5–4.6kg (considerable seasonal variation). (Goodman 2009)

Falanouc are adept earthworm hunters, Montagne d'Amber National Park.

DESCRIPTION & IDENTIFICATION A small to medium-sized carnivoran (larger than a domestic cat) with stocky body, small delicate head, prominent ears, distinctive elongated pointed snout and characteristic fat, tapered cylindrical tail. Pelage dense and soft. Upperparts vary from grey to sombre rufous-brown or light brown, sometimes with russet spots and tinges around thighs (western form). Underparts pale grey-brown. Head and tail grey. Tail large, broad at base and tapers to a point, covered with longer hairs, giving bushy appearance. May also show faint brown bands on flanks. Forepaws and claws formidable and well developed for digging (Albignac 1984). Gait is slow and 'sauntering' (Goodman 2009). Larger size and very distinct pointed muzzle should prevent confusion with other small carnivorans.

HABITAT & DISTRIBUTION Humid forests of east and dry forests of north and north-west, from around 50m to 1,600m. Throughout range, sparsely and patchily distributed but occasionally locally common. Previous suggestions of close association with riverine, swampy and marshy habitats (Albignac 1984, Goodman 2009) may be over-simplifications (Wampole *et al.* 2021).

Previously, two subspecies were recognised: Eastern Falanouc *E. g. goudotii* in eastern lowland rainforests, recorded from Andohahela region in south to Anjanaharibe-Sud and Marojejy Massif in north (Goodman 1996b, Dollar 1999, Goodman & Pidgeon 1999, Ross *et al.* 2020) and Western Falanouc *E. g. major* in Montagne d'Ambre, Ankarana, Dariana, Sambirano region in north-west, Mariarano Forest, Ankarafantsika, Tsiombikibo Forest and Soalala/Baly Bay region (Hawkins 1994, Goodman 2009, Evans *et al.* 2013, Mann 2015, Merson *et al.* 2018a, Ross *et al.* in press). Based on morphological and distributional evidence these taxa were elevated to species status (Goodman & Helgen 2010), but more recent genetic analysis does not support differentiation at any taxonomic level (Veron & Goodman 2018).

Throughout its range, populations threatened by habitat destruction and fragmentation, and more directly by localised hunting for food and forest intrusion by feral and non-native carnivorans, especially dogs and Small Indian Civet (Farris *et al.* 2020, Golden *et al.* in press). Introduction of toxic Asian Common Toad *Duttaphyrnus melanostictus* may also threaten Falanouc (Brown *et al.* 2016). IUCN Vulnerable.

Falanouc, Montagne d'Amber National Park.

BEHAVIOUR Madagascar's most specialised carnivoran. Elongated snout and tiny conical teeth (resembling an insectivore) have evolved to predate soft-bodied invertebrates, like earthworms and slugs on which Falanouc almost exclusively feeds (Albignac 1984, Rivera *et al.* in press). Other invertebrates and small vertebrates like frogs and chameleons are eaten occasionally (Dollar 1999b, Goodman 2009). Terrestrial: forages in shallow soil, leaf litter and rotten wood, scraping and digging up morsels using strong forepaws and long claws. Claws held above ground when walking.

During wet season, when food plentiful, Falanouc accumulates large amounts of fat (up to 800g/20% of mean body mass) in tail (Albignac 1984), aiding survival during austral winter (June–August) when invertebrate prey is scarce (Dollar 1999b).

Other information is scant and anecdotal. Falanouc appears to be mainly solitary (Dollar 1999b), although females with offspring have often been observed together (especially in June) (Farris 2021). Probably cathemeral/crepuscular, with most activity at night, but also occasionally active early morning and during day, with some seasonal variation (Goodman 2009, Gerber *et al.* 2012a, Farris *et al.* 2015a). Adults sleep in dens, either in existing burrows or holes/hollows at the base of trees with thick vegetation or under logs (Rivera *et al.* in press). Adults and subadults have been recorded climbing up to 1–2m in trees (Farris 2021) and sleeping on low tree branches (Goodman 2009).

Defends a large territory, with both males and females marking with scent glands around anus and neck, especially during the breeding season (Albignac 1984, Goodman 2009). Based on observations in captivity, courtship and mating occur between August and September; offspring are born in November. Litter size one or two (Albignac 1974). Gives birth to extremely well-developed, fully-furred young (pelage darker than adult, sometimes almost black) with open eyes, weighing around 150g. Can follow foraging mother within two days. Weaned after nine weeks. At around one month, infants able to climb: individuals observed climbing/resting on tree branch at 2m (Ravoahangy *et al.* 2011). Juveniles also observed sleeping in trees, but this behavior is unknown in adults (Rivera *et al.* in press). Females communicate with young via audible 'hiccup' vocalisations.

WHERE TO SEE Rare over most of its range. Shy, secretive and nocturnal, so a very difficult species to observe. In Montagne d'Ambre National Park, very occasionally seen at night near *Station Roussettes*.

Subfamily Galidiinae

The smaller 'mongoose-like' Malagasy carnivorans comprise the subfamily Galidiinae (Veron *et al.* 2021), which are derived from a common ancestor, i.e. monophyletic (Veron *et al.* 2017). Precisely how many species this group contains is debated: molecular genetics have drawn into question the validity of two taxa previously recognised as species, Grandidier's Vontsira *Galidictis grandidieri* and Durrell's Vontsira *Salanoia durrelli* (Veron *et al.* 2017). Here five species are recognised.

RING-TAILED VONTSIRA

GENUS *Galidia*

The genus *Galidia* contains a single species, *G. elegans*. Being largely diurnal, it is the most conspicuous native carnivoran, occurring in the majority of natural forest types, except the drier south-western deciduous forest and southern arid spiny forest (Farris & Goodman in press a).

RING-TAILED VONTSIRA
Galidia elegans   I. Geoffroy Saint-Hilaire, 1837

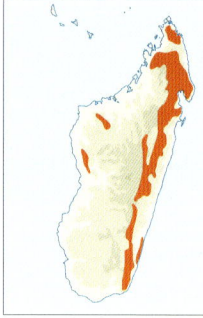

Other name: Ring-tailed 'Mongoose'

Malagasy name: Vontsira Mena

MEASUREMENTS
Head-body: 300–380mm.
Tail: 250–290mm. Weight:
males 900–1,085g, females
650–890g. (Goodman
2003d)

DESCRIPTION & IDENTIFICATION Upper parts, thighs and legs rich russet-chestnut. Head, throat and chin more olive-grey to olive-brown. Tail bushy and ringed with 5–7 (six most common) alternate bands of russet-chestnut and black. Head small, snout pointed, ears rounded. Relatively large feet, with well-developed foot pads and slight webbing between toes. Moderately sexually dimorphic; males slightly larger than females.

Three subspecies vary slightly: nominate eastern form, *G. e. elegans* is darkest with dark brown to black feet; northern *G. e. dambrensis* has paler chestnut upperparts and belly; and western *G. e. occidentalis* has light chestnut upperparts blending into dark belly, and legs and paws black (Goodman 2003d, Goodman 2009).

A small delicate carnivoran, 'mongoose-like' in shape. Might be confused with similarly diurnal Brown-tailed Vontsira *Salanoia concolor*, but more russet coat and distinctively ringed tail should prevent misidentification.

HABITAT & DISTRIBUTION Commonest and most widespread of Madagascar's small carnivorans (Goodman 2003b). Found widely in intact native forests, including littoral forest of east, north and west, from sea level to around 1,950m (Andringitra Massif) (Goodman 1996b). Absent from very dry forests of south-west. More abundant below 1,500m. Primary habitats preferred, but able to survive in secondary/partially logged forests and forest edge abutting cultivation (Goodman 2009, Kotschwar Logan *et al.* 2014). Particularly adversely affected by feral cat incursion into forest (Gerber *et al.* 2012b, Farris *et al.* 2015b). Evidence from Ranomafana National Park suggests over short time period (six years), Ring-tailed Vontsira has retracted from previous haunts at higher elevations (Beaudrot *et al.* 2018).

Three subspecies: eastern *G. e. elegans* throughout eastern rainforests from Marojejy area in north to Tolagnaro region in south, with a record from gallery forest at Berenty, 40km west of humid forest areas (Schnoell 2012); western *G. e. occidentalis* in deciduous forests between Manambolo River in south and Soalala in north, particularly around karst formations at Bemaraha and Namoroka; and northern *G. e. dambrensis* in moist forests of Montagne d'Ambre and drier forests around Ankarana, Daraina (Ross *et al.* in press) and between Ambanja and Ambilobe. Which form occurs in Sambirano further south is unclear (Farris & Goodman in press a). Subfossil remains of Ring-tailed Vontsira, thought to be c.500 years old, have been found in Ankilitelo Cave in the south-

Ring-tailed Vontsira: northern subspecies Galidia elegans dambrensis, *Ankarana Special Reserve*

west, well outside the species' current known range (Muldoon *et al.* 2009).

All populations throughout range threatened by habitat destruction and fragmentation, and by some localised hunting for food and forest intrusion by feral carnivorans (dogs and cats) (Farris *et al.* 2020, Golden *et al.* in press). Also certain body parts (e.g. the tail) used culturally by some tribal groups (Goodman 2003d). IUCN Least Concern.

BEHAVIOUR Primarily diurnal, with occasional activity at dusk and after dark (Goodman 2009, Farris *et al.* 2015a). Can be solitary but more often in pairs or family units of up to five (parents, with subadult/dependent offspring) (Goodman 2009, Farris & Goodman in press a). In rainforests, dens excavated at base of large trees or stream banks and have several entrances/exits (Dunham 1998). Also dens in fallen hollow trees. In dry forest zones with karst formations (Ankarana, Bemaraha), lairs are in rock crevices. Den sites changed frequently (every few days), to avoid predation and reduce parasites like fleas.

Very versatile, equally at home on the ground or in trees. Regularly climbs nimbly up trunks, along vines or branches. Seen in canopy >15m above ground, where it often rips open epiphytes when foraging (Goodman 2003d). Sometimes hunts cooperatively: observations of one individual climbing to remove sleeping mouse lemur from leaf nest, with second animal waiting on ground to dispatch the lemur

when it fell (Deppe *et al.* 2008). An adept swimmer which regularly hunts along forest streams, resting silently at edge before leaping into water to seize prey (Dunham 1998, Goodman 2003d).

Diet very varied and includes small lemurs, particularly mouse lemurs and dwarf lemurs, rodents (including non-native species), small tenrecs (streaked and shrew tenrec species) adult and young birds (mainly terrestrial species),

Ring-tailed Vontsira: eastern subspecies Galidia elegans elegans, *Marojejy National Park.*

Ring-tailed Vontsira can live in family units, Marojejy National Park.

eggs, smaller reptiles, frogs, invertebrates, aquatic invertebrates (crabs and crayfish) and even fish (Goodman & Pidgeon 1999, Goodman 2003d, Ryan 2003, Farris & Goodman in press a). Also record of single Ring-tailed Vontsira killing a 2m boa-type snake, then joined by two other individuals to consume it (Albignac 1973). Victims dispatched by bite to back of neck. Large eggs and snails broken in typical 'mongoose-like' fashion – holding item between forepaws and kicking it against a hard surface with hindfeet. Not averse to scavenging at forest research camps: can be very bold, even entering tents, and consumes non-natural foods (biscuits, chocolate, condensed milk etc.) (Farris & Goodman in press a). Around villages Ring-tailed Vontsira suffers retribution for predating chickens (Borgerson 2016) and is hunted for bushmeat (Golden 2009, Kotschwar Logan *et al.* 2014).

Family groups occupy territories of 20–25ha (Albignac 1984) with scent-marking important in defining and maintaining these. Scent smeared from large anal glands against rocks, tree trunks and branches (Dunham 1998). Can be vocal. Contact calls are high-pitched 'peeping' whistles that keep family together while foraging. Muffled 'meows' and soft 'mews' emitted during prey capture; louder shrieks and growls heard when disputes break out; alarm calls are low moans or grunts (Britt 1999).

Prior to copulation, males pursue females, and both animals scent-mark with increased intensity. When receptive, female lowers and quivers her rear body and repeatedly treads her hindfeet. Copulation is brief (10–30 seconds) but may be repeated 7–12 times in 15 to 80-minute period (Larkin & Roberts 1979, Albignac 1984). Gestation *c.*75–90 days (captivity), births occur between November and January. Litter size one; newborns weigh *c.*50g and are fully furred. Eyes open at 4–5 days. Female initially raises young alone in burrow and does not allow male to approach. After one month male allowed access to infants. Development slow: young take first steps at 12–14 days, can take small amounts of meat at one month, not fully weaned until 2–3 months. Physically mature at *c.*1 year, but do not leave parents, and attain sexual maturity, until 18 months to two years (Albignac 1984).

WHERE TO SEE Most conspicuous and regularly seen native carnivoran. Eastern form readily encountered in Ranomafana National Park and Marojejy National Park at *Camp Mantella* and *Camp Marojejia*. Also seen in Mantadia and Masoala National Parks. Northern subspecies seen in Montagne d'Ambre National Park and Ankarana Special Reserve; western form in southern portion of Tsingy de Bemaraha National Park.

BROWN-TAILED VONTSIRA

GENUS *Salanoia*

Salanoia is a monotypic genus, containing *S. concolor*, although a second species, *S. durrelli*, has previously been proposed (see below). It is perhaps the rarest and least-known member of Eupleridae, and much basic information on its behaviour and ecology is lacking.

BROWN-TAILED VONTSIRA

Salanoia concolor (I. Geoffroy Saint-Hilaire, 1837)

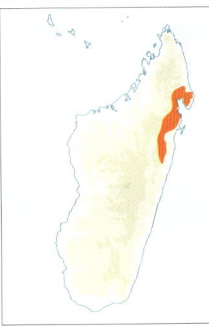

Other names: Brown-tailed 'Mongoose'

Malagasy name: Vontsira Boko

MEASUREMENTS
Head-body: 350–380mm. Tail: 160–200mm. Weight: 550–700g. (Albignac 1973, Goodman 2009, 2012)

DESCRIPTION & IDENTIFICATION Pelage short, dense and dark brown, flecked with paler guard hairs on back and flanks. Area below ears to around throat paler reddish-brown. Under chin and around mouth whitish-grey. Underparts reddish-brown. Ears relatively large and rounded. Muzzle thickly furred and sharply pointed. Snout protrudes over lower jaw. Whiskers short. Tail uniformly dark brown and bushy (Britt & Virkaitis 2003). When alarmed long tail hairs become erect. Claws long and relatively straight.

Based on geographic and apparent ecological separation, and morphometric and pelage differences, populations of *Salanoia* in reedbeds around Lac Alaotra have been described as a new species, Durrell's Vontsira *S. durrelli* (Durbin *et al.* 2010), but further investigation and genetic analysis suggest such differentiation is unwarranted (Veron *et al.* 2017). Hence, the genus *Salanoia* is treated as monotypic here.

A small delicate carnivoran, smaller and more gracile than similar Ring-tailed Vontsira. Also has more rounded ears and shorter tail with no banding.

HABITAT & DISTRIBUTION Prefers intact lowland rainforest (all records between near sea level and 680m, except Lac Alaotra at 750m). Occasional in secondary/degraded forest and cultivated areas adjacent to primary forest (Britt & Virkaitis 2003, Farris *et al.* 2012). Brown-tailed Vontsira disappears quickly from human-disturbed forests and is very susceptible to invasive feral carnivorans, especially cats (Farris *et al.* 2016). Also suffers incidental/opportunistic human hunting for food (Golden *et al.* 2014, Borgerson 2016). Restricted to north-east Madagascar (only area on island where lowland rainforest persists), from Masoala Peninsula and Makira region in north, to Zahamena and Betampona in south (Dollar 2000, Farris *et al.* 2012, Goodman 2013, Hawkins 2016a). Isolated population in reedbeds around southern portion of Lac Alaotra (Durbin *et al.* 2010).

Populations threatened by habitat destruction and fragmentation, by localised hunting for food and forest intrusion by non-native and feral carnivorans, e.g., Small Indian Civet, dogs and feral cats (Farris *et al.* 2017, 2020, Golden *et al.* in press). IUCN Vulnerable.

BEHAVIOUR Least-known Malagasy rainforest carnivoran. Strictly diurnal, activity peaks early in morning and late in afternoon (Farris *et al.* 2015b). Sleeps in burrows or hollow trees (Albignac 1973). Primarily terrestrial; appears to prefer ridgetops (90% of observations at Betampona); sometimes seen climbing to around 2m and occasionally to 5–10m (Britt & Virkaitis 2003).

Observed mainly in pairs (presumably male/female), solitary animals unusual; groups of three

Brown-tailed Vontsira, Betampona Reserve.

Brown-tailed Vontsira, Makira Natural Park.

to five presumed to be parents with offspring (Britt 1999, Britt & Virkaitis 2003). Mating season perhaps variable (Farris & Murphy in press) between August and October, gestation 74–90 days, single offspring born October to January. Infants left concealed while mother is away foraging.

Diet thought to be mainly insects/invertebrate-based, foraged from leaf litter; beetle larvae excavated from rotting wood using long straight claws (Britt & Virkaitis 2003). Also observed eating pill millipedes (Farris & Murphy in press). Generally silent when foraging but can be heard making soft throaty squeaks and growls (Britt & Virkaitis 2003). Nervous when encountered, growls loudly with alarm, sometimes rears onto hindlegs before fleeing. Also tail hair becomes erect and gait more stiff-legged (Goodman 2012).

Brown-tailed Vontsira is often sympatric with similar-sized diurnal Ring-tailed Vontsira; apparent microniche separation through differences in diet, subtle habitat preferences and activity periods (Britt & Virkaitis 2003, Farris et al. 2015a). Brown-tailed Vontsira similar in aspects of morphology and diet to Narrow-striped Boky *Mungotictis decemlineata* from dry forest and may be considered an ecological equivalent.

WHERE TO SEE Shy and secretive; sightings very rare. Most known localities (Betampona, Zahamena, Lac Alaotra and Makira) remote and difficult to reach.

Very occasional sightings on Masoala Peninsula along ridge trail connecting main tourism area with forest adjacent to Arol Lodge.

STRIPED VONTSIRAS
GENUS *Galidictis I.*

A genus containing two species, *G. fasciata* and *G. grandidieri*. Recent genetic analysis revealed only minor differentiation (based on mitochondrial and nuclear markers) between these, suggesting that subspecific treatment, *G. f. fasciata* and *G. f. grandidieri*, may be appropriate (Veron et al. 2017). Further research is required to substantiate this. Treated as species here. Both apparently occur at very low densities and are encountered infrequently (especially Broad-striped Vontsira) (Farris & Goodman in press b).

BROAD-STRIPED VONTSIRA
Galidictis fasciata   (J. F. Gmelin, 1788)

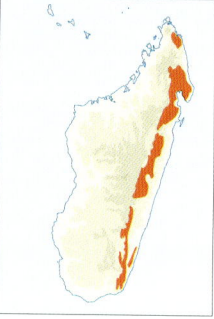

Other name: Broad-striped 'Mongoose' Malagasy name: Vontsira Fotsy

MEASUREMENTS
Head-body: 300–340mm. Tail: 250–295mm. Weight: 520–750g. (Goodman 2003e, 2009)

DESCRIPTION & IDENTIFICATION Base colour grey-beige extending onto legs and lower tail. Underparts paler creamy-white. Tail shorter than body length, bushy and creamy-white over distal two-thirds. Head grizzled grey-brown with cheeks, chin and throat noticeably paler. Several distinctive longitudinal dark brown stripes on flanks, from nape to base of tail, broader than grey-beige intervening areas. Muzzle short and slightly rounded; ears small. Digits and claws long, but webbing less developed than in Grandidier's Vontsira and Ring-tailed Vontsira (Goodman 2003e).

Most robust of native small carnivorans with conspicuous creamy-white tail and distinctive body markings. Much smaller than Spotted Fanaloka or non-native Small Indian Civet, and difficult to confuse with other species in range.

Broad-striped Vontsira, Tsitongambarika Forest.

HABITAT & DISTRIBUTION Eastern lowland and mid-altitude rainforests, and some northern dry forest areas, always at low densities. Majority of records below 700m, but has been seen up to 1,500m (Goodman & Pidgeon 1999). Known range extends from Andohahela in south to Marojejy and COMATSA corridor in north (Hawkins 2012), including Masoala Peninsula. New records from dry forests of Loky-Manambato Protected Area (near Daraina) extend the species' range (Ross *et al.* in press). Does not tolerate forest degradation and is negatively impacted by non-native and feral carnivorans (Small Indian Civet, dogs and cats) (Farris *et al.* 2015b). Broad-striped Vontsira is also hunted and eaten in areas adjacent to forest (Borgerson 2016, Hawkins 2016c). IUCN Vulnerable.

BEHAVIOUR Shy and secretive, occurring at very low densities. Primarily nocturnal, occasionally crepuscular, thus avoids competition with other similar-sized native carnivorans (Ring-tailed Vontsira and Brown-tailed Vontsira) (Gerber *et al.* 2012a, Farris *et al.* 2015a). One record of daytime activity (Ravoahangy *et al.* 2011). Sometimes seen singly, but more often in pairs (camera-trap images) with duos foraging together and digging cooperatively (Ravoahangy *et al.* 2011, Farris *et al.* 2015b). Occasionally seen around forest camps and sometimes raids camp food stores (Goodman 1996b).

Most records of the Broad-striped Vontsira are from camera traps, Ranomafana National Park.

Relative to body size, able to tackle largest prey of native small carnivorans: diet mainly small vertebrates, like rodents, reptiles and amphibians, and occasionally invertebrates. Probably also takes small lemurs (Goodman 2009). Forages on forest floor, often with tail held vertically (Goodman 2003e). Observed climbing up and down fallen logs and into lower branches of undergrowth to 1.5m (Barden *et al.* 1991).

WHERE TO SEE Very difficult. Occasionally seen in Ranomafana National Park, along road between park entrance and village of Ranomafana. Other potential localities are Makira, Masoala, Marojejy, Zahamena and Andohahela National Park.

GRANDIDIER'S VONTSIRA
Galidictis grandidieri   Wozencraft, 1986

Other names: Giant-striped Vontsira, Grandidier's 'Mongoose', Giant-striped 'Mongoose'

Malagasy name: Votsotsoke

MEASUREMENTS
Head-body: 380–465mm. Tail: 280–325mm. Weight: males: 1,450–2,350g (mean 1,650g), females: 1,225–1,625g (mean 1,400g). (Marquard *et al.* 2011)

DESCRIPTION & IDENTIFICATION Largest member of Galidiinae. Sexually dimorphic; male larger than female. Long, pointed snout. Upperparts grizzled beige-grey, with back and flanks marked with dark brown longitudinal stripes from neck to base of tail; intervening areas broader than stripes. Crown and nape darker grey blending into stripes. Underparts paler creamy-grey. Tail shorter than body length, off-white in colour and bushy. Feet are proportionately long with webbing between toes, which are long and do not retract (Goodman 2009).

Described in 1986: originally named *G. grandidiensis*, emended to *G. grandidieri* (Wozencraft 1986, 1987) and subsequently recognised as a species by most authorities (Garbutt 2007, Goodman 2009, 2012). Recent genetic analysis reveals close alliance with congener, Broad-striped Vontsira, which questions the validity of Grandidier's Vontsira as a distinct species, with suggestion that subspecies treatment is more appropriate (Veron *et al.* 2017, Farris & Goodman 2022b). However, based on broader factors such as larger size, unique habitat and ecology and allopatric distribution, Grandidier's Vontsira is maintained as a species here (see The Species Conundrum).

Large, and distinctively marked carnivoran, with prominent ears and pale bushy tail. Non-native Small Indian Civet is only broadly similar species in range, but is larger and has obviously ringed tail.

HABITAT & DISTRIBUTION Inhabits narrow band of spiny forest/bush formation south of Onilahy River, and north of Linta River, centred on exposed limestone Mahafaly Plateau in vicinity of Lac Tsimanampetsotse, at 35–145m (Mahazotahy *et al.* 2006, Marquard *et al.* 2011). This is driest region of Madagascar: <400mm of rain per year, daytime temperatures exceeding 40°C. Grandidier's Vontsira is commoner on western edge of plateau, where there is accessible water from aquifers along the

Grandidier's Vontsira, Tsimanampetsotse National Park.

limestone escarpment. Current range 440 km² (Mahazotahy *et al.* 2006) to 1,500km² (Marquard *et al.* 2011). Subfossil remains, including from Ankilitelo Cave in Mikoboka Mountains, north of the Onilahy River and some 180km north of Tsimanampetsotse (therefore outside current range), indicate species was probably more widespread (Goodman 1996a, Muldoon *et al.* 2009).

Population estimates vary between 2,650–3,540 individuals (Mahazotahy *et al.* 2006) and 3,100–5000 individuals (Marquard *et al.* 2011). Grandidier's Vontsira is the island's most threatened carnivoran (Cartagena-Matos *et al.* 2017) and one of the most geographically restricted and rare carnivorans in the world. Due to habitat loss, livestock grazing and intrusion by non-native and feral carnivorans, populations are certainly declining (Hawkins 2015c). IUCN Endangered.

BEHAVIOUR Strictly nocturnal and primarily terrestrial, but does climb in bushes and trees (Goodman 2003e, Andriatsimiety *et al.* 2009). Sleeps in natural cavity or crevice in peculiar fissured karstic limestone terrain that dominates its habitat (Wozencraft 1990). These crevices may be several metres deep and probably help cooling when daytime temperatures very high. Also recorded sleeping in tree holes (Goodman 2009). Emerges shortly after dusk, generally returning to den well before dawn. Does not use same burrow each night and different individuals may use same sleep site on successive nights.

Forages alone or in pairs, or occasionally larger groups up to five, mainly on the ground, in leaf litter, cracks and holes in limestone, hollow trees and beneath bark of fallen logs (Andriatsimiety *et al.* 2009). Covers up to 1.5km straight-line distance per night (Goodman 2009).

Diet dominated by invertebrates, especially hissing cockroaches (*Gromphadorhina* spp.) grass-hoppers and scorpions, with vertebrate prey like small lizards taken less frequently; tenrecs, (genera *Echinops*, *Geogale*) and mouse lemurs also recorded (Andriatsimiety *et al.* 2009). Even though far fewer vertebrates are eaten, they constitute *c.*60–80% of diet by biomass; a greater proportion of vertebrates is eaten in short wet season, presumably reflecting increased availability (Andriatsimiety *et al.* 2009).

Grandidier's Vontsira defecates at specific latrine sites, mostly on exposed rocks or similar prominent places, suggesting these are territorial markers (Wozencraft 1990). Wary when first approached, but very docile and soon becomes accustomed to human presence (Goodman 2003e).

Breeding regime unclear. Some evidence suggests no fixed reproductive season – lactating and pregnant females, and subadults at varying stages of maturation are known from different times of year (Goodman 2009). The high degree of climatic seasonality, and therefore prey base, would make this unusual. Counter to this, males show increased testes size in dry season (July–September) suggesting mating occurs at this time. Gestation period *c.*7–15 weeks with young therefore born at onset of wet season in December (Marquard *et al.* 2011).

WHERE TO SEE Known only from Tsimanampetsotse National Park, 80km south of Toliara. Best looked for in Euphorbiaceae scrub at base of escarpment on east side of lake, where locally common and easily seen at night.

NARROW-STRIPED BOKY
GENUS *Mungotictis*

The monotypic genus *Mungotictis* is one of two small native carnivorans in dry western forests; however the ranges of *Mungotictis* and Ring-tailed Vontsira do not overlap. Due to similarities in aspects of morphology and diet, genus *Mungotictis* may be considered an ecological equivalent of the rainforest-only Brown-tailed Vontsira.

NARROW-STRIPED BOKY
Mungotictis decemlineata (A. Grandidier, 1867)

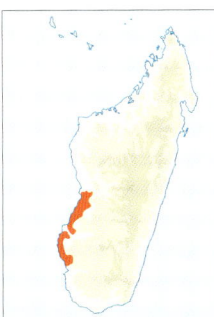

Other name: Narrow-striped 'Mongoose'

Malagasy name: Boky-boky

MEASUREMENTS
Head-body: 260–320mm. Tail: 190–240mm. Weight: males: 475–625g, females: 450–740g. (Woolaver 2006, Goodman 2009, Kappeler *et al.* in press b)

DESCRIPTION & IDENTIFICATION Pelage short and dense, longer under belly. Upperparts grizzled grey-beige fading to paler beige below. Back and flanks marked with 8–10 broadly spaced thin longitudinal dark stripes from nape to base of tail, often becoming indistinct. Tail bushy, brush-like and similar in colour to upper body but often greyer and more flecked, with alternate darker and lighter areas, giving a vaguely striped impression. Muzzle shortish and pointed, ears short and rounded. Legs short and delicate, feet webbed with long claws (Albignac 1976, Goodman 2009). Two subspecies, *M. d. decemlineata* and *M. d. lineata*, differentiated on basis of size and colour differences, and confirmed by genetic analysis (Veron *et al.* 2017): *M. d. lineata* differs from the nominate (described above) in being darker on the back with darker more distinct stripes and more russet on belly (Hawkins *et al.* 2000, Goodman *et al.* 2005d).

HABITAT & DISTRIBUTION Primary and fragmented deciduous forests of west and arid spiny forest of south-west, from sea level to *c.*400m (Woolaver 2006, Goodman 2009). Known range restricted to region between Tsiribihina River in north and Fiherenana River in south, inland possibly as far as Mahabo. *M. d. decemlineata* occurs from Tsiribihina River south to Mangoky River and is rare at southern extreme of deciduous forest (north of Mangoky River) (Woolaver *et al.* 2006). *M. d. lineata* occurs south of Mangoky River to Fiherenana River (Hawkins 2015d), and is extremely rare in spiny forest south of Mangoky River. Subfossil remains of *M. decemlineata*, *c.*500 years old, have been found in Ankilitelo Cave in south-west, some 150km south of the current known range (Muldoon *et al.* 2009). Habitat destruction and fragmentation are severe threats, as is predation by feral dogs. IUCN Endangered.

BEHAVIOUR Largely diurnal, terrestrial and arboreal (Razafimanantsoa 2003); more often encountered on ground, but is an accomplished climber. Prefers forest with dense undergrowth (Kappeler *et al.* in press b).

Lives in separate, same-sex social units. Female units are stable, matrilineal and generally comprise two adults, a juvenile and an infant of the dominant female (Schneider & Kappeler 2016, Schneider *et al.* 2016). Female members den together and aggressively defend territory boundaries from other female units, although females in neighbouring units are often related, suggesting females disperse less than males (Schneider *et al.* 2016). Males form less stable associations of 2–4 individuals, except during short mating season (July–October), when they are solitary (Schneider & Kappeler 2016). Social units are important so that learned behaviours are passed between individuals (Rasolofoniaina *et al.* 2020). In Kirindy Forest, population density *c.*20 individuals/km², including 15 adults (Schneider & Kappeler 2016).

Male social unit ranges (up to *c.*100ha) overlap significantly with one another and incorporate up to four female social unit ranges, which are smaller (*c.*35ha) but expand during the dry season when food is scarce (Schneider & Kappeler 2016). Narrow-striped Boky may be opportunistically predated by large raptors (Madagascar Harrier-Hawk, Henst's Goshawk), large snakes (Madagascar Ground Boa), Fosa and possibly feral dogs (Hawkins & Racey 2008, Kappeler *et al.* in press b). It is believed that their unusual social system helps reduce predation (Schneider & Kappeler 2016).

Narrow-striped Boky, Kirindy Forest.

Narrow-striped Boky often erects its bushy tail when alert, Kirindy Forest.

Also hunted for food around some villages (Goodman & Raselimanana 2003).

The vocal repertoire seems limited. After birth young emit a shrill call, which is similar to that of adult males when calling to females: the latter sounds like *bouk-bouk*, the Malagasy name, Boky-boky being onomatopoeic. When alert, often erects its bushy tail in a pulsing fashion, which clearly serves in visual communication.

Forages primarily in leaf litter and by digging in topsoil but also climbs and excavates in fallen decaying wood, particularly in wet season (Rasolofoniaina *et al*. 2019). Can cover in excess of 2km per day. Year-round, diet dominated by insects, insect larvae and other invertebrates, particularly during dry season with emphasis on giant cockroaches (*Blaberidae*), beetles (*Coleoptera*), grasshoppers and crickets (*Orthoptera*) (Rasolofoniaina *et al*. 2019). Diet slightly more varied in rainy season, includes small mammals (including tenrecs, rodents, small lemurs), reptiles, small scorpions, worms and other invertebrates. Land snails also feature during wet season; these are broken by being thrown into air or knocked against a tree (Razafimanantsoa 2003), or are left in direct sun forcing the occupant out to escape the heat (Rasolofoniaina *et al*. 2019). Same-sex social units may hunt larger prey cooperatively, including mouse lemurs, giant mouse lemurs and dwarf lemurs (Rabeantoandro 1997, Goodman 2003a).

In cool dry season (June–October) social units prefer to den at ground level in abandoned ant nests or collapsed burrows, then at onset of rainy season (November) switch to holes in fallen logs above ground or tree holes up to 10m off the ground. At no time are shelters fixed; sleep sites change regularly to avoid predation and reduce parasitic loads (Razafimanantsoa 2003).

Most mating occurs in August and September, but is flexible: females that lose their infant in first month, mate again (December–February) (Kappeler *et al*. in press b). When courting, males wait early in morning at den entrance of dominant female for her to emerge. Females can be vocal and initially aggressive towards males, but gradually calm and eventually the pair mate up to three times (Goodman 2009). Dominant female mates before other adult females in social unit. Gestation period *c*.90–105 days (Razafimanantsoa 2003). A single infant is born weighing *c*.50g, with ears and eyes open (Zahmel 2002). Young precocious, walking within a day, but mother may carry them in her mouth when moving between dens and sleep sites. Eat solids at 15 days, continue to suckle for two months. Agile climbers by 45 days, and can hunt for themselves at three months (Goodman 2009). Fewer than 30% of young survive their first year (Kappeler *et al*. in press b). Males disperse away from natal unit at *c*.2 years, females at around three years, or they remain in natal unit to begin breeding (Schneider & Kappeler 2016). If dominant female dies, she is replaced by new female from outside unit.

WHERE TO SEE Kirindy Forest, north-east of Morondava, is best place to see this species. Here it is locally quite common. During a one- or two-day stay the chances of success are very good. Alternatively, try Kirindy-Mitea National Park, *c*.100km south of Morondava.

Rats and Mice

Order Rodentia

Rodents comprise c.40% of all known mammal species, making the order Rodentia by far the most speciose order of mammals, with c.2,500 species in 34 families. They naturally occur on all continents, except Antarctica, but as natives they were absent from New Zealand and many other oceanic islands (Wilson *et al.* 2017). Evolutionary relationships within the order remain very much in question and are constantly being reassessed, resulting in regular taxonomic changes, with the descriptions of new species and previously recognised single forms being split into multiple 'cryptic' species.

Three broad divisions exist within Rodentia: the sciuromorphs (with squirrel-like skull morphology), the hystricomorphs (porcupine-like skull morphology) and suborder Myomorpha (mouse-like skull morphology). Only the latter, represented by a single family, Nesomyidae, is native to Madagascar. Three species from the family Muridae have been accidentally introduced by man, Black Rat *Rattus rattus*, Brown Rat *R. norvegicus* and House Mouse *Mus musculus* (see Non-native Species, p.367).

Pouched Rats, Climbing Mice and Fat Mice
Family Nesomyidae

The native rodents of Madagascar constitute a remarkable assemblage, which display a considerable diversity of body forms and lifestyles (Goodman *et al.* 2003b). On the face of it, their only common feature appears to be cohabiting the same island. Consequently, their evolutionary affinities have been contentious and previous taxonomies based primarily on morphology have been varied and confusing. Using morphological characteristics resulted in the different genera on Madagascar being aligned with widely different rodent groupings elsewhere in the world. This implied that rodents in Madagascar had diverse origins and must have been the consequence of multiple independent colonisations of the island.

However, recent molecular DNA analysis confirms that Malagasy rodents constitute a classic example of adaptive radiation derived from a single ancestral species (Jansa & Carleton 2003b, Poux *et al.* 2005) and, therefore, resemblances to rodents elsewhere are the result of convergent evolution. Nonetheless, in a broader sense Madagascar's rodents do exhibit affinities with other rodent groupings in Africa. As such they are placed in the subfamily Nesomyinae, within the family Nesomyidae, which contains five other subfamilies restricted to sub-Saharan Africa: Delanymyinae (Delany's Swamp Mouse), Mystromyinae (African White-tailed Rat), Petromyscinae (pygmy rock mice), Cricetomyinae (pouched rats and pouched mice) and Dendromurinae (climbing mice and fat mice).

Malagasy Rats and Mice Subfamily Nesomyinae

Molecular evidence points to a single rodent ancestor arriving on Madagascar by rafting across the Mozambique Channel from Africa 20–24 mya (Poux *et al.* 2005). Subsequently, a spectacular radiation has taken place, such that the endemic subfamily Nesomyinae currently contains 28 extant species in nine genera (Jansa & Carleton in press). A further three species, now extinct, are known only from subfossil remains (see The Extinct Mammals of Madagascar).

There are significant differences between genera, but close relationships within them. For an island of its size, Madagascar has fewer rodent species than might be expected (Borneo and New Guinea, both slightly larger than Madagascar, each support 55–60 species) (Carleton & Goodman 1998). There are two possible explanations for this apparent paucity. Either the rodent fauna of Madagascar is relatively impoverished or the inventory is incomplete. While it is probable new species await discovery (especially cryptic species), it seems unlikely rodent diversity on Madagascar matches that seen on Borneo and New Guinea (Goodman *et al.* 2003b).

In the relatively recent past two new genera have been discovered and described, *Monticolomys* and *Voalavo* (Carleton & Goodman 1996, 1998), as have a number of new species (Carleton's Tufted-tailed Rat *Eliurus carletoni*, Daniel's Tufted-tailed Rat *E. danieli*, Rock-loving Tufted-tailed Rat *E. tsingimbato* and Eastern White-tailed Mountain Mouse *Voalavo antsahabensis*), with further species likely awaiting description. The anticipated eventual total of endemic rodents is perhaps 30–35 forms (Goodman *et al.* 2003b).

Malagasy rodents are exceptionally diverse in their morphology and ecology: there are great differences in body form, size and behaviour. This striking variety is underlined by comparisons that have been drawn to their appearance and behaviour, some of the genera and species being likened to gerbils (*Macrotarsomys*), voles (*Brachyuromys*), arboreal dormice (*Eliurus*), squirrels (*Nesomys lambertoni*) and even lagomorphs (*Hypogeomys*) (Goodman & Monadjem 2017, Jansa & Carleton in press). Nevertheless, they do share certain features: all prefer native forest habitats (they rarely occur in deforested or degraded areas) and most appear to be predominantly nocturnal.

Madagascar's rodents are known from all native forest types, from lowland littoral forests to extremely wet rainforests, to high-mountain zones above the treeline to arid subdesert spiny forest. The humid forests of the east harbour the greatest diversity, with seven genera (*Brachytarsomys*, *Brachyuromys*, *Eliurus*, *Gymnuromys*, *Monticolomys*, *Nesomys* and *Voalavo*), whilst deciduous forests of the west are home to three genera (*Eliurus*, *Macrotarsomys* and *Hypogeomys*) and arid spiny forests of the south and south-west just two (*Macrotarsomys* and *Eliurus*).

Like the vast majority of native mammal species, the primary threat to the island's rodents is habitat destruction and fragmentation, an issue made more acute by the dependence of Nesomyinae on native forests. Madagascar has lost as much as *c.*80% of its original forest cover (Green & Sussman 1990), with perhaps a 70% decline since 1950 (Harper *et al.* 2007) as human populations have grown exponentially and subsistence agriculture has expanded.

In addition, native rodents suffer adversely from competition from non-native species, especially the Black Rat which invades deep into native forest zones. Not only do introduced rats out-compete native species for space, food and other resources, they also spread disease: all Nesomyinae species, especially those found above 800m are susceptible to 100% mortality from rodent-introduced plague – although these seem to be localised events (Kennerley 2016a). Native rodents also suffer direct predation from feral dogs and cats.

Unlike other native mammal groups, Nesomyinae suffer relatively little from bushmeat hunting. For the most part, rodent consumption does not seem to be culturally acceptable in Madagascar. There are exceptions: in some localities White-tailed Tree Rat *Brachytarsomys albicauda* is eaten and in the Andasibe region Eastern Red Forest Rat *Nesomys rufus* is hunted and consumed (Kennerley 2016a, Dolch 2021).

MALAGASY WHITE-TAILED TREE RATS
GENUS *Brachytarsomys*

A genus of two highly distinctive species of large nocturnal arboreal rats with a number of traits that betray their scansorial habits, most notably a long flexible tail for counterbalance, and exceptionally short but broad feet with narrow sharp claws. The genus *Brachytarsomys* has thick, soft woolly fur. The upperparts are brownish-grey to dark grey and the underparts grey-white to almost pure white. The tail is obviously longer than the head-body with a distinctive white tip. Ears are round and relatively tiny, hardly protruding above the surrounding fur (Carleton & Goodman 2003a, Carleton & Goodman in press a).

The two species are distinguished mainly by size, Hairy-tailed Tree Rat *B. villosa* being substantially larger, but also by the colour and shade of the upperparts and more subtle tail characteristics.

During the day, holes and hollows in standing trees (usually within 2.5m of the ground) are used as sleep sites. They are thought to feed mainly on fruit. Because of their nocturnal and arboreal habits, the Malagasy tree rats are not commonly encountered in the wild (Carleton & Goodman 2003a).

An extinct species of tree rat, *B. mahajambaensis* has been described from subfossil remains unearthed north of Mahajunga in the dry north-west. Aged at *c.*12,000 years old, *B. mahajambaensis* was smaller than any extant congener. Given the genus *Brachytarsomys* is today restricted to humid forest regions, this suggests the area of discovery was once much wetter and that western Madagascar has since undergone significant climate change (Mein *et al.* 2010).

WHITE-TAILED TREE RAT
Brachytarsomys albicauda Günther, 1875

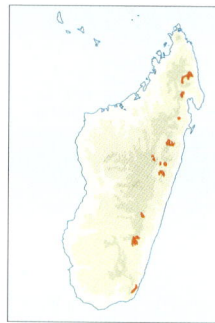

Other name: White-tailed Antsangy

Malagasy names: Antsangy, Angatra

MEASUREMENTS
Head-body: *c.*200–240mm.
Tail: *c.*210–260mm.
Weight: *c.*175–285g, mean 200–220g. (Carleton & Goodman 2003a, Goodman & Monadjem 2017, Carleton & Goodman in press a)

DESCRIPTION & IDENTIFICATION Pelage slightly shaggy, with woolly texture. Upperparts greyish-brown, more rufous on flanks, underparts and feet cream to off-white. Head more tawny to buffy-brown. Snout quite short, giving head rather blunt appearance. Ears relatively small, only just protruding above fur. Five rows of longish dark whiskers. Eyes largish and sometimes bulbous. Tail prehensile, appears mainly naked and scaly (but covered in short fine hairs); dark over most of length, but last 8–10mm are white. Claws prominent and hindfoot has extra-long fifth digit (Carleton & Goodman 2003a).

Typical 'rat-like' appearance, similar to red forest rats (genus *Nesomys*) but with much shorter snout, very pale underparts and exclusively arboreal habits (red forest rats are terrestrial). Can appear quite 'lemur-like' when moving along branches (Goodman & Monadjem 2017, Carleton & Goodman in press a). Distinguished from Hairy-tailed Tree Rat by smaller size, more rufous-brown upperparts and hairless tail (Carleton & Goodman 2003a).

HABITAT & DISTRIBUTION Native forest dependent: occurs in lowland and mid-altitude montane rainforest between 450 and *c.*1,900m (Soarimalala & Goodman 2011). Distribution appears patchy, being known from dispersed localities covering full extent of rainforest region (Goodman *et al.* 2013), from Marojejy and Anjanaharibe-Sud in north, to Tsitongambarika and Andohahela in south (Carleton

White-tailed Tree Rat, Tsitongambarika Forest.

& Schmidt 1990, Goodman & Carleton 1998, Goodman *et al.* 2001, Goodman *et al.* 2018) and may extend further south to Andohahela (Goodman *et al.* 1999a). Unlike most other rodents, White-tailed Tree Rats are hunted for food in some localities, probably because local people liken them to small nocturnal lemurs. IUCN Least Concern.

BEHAVIOUR A poorly known species. Nocturnal and probably wholly arboreal, moving freely on all aerial substrates from tree trunks to narrow branches and lianas, at all levels from understorey to high canopy (Carleton & Goodman in press a). They inhabit tree holes 2.5–7m above ground (Carleton & Goodman 2003a, Carleton & Goodman in press a), generally close to dense tangles of lianas and vines that allow easy access to canopy. Has been recorded sharing a tree hole with similarly nocturnal Hairy-eared Dwarf Lemur (Biebouw *et al.* 2009). If disturbed may appear at hole entrance and chatter. Diet mainly various canopy fruits, and seeds which may be cached in tree holes. Litters of six produced in captivity (Carleton & Goodman 2003a).

WHERE TO SEE Rarely seen well. Sometimes observed in Andasibe area, e.g. Mitsinjo Forest, or forests at Anjozorobe. Given potential confusion with small nocturnal lemurs, it is worth being vigilant on night walks in any rainforest reserve within the species' range.

HAIRY-TAILED TREE RAT
Brachytarsomys villosa Petter, 1962

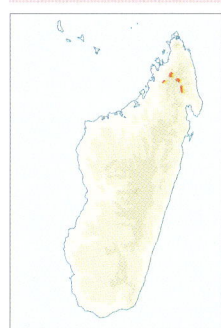

Other name: Hairy-tailed Antsangy

Malagasy names: Antsangy, Angatra

MEASUREMENTS
Head-body: 228–245mm. Tail: 260–272mm*. Weight: 236–350g, mean 290–300g. (Carleton & Goodman 2003a) (*two individuals)

DESCRIPTION & IDENTIFICATION Pelage soft with woolly texture. Upperparts, including head, grey-brown, blending to pale grey on flanks, then pale cream to white on underparts. Some show rufous tinges on flanks. Ears short, rounded and relatively small. Long prehensile tail very distinctive: bushy and densely covered in soft dark hairs of uniform length (6–8mm), with white tip (last 8–10mm) (Carleton & Goodman 2003a, Carleton & Goodman in press a). Larger than White-tailed Tree Rat with very obviously hairy tail.

HABITAT & DISTRIBUTION Mid- and higher-elevation montane rainforest at *c.*1,200–2,050m. Dependent on native forest. Currently known from only a handful of sites in northern mountains and highlands (Anjanaharibe-Sud, COMATSA Nord, Bemanevika, Tsaratanana) (Goodman *et al.* 2001b, Carleton & Goodman 2003a, Maminirina *et al.* 2008, Goodman & Raherilalao 2013). Future surveys may extend the known range. IUCN Vulnerable.

BEHAVIOUR Nocturnal and arboreal. Probably feeds on fruit and seeds. Male and female reproductive condition suggests breeding occurs in late October/ November (Carleton & Goodman 2003a). Females have three pairs of mammae, and litters up to six. Subadults observed in April (Goodman & Monadjem 2017). Other behaviour unknown, but probably similar to White-tailed Tree Rat.

Hairy-tailed Tree Rat, Anjanaharibe-Sud.

SLEEK-FURRED GROUND RAT

GENUS *Gymnuromys*

A distinctive monotypic genus. Since its discovery and description in 1896, Sleek-furred Ground Rat *G. roberti* has remained mysterious, with virtually nothing known of its behaviour and ecology, but surveys during the last 30 years have begun to shed light on the species' distribution and natural history (Carleton & Goodman 2003b). Within Nesomyinae, most closely related to tufted-tailed rats (genus *Eliurus*) and white-tailed mountain mice (genus *Voalavo*) (Jansa et al. 2009, Jansa & Carleton in press). Recent DNA analyses also allude to genetic divergence between populations in the northern highlands, which may lead to the recognition of greater taxonomic diversity in the future (Everson et al. 2020b, Carleton & Goodman in press b).

SLEEK-FURRED GROUND RAT
Gymnuromys roberti   Forsyth Major, 1896

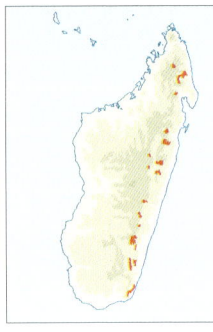

Other name: Robert's Forest Rat

Malagasy name: Voalavoanala

MEASUREMENTS
Head-body: 149–175mm.
Tail: 176–197mm. Weight: 98–155g, mean 120–140g.
(Carleton & Goodman 2003b, Carleton & Goodman in press b)

DESCRIPTION & IDENTIFICATION Medium-sized typical rat-like rodent with stout body and relatively long tail (110–115% head-body length). Pelage short, dense and sleek: upperparts dark grey to slate-grey, underparts grey-white with silvery sheen. Ears oval, moderately large and clearly protrude. Vibrissae dark and long (50–60mm). Tail sparsely haired and bicoloured: dark grey on top and pale grey to white below, with final 25–30% often white. No terminal tuft (Carleton & Goodman 2003b). Superficially resembles young Black Rat and possibly some tufted-tailed rats, particularly Major's Tufted-tailed Rat. *E. majori*. Fur of Major's Tufted-tailed Rat is woollier and tail appreciably longer (>120% head-body length) ending in bushy tuft (Carleton & Goodman 2003b).

HABITAT & DISTRIBUTION Lowland and montane forest locations along eastern flank of Central Highlands, from Anjanaharibe-Sud and Marojejy in north (Carleton & Schmidt 1990, Stephenson 1993, Goodman & Carleton 1998, Carleton & Goodman 2000) to Anosyenne Mountains in south (Goodman et al. 1999a). Recorded between c.500 and 1,800m (Carleton & Goodman 2003b), but most sites at c.900–1,600m, and nowhere is it common (Carleton & Goodman in press b). Never recorded outside native forest, but some records from secondary environments (Carleton & Goodman 2003b). Recorded in sympatry with red forest rats (*Nesomys* spp.) and several species of tufted-tailed rats (*Eliurus* spp.). IUCN Least Concern.

BEHAVIOUR Appears exclusively terrestrial and nocturnal. Probably forages for fallen seeds and fruits in leaf litter, among roots, trunks and fallen logs (Carleton & Goodman 2003b). Lives in burrows up to 1m deep, often under fallen logs. Tunnel relatively straight and ends in small chamber used for food storage: one contained 22 cached high-energy value Ramy nuts *Canarium madagascariensis*, 21 of which were gnawed open and cotyledons eaten (Carleton & Goodman 2003b). Breeding biology largely unknown; males suspected of being reproductively active in October–December (Carleton & Goodman 2003b). Litter size small. Probably suffers increasing

Sleek-furred Ground Rat, Ranomafana National Park.

competition from introduced Black Rat (Carleton & Goodman in press b).

WHERE TO SEE Nocturnal habits make observation difficult: chance sightings possible on night walks in locations like Marojejy, Andasibe-Mantadia and Ranomafana.

TUFTED-TAILED RATS

GENUS *Eliurus*

The tufted-tailed rats are Madagascar's most diverse group of rodents. On pristine Madagascar, the genus was probably ubiquitous and today they are found wherever there are significant patches of native/primary forest (Goodman *et al*. 2003, Carleton *et al*. in press b). Population densities are highest in pristine forests, although some species can tolerate a degree of forest degradation, e.g. Webb's Tufted-tailed Rat *E. webbi*. However, tufted-tailed rats soon disappear when disturbance becomes substantial (Ganzhorn *et al*. 2003).

They are strictly nocturnal, small to medium in size, and share various distinctive morphological features associated with a scansorial (climbing) way of life: most noticeable are broad hindfeet with a long outer digit and cushion-like plantar pads, and a long tail, which is not prehensile. In all species, tail length exceeds head-body length (*c*.115–130%). The tail is sparsely haired over its proximal portion and ends in a conspicuous terminal tuft or 'pencil' that covers the distal half to quarter of the tail (Carleton 2003, Carleton *et al*. in press b). In four species, the terminal tuft is white and particularly distinctive: Daniel's Tufted-tailed Rat *E. danieli*, Grandidier's Tufted-tailed Rat *E. grandidieri*, White-tailed Tufted-tailed Rat *E. penicillatus* and Tanala Tufted-tailed Rat *E. tanala*.

In all species the pelage is soft and fine, consisting of a thick coat of cover hairs, interspersed by longer, darker, coarse guard hairs. The dorsal pelage is usually longer and denser than ventrally. Overall, upperparts appear a shade of brown to brownish-grey, and underparts are generally creamy-grey to buff, with a quite abrupt transition between them. In two species (Major's Tufted-tailed Rat and White-tailed Tufted-tailed Rat), this is not always so and dorsal and ventral fur may be similar shades with no pronounced lateral line. Dark hairs around the eye often give the impression of an eye-ring, and facial vibrissae are well developed. The sexes are similar in size.

All tufted-tailed rats are adept climbers and are routinely found above ground among lianas, branches and vines. However, they are also seen much lower down among boulders, tree buttresses, fallen logs, tangled roots and even leaf litter.

At present 13 species are recognised: eight species in eastern rainforest regions and five in dry forests and arid scrub/spiny forest in the west and south (Carleton *et al*. 2001, Carleton 2003, Carlton & Goodman 2007, Goodman *et al*. 2009, Jansa *et al*. 2019). External differences between species principally involve size, relative tail length, colour and texture of the fur, and colour, extent and characteristics of the tail tuft. However, precise identification in the field can be challenging as visible differences between species are often subtle and the ranges of measurable variables often overlap. In most locations two or more species may occur sympatrically, and in rainforest locations this may be as many as 4–6 species, although there is elevational allopatry (Carleton *et al*. in press b).

The application of molecular genetic techniques has advanced the understanding of evolutionary relationships: The genus *Eliurus* has been provisionally divided into five species groups (Carleton & Goodman 2007, Jansa *et al*. 2019).

- *Eliurus antsingy* group (*E. antsingy* and *E. carletoni*)
- *Eliurus majori* group (*E. danieli, E. majori* and *E. penicillatus*)
- *Eliurus myoxinus* group (*E. minor* and *E. myoxinus*)
- *Eliurus petteri* group (*E. grandidieri* and *E. petteri*)
- *Eliurus tanala* group (*E. ellermani, E. tanala, E. tsingimbato* and *E. webbi*)

Nonetheless, uncertainty surrounding these relationships remains and the validity of certain species is questionable, e.g., White-tailed Tufted-tailed Rat may not be distinct from Major's Tufted-tailed Rat. Conversely, some existing species, e.g. Grandidier's Tufted-tailed Rat, Western Tufted-tailed Rat *E. myoxinus* and Webb's Tufted-tailed Rat may warrant further division and description of new taxa (Carleton *et al*. in press b).

Birds of prey and mammalian carnivorans prey heavily upon tufted-tailed rats. In some areas they constitute *c*.60% of the diet of some owls, including Madagascar Long-eared Owl, Madagascar Red Owl and Barn Owl (Goodman *et al*. 1993, Goodman & Thorstrom 1998). Madagascar Harrier-Hawk has been observed extracting tufted-tailed rats from tree holes and crevices with its long legs (Carleton *et al.* in press b). The remains of tufted-tailed rats have also been found in the scats of Fosa (Goodman *et al.* 1997, Hawkins 2003). In western forests, there are also records of predation by large snakes, e.g. Madagascar Hog-nosed Snake *Leioheterodon madagascariensis* and Madagascar Cat Snake *Madagascarophis colubrinus* (Randrianjafy 2003).

The nocturnal and arboreal habits of tufted-tailed rats make them a difficult group to observe in the wild. However, they can sometimes be seen on night walks in places like Andasibe-Mantadia, Ranomafana, Marojejy, Masoala, Daraina, Kirindy and Ankarafantsika. The majority of records are from specimens caught in traps placed in the lower branches of the forest understorey.

TSINGY TUFTED-TAILED RAT
Eliurus antsingy   Carleton *et al.*, 2001

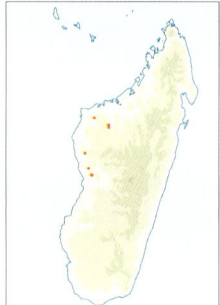

Malagasy names: Voalavo ala, Sokitralina

MEASUREMENTS
Head-body: 142–160mm. Tail: 153–195mm. Weight: 87–131g. (Carleton *et al.* 2001, Goodman *et al.* 2009)

DESCRIPTION & IDENTIFICATION A relatively large tufted-tailed rat. Pelage soft and fine. Upperparts drab brown to blackish-grey, underparts between white and grey-white. Head large and domed.

Tsingy Tufted-tailed Rat, near Daraina.

Hindfeet broad. Tail tuft well developed, covering distal 40–50% (longest hairs 12mm). Tuft darkens noticeably towards tip (Carleton *et al.* 2001).

HABITAT & DISTRIBUTION Known only from deciduous forests (fewer than ten locations) associated with limestone karst (*tsingy*) formations in centre-west, at *c.*100–200m, including Bemaraha and Namoroka National Parks (Carleton *et al.* 2001, Goodman & Raherilalao 2013). Recorded sympatrically with Western Tufted-tailed Rat and Rock-loving Tufted-tailed Rat. IUCN Data Deficient.

BEHAVIOUR Nocturnal and unlike congeners appears to be terrestrial some of the time. Found in limestone outcrops and associated forest. Has been caught in trees near rocky outcrops. Feeds on seeds and grains (Goodman & Monadjem 2017).

WHERE TO SEE Can possibly be seen on night walks in Tsingy de Bemaraha National Park.

CARLETON'S TUFTED-TAILED RAT
Eliurus carletoni   Goodman *et al.*, 2009

Malagasy names: Voalavo ala, Sokitralina

MEASUREMENTS
Head-body: 143–150mm. Tail: 164–183mm. Weight: 88–99g. (Goodman *et al.* 2009c)

DESCRIPTION & IDENTIFICATION A relatively large tufted-tailed rat. Fur fine and soft. Upperparts, including head and face, dark grey-brown (agouti); head and face paler brown in some individuals. Clear differentiation with white or greyish-white underparts. Ears relatively short. Tail longer than head-body length (115–120%), final third becoming increasingly covered in dark hairs to end in prominent tuft (hairs 10–13mm). In some individuals tail tuft is white.

HABITAT & DISTRIBUTION Restricted to far north, in dry forests associated with limestone karst (*tsingy*) formations (Ankarana) and similar dry forests of Analamerana and Loky-Manambato (Daraina) regions. Recorded at *c.*50–550m (Goodman & Raherilalao 2013, Rakotoarisoa *et al.* 2010, 2013a). Occurs in sympatry with Elleman's Tufted-tailed Rat, Lesser Tufted-tailed Rat and Western Tufted-tailed Rat. IUCN Least Concern.

BEHAVIOUR Nocturnal and scansorial. Known to climb in branches and among rocks in limestone formations. Presumed to feed primarily on seeds. Probably breeds at end of dry season, young born late November–December. In Ankarana *c.*50% of diet of Madagascar Red Owl *Tyto soumagnei* is *E. carletoni* (Cardiff & Goodman 2008).

WHERE TO SEE Reported as common at some localities, particularly in Loky-Mananbato (Daraina) area (Goodman & Raherilalao 2013). Any large tufted-tails rats (noticeably larger than Western Tufted-tailed Rat) seen on night walks in Ankarana, Analamerana and Daraina region are likely to be this species.

Carleton's Tufted-tailed Rat, Ankarana Special Reserve.

DANIEL'S TUFTED-TAILED RAT
Eliurus danieli   Carleton & Goodman, 2007

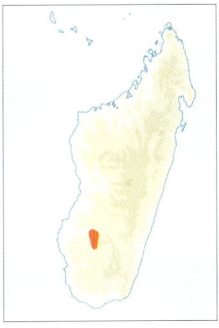

Malagasy names: Voalavo ala, Sokitralina

MEASUREMENTS
Head-body: 150–152mm. Tail: 179–195mm*. Weight: 91–100g. (Carleton & Goodman 2007) (*two individuals)

DESCRIPTION & IDENTIFICATION A relatively large tufted-tailed rat with most visually striking white tail tuft. Fur soft and fine. Upperparts grey to grey-brown with clear contrast versus greyish-white underparts. Ears relatively large and prominent. Tail *c.*120% head-body length and distinctive: initially appears bare (very fine dark hairs), with very dark hairs emerging halfway along its length. These become longer and change to long white hairs (12–15mm) over final one third. Overall effect is prominent white tuft covering final 30% which is accentuated by dense dark hairs preceding it (Carleton & Goodman 2007).

HABITAT & DISTRIBUTION Currently known only from sandstone Isalo and Makay Massifs in central south-west. Found in dry deciduous/local humid

Daniel's Tufted-tailed Rat, Isalo National Park.

forest/gallery forests in sandstone canyons between 650 and 700m (Carleton *et al.* in press b). Has also been found in adjacent human-altered habitat. Western Tufted-tailed Rat also recorded in Isalo Massif. IUCN Least Concern.

BEHAVIOUR Nocturnal and probably partially terrestrial. Specimens have been caught at base of sandstone rocky outcrops within forest.

ELLERMAN'S TUFTED-TAILED RAT
Eliurus ellermani   Carleton, 1994

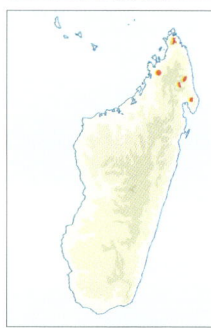

Malagasy names: Voalavo ala, Sokitralina

MEASUREMENTS
Head-body: 126–174mm.
Tail: 150–204mm. Weight:
55–120g. (Jansa *et al.* 2019)

DESCRIPTION & IDENTIFICATION Moderately large. Upperparts drab dark grey-brown, underparts dull creamy to off-white. Pelage longish and relatively coarse. Noticeably dark eye-ring and upper incisors are bright orange. Tail moderately long (115% head-body length) and tail tuft extends over final third. Tuft well developed (hairs 9–11mm long) and dark all the way to tip. Very similar to Tanala Tufted-tailed Rat: taxonomic validity may be questionable (Carleton & Goodman 2000).

HABITAT & DISTRIBUTION Known only from localities in north and north-east, ranging from

Grandidier's Tufted-tailed Rat, Anjanaharibe-Sud.

Montagne d'Ambre, through northern highlands to higher-elevation zones on Masoala Peninsula. Recorded in lowland and mid-elevation rainforest at *c.*400–1,570m, but mostly 800–1,200m (Carleton *et al.* in press b). Recorded sympatrically with Carlton's Tufted-tailed Rat, Grandidier's Tufted-tailed Rat, Major's Tufted-tailed Rat, Lesser Tufted-tailed Rat and Webb's Tufted-tailed Rat. IUCN Data Deficient.

BEHAVIOUR Nocturnal and scansorial. Other aspects unknown.

GRANDIDIER'S TUFTED-TAILED RAT
Eliurus grandidieri   Carleton & Goodman, 1998

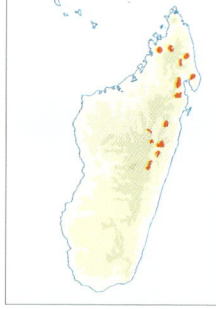

Malagasy names: Voalavo ala, Sokitralina

MEASUREMENTS
Head-body: 111–164mm.
Tail: 144–176mm. Weight:
42–62g. (Carleton & Goodman 1998, Carleton 2003)

DESCRIPTION & IDENTIFICATION Medium-sized tufted-tailed rat. Dorsal pelage sleek, fine-textured and sooty-brown to charcoal-grey, tending to be darker on back and more brown-grey on flanks. Underparts medium to pale grey; distinct but not contrasting with upperparts. Tail relatively long (130% of head-body length), tuft hair white, so distinctive, but not well developed, extending only over final 30% of tail. Muzzle proportionately longer than congeners. Ears also relatively long (Carleton & Goodman 1998).

HABITAT & DISTRIBUTION Known from numerous dispersed localities in mid-montane, upper montane and sclerophyllous forests of eastern-central and northern highlands. Recorded across wide elevational range, *c.*900–2.050m, but most records at 1,200–1,600m, where it appears particularly abundant (Carleton & Goodman 2000, Goodman & Carleton 1998). Southerly limit appears to be Tsinjoarivo area, northerly limits forest around Andapa Basin (Anjanaharibe-Sud and Marojejy) and Tsaratanana (Goodman & Carleton 1998, Goodman *et al.* 2000, 2013). Occurs sympatrically with Tanala Tufted-tailed Rat, Ellerman's Tufted-tailed Rat,

Major's Tufted-tailed Rat, Lesser Tufted-tailed Rat and Webb's Tufted-tailed Rat. (Carleton *et al.* in press b). IUCN Least Concern.

BEHAVIOUR Probably least scansorial humid forest tufted-tailed rat; climbs infrequently. Prefers environs of forest floor, around tree roots, fallen logs and rocks. Frequently trapped in front of hollows and tunnels, suggesting it may occupy burrows (Carleton & Goodman 2000). Breeding varies between locations and may be linked to elevation: in some areas females give birth in late August–September (Goodman & Monadjem 2017). Evidence from Anjanaharibe-Sud and Marojejy suggests breeding in October and November. Litter size up to three (Carleton & Goodman 2000, Goodman & Carleton 1998).

MAJOR'S TUFTED-TAILED RAT
Eliurus majori   Thomas, 1895

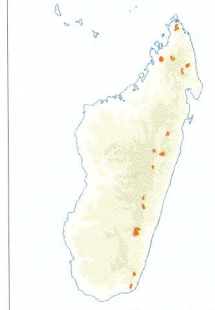

Malagasy names: Voalavo ala, Sokitralina

MEASUREMENTS
Head-body: 145–171mm. Tail: 170–207mm. Weight: 77–122g. (Goodman & Carleton 1996, Carleton 2003)

DESCRIPTION & IDENTIFICATION A relatively large *Eliurus*. Pelage soft, dense and noticeably woolly. Upperparts sombre slate-grey to blackish-brown, underparts mid to pale grey, but little or no contrast between them. Dark eye-ring often distinct. Tail relatively short (c.115% head-body length). Tuft less developed than in congeners, with a simple gradual increase in length and density of dark hairs towards tail tip. In some individuals tail tuft is white.

Dark coat, mainly uniform coloration and less developed tail tuft help prevent confusion with all congeners, except perhaps White-tailed Tufted-tailed Rat. Pelage colour and morphology of Major's Tufted-tailed Rat and White-tailed Tufted-tailed Rat are remarkably similar and their separate species status may be questionable (Carleton *et al.* in press b), but latter species has characteristic white tail tip.

HABITAT & DISTRIBUTION Restricted to mid-elevation and montane forests, and upland sclerophyllous vegetation (mossy forest) at elevations of c.875–2,500m (Carleton 2003, Carleton *et al.* in press b). Recorded from montane locations along extent of eastern highlands, from Montagne d'Ambre in north to Andohahela in south, including Tsaratanana, Anjanaharibe-Sud, Marojejy, Ambohimitambo and Andringitra (Raxworthy & Nussbaum 1994, Goodman & Carleton 1996, 1998, Goodman *et al.* 1999a, Carleton & Goodman 2000, Maminirina *et al.* 2008). Individuals from isolated population at Montagne d'Ambre are relatively small compared to those at other locations (Carleton & Goodman 2007). Recorded in sympatry with Grandidier's Tufted-tailed Rat, Lesser Tufted-tailed Rat, Ellerman's Tufted-tailed Rat, Tanala Tufted-tailed Rat and Webb's Tufted-tailed Rat only on Montagne d'Ambre (Carleton *et al.* in press b). IUCN Least Concern.

BEHAVIOUR Nocturnal and principally arboreal, with occasional forays to ground level. At mid-elevations (c.1,200–1,400m) most specimens are caught in vines, lianas or tangled lower branches. At higher altitudes (above c.1,600m), most still captured above ground, but some caught on forest floor (Goodman & Carleton 1996, 1998, Goodman *et al.* 1999a). Breeding and births probably occur between late October and December. Litter size up to five. Recorded predators include Fosa (Goodman *et al.* 1997).

LESSER TUFTED-TAILED RAT
Eliurus minor   Forsyth Major, 1896

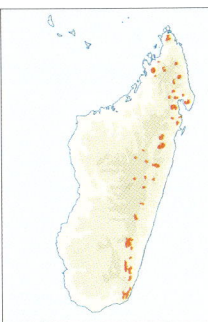

Malagasy name: Voalavo ala, Sokitralina

MEASUREMENTS
Head-body: 101–124mm. Tail length: 119–137mm. Weight: 22–50g. (Carleton 2003)

DESCRIPTION & IDENTIFICATION Smallest tufted-tailed rat. Upperparts light greyish-brown to rich cinnamon-brown, underparts pale grey to cream-beige. Upperparts obviously paler than upperparts: transition abrupt. Tail tuft dark brown to blackish and well developed, covering 50% of tail, increasing in density at tip. Ears proportionately short. Easily recognised by small size, longish tail (120% head-body length) and well-developed dark tail tuft. There is considerable variation within this taxon and Lesser Tufted-tailed Rat may, in the future, be separated into subspecies or even species (Carleton 2003).

HABITAT & DISTRIBUTION Extensive range over entire eastern rainforest belt, from Montagne d'Ambre in north to Anosyenne Mountains in south (Carleton 2003, Carleton et al. in press b). Most widely distributed and ecologically versatile tufted-tailed rat with greatest elevational tolerance of any Nesomyine rodent; from sea level to c.2,030m, but mid-elevations preferred (Goodman & Carleton 1996, 1998, Goodman et al. 1999a, Carleton & Goodman 2000). Also recorded in degraded and disturbed forests (Goodman & Monadjem 2017). Recorded sympatrically with most other rainforest tufted-tailed rat species and with Western Tufted-tailed Rat in transitional lowland deciduous forests in far north (Carleton et al. 2022b). IUCN Least Concern.

BEHAVIOUR Nocturnal and scansorial but utilises wide variety of terrestrial and arboreal microhabitats. Probably sleeps in ground dens but forages above ground among vines and lianas in understorey. Observed climbing to several metres (Goodman & Carleton 1996, 1998, Goodman et al. 1999a, Carleton & Goodman 2000). Probably breeds during wet season, young born in November–December. Litter size up to four. Recorded predators include Madagascar Long-eared Owl, Madagascar Red

Lesser Tufted-tailed Rat, Marojejy National Park.

Owl and carnivorans, Fosa and Spotted Fanaloka (Goodman et al. 1991, 1997, Thorstrom et al. 1997, Goodman & Thorstrom 1998).

WESTERN TUFTED-TAILED RAT
Eliurus myoxinus   Milne-Edwards, 1885

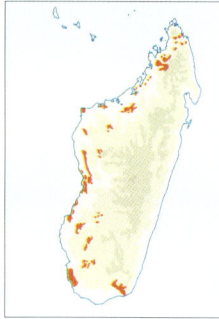

Other name: Milne-Edwards's Tufted-tailed Rat
Malagasy names: Voalavo ala, Sokitralina

MEASUREMENTS
Head-body: 117–136mm.
Tail length: 125–167mm.
Weight: 51–75g. (Carleton 2003)

DESCRIPTION & IDENTIFICATION Upperparts light greyish-brown to sandy-brown, with distinct abrupt transition to pale grey to creamy-grey underparts. Head stocky and squarish, upper incisors distinctive deep orange. Tail relatively short (105–110% head-body length), proximal third scaled, distal part covered in brown hair (12–15mm long), forming thick dark brush.

HABITAT & DISTRIBUTION Western deciduous forests and south-western and southern spiny forests between sea level and c.1,250m, and some dry and humid forests in far north between c.700 and 1,250m. Range extends from Tolagnaro in south-east, through all western forest areas to Ankarafantsika in north-west, then further north to Manongarivo, dry forests of Loky-Manambato Protected Area and drier transition forests on western edge of Marojejy (Goodman & Soarimalala 2002, Soarimalala & Goodman 2011, Goodman et al. 2013, Shi 2013). Seems commoner in deciduous forests than spiny forests (Carleton 2003). Tolerates forest fragmentation, but less tolerant of edge (Andriatsitohaina et al. 2020). Also recorded in disturbed habitats and in forest regenerating after fire (Randrianjafy 1993, 2003). Occurs sympatrically with Carleton's Tufted-tailed Rat and Lesser Tufted-tailed Rat in transitional dry-lowland to moist evergreen forests in north, and with Tsingy Tufted-tailed Rat and Rock-loving Tufted-tailed Rat in limestone karst areas in centre-west (Carleton et al. 2022b). Subfossil remains from Holocene (c.2,500–500 years BP) recorded from Andrahomana Cave, south-east Madagascar (Goodman et al. 2006), Ankilitelo Cave in south-west

(Muldoon *et al.* 2009), and caves at Beanka, in west (Nomenjanahary 2019). IUCN Least Concern.

BEHAVIOUR Nocturnal and scansorial. Occupies ground burrows and readily climbs through undergrowth, including up vertical tree trunks and along thin branches. Studies in Ankarafantsika and Kirindy Forest suggest males have mutually exclusive home ranges, while female home ranges are larger and overlap those of males (Ganzhorn 2003, Randrianjafy 2003). Presumed to feed on seeds and fruits. In Andohahela National Park feeds on fruits of locally endemic Three-cornered Palm *Dypsis decaryi* (Carleton 2003). Western Tufted-tailed Rat is possibly polygynous. Females appear solitary during breeding season but gregarious at other times (Randrianjafy 2003). Gestation period 24 days. Litter size up to four and in captivity females can have four litters per year, but this is unlikely in the wild. Sexual maturity may be reached within six to seven months (Randrianjafy *et al.* 2007). Known predators include Fosa and Barn Owl (Rasoma & Goodman 2007).

WHERE TO SEE Night walks in vicinity of Ankarafantsika National Park and Kirindy offer best chance. Often seen running along branches at all levels in canopy.

WHITE-TAILED TUFTED-TAILED RAT *Eliurus penicillatus* Thomas, 1908

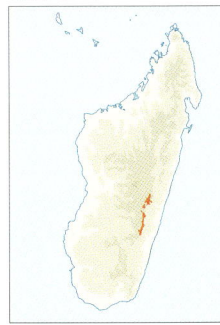

Other name: White-tipped Tufted-tailed Rat
Malagasy names: Voalavo ala, Sokitralina

MEASUREMENTS*
Head-body: *c.*150mm. Tail: *c.*175mm. Weight: *c.*100g. (Carleton & Schmidt 1990) (*single specimen)

DESCRIPTION & IDENTIFICATION Known only from a handful of specimens. Dorsal pelage brownish-grey to blackish-grey. Venter pale grey, with obvious demarcation between upper- and underparts. Tail tuft begins around halfway, the short white hairs becoming progressively longer and denser to form a conspicuous white tip. Very similar to Major's Tufted-tailed Rat and their species status may be questionable (Carleton *et al.* 2022b).

Western Tufted-tailed Rat, Ankarafantsika National Park.

HABITAT & DISTRIBUTION Known from only three localities within Marolambo protected area (Ampitambe Forest, vicinity of Ambositra and Ankerana, north-east of Fandriana) in central south-eastern montane forests between 900 and 1,670m (Carleton & Schmidt 1990, Carleton & Goodman 2007, Goodman *et al.* 2013). Several surveys further north and south have failed to locate additional populations: appears to be one of a handful of endemic rodents in east with restricted distribution. Other tufted-tailed rats recorded at these sites are Grandidier's Tufted-tailed Rat, Major's Tufted-tailed Rat, Lesser Tufted-tailed Rat and Tanala Tufted-tailed Rat (Carleton *et al.* 2022b). IUCN Endangered.

BEHAVIOUR Nocturnal and scansorial. Strict native forest dweller (Goodman & Raherilalao 2013).

PETTER'S TUFTED-TAILED RAT
Eliurus petteri   Carleton, 1994

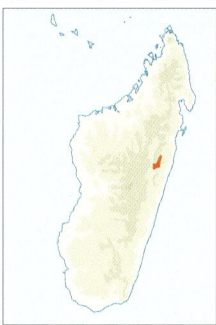

Malagasy names: Voalavo ala, Sokitralina

MEASUREMENTS*
Head-body: 130–136mm. Tail: 178–185mm. Weight: 74g. (Carleton 1994)
(*based on two specimens)

DESCRIPTION & IDENTIFICATION Moderately sized with very long tail, which is relatively longest of any tufted-tailed rat, *c.*135% head-body length. Pelage longish, soft with sleek appearance. Upperparts dark grey-brown, may vary from charcoal-grey to paler grey-brown. Underparts conspicuous bright white (distinguishes from all congeners). Distinct line between dorsal and ventral coloration. Tail tuft weakly developed, covering only final quarter of tail; tuft hairs light brown or greyish-brown, and relatively short (8–10mm).

HABITAT & DISTRIBUTION Restricted distribution: known only in lowland and mid-elevation rainforests in centre-east (Mantadia-Zahamena-Ankeniheny forest corridor), between *c.*430 and 1,200m (Rakotondraparany & Medard 2005, Goodman *et al.* 2013). Recorded sympatrically with Webb's Tufted-tailed Rat, Lesser Tufted-tailed Rat and Tanala Tufted-tailed Rat (Carleton *et al.* 2022b). IUCN Endangered.

BEHAVIOUR Nocturnal and scansorial. Diet likely to be seeds and fruits.

TANALA TUFTED-TAILED RAT
Eliurus tanala   Forsyth Major, 1896

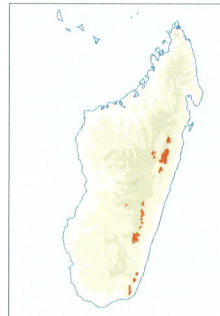

Malagasy names: Voalavo ala, Sokitralina

MEASUREMENTS
Head-body: 144–175mm. Tail: 163–210mm. Weight: 66–115g. (Jansa *et al.* 2019)

DESCRIPTION & IDENTIFICATION Distinctive large tufted-tailed rat, with relatively long tail. Upperparts dark greyish-brown, flanks slightly paler. Back noticeably darker in some individuals. Underparts white to creamy-grey, sometimes flecked with grey. Distinct line between dorsal and ventral coloration. Tail relatively long (120–130% head-body length) with short dark hairs covering terminal half, becoming progressively white, longer and thicker to form bushy conspicuous terminal tuft. Feet also white or pale grey.

HABITAT & DISTRIBUTION Humid lowland, mid-elevation, montane and sclerophyllous forest between *c.*400 and *c.*1,600m, with most records at 800–1,300m. Range covers centre-east and south-east regions, from vicinity of Lac Aloatra in centre-east to Andohahela in far south (Jansa *et al.* 2019). In some areas, Tanala Tufted-tailed Rat is common, but prefers elevations above *c.*1,100m; below this,

Tanala Tufted-tailed Rat, Tsitongambarika Forest.

E. webbi tends to be more common (Rakotoarisoa et al. 2013b, Goodman & Monadjem 2017). Recorded in sympatry with Grandidier's Tufted-tailed Rat, Major's Tufted-tailed Rat, White-tailed Tufted-tailed Rat, Lesser Tufted-tailed Rat and Webb's Tufted-tailed Rat (Carleton et al. 2022b). IUCN Least Concern.

BEHAVIOUR Nocturnal and scansorial. Utilises a variety of microhabitats including grassy glades, herbaceous growth, tangles of vines along watercourses, stands of tree-ferns and bamboo thickets. Collected in lower branches, lianas and vines, but also at ground level (often along watercourses), particularly at higher elevations (Goodman & Carleton 1996, 1998, Goodman et al. 1999a, Carleton & Goodman 2000). Diet comprises a variety of seeds and nuts, including Ramy nuts *Canarium madagascariensis*, which it is able to gnaw through to extract the endocarp. Reproduction during wet season, young born late November and December. Litter size two to four.

Rock-loving Tufted-tailed Rat, Tsingy de Bemaraha National Park.

ROCK-LOVING TUFTED-TAILED RAT *Eliurus tsingimbato* Jansa et al., 2019

Other name: Limestone Tufted-tailed Rat

Malagasy names: Voalavo ala, Sokitralina

MEASUREMENTS
Head-body: 144–175mm. Tail: 163–210mm. Weight: 66–115g. (Jansa et al. 2019)

DESCRIPTION & IDENTIFICATION A relatively large tufted-tailed rat. Fur fine and soft. Upperparts including head and face dark brown to mid-brown, grading into tawny along central flanks, with paler areas on upper hindquarters. Sharply demarcated from entirely creamy-white underparts, which extend to throat and chin and inner surfaces of hind- and forelimbs. Hairs on tail short and sparse except for tuft on distal 40%, which is moderate but distinct: brown in most individuals, with dark brown tip, but in some tuft has pale whitish band in middle portion. Forms a clade with Tanala Tufted-tailed Rat and Ellerman's Tufted-tailed Rat.

HABITAT & DISTRIBUTION Known from karst landscapes dominated by limestone *tsingy* formations in centre-west at elevations below c.300m. Recorded in Tsingy de Namoroka and Tsingy de Bemaraha National Parks, Beanka Forest and Forêt d'Antsahalaza (Jansa et al. 2019). In two forest types: in Namoroka, Bemaraha and Beanka recorded in dry deciduous forest zones dominated by limestone pinnacles with a profusion of lianas; in Forêt d'Antsahalaza dry deciduous forest grows on limestone bedrock with no *tsingy* formations and a noticeably less dense understorey, but vertical lianas and vines persist. Occurs sympatrically with Tsingy Tufted-tailed Rat and Western Tufted-tailed Rat and also Western Red Forest Rat *Nesomys lambertoni* in Tsingy de Bemaraha National Park (Jansa et al. 2019). IUCN Not evaluated.

BEHAVIOUR Nocturnal and scansorial. Animals collected at end of dry season (September–December) show signs of reproductive activity: females in oestrous, males with enlarged testes. This implies mating occurs from onset of wet season, with birth and nursing period coinciding with maximum fruit and seed abundance (Jansa et al. 2019).

WHERE TO SEE Possibly seen on night walks in Tsingy de Bemaraha National Park.

WEBB'S TUFTED-TAILED RAT
Eliurus webbi   Ellerman, 1949

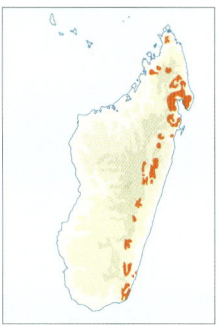

Malagasy names: Voalavo ala, Sokitralina

MEASUREMENTS
Head-body: 105–161mm. Tail: 150–183mm. Weight: 54–90g. (Goodman & Carleton 1996, Carleton 2003)

DESCRIPTION & IDENTIFICATION Moderately large but lighter build than other large congeners, Major's Tufted-tailed Rat and Tanala Tufted-tailed Rat. Pelage soft and quite long, particularly over dorsal region and flanks. Upperparts dark brown to blackish-brown, underparts dull pale grey with brown tints on sides; transition is abrupt. Regional variation in ventral pelage: individuals from isolated population on Montagne d'Ambre have near-white underparts. Tail relatively long (120–130% head-body length); tuft entirely dark brown to blackish-brown, and well developed (hairs 10–12mm long) covering distal two-fifths of tail. Proportionately smaller ears than other tufted-tailed rats (Jansa et al. 2019).

HABITAT & DISTRIBUTION Mainly in undisturbed lowland rainforest, including littoral forest along parts of east coast (Carleton 2003). Also in degraded habitats with non-native trees. Occurs from Montagne d'Ambre in north to Tolagnaro on southeast coast. Most localities between sea level and c.1,000m, and can be common at lower elevations; occurs sporadically to c.1,300m (Tsaratanana, Manongarivo, Anjozorobe-Angavo and Befotaka-Midongy du Sud) with exceptional records at 1,500m on Andringitra Massif (Pic d'Ivohibe) (Goodman & Carleton 1996, 1998, Goodman et al. 1999a, Carleton & Goodman 2000, Carleton et al. 2022a). Broadly sympatric with Tanala Tufted-tailed Rat, but prefers elevations below c.1,000m, whereas Tanala Tufted-tailed Rat prefers higher elevations. Also recorded sympatrically with Ellerman's Tufted-tailed Rat, Lesser Tufted-tailed Rat and Major's Tufted-tailed Rat (Montagne d'Ambre) (Carleton et al. 2022b). IUCN Least Concern.

BEHAVIOUR Nocturnal and primarily scansorial (Carleton 2003); observed and trapped in lianas and lower branches. Evidence suggests it may live in monogamous pairs occupying ground burrows up to 1m in depth. Inhabit chamber at rear of burrow, also used to store seeds (*Cryptocarya* sp. recorded). When exiting clogs burrow entrance with soil and leaf litter (Goodman 1994). Probably nests and forages at different levels in forest. Recorded sharing same burrow system as nesting Scaly Ground Roller *Brachypteracias squamiger* (Goodman 1994); ground-roller nest in principal tunnel of burrow system, while rats' food cache was at rear, so rats had to climb over birds' nest when entering and exiting. Time of breeding may vary across range (Goodman & Carleton 1996, 1998, Goodman et al. 1999, Carleton & Goodman 2000). Litter two to three. Predators include Madagascar Long-eared Owl and Madagascar Red Owl (Thorstrom et al. 1997, Goodman & Thorstrom 1998).

Webb's Tufted-tailed Rat, Masoala National Park.

WHITE-TAILED MOUNTAIN MICE
GENUS *Voalavo*

A recently described genus (Carleton & Goodman 1998) currently comprising two species: *Voalavo* species are characterised by their diminutive size (the smallest endemic rodent in rainforest regions) and very long tail (c.130% head-body length) (Jansa & Carleton 2003). They appear to occupy a similar niche to the big-footed mice (genus *Monticolomys*), although as far as is known the ranges of these two genera do not overlap. The genus *Voalavo* shows closest affinity to tufted-tailed rats (Soarimalala *et al*. in press). The name Voalavo comes from the common Malagasy term for rodent.

EASTERN WHITE-TAILED MOUNTAIN MOUSE
Voalavo antsahabensis Goodman *et al.*, 2005

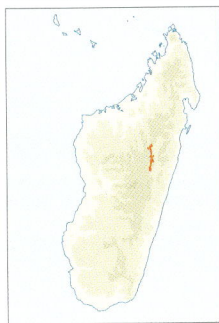

Other names: Anjozorobe Voalavo, Eastern Voalavo

MEASUREMENTS
Head-body: 85–100mm. Tail: 102–127mm. Weight: 18.5–26g. (Goodman *et al.* 2005c, Soarimalala *et al.* in press)

DESCRIPTION & IDENTIFICATION Fur soft, silky and dense. Upperparts mid-grey with brownish tones on neck and flanks. Underparts pale grey to off-white. Ears short and rounded. Tail long (c.120% of head-body length), naked and bicoloured: grey above, white below. Males have paired glands on chest, not seen in other Nesomyine rodents (Soarimalala *et al.* in press). Resembles small tufted-tailed rat, but is darker grey and has no tail tuft.

Sympatric with several other native rodents, all noticeably larger, e.g. tree rats (*Brachytarsomys* spp.), short-tailed rats (*Brachyuromys* spp.), tufted-tailed rats (*Eliurus* spp.), Sleek-furred Ground Rat and Eastern Red Forest Rat *Nesomys rufus* (Soarimalala *et al.* in press). Yet to be recorded in sympatry with similarly small Malagasy Mountain-dwelling Mouse *Monticolomys koopmani* (Soarimalala *et al.* in press).

HABITAT & DISTRIBUTION Restricted distribution: in montane humid forest, with abundant lianas, mosses and lichens, at 1,250–1,425m, and currently known only from Anjozorobe region in eastern Central Highlands (Goodman *et al.* 2005c, Soarimalala *et al.* 2007). IUCN Endangered.

BEHAVIOUR Nocturnal and both scansorial and terrestrial. Thought to occupy ground burrows but forages by climbing in undergrowth. Presumed to feed on seeds and fruits. Reproductive males have a chest gland that produces a distinctive odour. Litter size of two recorded (Goodman *et al.* 2005c, 2013, Soarimalala & Goodman 2011).

WHERE TO SEE Difficult to see. Sometimes seen in forests around Anjozorobe.

NORTHERN WHITE-TAILED MOUNTAIN MOUSE
Voalavo gymnocaudus Carleton & Goodman 1998

Other name: Northern Voalavo

MEASUREMENTS
Head-body: 80–90mm. Tail length: 107–126mm. Weight: 17–25.5g. (Carleton & Goodman 1998, 2000)

DESCRIPTION & IDENTIFICATION Fur soft, short and dense. Upperparts smoky-grey with brownish tones on flanks. Underparts from throat also grey but slightly paler than upperparts with no sharp contrast. Whitish-grey chin and throat. Vibrassae medium-long and white. Ears short and rounded. Tail long (135% of head-body), bicoloured (grey dorsally, white ventrally) virtually naked, except final third which is much paler and finely haired but not tufted. As with congener, males have pair of glands on chest, not known in other Nesomyine rodents (Soarimalala *et al.* in press). Resembles small tufted-tailed rat, but is darker grey and lacks discernible tail tuft.

Occurs sympatrically with several other Nesomyine species, e.g. tree rats (*Brachytarsomys* spp.), tufted-tailed rats (*Eliurus* spp.), Sleek-furred Ground Rat and Eastern Red Forest Rat *Nesomys rufus* (Soarimalala *et al.* in press). White-tailed

353

Northern White-tailed Mountain Mouse, Anjanaharibe-Sud.

mountain mice are considerably smaller. Not yet recorded sympatrically with similarly diminutive Malagasy Mountain-dwelling Mouse *Monticolomys koopmani* (Soarimalala *et al.* in press).

HABITAT & DISTRIBUTION Known only from montane, upper montane and sclerophyllous forest around Andapa Basin, specifically Anjanaharibe-Sud and Marojejy (Goodman & Carleton 1998, Carleton & Goodman 1998, 2000). Habitat typified by lush epiphytic growth with profuse moss coverage and regular drenching cloud cover. Recorded at *c.* 1,225–1,950m, and appears more abundant at higher elevations. IUCN Least Concern.

BEHAVIOUR Nocturnal, showing both scansorial and terrestrial behaviour. Appears to prefer environs near ground level with dense tangles of roots. Uses natural tunnels and runways between roots. Also trapped on branches 1.5m above ground (Goodman & Carleton 1998, Carleton & Goodman 2000). Chest glands on reproductive males produce distinctive musky odour, which may function to scent-mark trails and territory boundaries (Soarimalala *et al.* in press). Litter size of two or three (Goodman *et al.* 2013, Soarimalala & Goodman 2011). Skeletal remains found in pellets of Madagascar Long-eared Owl (Soarimalala *et al.* in press).

MALAGASY BIG-FOOTED MICE
GENUS *Macrotarsomys*

A genus containing three species of small forest mice that are strictly nocturnal and outwardly show close resemblance to some Old World gerbils (Muridae: Gerbillinae), particularly the genus *Gerbillus* from Africa, the Middle East and western Asia. All big-footed mice have soft fine textured fur and similarly coloured light brown dorsal and white ventral pelage. Tail length in all species is longer than head-body length, exceptionally so in one species, Long-tailed Big-footed Mouse *M. ingens*. Tail ends in a tuft (covering distal one third) – most pronounced in Petter's Big-footed Mouse *M. petteri*, least pronounced in Long-tailed Big-footed Mouse (but never as pronounced as in tufted-tailed rats). Within Nesomyine rodents, big-footed mice are most closely related to Malagasy Mountain-dwelling Mice genus *Monticolomys*, from high mountains of east and far north (Jansa *et al.* 2009, Steppan & Schenk 2017). Big-footed mice appear to be restricted to dry forests of western and southern Madagascar and can be quite common and readily observed in some areas.

WESTERN BIG-FOOTED MOUSE
Macrotarsomys bastardi
Milne-Edwards & G. Grandidier, 1898

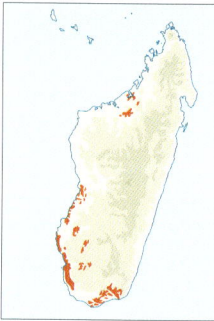

Other names: Bastard's Big-footed Mouse, Lesser Big-footed Mouse

MEASUREMENTS
Head-body: 78–88mm.
Tail: 104–123mm. Weight: 21–28g. (Carleton *et al.* 2022c)

DESCRIPTION & IDENTIFICATION Pelage soft and fine. Upperparts light brown to fawn with more greyish underfur showing through. Underparts and legs paler creamy or yellowish-white. Ears oval in shape, relatively large (20–25mm) and prominent. Eyes also relatively large. Tail long (*c.* 135% head-body length), quite stiff with slight dark terminal tuft (Carleton & Goodman 2003c). A small gerbil-like rodent with large ears and long tail. Can be confused only with its congener, Long-tailed Big-footed Mouse, which has proportionately longer tail. Western Big-footed Mouse is more likely to be seen on the ground.

HABITAT & DISTRIBUTION Deciduous forests and arid spiny forests with sandy soils from north-west (Anjajavy, Mariarano, Ankarafantsika) to south-west (Bezà Mahafaly) and southern tip of island, and east to foothills of Anosyenne Mountains near Tolagnaro (Carleton & Schmidt 1990, Goodman *et al.* 1999, Youssouf Jacky & Rasoazanabary 2008), although no records currently exist between Betsiboka and Tsiribihina Rivers, and further field surveys are required (Carleton *et al.* 2022c). Recorded from sea level to *c.*915m (Carleton & Goodman 2003c). Also able to survive in human-altered and degraded forests. Almost certainly suffers adverse competition from introduced House Mouse. IUCN Least Concern.

In the past two subspecies have been proposed: *M. b. bastardi* in open savanna and bush areas in central south-west, and *M. b. occidentalis* from Parcel 2 in Andohahela National Park in extreme south to deciduous forests of north-west. More recent analysis concluded such differentiation was unwarranted and they are no longer recognised (Jansa *et al.* 2008).

BEHAVIOUR Strictly nocturnal and largely terrestrial. Moves on forest floor, often in pairs, in typical gerbil-like fashion: a slightly erect posture, hopping with forelimbs off ground, using tail as counterbalance (Carleton & Goodman 2003c, Randrianjafy 2003). Hopping more pronounced when covering larger areas of bare ground. Diet consists of seeds, fruits, roots, some plant stems and perhaps the occasional insect (Ganzhorn 2003).

Pairs spend day in long burrows (up to 1.5m) excavated under large rocks or tree stumps (Ganzhorn 2003). These are shallow with inconspicuous openings that are backfilled and plugged after entering. The plugs are obvious in early morning when surroundings are damp and darker. Burrow terminates in small nest chamber lined with dry leaves and grass. Reproductive activity March to June, peaks April–May (Randrianjafy 2003). Litter size two or three (based on embryo counts), gestation period *c.*24 days (Carleton & Goodman 2003c).

Known predators include Barn Owl, scops owls (*Otus* spp.), White-browed Owl and Fosa (Dollar

Western Big-footed Mouse, Kirindy Forest.

et al. 2007, Rasoma & Goodman 2007, Goodman *et al.* 2014a).

WHERE TO SEE Readily seen on night walks in Kirindy Forest, near Morondava, and Ankarafantsika National Park. Disturbance of dry leaf litter often heard before animal is seen: tends to move in short bursts then pause.

LONG-TAILED BIG-FOOTED MOUSE *Macrotarsomys ingens* Petter, 1959

Other names: Ankarafantsika Big-footed Mouse, Greater Big-footed Mouse

MEASUREMENTS
Head-body: 115–150mm.
Tail length: 193–240mm.
Weight: 54–73g. (Carleton & Goodman 2003c, Carleton *et al.* 2022c)

DESCRIPTION & IDENTIFICATION Dorsal pelage soft, fine and pale sandy brown, ventral pelage creamy-white; transition abrupt. Ears large, oval and obvious. Tail exceptionally long (up to 170% of head-body length) and lightly haired. Hairs longer and denser at tail tip forming a modest tuft (not as well developed as tufted-tailed rats) (Carleton & Goodman 2003c). Broadly similar appearance to sympatric Western Big-footed Mouse, but appreciably larger with much longer tail. Length of tail should distinguish from similarly scansorial tufted-tailed rats.

HABITAT & DISTRIBUTION Very limited range: known only from dry deciduous forests of Ankarafantsika and immediate vicinity in north-west, at elevations of *c.*100–400m (Petter 1959, Carleton & Schmidt 1990). Possible sightings from Bongolava Massif need confirmation (Dollar & Kennerley 2016). Directly threatened by intrusion from introduced feral dogs and cats. IUCN Endangered.

BEHAVIOUR Nocturnal and probably solitary. Scansorial and rarely observed low down; generally seen among narrow branches and vines, often several metres above ground (Carleton & Goodman 2003c). Moves in kangaroo-like hops, bounding on disproportionately large feet and using tail as counterbalance. Often pauses and remains motionless if threatened. Lives in shallow burrows recognisable by small piles of soil outside entrance (Petter 1972). Entrance always backfilled and closed. Burrows often exceed 25cm deep and acts as thermoregulatory buffer: interior maintains ambient temperature above 22°C even in austral winter (Lobban *et al.* 2014). May reproduce year-round. Litter size two (based on embryo counts) (Randrianjafy 2003). Preyed upon by owls (possibly Barn Owl, scops owls (*Otus* spp.), White-browed Owl), Fosa and large snakes, e.g., Madagascar Ground Boa (Carleton & Goodman 2003c, Randrianjafy 2003, Dollar *et al.* 2007).

WHERE TO SEE Ankarafantsika National Park is only site where the species can be seen. On night walks it is worth searching narrow branches and tangles of vines; gives faint reflective eyeshine in torch beam.

PETTER'S BIG-FOOTED MOUSE
Macrotarsomys petteri

Goodman & Soarimalala, 2005

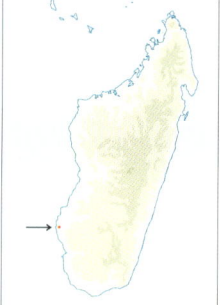

MEASUREMENTS*
Head-body: 150mm. Tail: 238mm. Weight: 105g. (Goodman & Soarimalala 2005). (*single specimen)

DESCRIPTION & IDENTIFICATION Dorsal fur short, soft and dense with agouti appearance. Flanks mid-brown, obvious dark brown dorsal ridge. Underparts buff-white, sharp transition on lower flanks. Tail very long (160% head-body length) and partially naked; dark over proximal half and off-white distally. Terminal tuft well developed, covering 25% of tail and increasingly bushy towards tip (Goodman & Soarimalala 2005). Large, stocky big-footed mouse (noticeably larger than congeners) with long hindfeet, short forelimbs, long ears (32mm) and long vibrissae (60+mm).

HABITAT & DISTRIBUTION Currently known only from Forêt d'Andaladomo within extensive Mikea Forest (elevation 80m), south of Lac Ihotry in south-west (Goodman & Soarimalala 2005), a unique zone of transition between deciduous forest and xerophytic spiny bush formations (Seddon *et al.* 2000). Subfossil remains have been collected at Andrahomana Cave in south-east and on Mahafaly Plateau in south-west, suggesting the species was once much more widespread (Goodman *et al.* 2006e, Muldoon *et al.* 2009). Occurs sympatrically with Western Big-footed Mouse. IUCN Data Deficient.

Long-tailed Big-footed Mouse, Ankarafantsika National Park.

BEHAVIOUR Nocturnal. Morphology suggests terrestrial behaviour. Single specimen trapped on ground (Goodman & Soarimalala 2005). Other aspects unknown.

MALAGASY MOUNTAIN-DWELLING MOUSE
GENUS *Monticolomys*

A monotypic genus characterised by its small size, dense pelage, rounded ears and relatively long tail (Carleton & Goodman 1996). Among Nesomyinae from eastern humid forest regions, only white-tailed mountain mice (*Voalavo* spp.) are similarly small, although, based on known ranges, these two genera are allopatric and apparently occupy a similar upper montane niche. The genus *Monticolomys* is more closely related to big-footed mice from western Madagascar, than to other species in eastern regions (Jansa et al. 1999, Jansa & Carleton in press).

Based on single molar, *Monticolomys* sp. has been identified in Late Pleistocene deposits at Belobaka Cave, east of Mahajanga (Gommery et al. 2018). This site is far outside the known distribution and habitat of extant populations and requires confirmation (Carleton et al. 2022a).

MALAGASY MOUNTAIN-DWELLING MOUSE *Monticolomys koopmani*
Carleton & Goodman, 1996

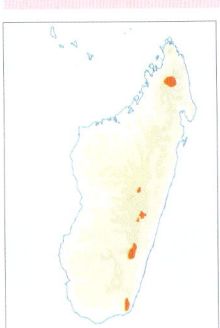

Other name: Koopman's Forest Mouse

MEASUREMENTS
Head-body: 84–101mm. Tail: 116–143mm. Weight: 18.5–27.5g. (Carleton & Goodman 1996, Goodman et al. 1999)

DESCRIPTION & IDENTIFICATION Pelage fine, soft and relatively thick. Upperparts sombre dark brown to brownish-grey with black guard hairs; underparts, including chin and throat, paler grey. Dorsal-ventral pelage transition indistinct. Ears short, rounded and densely furred both inside and out. Tail uniform grey, very long (about 135–140% head-body length) and covered in fine brown hairs. No terminal tuft (Carleton & Goodman 1996, Carleton et al. 2022a). Within its range and ecotone, small size and mouse-like appearance should distinguish it from other rodents. Only Lesser Tufted-tailed Rat is similarly diminutive, but Malagasy Mountain-dwelling Mouse is lighter in build and has a proportionately longer tail, which lacks a terminal tuft.

HABITAT & DISTRIBUTION Primarily in upper montane and sclerophyllous forests up to c.2,200m which are regularly covered in cloud and mist, often dominated by dense bamboo and similar species, encrusted with mosses and epiphytes. Also recorded in mid-elevation forest down to c.900m. Currently known from several widely dispersed high-mountain locations from Tsaratanana in far north to Ankaratra in centre-east, to Andringitra and Andohahela in south-east and south (Carleton & Goodman 1996, Goodman et al. 1999a, Maminirina et al. 2008, Goodman et al. 2013). Not yet documented in high-elevation zones of Anjanaharibe-Sud and Marojejy, where similarly small Northern White-tailed Mountain Mouse occurs. This species occurs above treeline and can be common in degraded montane grasslands; it returns quickly to such habitats after fire damage (Rasolonondrasana & Goodman 2006). IUCN Least Concern.

BEHAVIOUR Nocturnal. Exhibits terrestrial behaviour at times, but also an adept climber and caught on narrow lianas c.2m above ground (Carleton & Goodman 1996, Goodman et al. 1999a). Long tail and enlarged claws are clear adaptations to climbing. Diet apparently dominated by seeds and some fruits (Rasolonondrasana & Goodman 2006). Breeding

Malagasy Mountain-dwelling Mouse, Andringitra National Park.

likely takes place in wet season, with litter sizes up to three (based on embryo count) (Carleton *et al.* 2022a). Recorded predators include Madagascar Red Owl (skeletal remains found in pellets on Mount Papango, Befotaka-Midongy du Sud) (Carleton *et al.* 2022a). Other owls and small carnivorans also likely feed on this species.

GIANT JUMPING RAT
GENUS *Hypogeomys*

A distinctive genus containing a single extant species, the Giant Jumping Rat *H. antimena*; this splendid creature is the largest living rodent on Madagascar. Described in 1869 by Alfred Grandidier, the Giant Jumping Rat was the first known native rodent from the island. In evolutionary terms, the genus *Hypogeomys* is most closely related to short-tailed rats (*Brachyuromys* spp.) and forest rats (*Nesomys* spp.) within the endemic subfamily Nesomyinae (Jansa *et al.* 2009, Steppan & Schenk 2017). A second larger species. *H. australis*, is known from the Holocene subfossil record, *c.*4,400 years ago, and is thought to have survived until around 1,500 years BP (Goodman & Rakotondravony 1996, Burney *et al.* 2004) (see The Extinct Mammal Fauna).

GIANT JUMPING RAT
Hypogeomys antimena   A. Grandidier, 1869

Malagasy names: Vositse, Votsovotsa, Votsotsa

MEASUREMENTS
Head-body: *c.*285–345mm.
Tail length: *c.*200–255mm.
Weight: *c.*1.0–1.3kg.
(Sommer & Carleton in press)

DESCRIPTION & IDENTIFICATION Fur short and dense. Upperparts medium to dark brown or light brown-grey, with reddish tones in some. Head usually darkest area. Underparts paler, creamy-grey or off-white. Sometimes has darker V-shaped area on top of nose. Snout blunt and ears very conspicuous (50–65mm long). Tail thick and muscular, dark and covered in short stiff hairs. No sexual dimorphism. Juvenile is greyer above, paler below (Sommer

2003a, Sommer & Carleton in press). A large, rotund rabbit-like rodent with a blunt snout, conspicuous large ears and elongated hindfeet.

HABITAT & DISTRIBUTION Now restricted to narrow coastal zone of deciduous forest on sandy soils interspersed with baobabs (*Adansonia* spp.), characterised by permanent dry leaf litter (Cook *et al.* 1991), between 40 and 100m, north of Tomitsy River and south of Tsiribihina River (Sommer 2003a). Found only in two isolated areas (separated by Mandroatra River) totalling <200km² (Sommer *et al.* 2002). In good habitat (high canopy well away from forest edge), population densities reach *c.*50 individuals/km², but in most areas much lower (Sommer & Hommen 2000). Total population estimated to be *c.*36,000 individuals (Young *et al.* 2008). Sympatric with two other Nesomyine rodents, Western Tufted-tailed Rat and Western Big-footed Mouse, both of which are significantly smaller.

Historically much more widespread: only 1,400 years ago range extended from Central Highlands well south and south-west (Lac Tsimanampetsotse region) and this persisted until the 14th century (*c.*1350) (MacPhee 1986, Goodman & Rakotondravony 1996, Muldoon *et al.* 2009). Climate change, leading to aridification, and extensive forest clearance have resulted in massive retraction. Giant Jumping Rats continue to be threatened by habitat loss through forest clearance, charcoal production, burning, and incidental hunting with dogs (targeting Tenrecinae spp.). Feral dogs and cats are also a major threat (Sommer & Hommen 2000). IUCN Endangered.

BEHAVIOUR Strictly nocturnal, terrestrial and monogamous, the latter an extremely rare trait in rodents (Sommer 1997) and known in less than 5% of all mammals (Sommer 2003b, Sommer & Carleton in press). Pair bonds last more than one breeding season and persist outside this, normally until one mate dies or is predated (Sommer 1997, 2001, Sommer & Tichy 1999). Predation rates high, mainly by Fosa and Madagascar Ground Boa (Rasoloarison *et al.* 1995, Sommer 2001). After a death, remaining animal normally develops new pair bond within a few days or weeks (Sommer 1997). Females always keep burrow and territory until new bond forms; males generally maintain existing territories, but sometimes move to begin partnership with recently widowed female (Sommer 2000). Infanticide does not occur: after taking over new burrow, males tolerate infants sired by previous male (Sommer 2003a).

Moves on all fours with ambling gait or kangaroo-like hops on hindlegs, particularly when startled or nervous (Cook *et al.* 1991, Sommer 1996, 1997). Stout tail can be used as counterbalance or 'prop' (fifth limb) during locomotion. Forages, alone or in pairs, on forest floor for fallen fruit, including baobab fruits, seeds and succulent leaves. Also strips bark from saplings and digs for roots and tubers. Food held in forepaws and manipulated while sitting semi-upright (like a rabbit) (Sommer 2003a).

Social communication appears complex. When active at night, individuals and pairs tap the ground with hindfeet, often with head and ears cocked. Audible calls are variable and include a rolling whirr, *brou-brou-brououou* and *kouitsch-kouitsch*.

Family unit occupies territory of 3–4ha (larger in dry season when food is scarce), marked with urine, faeces and scent gland secretions at specific latrine sites important in demarcating territorial boundaries (Sommer 2003a). Home ranges of neighbouring pairs are mutually exclusive and defended year-round.

Family burrow sited on slightly elevated area of bare soil. Adults and offspring spend day in burrow: provides protection against predation, sanctuary from high daytime temperatures and heavy rain, and security for raising offspring. Burrow is complex of tunnels (c.45cm diameter) up to 5m across that has between one and seven entrances; only one to three in use simultaneously, others blocked by soil and leaves. Entrances in use are 'plugged' with barrier of soil, at depth of c.50cm that must be excavated to allow passage in and out (Sommer 1996). At night, individuals, including recently independent males, use unoccupied burrows as temporary refuges. After death of a resident, or burrow abandonment, immigrants readily occupy empty burrows (Sommer 2000, Sommer & Hommen 2000). Digging completely new burrow is rare: existing burrows used continuously across many generations, with records showing occupancy lasting 25 years and more (Sommer & Carleton in press).

In rainy season (late November–March), when food resources optimal, a pair produces only one or two (exceptionally three) offspring (an inordinately low annual breeding rate for a rodent) (Sommer & Carleton in press). This corresponds to a reduction in parents' home range. Gestation period c.130 days (Veal 1992). Young remain in burrow for first 4–6 weeks; when first venturing out stay close to entrance, and do not emerge regularly until

Giant Jumping Rat, Kirindy Forest.

4–8 weeks. Subadults attain weight less than 1kg in first year; adult weight reached during second year (Sommer 1996, 2000). Female offspring remain with parents for two to three years and then leave to breed; males leave at *c*.1 year. Males sometimes establish a territory and mate within a short period, but normally this takes two years (Sommer 2001).

WHERE TO SEE Giant Jumping Rats are a key attraction at Kirindy Forest, 60km north-east of Morondava. Here they can be encountered at night in areas of main trail network, sometimes near central camp area. It is often well into the night before they become active and success may require persistent searching over two or three nights. Probably more active on moonless nights. Kirindy is arguably the best site for nocturnal animals on the island. Also look for Western Big-footed Mouse and several lemur species.

SHORT-TAILED RATS
GENUS *Brachyuromys*

The genus *Brachyuromys* shows remarkable resemblance in body form, particularly small ears and short tail, and ecology, to some Holarctic voles (subfamily Arvicolinae), and are sometimes referred to as Malagasy voles or vole rats. However, they are not closely related, the superficial similarity being an example of evolutionary convergence, rather than phylogenetic affinity (Jansa *et al.* in press). Amongst endemic Malagasy rodents they appear most closely aligned to forest rats (genus *Nesomys*) (Jansa and Carleton in press).

Their body is compact, with a broad head and inconspicuous, small rounded ears and small eyes. Legs are short with relatively narrow hindfeet. Tail is noticeably shorter than head-body length (*c*.50–60%). Pelage is fine, but dense and soft in texture. Dorsally, short-tailed rats are rich reddish-brown or yellowish-brown with a darker undercoat visible beneath. Venter similarly coloured with no obvious transition. Uniformly dark tail is sparsely haired on the dorsal surface, but has a dense covering on the lower surface (Jansa & Carleton 2003a, Jansa *et al.* in press).

They inhabit marshy areas or abandoned rice paddies on edge of native forests or moist meadows at high elevations (above *c*.2,000m) above the treeline, especially areas of matted grasses and reeds (Jansa & Carleton 2003a, Jansa *et al.* in press).

Beneath such vegetation they construct networks of runways, the vegetation often being so dense that no sunlight penetrates. Within these 'tunnel' systems, they appear to be active at all hours, day and night, feeding predominantly on grass. Short-tailed rats may also live in the same areas as shrew tenrecs (*Microgale* spp.), tufted-tailed rats (*Eliurus* spp.) and introduced rats (*Rattus* spp.).

Short-tailed rats may provide one of the few examples in which a native mammal benefits from human-induced habitat alteration. Abandoned rice paddies appear to create ideal habitat, but they may still suffer competition from introduced Black Rats: in studies in Ranomafana area, short-tailed rats were trapped only in centre of disused rice fields, whereas Black Rats dominated the edges (Jansa & Carleton 2003b).

The fossorial nature of these species makes observations extremely difficult, unless they are trapped.

GREGARIOUS SHORT-TAILED RAT
Brachyuromys ramirohitra   Forsyth Major, 1896

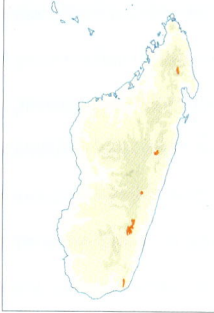

Other names: Large Short-tailed Rat, Greater Short-tailed Rat

Malagasy name: Voalavo

MEASUREMENTS*
Head-body: 165–169mm. Tail: 110–117mm. Weight: 117–129g. (Jansa *et al.* in press) (*based on two specimens)

DESCRIPTION & IDENTIFICATION Pelage thick and soft. Upperparts greyish-brown to reddish-brown with blackish tones. Underparts more greyish-beige but no obvious distinction. Snout blunt, head broad and rounded. Tail short and dark with paler underside. This is the larger short-tailed rat, easily differentiated from its smaller congener by size, slightly longer tail and more reddish coat (plus dental differences).

HABITAT & DISTRIBUTION Prefers mid-elevation, montane and sclerophyllous forest between *c*.900 and *c*.2,000m. Known from widely scattered localities in south-eastern, central and northern highlands, including Ampitambe (the type locality), Amboasary, Andringitra, Marolambo Massif and Anjanaharibe-

Sud (Carleton & Schmidt 1990, Goodman & Carleton 1996, Jansa & Carleton 2003a, Goodman *et al.* 2013, Kennerley 2016b). In south, occurs in close proximity to congener Betsileo Short-tailed Rat *B. betsileoensis*, although actual zone of sympatry appears limited. Gregarious Short-tailed Rat prefers montane and sclerophyllous forest from 1,200m to *c.*2,000m (Goodman & Rasolonandrasana 2001), corresponding to a lower elevational zone than preferred by its congener (Jansa & Carleton 2003a). IUCN Least Concern.

BEHAVIOUR Terrestrial and probably cathemeral: activity recorded around dusk and dawn, as well as during darkness. Specimens often captured on ground near extensive natural tunnel networks associated with tangled roots (Goodman & Carleton 1996). Presumed to be herbivorous, probably feeding on grasses and similar vegetation. Males trapped in October in Marolambo and Ambositra-Vondrozo protected areas had enlarged testes (Jansa *et al.* in press). Litter up to two (based on embryo counts). Predated by Fosa – skull remains found in scats in grassland at elevations above *c.*2,000m in Andringitra (Goodman *et al.* 1997).

Gregarious Short-tailed Rat, Andringitra National Park.

Betsileo Short-tailed Rat, Andringitra National Park.

BETSILEO SHORT-TAILED RAT
Brachyuromys betsileoensis   (Bartlett, 1880)

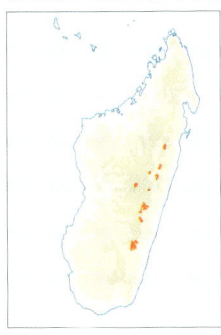

Other names: Small Short-tailed Rat, Lesser Short-tailed Rat

Malagasy name: Voalavo

MEASUREMENTS
Head-body: 140–163mm.
Tail length: 83–97mm.
Weight: 86–144g. (Jansa & Carleton 2003a, Goodman & Monadjem 2017, Jansa *et al.* in press)

DESCRIPTION & IDENTIFICATION Very similar to Gregarious Short-tailed Rat, but on average noticeably smaller, with proportionally shorter tail. Dorsal and ventral pelage plain brown to greyish-brown without rich reddish tones of its congener.

HABITAT & DISTRIBUTION Montane grassland, marshes, upper montane and mossy sclerophyllous forest, from *c.*900m up, but mainly above 1,900m, and in heathland above treeline to *c.*2,600m. Also in open grassland outside forest, where sometimes common (Langrand & Goodman 1997) and in degraded habitats, including abandoned rice paddies, and adjacent to habitation well away from native formations in centre-east (Randriamoria *et al.* 2015, Randriamoria 2017). Recorded at several locations in Central Highlands, from Lac Alaotra area south to Andringitra Massif (Carleton & Schmidt 1990, Goodman & Carleton 1996, Pidgeon 1996, Ramanana 2010). IUCN Least Concern.

BEHAVIOUR Terrestrial, and active day and night, often early morning and around dusk, sometimes by day, as well as at night (Jansa *et al.* in press).

Utilises network of tunnels through dense matted meadow vegetation. Probably feeds on grasses and similar vegetation. Reproductively active during July in Ranomafana: males with enlarged testes, females pregnant or lactating (Jansa et al. in press). Litter size two (based on embryo counts). Eaten by Fosa – skeletal remains found in scats in heathland at elevations above c.2,000m in Andringitra (Goodman et al. 1997).

RED FOREST RATS

GENUS *Nesomys*

The genus *Nesomys* contains three species of medium-sized, comparatively thick-set, diurnal, terrestrial and forest-dwelling rats. All are chestnut-brown or reddish-brown, with tail length slightly less than head-body length (Carleton et al. 2022d). Second largest endemic rodents, only Giant Jumping Rat is larger.

Diurnal rodents are rare in tropical forests: such behaviour probably developed in red forest rats because of reduced predator pressure in Malagasy forests (compared to other tropical forests). Diurnal activity may also reduce competition with other Nesomyine species that are primarily nocturnal. Within Nesomyinae, red forest rats are most closely related to short-tailed rats (*Brachyuromys* spp.) (Jansa & Carleton in press, Jansa et al. in press).

Subfossil remains of an extinct species, *N. narindaensis*, described from c.12,000-year-old deposits north of Mahajunga in the dry north-west. This species was larger than any extant *Nesomys*. The genus is more normally associated with humid

forest regions, and it may be that the area of discovery was once much wetter than it is today (Mein et al. 2010).

EASTERN RED FOREST RAT
Nesomys rufus   Peters, 1870

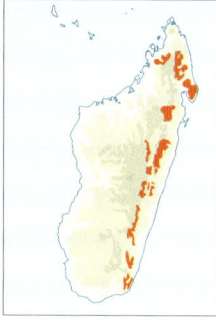

Other names: Red Forest Rat, Rufous Forest Rat, Rufous Nesomys

MEASUREMENTS
Head-body: 163–198mm. Tail: 139–180mm. Weight: 122–205g; female mean 150g, male mean 175g. (Ryan 2003)

DESCRIPTION & IDENTIFICATION Pelage longish, soft and smooth. Upperparts reddish-brown, flecked with darker brown and darker guard hairs. Around cheeks, flanks and rump may be hint of reddish-chestnut. Underparts slightly paler rufous-brown; chin and throat can be whitish. Ears prominent (20–25mm) and rounded, whiskers fine and black. Tail medium-long (90% head-body length), narrow, moderately covered in hair, sometimes white at tip, but no suggestion of a terminal tuft. A medium-sized, robust forest rat, bearing passing resemblance to Brown Rat *Rattus norvegicus*, but appreciably smaller and reddish in colour.

HABITAT & DISTRIBUTION Throughout mid- and higher-elevation rainforests, at c.650–2,300m, although uncommon above 1,900m. Range extends from Tsarantanana Massif and Sambirano region (Manongarivo) to mountains around Andapa Basin, south to Andohahela region (Goodman & Carleton 1996, 1998, Carleton & Goodman 2000, Goodman & Soarimalala 2002). Can tolerate minor habitat disturbance and degradation (Irwin et al. 2010b): in some areas, e.g. Anjozorobe in centre-east, has been recorded outside forest in agricultural areas where it feeds on crops (Ryan 2003). At some localities within 600–1,000m elevational band (e.g. Masoala, Mantadia, Ranomafana), Eastern Red Forest Rat is sympatric with Lowland Red Forest Rat *N. audeberti* (Carleton et al. 2022d). IUCN Least Concern.

BEHAVIOUR Strictly terrestrial and diurnal, with some crepusciular behaviour. Activity begins prior

Red Forest Rat, Andasibe-Mantadia National Park.

to sunrise, males often before females (Ryan *et al*. 1993). Foraging more intense in early morning and late afternoon, and to lesser extent around midday. Average daily distance travelled *c*.400m. Forages in dense vegetation, leaf litter and around dead logs. Diet mainly nuts and seeds, including Ramy *Canarium madagascariensis* and fallen fruits (e.g. *Cryptocarya*). Fat rich seeds like *Canarium* are cached for later consumption (Goodman & Sterling 1996), and therefore may play a role in seed dispersal (Razafindratsima 2017). An adept scavenger around research camps and associated rubbish pits (Goodman & Carleton 1996).

Males and females occupy home ranges of *c*.0.5ha (male's generally slightly larger), which appear constant between years (Ryan *et al*. 1993). Home ranges non-exclusive and may be shared with several other individuals. Considerable overlap between ranges of males and females and between neighbouring individuals of same sex.

Home range contains a number of dispersed burrows or dens, all in regular use, although one is often preferred. Burrows generally under fallen logs and brush piles; each has several entrances (Ryan *et al*. 1993). They are multi-chambered and connected via 1–2m tunnels: upper chamber lined with freshly clipped grass and used as food storage cache, deeper chambers lined with deep beds of shredded palm fronds or similar, and used for sleeping (Ryan *et al*. 1993). Occasionally males and females share a den.

Little known about reproduction. In higher-altitude areas, reported to be reproductively active mid-October to late December (Goodman & Carleton 1996, 1998, Goodman *et al*. 1999). At lower elevations breeding activity observed between July and early September (Ryan *et al*. 1993). Litter size up to four (based on embryo counts) (Carleton *et al*. 2022d).

Predated especially by Ring-tailed Vontsira, also Fosa, diurnal raptors, e.g. Henst's Goshawk, Madagascar Harrier-Hawk and Madagascar Buzzard, and large snakes (e.g. boas, *Acrantophis* and *Sanzinia*) (Ryan 2003, Carleton *et al*. 2022d).

WHERE TO SEE Encountered in many eastern rainforest parks. Often seen in Ranomafana National Park, particularly in late April and early May, when fallen Strawberry Guava *Psidium cattleianum* fruits plentiful on forest floor. Andasibe-Mantadia and Marojejy National Parks also good places.

LOWLAND RED FOREST RAT
Nesomys audeberti (Jentink, 1879)

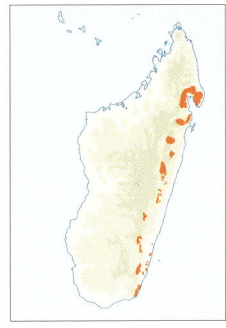

Other names: Audebert's Forest Rat, White-bellied Forest Rat, White-bellied Nesomys

MEASUREMENTS
Head-body: 178–213mm. Tail: 167–210mm. Weight: 155–235g; female mean 210g, male mean 225g. (Ryan 2003, Carleton *et al*. 2022d)

DESCRIPTION & IDENTIFICATION Very similar to Eastern Red Forest Rat, but generally larger. Upper pelage reddish-brown flecked with darker guard hairs. Throat, chest and belly very pale, often off-white. Tail covered in dense hair and often has white tuft at tip. Snout elongate; ears long and prominent (Ryan 2003). Large robust forest rat, with overall similarity to Brown Rat, but obviously reddish in colour with distinctive white underparts.

HABITAT & DISTRIBUTION Prefers lowland and mid-elevation rainforest, including littoral forest, from sea level to *c*.1,000m; mostly below 800m. Patchily distributed within range which extends from Masoala Peninsula in north to region of Tsitongambarika, in Vohimena Mountains in south (Carleton & Schmidt 1990, Goodman *et al* 1999a, Andrianjakarivelo *et al*. 2005, Rakotomalala *et al*. 2007). Sympatric with Eastern Red Forest Rat at some localities between 800 and 1,000m, e.g. Masoala, Mantadia, Ranomafana (Ryan 2003). IUCN Least Concern.

Lowland Red Forest Rat, Masoala National Park.

BEHAVIOUR Terrestrial, foraging on forest floor for seeds and fallen fruits, but also capable of climbing among fallen logs. Night spent in burrow, located under tangle of roots, a fallen log or similar. Burrows similar to Eastern Red Forest Rat: multi-chambered with several entrances, deeper chambers used for sleeping. Activity begins around sunrise and continues through morning. Declines after midday and rats return to burrow before dark. Male home range up to 1.4ha: males appear more active than females, whose home ranges are smaller, c.0.5ha (Ryan 2003). Home ranges non-exclusive. Daily travel no more than 400–500m (Ryan 2003). Reproductive activity recorded July and August, with young born c.6 weeks later. Litter size of two suspected. Main predators probably similar to Eastern Red Forest Rat, i.e. Eupleridae carnivorans, especially Ring-tailed Vontsira, diurnal raptors and large snakes.

WHERE TO SEE Can be encountered at lower elevations in Ranomafana National Park, where it is sympatric with Eastern Red Forest Rat. Forests at Lohatrozona and Tampolo on Masoala Peninsula are also likely places.

Western Red Forest Rat, Tsingy de Bemaraha National Park.

WESTERN RED FOREST RAT
Nesomys lambertoni   G. Grandidier, 1928

Other names: Lamberton's Forest Rat, Tsingy Red Forest Rat, Western Nesomys

MEASUREMENTS*
Head-body: 189–227mm. Tail: 160–191mm. Weight: 225–243g. (Goodman & Schütz 2003) (*two specimens)

DESCRIPTION & IDENTIFICATION A large red forest rat, with distinctive 'bushy' tail that looks vaguely 'squirrel-like' (Goodman & Monadjem 2017). Body similar to Eastern Red Forest Rat, but dorsal pelage darker reddish-brown, underparts paler brown, with obvious orange tinge. Snout and ears long. Tail long and conspicuously hairy (dark) with no obvious tuft at tip.

HABITAT & DISTRIBUTION Very limited range: known only from deciduous forests associated with limestone karst (*tsingy*) formations (up to c.100m), south of Antsalova in Tsingy de Bemaraha National Park. Can appear common in some areas. IUCN Endangered.

BEHAVIOUR Terrestrial and probably diurnal. Observed foraging on forest floor among leaf litter between limestone karst formations. Also climbs over smaller rocky formations and eats seeds and fallen fruits.

WHERE TO SEE Readily seen in forests within Tsingy de Bemaraha National Park, north of Manambolo River. Look in more shaded forest areas, particularly around fallen rocks and boulders, adjacent to damp areas.

Non-native Mammals

Unlike many other tropical islands, Madagascar has not (yet) suffered a major intrusion of species, accidentally or intentionally, introduced by man. Nonetheless, a number of non-native mammal species are now established. These include species that live as human commensals, e.g. Black Rat *Rattus rattus*, Brown Rat *Rattus norvegicus*, House Mouse *Mus musculus* and Musk Shrew *Suncus murinus*, some of which have penetrated native forest areas. Others include domesticated species like dogs *Canis familiaris*, cats *Felis silvestris*, goats *Capra hircus*, sheep *Ovis aries*, pigs *Sus scofa*, rabbits *Oryctolagus cuniculus domesticus* and zebu cattle *Bos taurus*. There are also introduced wild species that have become naturalised, including the ungulates Fallow Deer *Dama dama*, Rusa or Timor Deer *Cervus timorensis* and Bush Pig *Potamochoerus larvatus*, plus one carnivoran, the Small Indian Civet *Viverricula indica* (Goodman *et al.* 2003b, Goodman & Soarimalala in press).

Non-native Rodents

Three rodents have been unintentionally introduced to Madagascar. Black Rat, Brown Rat and House Mouse have spread to most parts of the planet and, with the exception of humans, must be among the most widely distributed and prolific mammals. It is not known when the first introductions to Madagascar occurred, but initial colonisations may have been when the first humans arrived around 2,000–3,000 years ago (Duplantier & Duchemin 2003). Since then, other waves of introduction may have taken place and the frequency of these has probably increased in recent times.

A fourth rodent, the Guinea Pig *Cavia porcellus* (family Caviidae) was intentionally introduced to the island as a farm animal for meat production and as a household pet. They have not proved popular as a food source. There are no free-living populations and no negative ecological consequences for endemic species (Goodman & Soarimalala in press, Ramasindrazana *et al.* in press).

Old World Rats and Mice Subfamily Murinae

Due to their ability to survive and adapt to change, capacity to reproduce quickly, and history of spectacular diversification, this large subfamily is regarded as one of the most successful groups of mammals: the subfamily Murinae is naturally distributed throughout the Old World south of the Arctic Circle (but not in Madagascar) with approximately 656 species and at least 135 genera (Denys *et al.* 2017).

OLD WORLD RATS

GENUS *Rattus*

Rattus is one of the largest mammalian genera, containing 64 species found throughout the Old World (Denys *et al.* 2017). The two species introduced to Madagascar, Black Rat and Brown Rat, are the most familiar because of their constant association with humans. However, they are not representative of the genus as a whole: most *Rattus* species are more specialised, prefer natural forests, have restricted ranges and often avoid habitation, and all are omnivorous and eat a wide variety of animal and plant matter. Seeds, grains, nuts, fruits and vegetables are preferred by most species.

The earliest records of the genus *Rattus* in Madagascar are somewhere between the 11th and 14th centuries from a site near Ambanja associated with a Middle Eastern/East African maritime trading network (Rakotozafy 1996), although an even earlier arrival on Madagascar is suspected. Initially it seems likely that they were largely confined to areas associated with humans, and their spread beyond into native habitats has been comparatively recent, i.e. within the last century (Randriamoria *et al.* in press). Subsequently, *Rattus*, especially Black Rat, has spread dramatically throughout the island and into all native habitats, with greater densities in eastern humid areas (Duplantier & Duchemin 2003).

There is considerable dietary overlap between the endemic Nesomyine rodents and the genus *Rattus*, and there is increasing evidence that they are displacing endemic taxa from some native forests via resource competition and direct predation (Goodman 1995, Jansa & Carleton 2003a, Ryan 2003). An increasing *Rattus* population may also have a devastating effect on endemic birds, particularly ground-dwellers and nesters such as mesites (Mesitornithidae) and ground rollers (Brachypteraciidae).

BLACK RAT
Rattus rattus
Linnaeus, 1758

Other names: Common House Rat, House Rat, Ship Rat, Roof Rat

Malagasy name: Voalavo

MEASUREMENTS*
Head-body: *c.*130–180mm.
Tail: *c.*180–210mm.
Weight: 100–185g. (Denys *et al.* 2017) (* Africa specimen measurements)

DESCRIPTION & IDENTIFICATION Typical, but highly variable, slender *Rattus*. Pelage appears coarse due to long dark guard hairs, but sleek. Upperparts dark slate-grey (almost black, more common in urban areas) to greyish-brown (especially in rural areas), underparts and throat paler. Ears relatively long, pinkish and virtually hairless. Dark eyes prominent. Vibrissae long and stiff. Feet broad and long. Tail unicoloured, long (95–120% head/body), thin and lacks hairs. Taxonomy of *R. rattus* and relatives currently in flux. May represent a complex of 4–6 species. Immature Black Rats could be confused with any endemic species, particularly Sleek-furred Ground Rat and possibly red forest rats (*Nesomys* spp.) or the larger tufted-tailed rats (*Eliurus* spp).

HABITAT & DISTRIBUTION Originally native to southern Indian Subcontinent (Alpin *et al.* 2011), and now introduced worldwide. Black Rats in Madagascar conform to the most widespread lineage, spreading originally from India to Europe with early human migrations, then to Africa with European expansion. In Madagascar probably derived from two independent introductions, one to north, the other to south, and both probably from same ancestral source on the Arabian Peninsula (Ramasindrazana *et al.* in press, Randriamoria *et al.* in press).

Populations may have remained local until late 19th century. By mid-20th century probably widespread, including native forest areas. Today widely distributed throughout the island, in virtually all natural and anthropogenic habitats; commoner in humid eastern areas, than in west and south. Associated with rural villages, degraded areas and wherever native forests have been replaced by agriculture. Also penetrates deep into native forest and across elevational zones to over 2,500m, as well as offshore islands like Nosy Mangabe (Goodman *et al.* 2003b, Ramasindrazana *et al.* in press, Randriamoria *et al.* in press). Where Black Rat has colonised native habitats, there is a corresponding decline in endemic Nesomyine rodents which are strictly forest-dwelling. In some eastern humid native forests Black Rats are commoner than endemic rodents, with highest densities at mid-elevations; less common in drier regions of west and south-west, probably due to preference for water sources and damp habitats (Ramasindrazana *et al.* in press, Randriamoria *et al.* in press).

BEHAVIOUR Predominantly nocturnal/crepuscular. Extremely generalist omnivore: diet consists mainly

Black Rat.

of vegetables, fruits, seeds, grains and nuts. Also eats invertebrates, eggs of birds and reptiles, including endemic species, and small reptiles and amphibians. An adept scavenger that also feeds on human refuse (Denys *et al.* 2017).

Territorial, defending its entire home range, which is usually small (especially in urban environments). Two or more females are generally subordinate to a larger dominant male, but are themselves dominant over other group members.

When conditions are favourable, Black Rat can breed year-round (Duplantier & Rakotondravony 1999): females can breed three to nine litters per year with up to ten pups per litter. Sexual maturity can be reached by 68 days, gestation period 20–22 days. Such prolific reproduction theoretically means a pair can have hundreds of descendants in a year and millions within three years; however, on average a single female produces no more than 25 per year (Randriamoria *et al.* in press). In the wild mean longevity one year, with 90–95% annual mortality rate.

Now widespread in many rainforest areas and probably a severe competitor to sympatric endemic rodents: broad dietary overlap between Black Rat and Nesomyine species, and direct predation may also occur (Dammhahn *et al.* 2017, Ramasindrazana *et al.* in press). In some isolated forest patches, native species have been extirpated already, whilst in other native forests Black Rat may be the most abundant rodent species. Endemic rodents suffering most directly are short-tailed rats (*Brachyuromys* spp.) (Jansa & Carleton 2003a), and, to a lesser extent, red forest rats (*Nesomys* spp.) and tufted-tailed rats (*Eliurus* spp.). Black Rat is eaten by Fosa (Goodman *et al.* 1997b, Hawkins 2003).

Black Rat is a major cause of public health issues, being a significant reservoir for diseases including bubonic plague. Also a major crop pest and causes extensive damage to standing and stored rice stocks, also cassava, sweet potatoes, tomatoes, corn and beans (Duplantier & Rakotondravony 1999, Ramasindrazana *et al.* in press, Randriamoria *et al.* in press).

WHERE TO SEE Often seen in towns and villages around rubbish and where rice and grain is stored. Also often seen around more permanent forest camps.

BROWN RAT
Rattus norvegicus
Berkenhout, 1769

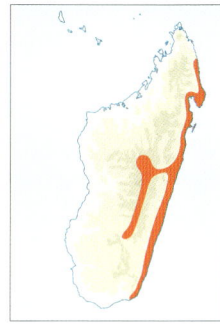

Other names: Common Rat, Sewer Rat, Domestic Rat, Norway Rat

Malagasy names: Voalavo, Voalavo siedoa

MEASUREMENTS*
Head-body: *c.*160–290mm. Tail: *c.*120–250mm. Weight: *c.*200–540g. (*worldwide specimen measurements)

DESCRIPTION & IDENTIFICATION Typical, variable, robust, large *Rattus* species. Pelage long, shaggy and coarse. Dorsal areas various shades of brindled brown-grey, overall appearance from mid-brown to dark grey. Ventral areas paler grey, sometimes creamy-white on throat and chest. Ears short, hairy and partially obscured. Vibrissae long and stiff. Feet broad, hindfeet longer. Tail relatively short (*c.*80–95% head/body), thick, lightly haired and distinctly bicoloured, dark brown on top, paler brown below. Overall, larger and more robust than Black Rat. Less genetic and morphological variation in Brown Rat than Black Rat (Denys *et al.* 2017).

HABITAT & DISTRIBUTION Probably originated in South-East Asia (Zeng *et al.* 2018), but previously thought to have originated in northern China and Mongolia (Aplin *et al.* 2003): had spread throughout Asia, Middle East, Africa and Europe by 16th century. Historically found in primary forest and fields, now all kinds of disturbed habitats, including almost all major

Brown Rat.

cities, agricultural land and arable areas, throughout its human-expanded distribution (Denys *et al.* 2017). More common in temperate colder habitats, less so in warmer tropical environments.

In Madagascar, has most restricted range of introduced rodents; limited mainly to ports and larger towns and villages in eastern lowlands and Central Highlands to 1,500m (Duplantier & Duchemin 2003, Soarimalala & Goodman 2011). Also trapped well away from habitation in some areas, particularly on coast. Earliest record (museum specimen) in Madagascar is 1929 (Ramasindrazana *et al.* in press). Timing of arrival in Central Highlands unknown, but thought to be relatively recent, perhaps in last 50–70 years: first records in Antananarivo early 1980s (Randriamoria *et al.* in press). Subsequent colonisation of suitable areas has been dramatic; in Antananarivo now more than 90% of all rodents trapped (Duplantier & Duchemin 2003, Ramasindrazana *et al.* in press). Prefers damp habitats, with regular access to drinking water. Does not appear to penetrate native forest areas to any extent (Soarimalala *et al.* 2019, Ramasindrazana *et al.* in press).

BEHAVIOUR Nocturnal with primary activity before midnight, although daytime activity is not uncommon. Mostly terrestrial; rarely climbs. Excellent swimmer, so also occurs in semi-aquatic environments, thriving in sewers. Opportunistic omnivore, feeding on a variety of seeds, nuts, fruits, invertebrates and smaller vertebrates. Also eats any foodstuff associated with humans. In native forests may also predate eggs and chicks of native birds. Rarely competes directly with Black Rat as generally more terrestrial, whereas Black Rat is arboreal and scansorial.

In urban environments range is small, rarely moving more than 70–80m. In open areas may travel 3–4km per night. Colonies develop from pair or single pregnant female. Very large aggregations the result of several smaller units (clans) coming together. Reproduces year-round: when conditions favourable up to five litters per year. Gestation *c.*21 days, with 3–7 young per litter (Denys *et al.* 2017).

Due to scarcity in native forests, impact of Brown Rat on endemic species is difficult to assess and probably low: certainly the impact of Black Rat on endemic Nesomyine species is far more serious (Ramasindrazana *et al.* in press, Randriamoria *et al.* in press).

WHERE TO SEE Can be seen around waste ground, sewers and refuse tips in urban areas.

OLD WORLD MICE
GENUS *Mus*

A large genus containing *c.*40 species distributed throughout the Old World (Denys *et al.* 2017). Most dwell in natural habitats like forests and savannas, and have restricted distributions, but House Mouse *Mus musculus* and a few closely related species have spread dramatically in association with man. The earliest records on Madagascar date from an Islamic port south of Ambanja between the 9th and 10th centuries (Radimilahy 1998) and suggest colonisation from the Middle East via Arab trading routes (Hutterer & Tranier 1990, Ramasindrazana *et al.* in press).

HOUSE MOUSE
Mus musculus   Linnaeus 1758

MEASUREMENTS
Head-body: *c.*70–100mm.
Tail: *c.*65–105mm. Weight: *c.*12–35g.

DESCRIPTION & IDENTIFICATION Medium-sized *Mus* species. High variable in size and colour depending on environment. Generally, upperparts uniform greyish-brown, underparts paler. Ears rounded and relatively small. Eyes small. Tail prominently ringed, moderately long, usually equal to head-body length. In Madagascar represented by subspecies *M. m. castaneus*, which also occurs in Kenya, Pakistan, India and South-East Asia (Denys *et al.* 2017). Smaller than majority of endemic rodents: Only white-tailed mountain mice (genus *Voalavo*), Malagasy Mountain-dwelling Mouse and Western Big-footed Mouse are similarly diminutive.

HABITAT & DISTRIBUTION Evolutionary origins of House Mouse probably Indian Subcontinent; naturally occurs throughout Eurasia and many parts of Africa, and now introduced globally to all regions except Antarctica (Denys *et al.* 2017). In Madagascar, found throughout the island in association with habitation from rural villages and agricultural areas to major towns and cities. Also occurs in native habitats including montane zones (Ramasindrazana *et al.* in press).

BEHAVIOUR Terrestrial; mainly crepuscular and nocturnal. Omnivorous, with preference for invertebrates, cereals, seeds and other plant matter. Home range varies with habitat and locality: in farmyards, buildings or similar may be as small as 4m², while field-dwelling populations may be semi-nomadic and cover 1–2km². Socially flexible: in commensal populations females non-aggressive towards each other, in other situations there is high degree of intolerance (Denys *et al.* 2017). Lives in colonies or extended family clans, with each colony dominated by single male (gains more mating success than subordinate males). Able to breed year-round, but reproduction reflects resource availability. Litter size and frequency reduce in winter, peak in spring/ summer. Gestation *c.*20–25 days. Can be prolific,

House Mouse.

with 5–10 litters per year, each containing between four and eight young.

WHERE TO SEE Often seen in villages, especially where there are grains or cereals.

Non-native Carnivorans

Three species of carnivoran have been introduced to Madagascar: the domestic dog *Canis familiaris*, feral cat *Felis silvestris* and Small Indian Civet *Viverricula indica*. The impact of all three species is significant, not only in terms of direct competition and exclusion of native carnivorans (Goodman *et al.* 2003, Farris *et al.* 2015b), but also more broadly via predation on many other native faunal groups, e.g. lemurs, rodents and ground-nesting birds (Brockman *et al.* 2008, Murphy 2015, Farris *et al.* 2017b, Merson *et al.* 2019b, Murphy *et al.* 2019). In one study in Makira Natural Park in the north-east, the populations of four out of six resident native carnivorans, Falanouc, Ring-tailed Vontsira, Broad-striped Vontsira and Brown-tailed Vontsira decreased by at least 60% over a six-year period as a consequence of increases in non-native carnivoran populations (Farris *et al.* 2017b).

True Civets Family Viverridae, Subfamily Viverrinae

Restricted to the Old World, the Viverridae is one of the most diverse carnivoran families and includes the civets, genets, oyans and their allies. Currently 34 species are recognised, although this is likely to change with further taxonomic research and revision (Jennings & Veron 2009). The terrestrial civets form the subfamily Viverrinae, which contains six species in three genera, *Viverricula*, *Civetticis* and *Viverra*. They are distributed through Africa and Asia (Jennings & Veron 2009).

SMALL INDIAN CIVET
GENUS *Viverricula*

Viverricula is a monotypic genus whose single representative naturally occurs across much of the Indian Subcontinent and South-East Asia.

SMALL INDIAN CIVET
Viverricula indica (É. Geoffroy Saint-Hilaire, 1803)

Other name: Lesser Oriental Civet

Malagasy name: Jaboady

MEASUREMENTS
Head-body: 485–680mm.
Tail: 300–430mm. Weight: 2–4kg. (Jennings & Veron 2009).

DESCRIPTION & IDENTIFICATION Taxonomy debated, with as many as 11 subspecies recognised (Jennings & Veron 2009), although more recent research suggests only four are valid (Gaubert *et al.* 2017). In Madagascar, nominate *V. i. indica* occurs (Murphy & Farris in press).

Pelage short and coarse. Muzzle short and ears close-set. Upperparts brownish-grey to tawny or beige, underparts slightly paler and greyer. Head and rounded face slightly darker with few markings, except dark areas around eyes. Forequarters covered in small dark spots, which gradually increase in size towards rear and may fuse to form 6–8 longitudinal stripes on back and flanks (effect more obvious on back/upper flanks than lower flanks). Tail has six to nine dark rings, broader than intervening spaces, with whitish tip. Legs

Small Indian Civet.

and feet very dark brown-grey, sometimes almost black. Gait digitigrade (walks on its digits).

A small terrestrial civet with conspicuous body markings. Larger than all but two of native carnivorans, Fosa and Falanouc. Confusion between smaller individuals and Spotted Fanaloka possible: stockier build, shorter muzzle, broader, more rounded ears (rather than narrower and more triangular), flank markings, dark legs and feet, and boldly ringed tail of Small Indian Civet should serve to differentiate.

HABITAT & DISTRIBUTION Originates from Indian Subcontinent including Sri Lanka, Indochina, mainland South-East Asia, and islands of Hainan, Taiwan, Sumatra and Java. Its presence on Indonesian islands of Bali, Bawean, Kangean, Lombok and Sumbawa may be due to introduction (Jennings & Veron 2009). Other than Madagascar, Small Indian Civet has also been introduced to Zanzibar and Mafia Islands (off Tanzania) and the Comoros Islands and Socotra in the Indian Ocean (Choudhury *et al.* 2015).

When the introduction to Madagascar occurred is uncertain (Goodman *et al.* 2003), although it may have arrived from the Indo-Malay region (probably sometime after 900 CE) as a consequence of 'wild farming' and the trade in 'civet oil', extracted from the perineal glands and used to produce perfume during the 'Islamic Golden Age' (Gaubert *et al.* 2017, Murphy & Farris in press). Skeletal remains, almost certainly of Small Indian Civet, found in archeological remains dated 1030 BP recovered at Mahilaka, an Islamic trading port in north-west Madagascar (Rakotozafy 1996). In addition, Small Indian Civet remains have been found at various other paleontological and archaeological sites around the island (Murphy & Farris in press). The species' subsequent gradual spread from north to south across the island mirrored that of humans.

Today in Madagascar, widely distributed except in arid south and south-west, and recorded in most major natural and anthropogenic habitat types (Goodman 2012). Probably prefers lower elevations, but recorded as high as 1,500–2,000m (Jenkins & Carleton 2005). Small Indian Civet can penetrate intact native forest, but more often is associated with degraded and secondary forests and cultivated areas (Gerber *et al.* 2012, Farris *et al.* 2015a, 2017a,b, Murphy & Farris in press). Regularly found in and around human settlements and agricultural areas including the Central Highlands, where seen considerable distances (many tens of kilometres) away from forest and native habitat (Murphy & Farris in press). Trapped and killed as a pest and hunted for food (Golden 2009, Borgerson 2016).

BEHAVIOUR Usually solitary and nocturnal, but may be active by day in undisturbed areas. Terrestrial, but capable of climbing. Normally sleeps in shallow burrows that it excavates, or in clumps of dense vegetation. In urban areas, frequents disused buildings and drainage ditches. Virtually nothing known about breeding in Madagascar. Males and females probably pair during austral winter when mating occurs. Gestation period c.67 days. Litter of two to five born between September and December in a burrow or equally safe area on the ground.

Diet generalist and omnivorous: feeds on fruits, roots, carrion, insects, other invertebrates and small vertebrates, probably including rodents, tenrecs and small lemurs (mouse lemurs and dwarf lemurs) and possibly larger species including Ring-tailed Lemur (Goodman *et al.* 1993c, Goodman 2003a). In rural areas regularly predates domestic poultry (Goodman 2009) and in urban areas often scavenges around refuse.

Presence of Small Indian Civet negatively impacts native carnivoran species and communities. There is considerable niche overlap, particularly in diet, with Spotted Fanaloka, which suffers accordingly and is consequently excluded from forest: Spotted Fanaloka occupancy can decrease by 40% at sites where Small Indian Civet is present (Farris *et al.* 2021).

WHERE TO SEE Because of association with human settlements, may be encountered at night patrolling perimeter of rural villages and field camps, or urban

Small Indian Civets are widely killed as pests and hunted for food.

fringe areas, especially where there is rubbish. Often seen in Berenty Private Reserve, especially in October–December when they predate fallen eggs and chicks below colonies of breeding Cattle Egrets *Bubulcus ibis*.

Domestic Carnivorans

Domestic/feral dogs and cats in and around native forest and fragmented forest areas negatively impact all native carnivorans, via direct predation, competition for resources, disease transmission and by forcing alteration in regular activity patterns of native species (Farris *et al.* 2015a, 2017b, 2020, 2021).

DOGS

The domestic dog *Canis familiaris* occurs in close association with man throughout the island and in places is used to hunt native species such as Common Tenrec and Greater Hedgehog Tenrec. It is probable that some dogs live a largely feral existence at the edge of native forest (within 2km of a village) and prey on some endemic mammals, including lemurs, small mammals and birds (Goodman *et al.* 2003b). Because activity periods overlap considerably, the presence of feral dogs (which are largely diurnal) negatively impacts the occurrence of Ring-tailed Vontsira and

Pack of feral domestic dogs, Andasibe-Mantadia National Park.

Brown-tailed Vontsira, while it not only adversely affects Fosa occurrence, but also causes a shift in activity patterns, with the result that Fosa becomes much more nocturnal (Farris *et al.* 2015a, 2017a).

FERAL CATS

Both domestic cats, i.e. those belonging to a local household and largely confined to villages, and feral cats, i.e. those without owners that forage independently within forested areas, have been documented throughout Madagascar (Goodman 2012). Feral or so called 'forest cats' survive in many forests island-wide, but appear more common in drier western regions (Goodman *et al.* 2003b). The majority broadly resemble European Wild Cats *Felis silvestris* or African Wild Cats *F. lybica*, with 'tabby-like' pattern and coloration, and a tail with dark rings and a dark tip. They are distinctly larger, often around twice the size, of a typical domestic cat (Goodman 2012). In 1792 a cat from the island was

Feral cat in western dry forest.

described as *Felis catus madagascariensis* (Kerr 1792), the author noting that, 'this is a beautiful variety, which inhabits the island of Madagascar'; it is now considered synonymous with the 'wild cat' form, *F. silvestris* (Driscoll *et al.* 2007). They are well known in rural areas as they often predate poultry from villages. Malagasy names include *ampaha* and *kary* and distinguish them from the island's pet cat population.

In parts of the north-east (Masoala Peninsula and Makira Natural Park) an all-black form (occasionally with faint darker stripes) occurs, known as *fitoaty*, which is described as larger than 'regular' tabby variants (Borgerson 2013, Farris *et al.* 2016a). Anecdotal and camera-trap evidence suggests these cats have longer legs and more muscular hindlegs than tabby forms, and are comparable in height to the largest native carnivore, the Fosa. Indeed, local people describe the *fitoaty* as being larger and heavier than Fosa and sometimes refer to it as the 'black fosa'. Studies also suggest the two forms have different ecological preferences, with tabby forms occurring in and around secondary and degraded forests in the vicinity of villages, whereas the black variant primarily occupies intact forest and rarely ventures close to forest edges and villages (Farris *et al.* 2016a).

The origin of feral cats in Madagascar has been the subject of much speculation. It is likely they were introduced in both domestic and 'wild' forms, although when is unclear; remains have been found in subfossil deposits, but these are thought to be relatively recent (Goodman *et al.* 2003). Molecular analysis suggests they originated from areas around the Arabian Sea in the vicinity of United Arab Emirates, Oman and Kuwait (with earlier genetic influences from India and Pakistan), and probably arrived with Arab traders, who sailed along the East African coast to Zanzibar, the Comoros Islands and Madagascar (Sauther *et al.* 2020). It is probable that both cats and the Small Indian Civet arrived at a similar time (between the 8th and 13th centuries) via the same trade routes.

Feral cats pose a significant threat to a wide variety of native species due to their adaptability, elusive nature, and efficient, generalist hunting behaviour. Their negative impact is considerable, both spatially and temporally. Via direct competition for resources and subsequent exclusion, their presence in forests is particularly detrimental to Spotted Fanaloka, Ring-tailed Vontsira, Brown-tailed Vontsira and Broad-striped Vontsira (Barcala 2009, Gerber 2012, Farris *et al.* 2015a,b, 2016b, Rasambainarivo *et al.* 2017, Merson 2018, Merson *et al.* 2018b). A study in the north-east over a six-year period found a significant increase in feral cat populations contributed to a *c.*60% decline in those of smaller native carnivorans (Farris *et al.* 2017b).

Feral cats are known predators of large (sifakas), medium ('true' lemurs) and small (mouse lemurs)

Feral cat predates male Red-fronted Brown Lemur, Andranomena Forest.

lemurs (Brockman *et al.* 2008, Merson *et al.* 2019), endemic rodents (Farris *et al.* 2017b), endemic birds (Murphy 2015, Murphy *et al.* 2019) and endemic reptiles (Merson *et al.* 2019).

Non-native Insectivores

Insectivore is an informal term used in connection with members of the order Eulipotyphla, which have a relatively generalised diet dominated by invertebrates. The order includes hedgehogs and gymnures (family Erinaceidae) and shrews (family Soricidae).

Shrews Family Soricidae

The shrews (Family Soricidae) represent the fourth largest mammal family, currently numbering 448 species in 26 genera (Burgin & He 2018). Soricidae includes one of the world's smallest terrestrial mammal species, Etruscan Shrew *Suncus etruscus*. Their distribution is extensive, on all major landmasses with the exception of Arctic islands, Greenland, Iceland, the West Indies, New Guinea, Australia, New Zealand and some Pacific islands. In the New World, they have spread as far as northern South America.

Some members of the endemic family Tenrecidae, particularly shrew tenrecs (genera *Nesogale* and *Microgale*) and Soricidae show obvious similarities in body form, lifestyle, behaviour and ecology, and provide excellent examples of convergent evolution (Goodman *et al.* 2003).

Shrews are all small, short-legged insectivores, with elongated pointed snouts. Their pelage is short and dense. Upperparts are generally grey to grey-brown, becoming paler below. Eyes are very small and often almost hidden, whilst the vibrissae (whiskers) are long and prominent, and the ears small.

Two species of shrew in the genus *Suncus*, the musk or pygmy shrews, occur on Madagascar (Goodman 2003b), both a consequence of introductions, although the precise situation regarding the Etruscan Shrew is uncertain.

MUSK SHREWS
GENUS *Suncus*

The resemblance of the genus *Suncus* to native shrew tenrecs (*Nesogale* and *Microgale* spp.) is striking and confusion may occur if only brief views of an animal in the field are possible. On closer examination differences are significant. The dentition of true shrews is reduced (no more than 32 teeth) compared to shrew tenrecs. Also, unlike shrew tenrecs, musk shrews do not possess a cloaca.

ASIAN MUSK SHREW
Suncus murinus   (Linnaeus, 1766)

Other names: Asian House Shrew, Indian Musk Shrew, Grey Musk Shrew

Malagasy names: Voalavonarabo, Voalavo fotsy

MEASUREMENTS*
Head-body: 90–160mm. Tail: 45–110mm. Weight: *c.*25–140g. (Burgin & He 2018) In Madagascar: females 30–35g, males 45–50g. (* worldwide specimen measurements)

DESCRIPTION & IDENTIFICATION A large shrew with highly variable body size and very strong, distinctive odour. Male larger than female. Pelage short, fine and velvet soft. Two colour morphs: one

Asian Musk Shrew, Berenty Private Reserve.

has pale grey upperparts, slightly paler underparts and a whitish tail; the other has dark greyish-brown upperparts, paler grey underparts and dark grey tail. Both occur on Madagascar. Pale form is more common. Many introduced populations on different islands around the world, and these subpopulations have now diverged considerably in morphology, physiology and size, hence subspecific differentiation may apply. In Madagascar individuals are relatively small.

HABITAT & DISTRIBUTION Originates in Asia where widespread: natural distribution from Japan to Pakistan. All records further west are probably introductions by humans (Hutterer & Tranier 1990). Now widely distributed in Arabia, East Africa, the Mascarene Islands and Madagascar.

May have arrived initially in Madagascar via African and Arabian trade routes: the Malagasy name *Voalavonarabo*, which means 'Arab mouse' alludes to this (Ramasindrazana *et al.* in press). More recently, this species has been accidentally introduced to other parts of the world via freight on ships, etc., with subsequent waves of introduction to Madagascar (Hutterer & Tranier 1990). The Malagasy population derived from several separate colonisations involving different founder stock localities: some probably came from Arabian Peninsula via East Africa, others directly from South-East Asia. First record in Madagascar was in 1858 (Hutterer & Tranier 1990).

Today Asian Musk Shrew occurs as a human commensal in cities, towns and villages in both eastern humid regions and dry western ones. In east occurs in native forests, including secondary and degraded zones, and in and around agricultural areas, including rice fields (Duplantier & Duchemin 2003, Soarimalala & Goodman 2011, Ramasindrazana *et al.* in press). Also occurs in dry deciduous forest, but appears absent from south and south-west spiny thicket (Goodman *et al.* 2002, Soarimalala 2008). Recorded from near sea level to *c.*1,600 m (Ramasindrazana *et al.* in press).

BEHAVIOUR Lives as human commensal throughout range (Hutterer & Tranier 1990) and found in man-made habitats, cultivation and around habitation. Nocturnal and solitary. Vocal repertoire of high-pitched chirrups, clicks and buzzes associated with aggressive behaviour. Day spent in burrow or inside buildings.

Diet omnivorous, but voracious hunters, eating a variety of insects, other invertebrates and small vertebrates, as well as fruits, other plant material and edible human refuse.

May breed throughout year in Madagascar, with peak in October–December. Gestation 30 days, litter size 2–6, three average. Weaning between 17 and 20 days and apparent sexual maturity is reached after c.36 days.

There appears to be no correlation between high densities of Asian Musk Shrew and reduced diversity in shrew tenrecs and other tenrecs (Goodman & Rakotondravony 2000).

WHERE TO SEE May be encountered at night in most areas around human settlements.

Etruscan Shrew.

ETRUSCAN SHREW
Suncus etruscus   (Savi, 1822)

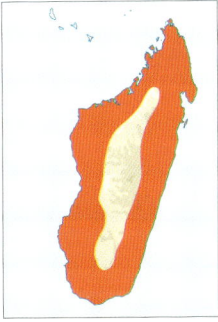

Other names: Pygmy Musk Shrew, Common Dwarf Shrew, Pygmy White-toothed Shrew

Malagasy name: Voalavo fotsy

MEASUREMENTS
Head-body: c.39–51mm. Tail: 31–35mm. Weight: 1.8–3.2g, mean 1.7g. (Goodman *et al.* 1999, 2003b)

DESCRIPTION & IDENTIFICATION Pelage short, soft and greyish-brown. No clear demarcation between upperparts and underparts. Ears rounded and prominent. Tail usually longer than 60% head-body length. Smallest shrew and smallest terrestrial mammal in Madagascar.

HABITAT & DISTRIBUTION Widespread throughout the Old World, from South-East Asia, Indian Subcontinent, Central Asia, Middle East, Europe and North Africa. Populations in Madagascar previously described as endemic species *S. madagascariensis*, or subspecies *S. e. madagascariensis*. Recent phylogenetic analysis confirms such differentiation is invalid and Malagasy populations conform to *S. etruscus* (Omar *et al.* 2011, Sheftel 2018).

In Madagascar widely distributed including eastern, western and southern regions, from sea level to c.1,580m. Occurs in villages and rural communities, and various native forest formations, where often sympatric with Asian Musk Shrew (Ramasindrazana *et al.* in press). Distinctly more common in drier west and south than in humid east: in Andohahela National Park, not recorded in rainforest (Parcel 1), but commonly encountered in disturbed spiny forests (Parcel 2) (Goodman *et al.* 1999b).

BEHAVIOUR Probably solitary and nocturnal. Diet primarily small insects and other small invertebrates. Litter size up to five (Goodman *et al.* 2003b).

Hoofed Mammals

Although there are no extant native ungulates in Madagascar, historically there were three species of dwarf hippopotamus that are now extinct (see Extinct Mammals). At various times, seven species of hoofed mammal have been introduced, four of them domesticated species – goats, sheep, pigs and zebu cattle, and three are wild – Bush Pig, Fallow Deer and Javan or Timor Deer.

Wild Pigs Family Suidae

An Old World family currently containing 18 species in six genera (Meijaard *et al.* 2011). Two species occur on Madagascar, the domestic pig *Sus scrofa domesticus* and feral Bush Pig *Potamochoerus larvatus*.

There is conjecture as to how Bush Pig arrived on the island. Some authorities have suggested the species arrived naturally and should be considered native (Andrianjakarivelo 2003). On mainland Africa, Bush Pigs are known to venture frequently into extensive papyrus beds, which may become detached, float downriver and out to sea. In this manner (sweepstake dispersal) Bush Pigs might have reached Madagascar independently of humans (the same way it is assumed the now extinct dwarf hippopotamus also reached the island). Alternatively, the majority view, followed here, is that they were introduced and have subsequently become naturalised.

Introductions may have coincided with the arrival of the first humans around 2,000 years ago, in which case it is supposed Bush Pigs were, at least partially, domesticated before they reached Madagascar and there is evidence supporting this in some parts of Africa (Andrianjakarivelo 2003).

There are no known remains in subfossil deposits and the earliest known written reference is from the mid-17th century (Goodman *et al.* 2003b). The lack of evidence until recent times suggests the species is a relatively recent addition to the Malagasy fauna, adding further weight to the theory that the species is introduced. It is nonetheless an important component of the island's biota, not only ecologically, but also socially and economically, so is afforded more extensive treatment here.

BUSH PIGS

GENUS *Potamochoerus*

A genus containing two species, Red River Hog *P. porcus* and Bush Pig *P. larvatus*. They are distributed through equatorial, southern and eastern sub-Saharan Africa: Red River Hog being found in tropical rainforests of West and Central Africa, from Senegal and Guinea-Bissau to northern Uganda and eastern Democratic Republic of Congo, and as far south as northern Angola; Bush Pig's range is primarily limited to East and south-east Africa and Madagascar (Meijaard *et al.* 2011).

The Malagasy people generally consider Bush Pigs as harmful animals. The devastation they cause is very apparent in agricultural areas, where they can quickly destroy crops of cassava, sweet potatoes, corn, sugarcane and even rice (Andrianjakarivelo 2003).

Male Bush Pig.

BUSH PIG

Potamochoerus larvatus (F. Cuvier, 1822)

Malagasy names: Lambo, Lambodia, Lamboala, Antsanga

MEASUREMENTS*
Head-body: 900–1,200mm. Tail: 300–400mm. Shoulder height: 600–800mm. Weight: females up to 55kg, males up to 70kg. (Paulian 1984) (*estimated measurements for Malagasy specimens)

DESCRIPTION & IDENTIFICATION A compact-bodied pig with short legs, rounded back, long snout and shaggy appearance. Hair covers entire body, including head and face, long and bristly but relatively sparse. Distinct dorsal crest from between ears down length of spine. Ears tufted with extended pinnae. Cheeks of males in particular bushy and beard-like. Males have small tusks, barely visible but used for defence. Coloration variable: generally face and head predominantly grey, extending over nape and part-way down dorsal crest. A darker broad band surrounds muzzle. Upperparts reddish-brown to greyish-brown, and grade to darker grey-brown underparts and limbs. Populations from rainforest regions appear larger and greyer, while those from the west are smaller and distinctly reddish (Goodman & Andrianjakarivelo in press). The largest free-living mammal on the island and cannot be confused with any other species.

HABITAT & DISTRIBUTION Adaptable. In Madagascar, occurs in all major habitats, from rainforest to dry deciduous and spiny forests and even open savanna, from sea level to above treeline at 2,000m, e.g. Andringitra Massif (Goodman & Andrianjakarivelo in press). Its presence depends more on water and food availability and less on vegetation cover, so it occurs in many secondary and degraded habitats, and cultivated areas, providing hunting levels are not high (Vercammen et al. 1993, Andrianjakarivelo 2003). Reaches highest densities in protected areas with little hunting pressure. Probably absent from deforested Central Highlands and close to major urban areas.

On mainland Africa three subspecies: *P. l. hassama* in East Africa, including eastern Democratic Republic of Congo, as far north as Ethiopia, *P. l. koiropotamus* on lower Congo River (left bank), Angola, southern Democratic Republic of Congo, Zambia, Malawi, Mozambique, Zimbabwe, Botswana, Swaziland and South Africa, and *P. l. somaliensis* on Tana, Juba and Wabe Shebelle Rivers in north-east Kenya and Somalia (Meijaard et al. 2011).

Previously there has been much debate surrounding the occurrence of the genus *Potamochoerus* on Madagascar: either it colonised naturally in recent geological history or was introduced by Bantu people from Africa between 1,000 and 500 years BP (Pierron et al. 2017). Hence, taxonomy of Malagasy populations has been contentious, where two subspecies have been recognised: nominate *P. l. larvatus* (described by Cuvier in 1822) from western Madagascar and the Comoros Islands, and *P. l. hova* in eastern regions. Such taxonomy alludes to the species being native, or perhaps being introduced on two separate occasions from different source populations (Meijaard et al. 2011). Recent genetic studies confirm close affinities with African populations indicating a single introduction event and conclude that subspecific division of eastern and western populations is unjustified (Lee et al. 2020).

BEHAVIOUR Lives in groups of 2–10 individuals, but 4–6 average in Madagascar. Groups contain a dominant male and female, the only breeders within a group. Predominantly nocturnal, possible tendency towards diurnal activity in remote areas away from human disturbance. Day spent in self-excavated burrows or similar natural secluded hideaways like dense bushes, beneath large root systems, rock crevices or caves, becoming active only at sunset (Goodman & Andrianjakarivelo in press). Average home range 4–10km² (Andrianjakarivelo 2003).

Births may occur year-round, but mainly in October–December (early wet season), gestation c.120–127 days, litter size 1–4, with 3–4 most common. Newborn weighs 700–800g. Weaned at 2–4 months (Vercammen et al. 1993, Meijaard et al. 2011). Female constructs deep nest of vegetation to protect piglets, which are given considerable care and attention, not only by sow but by dominant boar. Dispersal from family group begins at c.8 months (Goutard 1999) and sexual maturity reached after c.18–21 months.

Omnivorous and very adaptable. Diet varies with season and food availability, mainly roots, tubers, rhizomes, berries, fallen fruit and some grass, but also animal matter like reptile eggs, small vertebrates, a variety of invertebrates, carrion and excrement (Andrianjakarivelo 2003). May take advantage of fruit discarded by lemurs feeding in the canopy, and be an important seed disperser. When feeding, powerful snout used as a plough to dig up subsoil matter; tusks also used to help dig deeper. This is a highly destructive method of feeding; in short periods a group can strip an area of most plants, and is also capable of inflicting considerable damage to crops.

Has been an influential factor contributing to rapid decline of some native tortoise species. For instance, nests of Ploughshare Tortoise or Anogonoka *Geochelone yniphora* (one of world's most endangered tortoises, restricted to area around Soalala in western Madagascar) are regularly predated (both eggs and hatchlings) (Pedrono & Smith 2003). It is likely that other endemic reptiles and possibly ground-nesting birds suffer in a similar fashion.

Smaller individuals and juveniles may be predated by Fosa (Rasoloarison et al. 1995). Widely hunted by humans using dogs, spears, firearms, and also sophisticated traps and snares (Reuter et al. 2016b, Golden et al. in press). An important source of bushmeat and regarded as a commodity in both rural areas and urban centres (Golden et al. in press). For example, on Masoala Peninsula, Bush Pig meat represents c.65% of wildlife biomass consumed

(Golden et al. 2014, Borgerson et al. 2019). In parts of west (north of Mahajanga, south to Morondava) commercial Bush Pig hunting occurs, with foreign and Malagasy sport hunters organising shooting trips (Golden et al. in press).

WHERE TO SEE Widespread and common, but rarely seen due to shy and mainly nocturnal habits. Indications of their presence often seen in forested regions, e.g. large excavated areas on forest floor, droppings and hoof prints. Occasionally seen on forest walks at night, but soon disappears if startled.

Domestic Stock

There are no firm records as to when the various species of farm stock were introduced. The bone remains of goats, sheep, pigs and zebu cattle have been unearthed in village middens at Mahilaka, a port dating from 1,050–1,400 CE (Rakotozafy 1996).

The Malagasy cow or zebu (omby in Malagasy) is an descendant of the Asian cow Bos indicus. It is unclear if they arrived directly from Asia or via Africa (Souvenir Zafindrajaona & Lauvergne 1993), although either way this may have been via the Comoros Islands (Brucato et al. 2018). Feral zebu live in, and around, native forests in many parts of the island, including within protected areas. It is frequent to come across their dung pats on forest trails in parks such as Mantadia, Ranomafana and Masoala, even if the animals themselves are not seen. Their impact on the forest environment and how they affect native species is unclear. Of greater concern are the indirect threats caused by the practice of regularly burning pasture to encourage fresh growth on which zebu can graze. This is massively detrimental to soil structure and quality, and with time renders the soil sterile. In addition, these fires often encroach into forests where their affect on native wildlife can be catastrophic.

Small numbers of horses and donkeys have also been introduced in more recent times. The first record on Madagascar of the domestic horse Equus caballus was in 1817, when Robert Farquhar, governor of Mauritius, gifted animals to King Radama (Goodman & Soarimalala in press). Populations were augmented to a few thousand during the French colonial era (Deschamps 1960), but have subsequently reduced to no more than 300 individuals in 2010. Donkeys E. asinus have also been imported at various times, but today persist in just a handful of places (Goodman & Soarimalala in press).

Deer

Between 1928 and 1932 two species of deer were introduced, Fallow Deer and Rusa or Timor Deer (Lever 1985). Fallow Deer were first released near the Ankaratra Massif, in the Central Highlands, and Rusa Deer initially released near Station Forestière d'Analamazaotra (Périnet). Even though there is little forest remaining in Ankaratra and hunting pressure from local communities is high, a small population of Fallow Deer persisted until at least a few decades ago (Goodman et al. 2003, Goodman & Soarimalala in press). Populations of Rusa Deer, on the other hand, perhaps survived into the early 1960s but have since been hunted out (Lever 1985).

Threats to Madagascar's Mammals

The unique biodiversity of Madagascar is under siege. It is being bombarded by numerous and ever-increasing threats, primarily human in origin, that massively impact the long-term survival prospects of the island's fauna and flora, with mammal species among the most vulnerable. Madagascar is one of the world's poorest countries and most environmental problems are intertwined with the socio-economic issues facing the Malagasy people. Between 1970 and 2021, the population rose from c.6.5 million to c.28.5 million, with >60% living a subsistence agricultural lifestyle in rural communities that depend heavily on the land and native forests for their livelihood. World Bank figures (2021) suggest c.70% of the population live on less than US$2 a day and c.60% on less than US$1.25.

Humans are comparative newcomers to Madagascar: indeed the island was among the last places on Earth to be settled. Although early human colonisation may date back 2,000–3,000 years (Dewar *et al.* 2013, Hansford *et al.* 2018), archaeological evidence suggests the first sustained colonisation did not occur until around 1,500 years ago (Cox *et al.* 2012), from South-East Asia and mainland Africa (Hurles *et al.* 2005), with the process of initial colonisation still poorly understood (Tofanelli *et al.* 2009). The early colonists brought with them a culture with rice and zebu cattle at its core. Neither can be raised in dense forests, so trees were felled and the undergrowth burned to clear the land.

During the late 18th and early 19th centuries, King Andrianampoinimerina punished subjects who destroyed the forests, but the practice continued. Later, in 1883, the missionary James Sibree (after whom Sibree's Dwarf Lemur is named) commented, 'again we noticed the destruction of the forest and the wanton waste of trees' (Austin & Bradt 2020). It is bewildering to contemplate that in the relatively short time that our species has been resident, c.80% of the original native forest cover has been lost, with the majority in recent times as the population has expanded exponentially (Green & Sussman 1990): since c.1950 there has probably been a c.45% decline in forest cover (c.37% between the early 1970s and 2014) (Vieilledent *et al.* 2018). Lowlands have been most heavily impacted; in eastern rainforest regions there are now very few forests remaining below 800m (Harper *et al.* 2007). In addition, there

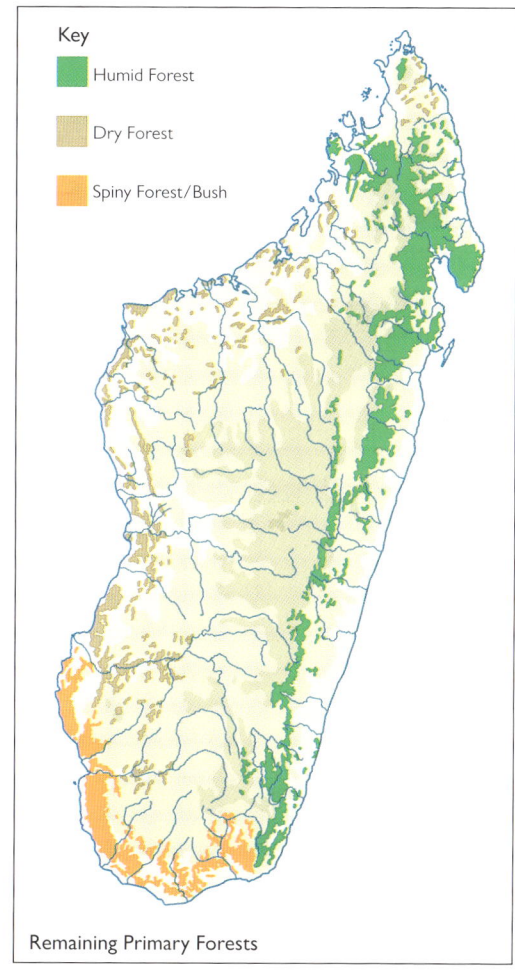

Key

Humid Forest

Dry Forest

Spiny Forest/Bush

Remaining Primary Forests

Vast swathes of forest are felled and burned annually.

Slash and burn (tavy) cultivation: Betsimisaraka woman planting hill rice after forest felling and burning, near Andasibe-Mantadia National Park.

has been large-scale drainage of lakes, marshes and wetlands as they are converted to rice paddies, with corresponding impacts on species adapted to these environments, for example the Lac Alaotra Reed Lemur and birds like Madagascar Pochard *Aythya innotata* and Madagascar Fish Eagle *Haliaeetus vociferoides*.

As measured in 2014, native forests cover *c.*8.9Mha, *c.*15% of Madagascar's surface area, comprising *c.*4.4Mha of humid rainforest, *c.*2.6Mha of dry deciduous forest, *c.*1.7Mha of spiny forest and *c.*177,000ha of mangroves (Vieilledent *et al.* 2018).

As if the loss, fragmentation and degradation of forests were not enough, many of Madagascar's mammals (and other species) face more direct threats from hunting, illegal trade, and competition from invasive non-native species. A synopsis of the major threats and their implications is outlined here.

Habitat Loss and Fragmentation

Habitat loss as a result of deforestation is arguably the single most severe issue facing the island's non-volant mammals (Steffens *et al.* in press). More than 90% of Madagascar's native land vertebrates, especially mammals, live exclusively in forest and woodland habitats (Goodman *et al.* 2003) and will not cross even narrow intervening areas in search of food, new territories or mates. For example, most lemur species, including the partially terrestrial Ring-tailed Lemur, are particularly sensitive to deforestation, as they require native forest for food and shelter, although some are capable of adapting to slightly degraded and secondary forest (Eppley & Goodman in press). Hence the impact of deforestation and forest fragmentation in terms of both overall numbers and species diversity is far-reaching and often catastrophic.

Given the extent of deforestation, it is remarkable that Madagascar has been largely spared the ravages of extensive industrial-scale timber extraction operations or clear felling for oil palm plantations seen in many other tropical regions. Forest clearance is primarily driven by local-level village decisions to obtain land that ultimately has short-term agricultural productivity. Traditional subsistence swidden agriculture involves slash-and-burn (*tavy*) cultivation, where small patches of forest are felled and burned to clear the land, which is subsequently planted with crops (Styger *et al.* 2007) – mainly hill rice or manioc, but also coffee, coconuts, lychees, cloves, vanilla and other spices. The soil cannot sustain food crops like hill rice for more than two or three years; without protection of the trees above it is quickly eroded and washed away, especially on slopes, often resulting in the formation of huge gullies, known as *lavaka*. Consequently, rivers and streams become clogged with silt. Within a short time, the cleared land is reduced to a sterile and barren moonscape covered in coarse grasses (that are often burned annually, further undermining soil structure and quality), so farmers move on to clear more forest, and the sad destructive cycle is reinforced and perpetuated.

In addition to subsistence agriculture, the production of charcoal contributes further to deforestation. Charcoal is the primary cooking fuel in rural and urban areas. Relatively small quantities are consumed in rural zones close to forests where the charcoal is produced: the primary driver of the market is consumption in urban areas (Minten *et al.* 2013, Randriamalala *et al.* 2017).

The local scale of deforestation is reflected in the configuration of remaining forests: instead of large cleared areas interspersed by large intact tracts, deforestation in Madagascar tends to be more of an intricate patchwork consisting of many small cleared areas. The result is a colossal degree of fragmentation with in excess of 80% of remaining forest within 1km of a forest edge (Harper *et al.* 2007) and *c.*50% within 100m (Vieilledent *et al.* 2018).

There is a direct relationship between fragment size and species diversity: as forest blocks become smaller, the number of species they support diminishes, i.e. in smaller patches some species become locally extinct (Ganzhorn 1998). Forest fragmentation inhibits gene flow through and between populations, thereby reducing genetic diversity via inbreeding depression, which in turn reduces fitness. This has been shown in several lemur species, e.g. Milne-Edwards's Sportive Lemur (Craul *et al.* 2009), ruffed lemurs (*Varecia* spp.) (Razakamaharavo *et al.* 2010, Baden *et al.* 2013, Holmes *et al.* 2013) and mouse lemurs (*Microcebus* spp.) (Olivieri *et al.* 2008, Radespiel *et al.* 2008). Fragmentation also reduces floral diversity, which correspondingly depresses food plant diversity for folivorous and frugivorous lemurs. Species that require a broad selection of leaves and other foodstuffs in their diet, like Diademed Sifakas, are particularly impacted (Irwin *et al.* 2010). Conversely, species are occasionally able to persist and even flourish in fragmented areas, especially if foods on which they rely are not negatively impacted. Isolated populations of Greater Bamboo Lemur have survived in fragmented habitats as their primary food source, large-culmed bamboos, often thrive following deforestation (Olson *et al.* 2013). Similarly, deforested areas in lowlands of the east are often replaced by stands of Traveller's Trees *Ravenala madagascariensis* (Strelitziaceae) and *Typhonodorum* sp. (Araceae), which are favoured roosting sites for the endemic Eastern Sucker-footed Bat, which can become locally common (Goodman 2019).

Fragmentation significantly increases 'edge effects' (interactions that occur at the boundary

Eroded gullies called lavaka are now a feature of many cleared areas in the Central Highlands.

of two or more habitats), where peripheral forest vegetation suffers changes in community structure because of factors like microclimatic alteration (changes in sunlight, humidity, air movement, soil temperature, etc.) which then compromises forest quality. Furthermore, forest edges allow for the invasion of both plant and animal species that either out-compete or directly impact forest obligates. For example, the main road (*Route Nationale* 4), that bisects Ankarafantsika National Park and effectively acts like a 'forest edge' limits the dispersal and movements of female Golden-brown Mouse Lemurs that do not cross the road, whereas males appear less affected (Ramsay *et al.* 2019, Steffens *et al.* in press). Also, native carnivorans tend to avoid spaces within 500m of forest edge, particularly smaller nocturnal species like Spotted Fanaloka, Falanouc and Broad-striped Vontsira, often because these areas have higher densities of free-ranging domestic dogs, feral cats and non-native Small Indian Civets which compete directly (Farris *et al.* 2015b, 2017b, 2020, Merson *et al.* 2019b, Murphy *et al.* 2019). These invasive species also cause some native carnivorans, e.g. Fosa, to alter their normal daily activity patterns (Gerber *et al.* 2012a, Farris *et al.* 2015a, Merson *et al.* 2018). Introduced rodents, most notably Black Rat, invade deep into forest zones, adversely affecting native rodents, endemic birds and other indigenous species (Goodman 1995).

Habitat Degradation

Even when considering the most remote blocks of forest, there are very few remaining areas that can truly be considered 'primary', in that they have avoided any human-induced degradation (Goodman & Jungers 2014): virtually every corner of the island has been influenced by humans at some stage. For example, pottery and tombs dating back *c.*650 years have been found in remote rainforests in Andohahela National Park, in areas with no trails and where people from the nearest villages rarely venture (Goodman & Rakotoarisoa 1998). Within all remaining native forest blocks, even those in protected areas, some form of human degradation continues and is often clearly evident, potentially further diminishing the forest's capacity to support mammals and other native animals.

A variety of forest products are regularly extracted, including timber, tree-ferns for modification to plant containers, and wild food plants. The selective logging of hardwoods like rosewood (*Dalbergia* spp.) and ebony (*Diospyros* spp.) is the most pervasive form of degradation, largely fuelled by demand from urban areas and international markets (Schwitzer *et al.* 2013). The vast majority of such extraction is illegal and its incidence increases dramatically at times of political instability, e.g. in 2009, when rosewood exports jumped to 35,000 tonnes, from 13,000 tonnes in 2008 (Randriamalala & Liu 2010).

Although such logging is 'selective' as far as the species being extracted is concerned, its broader effects on the forest environment are more far reaching. Dense hardwoods like rosewood are often rafted down rivers from remote forests, with trees of less dense 'lighter' species of wood (e.g. *Dombeya* spp.) being felled to construct the rafts: for every rosewood tree approximately five 'raft wood' trees are felled (Randriamalala & Liu 2010). In addition, it is often necessary to remove a number of other mature trees immediately adjacent to the target tree simply to gain access. Therefore, the harvesting of a single hardwood tree can result in a far greater quantity of other native trees being lost and the forest being damaged more widely (Schwitzer *et al.* 2013).

Such selective logging can be particularly detrimental to mammals, especially certain lemurs. The most negatively impacted are ruffed lemurs (*Varecia* spp.) that rely heavily on large-crowned fruiting trees at the core of their territories (Vasey 2000a), which are often the first to be removed by loggers. Ruffed lemurs also feed on targeted hardwoods like ebony and rosewood as well as 'raft wood' trees. Species like Indri, Diademed Sifaka and Milne-Edwards's Sifaka feed on rosewoods (Powzy and Mowry 2003, Arrigo-Nelson 2006). The leaves and unripe fruits of ebony are also eaten by Diademed Sifaka and Silky Sifaka (Irwin 2006, Patel 2011).

Mining Operations

Madagascar has extensive deposits of minerals, fossil fuels, gemstones and precious metals. In some cases, where these are especially valuable and considerable in extent they are being exploited on an industrial scale, whilst the majority is accessed largely by small-scale, artisanal operations. Both approaches can have significant impact on the environment and species inhabiting the affected areas.

Large-scale operations include the QIT Madagascar Minerals/Rio Tinto ilmenite and titanium dioxide operation in coastal areas near Tolagnaro (Mandena, Sainte Luce and Petriky) in the extreme south-east, the Ambatovy nickel and cobalt mine near Andasibe in the central-eastern rainforest corridor, and the Tantalum Rare Earth Malagasy (TREM) project on the Ampasindava Peninsula in the north-west. Further exploration is likely to result in an expansion of industrial-scale mining for chromium, rutile, zircon and vanadium. The island also harbours oil deposits that have previously been uneconomic to exploit, but this is changing. The main field of Tsimiroro extracts heavy oil suitable for power generation and road construction, and some authorities believe the island's reserves could hold the equivalent of at least 1,500 billion barrels (Euro News 2016).

Habitat destruction, which decreases the size and increases fragmentation of forest areas, is the most harmful aspect of large-scale mining, but the impact on various mammal species varies. The most vulnerable and impacted tend to be larger species that have more extensive home ranges, e.g. large lemurs like Indri, sifakas (*Propithecus* spp.) and ruffed lemurs

Illegal rosewood stockpiled in Maroantsetra.

Artisanal gold mining near Dariana.

(*Varecia* spp.), and carnivorans like Fosa. Species with specialist diets are also especially vulnerable, particularly if their foodstuffs are compromised, as are species most susceptible to 'edge effects'.

Established mining operations in theory have well-developed environmental programmes running in tandem with extraction. The companies work with various international conservation agencies on achieving 'net positive impact', via reforestation projects and funding new conservation initiatives to offset environmental degradation the mining operations cause (Ganzhorn *et al.* 2007, Reuters Events 2014). This includes for instance, the translocation of some lemurs to nearby refuge forests and also constructing 'primate bridges' to facilitate movement between forest patches fragmented by roads: as many as six lemur species have been recorded utilising such bridges (Mass *et al.* 2011).

The impact of large-scale mining generally reaches far beyond its immediate vicinity and there are always downsides and trade-offs, not only in terms of the direct negative impact on the environment, but also the indirect ways local culture and economies are affected. As part of the QIT Madagascar Minerals project a brand new deep-water port (Ehoala) was constructed near Tolagnaro to allow the direct export of minerals. According to a commissioned report, 'the construction of Ehoala

port, the pollution of rivers and the ecosystem changes caused by the mining activities have caused great harm and significantly reduced fishing' (Franchi *et al.* 2013). In addition, according to local people, the cost of living in the region tripled after the start-up phase of the mine, yet only a small proportion of the population were directly employed and so benefitted from increased incomes. There are positives too, as the mining company built new water wells for local communities and a hospital to serve the area.

Small-scale artisanal mining is primarily concerned with gold, precious and semi-precious stones (Madagascar is well-known for high quality rubies and sapphires) and crystals. These invoke different dynamics and pressures on the environment. This sector represents an important source of employment, with more than 500,000 people involved countrywide (EITI 2021), resulting in an illicit gem trade worth perhaps more than US$150 million per year (Tullis 2019). Many of the most productive precious metal and gem-rich areas are contiguous with remaining forest zones, including officially protected areas that support the highest levels of biodiversity. As such most operations are illegal (Duffy 2005, 2007).

Artisanal mining impacts the environment both directly and indirectly. Accessing the mining areas

Golden-crowned Sifaka fed by locals, Andranotsimaty near Daraina.

often requires the felling and clearing of trees leading to deforestation, with wood being used to construct mine shafts or other structures relating to extraction. The discovery of gems or gold generally results in the sudden influx of hundreds, or even thousands, of migrant miners into a small area that was previously inhabited by few or no people. This massively increases demands on the forest, as trees are cut to build makeshift shelters and produce charcoal, and streams are diverted to provide water supplies. With limited options for food provision, miners turn directly to the forest and the incidence of hunting, especially of lemurs, carnivorans and other species, increases dramatically. For example, since 2010 there has been a huge influx of miners from other parts of Madagascar into the protected Ankeniheny-Zahamena (CAZ) forest corridor that links Mantadia and Zahamena National Parks. After the discovery of sapphires, several illegal mines began operating to supply dealers in Sri Lanka (who dominate the Madagascar markets). There have subsequently been significant increases in deforestation and wild animal (especially lemur) hunting in these areas (Tullis 2019).

Perhaps nowhere better encapsulates the social issues and conservation challenges posed by artisanal mining operations than the Daraina region in the far north. Located south of the Loky River and north

of the Manambato River, this area is of significant biological importance, being home to two locally endemic lemurs, the Critically Endangered Golden-crowned Sifaka and Endangered Daraina Sportive Lemur, plus six other lemur species, including Aye-aye, and carnivorans like Fosa and Broad-striped Vontsira. The remaining deciduous and semi-evergreen forests are highly fragmented and initial efforts to formally protect the area encountered many obstacles because the region also supports one of the largest artisanal gold mining operations in Madagascar.

Visitors to the forests near Daraina, in search of Golden-crowned Sifakas and other lemurs, can see the issues laid bare before them. In some areas, the forest floor is pockmarked by mining excavations and deep shafts, rendering it a denuded moonscape, and yet in the trees above lives one of the world's rarest and most beautiful primates. Local traditions and beliefs dictate that the sifakas are sacred and it is taboo (*fady*) to hunt them. Indeed, around some villages the relationship is more intimate as the sifakas are fed. It is a striking, touching paradox.

The broader region has now received protection (Loky-Manambato Protected Area), but gold mining operations continue to expand as excavation techniques become more sophisticated. Furthermore, itinerant miners do not respect local taboos and hunt the sifakas and other lemurs. The construction of a Chinese-backed gold mining outpost suggests that this threat will be exacerbated (Semel *et al.* 2020). In addition, the reconstruction of *Route Nationale* 5a, between Iharana (Vohemar) and Ambilobe, with Chinese funding, will almost certainly lead to a further influx of mine workers and an increase in hunting and resource extraction across the region.

Hunting and Bushmeat

It is likely that one of the significant drivers of extinction of Madagascar's extraordinary Holocene megafauna, including giant lemurs, elephant birds and dwarf hippopotamuses, was over-hunting by early human colonists (perhaps along with climate change and drought) (see The Extinct Mammals of Madagascar). Evidence of prehistoric hunting and butchery of terrestrial vertebrates dates back perhaps 2,000–3,000 years (Crowley 2010, Godfrey *et al.* 2019). In short, the Malagasy have a long history of hunting wild animals for food. It is perhaps surprising, therefore, that until recently the hunting of mammals and other endemic fauna has been largely overlooked as a major threat to the extant fauna, with deforestation and expanding swidden agriculture generally perceived as applying the main pressures, in conjunction with rapid climate change (Green & Sussman 1990, Morelli *et al.* 2020).

However, more recently hunting has been revealed as a considerable threat, including within relatively intact forests, although the significance of its impact, especially with respect to conservation, has often been largely overlooked (Golden 2009, Golden *et al.* in press). Throughout the island, all manner of wildlife, especially larger mammals, are hunted for local consumption. This high level of hunting, in combination with habitat loss, has resulted in dramatic reductions in wildlife populations across Madagascar (Golden *et al.* in press). Madagascar has the largest number of endemic mammals (46 species) threatened by hunting in the world (Ripple *et al.* 2016).

In many other areas in the tropics, e.g. Africa and South-East Asia, hunting is practiced only by a skilled fraction of the population in order to supply wild meat (bushmeat) to villages and urban areas. The day-to-day food needs of the majority of people are largely met by provision from agricultural practices: as such, bushmeat is considered a luxury. In Madagascar the dynamic is different as the majority of wild meat harvest is at a subsistence level, driven by poverty and the simple need for protein (food insecurity) (Schwitzer *et al.* 2013), although wild meats are widely considered inferior in taste and not preferred to domestic meat (Reuter *et al.* 2016b, Merson *et al.* 2019a). For example, in the Makira and Masoala regions in the north-east around 50% of the population hunt and eat lemurs (Golden *et al.* 2014), and on the Masoala Peninsula *c.*90% eat lemur meat at some point during their life (Borgerson *et al.* 2016).

However, some mammals, in particular fruit bats, tenrecs and Bush Pigs have a more commercial aspect and are often transported further afield for sale to urban markets (Reuter *et al.* 2016c): the majority of people in urban areas consume wild meat (generally tenrecs or fruit bats) at some point (Reuter *et al.* 2016b). There is also evidence that a luxury bushmeat trade, targeting endangered lemur species, develops after political unrest and the breakdown of basic legal infrastructure (as in 2009) (Barrett & Ratsimbazafy 2009, Schwitzer *et al.* 2014), but this diminishes thereafter with time as stability is restored (Reuter *et al.* 2016c).

Despite all lemur species being protected by law, hunting is widespread across forest areas throughout the island (Golden *et al.* in press). Lemur hunting is particularly detrimental, as it tends to target larger, reproductively active, adults that are crucial to maintaining population stability (Schwitzer *et al.* 2013). Larger species only reach sexual maturity at a late age, have long gestation periods, long lifespans and small litter sizes; life history traits that make them extremely susceptible to rapid depletion. As

Madagascar Flying Fox for sale in regional market in south east.

Hunted aye-aye being cooked and butchered, Makira region.

such they can withstand only very low levels of hunting pressure (Golden 2009).

Traditionally, a variety of lemur species has effectively been protected by local prohibition beliefs and taboos (known as *fady* in Malagasy culture), for example Indri in parts of its range that overlap with the some of the Betsimisarika ethnic group (Britt *et al.* 1999, Powzyk & Thalmann 2003, Dolch 2021), Milne-Edwards's Sifaka and woolly lemurs in the Ranomafana National Park area (Jones *et al.* 2008), Decken's Sifaka and Crowned Sifaka in central-western regions (Durbin *et al.* 2003), Verreaux's Sifaka in the Isalo and Bezá-Mahafaly areas (Hawkins 1999, Louden *et al.* 2006), Perrier's Sifaka in the Analamerana region (Anania *et al.* 2018), Golden-crowned Sifaka in the vicinity of Daraina (Vargas *et al.* 2002) and Ring-tailed Lemur with some Malagasy ethnic groups (Mahafaly and Tanala) in the south-

west (Louden *et al.* 2006): generally the beliefs are linked with notions that these lemurs embody the spirits of dead ancestors, so hunting them is forbidden. However, as increasing numbers of young in-country migrants have moved around the island (e.g., because of population influxes caused by illegal gold and gem mines), local traditions and beliefs have been broken down (Lingard *et al.* 2003, LaFleur 2013, Golden & Comaroff 2015), and hunting these species has become culturally acceptable and more widespread.

In eastern regions, lemur hunting is mainly practiced during the cyclone season (February–April) and represents an important source of protein at times when the availability of domestic meat is more limited (Golden *et al.* 2019). On the Masoala Peninsula nine of the ten species are hunted, with the larger species, Red Ruffed Lemur and White-fronted Brown Lemur, especially prized (Borgerson *et al.* 2016). West of Masoala, hunting is equally prevalent in the Makira region, where species such as Black-and-white Ruffed Lemur, White-fronted Brown Lemur, Eastern Grey Bamboo Lemur and even Aye-aye are eaten. Here, Indri is also hunted for pelts, which are worn as clothing (Golden 2009). In the Andasibe area (in the vicinity of arguably the islands' premier protected rainforest), a high proportion of local people eat various lemurs including Diademed Sifaka, Brown Lemur, Eastern Woolly Lemur and Indri (Razafimanahaka *et al.* 2012, Randriamamonjy *et al.* 2015). Hunting is also prevalent in south-eastern forests, with large species like Milne-Edwards's Sifaka targeted in the Ranomafana area (Dunham *et al.* 2008). In littoral forests of the extreme south-east near Sainte Luce, Red-collared Brown Lemur and even small nocturnal species like dwarf lemurs and mouse lemurs are eaten (Randriamanalina *et al.* 2000). Dwarf and mouse lemurs are often hunted in the cooler dry season when they are excavated while dormant or hibernating in tree holes (Bollen & Donati 2006).

The picture is similar in drier forests on the west side of the island. The hunting of Brown Lemur and Coquerel's Sifaka is rife in and around Ankarafantsika National Park, with significant numbers being taken by raffia harvesters (Garcia & Goodman 2003, Salmona *et al.* 2014). Once protected by local taboos (*fady*), both Decken's Sifaka and Crowned Sifaka are now widely hunted in central-western regions (King *et al.* 2014, Rakotonirina *et al.* 2014). Also in the centre-west, in the Kirindy-Mitea area Verreaux's Sifaka, Ring-tailed Lemur, Red-fronted

Brown Lemur and sportive lemurs are hunted and eaten (Goodman & Raselimanana 2003).

Hunting in arid regions of the south-west also occurs. Traditional taboos (*fady*) held by several ethnic groups in southern Madagascar have previously protected lemurs from hunting pressure. Such beliefs are now breaking down due to an influx of migrant workers, primarily from artisanal mining industries (La Fleur 2013). In many areas, larger species like Verreaux's Sifaka and Ring-tailed Lemur have already disappeared (due to habitat destruction and over-hunting), so smaller lemurs like sportive lemurs, dwarf lemurs and mouse lemurs are now taken (Gardner & Davies 2014). In areas where Ring-tailed Lemurs persist they may be killed and smoked, and the meat transported to markets in Toliara for sale (Sauther *et al.* 2013, La Fleur 2013).

The techniques employed for hunting lemurs vary. The use of firearms is infrequent, as it is limited to relatively few with the financial capability, generally from urban areas (Reuter *et al.* 2016c). Most active hunting is with slingshots, spears, blowpipes or with dogs. Some nocturnal tree-hole dwelling species are reached by felling the tree (Reuter *et al* 2016c). However, the majority of lemur harvesting (and other mammals) is done passively using traps and snares, often called *laly* (Golden *et al.* in press)

or *fandrika* in the Andasibe area (Dolch 2021). When lemur hunting, two types of snare trap are generally used: both require a small area of forest to be cleared, which is then spanned with a single narrow wooden trunk, forcing any lemur to cross. Sometimes these bridges, set with snares, simply connect forest fragments and are not baited. They are called *laly lava* and capture any lemur species that makes use of the bridge. Alternatively, the wooden trunk bridge connects the forest to a specific fruiting tree, or is baited with bananas, so the trap targets frugivorous species. Such traps are called *laly totoko*. Snares and traps are generally made on site from forest products (Golden *et al.* in press).

While the vast majority of lemur hunting appears to be at subsistence level, there is increasing evidence of organised commercial or semi-commercial trapping to supply an illegal bushmeat trade (Golden *et al.* in press): animals specifically hunted for sale are captured using more efficient techniques, generally by hunters from urban areas (Reuter *et al.* 2016). For example, in Loky-Manambato Protected Area in the vicinity of Daraina in 2016, authorities seized the corpses of many tens of Crowned Lemurs, apparently destined for illicit markets c.200km away in Sambava (Goodman *et al.* 2018). Fruits were poisoned to drug and capture the animals alive,

Confiscated Crowned Lemur corpses (some charred) with arrested hunters, near Daraina.

Most lemur hunting is with snares known locally as laly.

with some lemurs immediately killed and dried to produce local biltong (called *kitoza*) and others fed to fatten up before sale. The market value (in 2016) of one lemur was *c*.3,000 Ariary (US$5) (Golden *et al*. in press). There are similar cases at other sites in the far north-east and north, including examples of the Critically Endangered Golden-crowned Sifaka entering the trade. Loky-Manambato Protected Area has a high density of artisanal gold mining operations (see Mining Operations) with incoming migrant miners providing a potential market for bushmeat. Furthermore, in 2019 construction began on a new road (part-funded by China) linking Iharana (Vohemar) and Ambilobe, and passing through Daraina, with a corresponding influx of hungry workers. Once complete this road will facilitate easier access to forest areas for hunters, and swifter transfer of bushmeat to markets.

Lemurs are not the only mammals to be impacted by hunting, in fact most other mammal groups, both endemic and introduced are also sought. In many areas the endemic carnivorans (family Eupleridae) and non-native Small Indian Civet are targeted. The larger endemic species (genera *Cryptoprocta*, *Fossa* and *Eupleres*), are protected by law so hunting is illegal (except when they inflict economic damage, e.g. poultry predation), whereas the smaller endemic carnivorans (genera *Galidia*, *Galidictis*, *Salanoia* and *Mungotictis*) are considered game species and can be legally hunted in season (1 April–30 June). Carnivorans are caught using dogs or with snares. Snares are set along forest trails, paths leading

to water sources or to domestic poultry pens. Alternatively, baited noose traps called *antombato* are set on branches (Golden *et al*. in press).

Hunting carnivorans is widespread in many rainforest regions, although the primary driver in some areas may not be food provision, but instead to reduce conflict and protect domestic animals: consumption of these carnivorans, including feral cats, is therefore secondary (Reuter *et al* 2016b). Snares, called *fandrikan-dia* in Malagasy, are widely employed. In Makira and the Masoala Peninsula in the north-east, the most heavily hunted species are the non-native Small Indian Civet, Ring-tailed Vontsira and Fosa, although other endemic species like Spotted Fanaloka, Falanouc, Broad-striped Vontsira and Brown-tailed Vontsira are also taken opportunistically (Golden 2009, Farris *et al*. 2015b, Borgerson 2016). A similar picture repeats in many areas further south, including around Andasibe, where Fosa, Spotted Fanaloka, Falanouc and Ring-tailed Vontsira are hunted (Jenkins *et al*. 2011, Dolch 2012, 2021, Randriamamonjy *et al*. 2015). In the west, carnivoran hunting has been documented around Ankarafantsika National Park and in the Menabe and Kirindy-Mitea regions in the centre-west (Garcia & Goodman 2003, Goodman & Raselimanana 2003, Randrianandrianina *et al*. 2010).

Spiny tenrecs (subfamily Tenrecinae) are the most commonly hunted and widely consumed endemic mammals across the island (Golden *et al*. in press). They are game species, so can be legally hunted (without the use of dogs) in season (1 April–31 May). The smaller furred tenrecs (subfamily Oryzorictinae) and large-eared tenrecs (subfamily Geogalinae) are almost never hunted, probably because of their mouse-like appearance, small size and musky taste (Dolch 2021).

In eastern rainforests the Tailless Tenrec, Greater Hedgehog Tenrec and, to a lesser extent, Lowland Streaked Tenrec comprise the majority of wildlife consumed (Golden *et al*. in press). For example in the Makira and Masoala Peninsula region in the north-east, >90% of households across *c*.40 communities consumed tenrecs, accounting for 75% of total wildlife biomass eaten (Golden *et al*. 2014). Most tenrec hunting utilises dogs (primarily at dusk or at night) and takes place outside the legal season (November–May), when availability and consumption of meat from domestic animals and fish is reduced (Golden 2009, Golden *et al*. 2019). In all rainforests further south, the picture is similar, with spiny tenrecs being hunted and eaten by a high proportion of the rural population and constituting 50% or more of the wild animal biomass consumed (Golden *et al.* 2014, Randriamamonjy *et al.* 2015, Borgerson *et al.* 2018).

Similarly, in western Madagascar both Tailless Tenrec and Greater Hedgehog Tenrec are hunted with dogs and eaten by a large proportion of the rural population (Garcia & Goodman 2003, Borgerson *et al.* 2018). In the Menabe region of the centre-west, spiny tenrecs (subfamily Tenrecinae) provide one of the most preferred meats (Merson *et al.* 2019a). In spiny forests of the extreme south-west, three species Tailless Tenrec, Greater Hedgehog Tenrec and Lesser Hedgehog Tenrec are commonly hunted and eaten (Gardner & Davies 2014).

Madagascar Flying Fox is legally protected (Dolch 2021), but all other bats are considered game species. Other fruit bats (*Eidolon* and *Rousettus*) can be legally hunted between 1 May and 1 September, insectivorous bats from 1 February to 1 May. However, throughout Madagascar, larger bats, particularly the three species of fruit bat (family Pteropodidae) including Madagascar Flying Fox, are hunted and eaten, but smaller insectivorous bats are also widely harvested, with species like Commerson's Leaf-nosed Bat taken in their tens of thousands annually (Goodman 2006, Goodman *et al.* 2008c, Jenkins & Racey 2008, Golden *et al.* 2014).

The three fruit bat species are widely distributed and the hunting techniques employed across different regions are broadly similar. Madagascar Flying Foxes are mainly hunted with nets and slingshots at tree day-roosts, although a particularly devastating technique involves felling an entire roosting tree and then beating the bats to death with sticks (Andrianaivoarivelo *et al.* in press). Consequently most roosts are now in remote locations away from immediate human disturbance. All three fruit bat species are taken using similar techniques while they feed in trees at night, but smaller insectivorous bats are targeted more opportunistically when in flight by throwing sticks and rocks, or at their day-roosts in caves or on trees (Golden 2005, Jenkins & Racey 2008, Andrianaivoarivelo *et al.* 2011, Reuter *et al.* 2016c).

While fruit bats are eaten widely across the island, hunting is seasonal and overall they constitute a relatively small proportion of wildlife biomass consumed (Golden *et al.* 2014, 2019, Borgerson 2016, Borgerson *et al.* 2019). However, in some south-eastern towns, e.g. Farafangana, there is occasional large-scale commercial fruit bat hunting, with bats appearing as *plat du jour* in small roadside restaurants (*hotely*) (Racey *et al.* 2010). Also, there may be a greater degree of fruit bat consumption in some western areas, for instance the Menabe region (Jenkins & Racey 2008).

In the far south-west, on the Mahafaly Plateau, including in and around Tsimanampetsotse National Park, smaller species like Commerson's Leaf-nosed Bat are actively hunted and eaten (several thousand individuals per year), and other insectivorous species like Glen's Long-fingered Bat and Rufous Trident Bat are taken incidentally or opportunistically (Goodman 2006, Goodman *et al.* 2008c). Previously, these bats were largely protected by taboos (*fady*) relating to the sacred caves in which they roost. However, traditional beliefs have largely broken down, and bat consumption is now widespread (Fernández-Llamazares *et al.* 2018).

Unlike other native mammal groups, the endemic Nesomyine rodents seemingly suffer relatively little impact from bushmeat hunting. The consumption of rodents is by and large culturally unacceptable in Madagascar, although there are exceptions. In some localities White-tailed Tree Rat is eaten, perhaps because it is regarded as comparable to small nocturnal lemurs (Kennerley 2016a) and in the Andasibe region Eastern Red Forest Rat is hunted (Dolch 2021).

Of the various non-indigenous mammals introduced to Madagascar, three species are of significance in the context of hunting and bushmeat consumption: feral cats, Small Indian Civet (see

All three fruit bat species, especially Madagascar Flying Fox are widely hunted and eaten across the island.

Tailless Tenrecs are a 'game' species, so can be legally hunted.

hunted with dogs, spears and firearms, and also sophisticated methods of trapping and snaring (Reuter *et al.* 2016c, Golden *et al* in press). Snares, made of rope or metal cable, called *fandrikandia*, are placed on forest trails used repeatedly by pigs and are set directly over an existing pig footprint. The noose is then disguised with twigs and leaves, and triggered by the heavy footfall of a passing animal (Golden *et al* in press). One animal provides a lot of meat, so portions of the carcass may be smoked, sold or gifted to friends (Adrianjakarivelo 2003, Golden 2005, Gardner & Davies 2014, Reuter *et al.* 2016c).

Furthermore, in the west of the island (north of Mahajanga, south to Morondava) there is commercial hunting of Bush Pigs, which includes foreign and Malagasy sport hunters taking part in organised shooting trips; forest sites are first baited by local people, who receive a portion of the meat after a successful hunt. In addition, large commercial operations, e.g. shrimp farms, engage hunters to supply Bush Pigs to local markets, to be subsequently purchased by and provide meat for factory workers (Golden *et al.* in press).

How sustainable is bushmeat hunting of mammals on Madagascar? This is easier to assess in some groups than others. Current rates of lemur harvest in many areas, especially larger species (perhaps excepting mouse lemurs and dwarf lemurs) are certainly unsustainable (Dunham *et al.* 2008, Golden 2009, Brook *et al.* 2019). In any given forest, lemur species density is inversely related to its body size and hunting targets larger-bodied species. Therefore, present rates of hunting will almost certainly result in local extirpations (Golden *et al.* in press). Similarly, carnivoran harvesting in the Makira and Masoala regions is deemed unsustainable (Golden 2009, Farris *et al.* 2015), and the situation is likely to be similar in other parts of the island. Low fecundity (one offspring per year) and high levels of harvesting in all three fruit bat species suggest significant declines are likely (Jenkins & Racey 2008), and the rates of depletion in some insectivorous bats in south-western regions exceeds recruitment, making local extirpations probable (Goodman 2006).

The only endemic mammal group where hunting levels may be sustainable are the spiny tenrecs (family Tenrecinae), due to the extremely high fecundity of the principal target species (genera *Tenrec* and *Setifer*) (Annapragada *et al.* 2021), although pregnancy and birthing in tenrecs overlaps with peak hunting periods, which compounds the negative impact (Borgerson 2016).

above) and Bush Pig. Bush Pig is widespread and common across the island and is the largest free-living mammal (55–70kg). Being an introduced species, it is legal to hunt and is of particular importance in the provision of bushmeat across both eastern and western regions. For instance, on the Masoala Peninsula Bush Pig meat represents *c.*65% of wildlife biomass consumed (Golden *et al.* 2014, Borgerson *et al.* 2019) and it is one of the most targeted animals in the Menabe region of the centre-west (Randrianandrianina *et al.* 2010) and Ranobe region of the south-west (Gardner & Davies 2014). Across the island, Bush Pig is regarded as a commodity and is probably the most marketed bushmeat species in both rural areas and urban centres (Golden *et al.* in press). While other wild meats are generally cheaper than domestic alternatives, wild pork is comparable in price to zebu (beef) (Reuter *et al.* 2016c).

Being so large, Bush Pigs are the only mammal that is potentially dangerous, so they are actively

Despite not being preferred, wild meat is critical in human nutrition and health in many parts of the island (Golden *et al.* 2011, 2019, Borgerson 2019) and its consumption is nearly always associated with poverty (e.g., Borgerson *et al.* 2016, Golden *et al.* 2016). Current rates of hunting, in combination with contracting habitat due to deforestation and degradation, will probably result in local species extirpations. Therefore, a 'perfect storm' scenario is being played out, whereby rural populations are bound into a downward spiral of self-depletion of their nutritional resources that will impact their health and livelihoods, and simultaneously cause the extinction of numerous endemics (Golden *et al.* in press).

The Illegal Pet Trade

In addition to hunting pressure, the live capture of wildlife in Madagascar poses an ever-increasing threat. This illegal trade is particularly well established for amphibians and reptiles, which are transported via well-organised networks to supply massive illicit international markets (Andreone *et al.* 2005). However, the situation pertaining to the island's mammals, particularly lemurs, has until recently been unclear.

Although the capture and sale of lemurs is illegal both domestically and internationally, there is an established culture of keeping lemurs as pets across the island (Reuter & Schaefer 2017a). Studies to quantify this have revealed startling figures: an estimated 33,000 lemurs, of a variety of species, are kept illegally in Madagascar, with c.55% of these in urban households (Reuter *et al.* 2016a, 2019).

In contrast to the well-established networks of collectors, middlemen and dealers involved in the amphibian and reptile trade, there is little evidence the live capture of lemurs involves such dedicated, organised supply chains. Indeed, most lemurs are captured by their subsequent owners, with smaller numbers bought locally or exchanged as gifts (Reuter *et al.* 2016a).

While species representing 13 genera have been recorded as pets (Reuter *et al.* 2019), the wild capture of lemurs disproportionately affects some species (Reuter & Schaefer 2017a, LaFleur *et al.* 2018, 2019). By far the most popular is the Ring-tailed Lemur, which accounts for approximately half: in a survey of 12 urban areas within or near the species' range, more than 3,000 individuals* were located (potentially more than now remain in the wild) (Reuter *et al.* 2019). Even relatively modest levels of live capture will probably result in substantial local population declines, further exacerbating the already perilous situation facing Ring-tailed Lemurs from the tandem threats of bushmeat hunting and habitat destruction (LaFleur *et al.* 2017, 2018). Diurnal group-living species are easier targets for capture and also tend to be frugivorous, making them easier to maintain in captivity. As such, the other principal target species for the pet trade are 'true' lemurs (*Eulemur* spp.), Black-and-white Ruffed Lemurs and bamboo lemurs (*Hapalemur* spp.) Species like sifakas are rarely kept (outside legal facilities), as they are very difficult to house and maintain in captivity (partly because of very specific folivorous diets that cannot be accurately replicated) and because they are protected by local taboos (*fady*) in some areas (Jones *et al.* 2008, Reuter *et al.* 2019).

Once in captivity, lemurs are rarely kept or fed appropriately. Most are caged or tethered on short ropes or chains, or when young carried around openly as 'fashion accessories'. In captivity lemurs are rarely given foods consistent with their natural diets: most, irrespective of species, are offered rice and bananas, and suffer poor health as a consequence (Reuter & Schaefer, 2016). Many survive only a few months, due to a combination of inappropriate nutrition and horrendous conditions, or because they are too young, having been captured before they were weaned (Reuter *et al.* 2016a, 2019, LaFleur *et al.* 2019): approximately one third of attempts to own a pet lemur result in its premature death (Reuter *et al.* 2019).

The two principal motivations for keeping lemurs are either as personal pets or for financial gain by hotels aiming to attract more affluent tourists. There is little evidence they are kept as a back-up source of food security (Reuter & Schaefer 2017b). Also, pet lemurs are rarely bred in captivity, so all pet lemurs originate from wild sources (Reuter *et al.* 2016a).

Approximately 50% of pet lemurs are kept in private households, to convey social status, to 'save' the animal from being killed for bushmeat and eaten, or for personal gratification and preference

---

* Many individuals have been confiscated by the authorities and transferred to the Reniala Lemur Rescue Centre, near Mangily north of Toliara. Here they are housed in semi-wild enclosures, for recuperation and recovery to full health, with the aim of reintroduction to the wild when possible (LaFleur *et al.* 2015).

(Reuter *et al.* 2018). Better-quality hotels and other tourism facilities account for the other 50%. Hotels advertising captive lemurs, either via their websites or on social media, charge significantly higher rates than those that do not (Reuter & Schaefer 2017b). In addition, many charge extra fees to interact with or be photographed with the lemurs, a demand fuelled by both foreign and Malagasy tourists seeking cute lemur 'selfies', often for their own social media accounts. Consequently, this can lead to other hotels procuring pet lemurs of their own, under the perception this will increase their own standing and income-generating capabilities (Reuter *et al.* 2019).

Furthermore, some tourism facilities legally keep lemurs including sifakas and even Indri and Aye-aye, as they are particularly charismatic (Reuter & Schaefer 2017c, LaFleur *et al.* 2020). Inappropriate direct tourist interaction with these animals, including feeding, is often encouraged. Despite being legal facilities, best practice standards for captive care are generally not maintained and animals are often fed with non-natural detrimental foodstuffs, like bananas. Facilities purport only to house animals confiscated from illegal facilities; however, in some cases illegal wild capture occurs to replace particularly attractive or charismatic individuals that have died in care (LaFleur *et al.* 2020). Any tourist who visits these facilities is adding credence to their *modus operandi*. This is strongly discouraged and all visitors to Madagascar should concentrate their efforts on seeing lemurs and other endemic fauna in the wild.

There is a clear correlation between well-known tourism locations and the prevalence of visible 'on show' pet lemurs, and between rates of pet lemur ownership and social media activity and an area's popularity (Reuter & LaFleur 2019, Reuter *et al.* 2019). Social media and online images strongly influence peoples' perception of endangered species: any image of a pet lemur ultimately leads to more people wanting a pet lemur. Although they are probably unaware of the negative impact, tourists are unwittingly fuelling demand. Foreign tourists, in particular, have a responsibility to set appropriate standards and trends. If they are seen posing with lemurs sitting on their shoulder and gleefully posting images on social media, it conveys a message of acceptability and authenticity, which encourages people in Madagascar to follow suit, adding to local demand for lemur capture from the wild, and ultimately increasing the likelihood of extirpations of the most popular species.

Intrusion by Non-native Species

Although Madagascar has not (yet) suffered major intrusion from non-indigenous species (either deliberately or accidentally introduced by humans), several non-native species do have a serious negative impact on some endemic mammal communities and species.

Black Rats penetrate deep into remaining native forest areas and compete for resources with endemic Nesomyine rodents (Goodman 1995). Species suffering most directly are short-tailed rats (*Brachyuromys* spp). (Jansa & Carleton 2003a) and also red forest rats (*Nesomys* spp). and tufted-tailed rats (*Eliurus* spp.), as well as endemic ground- and low-nesting birds, whose eggs and chicks are predated, and smaller reptiles and amphibians that are also eaten.

The three non-native carnivorans, domestic dog, feral cat and Small Indian Civet, all impact native Euplerid carnivorans, by competing for space and resources (Goodman *et al.* 2003, Farris *et al.* 2016b). These three carnivorans are also opportunistic predators of many native species.

The presence of feral cats is particularly detrimental to populations of Spotted Fanaloka, Ring-tailed Vontsira, Brown-tailed Vontsira and Broad-striped Vontsira (Gerber 2012, Farris *et al.* 2015a,b, 2016b, Rasambainarivo *et al.* 2017, Merson *et al.* 2018) and they are known predators of various lemurs (Brockman *et al.* 2008, Merson *et al.* 2019) as well as endemic rodents and reptiles (Murphy 2015, Farris *et al.* 2017b, Merson *et al.* 2019, Murphy *et al.* 2019).

There is considerable niche overlap, especially in diet, between the Small Indian Civet and endemic Spotted Fanaloka, which suffers accordingly and can be excluded from forest areas: Spotted Fanaloka occupancy decreases by *c.*40% at sites where Small Indian Civet is present (Farris *et al.* 2021).

For more detailed discussion of the impact of introduced, invasive species, see Non-native Mammals.

Conservation and the Island's Protected Areas

Conserving Madagascar's biodiversity is of the utmost importance. Not only are the vast majority of species, especially the mammals, endemic, but also they belong to ancient lineages that first became isolated millions of years ago. Therefore, they comprise taxonomic groupings (genera, subfamilies, families) without close relatives elsewhere. Having large numbers of higher-level taxonomic groups clustered on one island makes significant swathes of evolutionary history extremely vulnerable: extinction on Madagascar equates to global extinction of this evolutionary legacy.

The majority of Malagasy mammals (and other animal species) are native forest dwellers and are unable to survive if forests are degraded or removed. There are exceptions. Some mouse lemurs (*Microcebus* spp.), giant mouse lemurs (*Mirza* spp.) and dwarf lemurs (*Cheirogaleus* spp.) persist in partially degraded or secondary forests, and even foray into crop plantations. Greater Bamboo Lemur appears to thrive (relatively speaking) in some degraded areas, where large-culmed bamboos, the lemur's main food source, subsequently proliferate. Similarly, both species of sucker-footed bat (*Myzopoda* spp.), especially Eastern Sucker-footed Bat, appear to have benefitted from deforestation, which has led to an abundance of Traveller's Trees *Ravenala madagascariensis* (Strelitziaceae) in which it preferentially roosts. Also, several species of tenrec are able to survive in more open and human-altered habitats including grasslands and rice paddies; these include all five species of spiny tenrec (subfamily Tenrecinae), Least Shrew Tenrec and Mole-like Mole Tenrec. And, among rodents, Betsileo Short-tailed Rat has adapted to more open marsh environments. Even the island's largest carnivoran, Fosa, has been recorded in open non-forested habitats, although these are almost certainly animals that are not resident, but rather moving between forested areas.

Nonetheless, these few exceptions aside, the survival of the overwhelming majority of Madagascar's mammals is inextricably linked to native forests and their continued survival is dependent on the maintenance and conservation of forest ecosystems. Given that so much native forest has already disappeared, deforestation continues at an alarming rate, and other threats like hunting and collection for the pet trade add further pressures, there is no disguising the fact that the long-term future for the island's mammals (and other native species) appears bleak.

Unsurprisingly, where threats to survival are concerned, Madagascar's 'flagship' primates tend to hog the limelight. Scrutiny of their current conservation status, the Red List of Threatened Species published by the International Union for the Conservation of Nature (IUCN), makes for grim reading. Of the 108 described species only two – Grey Mouse Lemur and Grey-brown Mouse Lemur – are assessed as Least Concern, meaning they are not in imminent threat of extinction, whereas 24 species are regarded as Vulnerable, 47 species are listed as Endangered and 33 species deemed to be Critically Endangered (one species is described as Data Deficient and one recently described species has yet to be evaluated). Put another way, 104 or 96% of lemur species are threatened with extinction.

Species facing the most severe levels of threat are those with already extremely restricted ranges or those living in the most severely depleted habitats. For instance, species with massively restricted ranges include the Alaotra Reed Lemur whose reedbed habitat around Lac Alaotra encompasses no more than 220km^2, and Perrier's Sifaka, which may number no more than 500 individuals in the highly fragmented deciduous forests of the far north. Low-elevation rainforest is perhaps the habitat type under greatest threat and also supports several of the largest lemurs, such as the Indri, Diademed

*Deforestation is especially acute in Menabe in the centre west, where baobabs (*Adansonia spp.*) are left 'stranded'.*
Species like Giant Jumping Rat, Fosa and Madame Berthe's Mouse Lemur are particularly affected.

Sifaka, Black-and-white Ruffed Lemur and Red Ruffed Lemur, which are all regarded as Critically Endangered. How tragic it would be if eastern regions no longer echoed to the song of the Indri and this magnificent creature were lost to future generations forever.

Looking beyond the lemurs, only one of the island's endemic carnivorans, Ring-tailed Vontsira, is assessed as Least Concern, although its populations have undoubtedly fallen significantly since the species was last assessed, and this categorisation is probably inaccurate. (Near Threatened or Vulnerable would perhaps be a more accurate reflection.) Five carnivoran species are regarded as Vulnerable and two as Endangered. Being the island's largest carnivoran, Fosa requires a large home range (containing a suitably healthy prey-base). It is an adaptable creature and survives in all native forest and habitat types where there is prey, with the exception of very dry spiny forest in the south and south-west. However, because forests are so depleted and fragmented, there may now only be four intact forest blocks (two in eastern rainforests, two in western dry forests) substantial enough to support more than 300 adults in each. The island-wide population is c.8,500 individuals at best and probably considerably smaller than this (Gerber *et al.* 2012b).

Even among rodents, a group generally noted for their resilience and abundance, six of 28 native species are regarded at Endangered, although one of these, the Giant Jumping Rat is perhaps under more imminent threat than its official categorisation suggests: due to massive levels of deforestation and fragmentation, its range is now tiny covering no more than 200km², and ironically its adult breeding population is being heavily reduced by predation from the Fosa (Goodman & Soarimalala in press).

So what can be done to 'stem the tide' and reduce the rate of forest depletion and biodiversity loss? The Malagasy government, working in conjunction with several international conservation organisations and funding agencies (e.g. Conservation International, Duke Lemur Center, Durrell Wildlife Conservation Trust, German Primate Centre, Madagascar Fauna & Flora Group, Missouri Botanical Gardens, the Peregrine Fund, Wildlife Conservation Society and the World Wide Fund for Nature), has now implemented numerous projects that integrate the conservation of biodiversity with the development of local communities. A number of fundamental principles underpin these projects.

To devise sustainable (long-term) economic alternatives to forest clearance it is vital to understand the inter-relationships among the fauna and flora within the forest. To this end, biodiversity research, involving a wide spectrum of taxonomic groups, focuses on how each species fits into the forest community as a whole and its interdependence on other species. This information can then be integrated with indigenous knowledge from the local communities.

Armed with such knowledge, it is possible to devise, implement and support sustainable agricultural techniques sympathetic to the local environment. For example, in rainforest areas alternatives such as aquaculture, apiculture and horticulture of important native species have been encouraged, whilst attempting to find ways of maximising production from existing land under rice cultivation (by application of fertilisers, using more suitable seed varieties, and improved irrigation systems). This then reduces the need to expand *tavy* cultivation and maintains the integrity of the remaining forests and watershed.

Healthcare is also a vital component. Local communities generally acknowledge this to be their primary concern. It involves improved preventative measures, sanitation and family planning, and wherever possible incorporates both traditional medicines based on forest products (those that can be cultivated or sustainably harvested) as well as modern Western techniques. If improved health results in better child survivorship, then there is less incentive to have a large family, eventually leading to a slowing of population growth.

Of course, education and training run hand in hand with all of these measures. Education and conservation awareness programmes are designed to increase literacy and provide access to educational facilities for local residents. Local people also work closely with both Malagasy and Western scientists to gain a better understanding of the precarious nature of their environment: some become full-time researchers and others park officials.

Once all of the above are in place, the remaining piece of the jigsaw can be slotted into place – ecotourism. The recreational use of protected areas, by both Malagasy nationals and foreign visitors, is core to many projects, and creates more jobs for local people and increases revenue flow into the community. Local people, who may have once relied on harvesting forest products for their livelihood, are able to earn a living in connection with the park or reserve. Furthermore, tourists require accommodation and guides, so employment is created in local hotels and other secondary service industries, and foreign revenue further flows into local economies.

This is not to claim that ecotourism is a panacea for Madagascar's environmental and biodiversity issues. It is certainly not. There is always a great danger of cultural identity and integrity being eroded when indigenous communities are exposed to more materialistic foreign cultures. However, if the approach is cautious and the system carefully managed, there is every reason to believe that ecotourism has a vital role to play in the conservation of Madagascar's biological riches.

Parks and Reserves

The first efforts to legally protect the island's biological heritage came in 1927, when the French colonial government created ten reserves. For years the system was under-funded, inadequately manned and stagnant, offering only cursory protection against habitat destruction, disturbance and hunting. This began to change in 1985, when Madagascar hosted a major international conference for conservation and development. The government, in partnership with the World Wide Fund for Nature (WWF), evaluated all of the nation's protected areas (37 at the time, covering just 2% of the island's land area) and worked towards developing a strategy that would provide people living adjacent to reserves with viable alternatives and the economic means to halt forest exploitation.

In 2003, these initiatives were given further impetus when the then president, Marc Ravalomanana announced at the Fifth World Parks Congress his government's intension to triple the area of Madagascar's forests under protection. In what

Ecotourism is an important component in the conservation of Madagascar's biodiversity. Photographing a female Red-fronted Brown Lemur, Kirindy Forest.

became known as 'The Durban Vision' he said, 'We can no longer afford to watch our forests go up in flames, nor let our many lakes, marshes and wetlands dry up, nor can we inconsiderately exhaust our marine resources. This is not just Madagascar's biodiversity, it is the world's biodiversity. We have the firm political will to stop this degradation.'

A programme was launched to re-evaluate existing reserves and identify new areas in need of protection, and to create habitat and wildlife corridors that would re-connect existing forest areas that had become isolated. The new and existing protected areas have been assimilated into the System of Protected Areas of Madagascar (SAPM) which has three major objectives: to conserve the island's unique biodiversity; to conserve its cultural heritage; and to enable sustainable resource use to help alleviate poverty.

Madagascar now has an extensive network of protected areas, managed and administered by Madagascar National Parks (MNP) (www. parcs-madagascar.com). There are now c.50 or so major protected areas that fall into three main categories – national parks, special reserves and strict nature reserves – administered and managed

Madagascar now has 50 or so major protected areas. Zombitse-Vohibasia National Park is an isolated area of forest in the south-west that is home to a number of locally endemic species.

in slightly different ways. Only national parks and special reserves are open to tourists, although as far as visitors are concerned there is no perceptible difference. Some of these remain a considerable distance 'off the beaten track' and have yet to be developed for tourism. Nevertheless, an increasing number do have infrastructure at varying levels that facilitate tourism, with visits requiring differing levels of planning.

Among parks and reserves that are accessible, the norm is that only limited areas, constituting a small proportion of the park's total area, have been developed for tourism (helping to minimise some of the latter's negative impacts). Within these areas, a network of trails take visitors to the main areas of interest and it is compulsory to be accompanied by an official park guide.

In addition to the protected areas run by MNP, there are an increasing number of community-managed conservation projects and reserves (V.O.I.) around the island. Some of these have been established independently, others with help and collaboration from international conservation organisations; some are large and ambitious in extent, others are more modest but nonetheless rewarding. They have an important role to play in helping conserving the island's biodiversity and it is vital that both local and foreign visitors support them as much as possible.

Near pristine mid-elevation rainforest interior in the heart of Andasibe-Mantadia National Park.

Important Mammal Watching Sites

Madagascar has some of the best tropical forest environments in the world for watching wildlife, mainly because much of it is relatively accessible and easy to track down, but also because the forests are safe in which to wander around. Mammal watching in most tropical forests (particularly rainforests) can be challenging, as the canopy is high and species are elusive. In Madagascar the contrary is more often true, especially with respect to the island's endemic primates, that can often be viewed at relatively close quarters in locations where they have become habituated.

There is no doubting that the majority of mammal watching in Madagascar is primarily directed towards the island's lemurs; as such a list of 'Important Mammal Watching Sites' could just as easily be titled 'Important Lemur Watching Sites'. Nonetheless, other types of mammals can sometimes be encountered more 'opportunistically' in many locations: during daylight this is mostly limited to the diurnal carnivorans (e.g. Fosa, Ring-tailed Vontsira, Narrow-striped Boky), streaked tenrecs (*Hemicentetes* sp.), red forest rats (*Nesomys* spp.) and some roosting bats. All other native 'non-lemur' mammals are primarily nocturnal, so are harder to see and encountered only from time to time during night walks.

Madagascar is arguably the best place in the world for conducting night walks as there are so many visible nocturnal species, especially lemurs, as well as frogs, reptiles, including sleeping chameleons, and small mammals. Frustratingly, rules imposed by Madagascar National Parks currently prevent night walks in many of the island's protected areas (hopefully this situation may change in the future), but night walks are often possible in neighbouring forests outside protected zones, including in more formalised community-managed reserves.

The sites selected here include national parks, special reserves, community conservation projects and private reserves. They have been chosen firstly because they offer very good wildlife, particularly mammal-watching experiences, and secondly because they represent a cross-section of different habitat types and therefore cover much of the breadth of mammal diversity. Most of the selected sites are also accessible for both organised groups and independent

Indri watching, Association Mitsinjo Forest Reserve.

travellers, but some more 'off-the-beaten-track' locations might suit only the more adventurous.

Each site has been given a star rating, to indicate its relative stature and importance in terms of mammal diversity and the ease with which these mammals might be seen. There are three categories:

Three Star ***

These are sites that should be prioritised by the first-time visitor to Madagascar or those wanting to maximise the number of species they see. These locations offer the maximum diversity of mammals that might be seen in the shortest period. The minimum recommended stay is two nights, but most

Map Key

Eastern Rainforest Areas

1. *Andasibe-Mantadia National Park and Vicinity ****
2. *Anjozorobe-Angavo Forest Corridor ****
3. *Ankeniheny-Zahamena Forest Corridor (CAZ) **
4. *Antanetiambo Nature Reserve ***
5. *Farankaraina Reserve ***
6. *Lac Alaotra: Park Bandro ***
7. *Makira Natural Park **
8. *Manombo Special Reserve **
9. *Marojejy National Park ***
10. *Masoala National Park ****
11. *Montagne d'Ambre National Park ****
12. *Nosy Mangabe Reserve ***
13. *Ranomafana National Park ****
14. *Tsinjoarivo **
15. *Tsitongambarika Protected Area **
16. *Analamerana Special Reserve and Andrafiamena-Andavakoera Protected Area ***

Northern and Western Deciduous Forest Areas

17. *Anjajavy Private Reserve ****
18. *Anjiamangirana Classified Forest **
19. *Ankarafantsika National Park (Ampijoroa Forestry Station) ****
20. *Ankarana Special Reserve ****
21. *Antrema Reserve (Katsepy Forest) **
22. *Kirindy Forest ****
23. *Kirindy-Mitea National Park (Kirindy Mité) **
24. *Loky-Manambato Protected Area (Daraina) ***
25. *Mahavavy-Kinkony Wetland Complex (including Tsiombikibo Forest and Anjahamena) ***
26. *Mariarano Forest **
27. *Sahamalaza-Iles Radama National Park ***
28. *Tsingy de Bemaraha National Park ****
29. *Tsingy Mahaloka **
30. *Zombitse-Vohibasia National Park ***

Southern Areas including Xerophytic Spiny Forests

31. *Anja Community Reserve ****
32. *Andohahela National Park ***

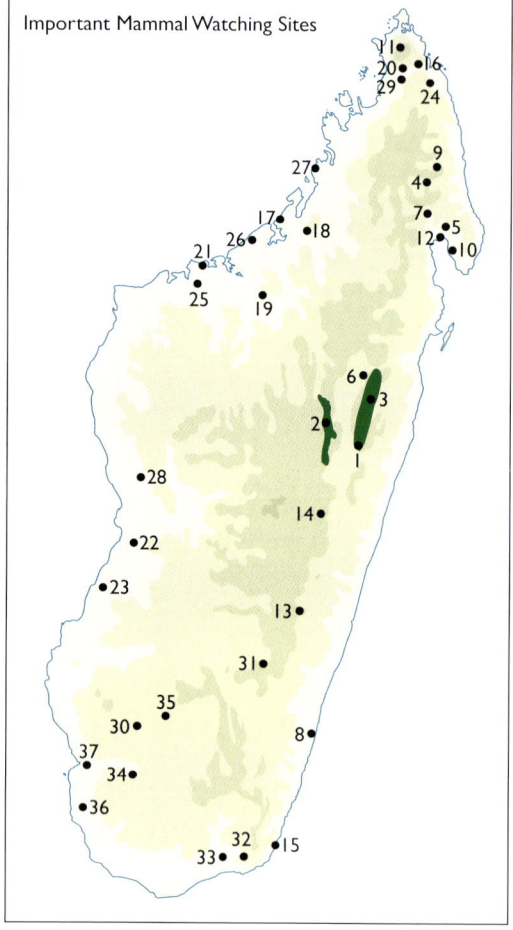

Important Mammal Watching Sites

33. *Berenty Private Reserve ****
34. *Bezà-Mahafaly Special Reserve **
35. *Isalo National Park ***
36. *Tsimanampetsotse National Park ***
37. *Tsinjoriake and Amoron'i Onilahy Protected Areas **

warrant stays of at least three or four nights. These locations include some of the island's best-known protected areas.

Two Star **

These locations may be slightly removed from regular tourist circuits so are suited to those who have more time at their disposal or want to see a particular species that perhaps has a very limited range. Given that getting to these sites may be more involved and take longer, a minimum recommended stay is three nights although some may justify longer. Alternatively any sites in this category that are

accessible might be more limited in the mammal watching opportunities they offer.

One Star *

Sites with one star are primarily aimed at the seasoned mammal watcher, who will be visiting Madagascar for perhaps the second or third time. These sites may offer the chance to see a particularly unusual species or one with a very restricted range, or they may offer opportunities to see more easily seen species but in an unusual location or situation. Most sites in this category will have little in the way of infrastructure and few facilities.

Each site account contains a list of the main mammal species that have been recorded at the location. Species in **bold** indicate that observations can be especially rewarding, and generally correspond to the recommended sites suggested in the 'Where to See' sections of the species accounts.

Eastern Rainforest Areas

The eastern rainforests are perhaps the most exciting region for naturalists as they support the greatest diversity of animal and plant species. Around 80% of species are endemic to this habitat. It can also be the most infuriating, as animals, particularly mammals, can be challenging to find. However, several key parks and community reserves in widely spread localities are accessible and some of the wildlife, especially lemurs, is habituated to varying degrees (from extremely well habituated to wary, but moderately tolerant). Visits to three or four of these areas would offer ample opportunities to see a wide cross-section of species.

Andasibe-Mantadia National Park and Vicinity ***

Andasibe is arguably Madagascar's premier rainforest location, with Andasibe-Mantadia National Park and several NGO and community conservation reserves in the immediate vicinity. In combination, these offer opportunities to see an exceptional diversity of diurnal, cathemeral and nocturnal lemurs, and some of the island's other mammals too.

This is THE place to see Indri, with habituated groups in Analamazaotra Special Reserve and Mitsinjo Forest Reserve. Also regularly seen in Mantadia National Park, but can be more challenging to locate in these extensive forests. Mantadia National Park is excellent for Diademed Sifaka, which has also been reintroduced to Analamazaotra Special Reserve.

Location and Access: The village of Andasibe (= Périnet) is 23km east of Moramanga and c.140km east of Antananarivo (3–4 hours by road). Andasibe-Mantadia National Park consists of two discrete sections: the smaller, Analamazaotra Special Reserve, is adjacent to Andasibe (park entrance 1.5km from the village), whilst Mantadia National Park is around 12–18km north of Andasibe along a poorly-maintained dirt road (4×4 required, 1.5-hour drive).

Habitat and Terrain: Analamazaotra Special Reserve is a reasonable example of mid-altitude montane rainforest covering 810ha. It was selectively logged in the 1960s so many of the largest trees have been removed and canopy rarely exceeds 20–25m. Dominant trees include *Tambourissa*, *Symphonia*, *Dalbergia* and *Weinmannia* spp. Reserve is centred on a small lake (Lac Vert), with moderately steep forested slopes around it. Mean annual rainfall is 1,700mm.

Mantadia National Park is excellent primary mid-altitude montane rainforest, covering >10,000ha. The valley bottoms resemble lowland forest, with numerous huge trees and the canopy averages >25–30m. Understorey dominated by tree-ferns (*Cyathea* spp.) and *Pandanus* spp. near watercourses; orchids and other epiphytes are common. A hilly area with little flat ground, other than in main valley bottom. Some slopes steep and difficult. Rainfall averages 1,500–2,000mm per year.

KEY SPECIES

LEMURS: **Indri**, **Diademed Sifaka**, **Black and white Ruffed Lemur** (Mantadia only), **Brown Lemur**, Red-bellied Lemur, **Eastern Grey Bamboo Lemur**, Weasel Sportive Lemur, **Eastern Woolly Lemur**, Furry-eared Dwarf Lemur, Hairy-eared Dwarf Lemur, Goodman's Mouse Lemur and **Aye-aye**.

CARNIVORANS: Spotted Fanaloka, Fosa, Falanouc and Eastern Ring-tailed Vontsira (all primarily in Mantadia).

TENRECS: Tailless Tenrec, **Greater Hedgehog Tenrec**, **Lowland Streaked Tenrec**, Mole-like Mole Tenrec, Dobson's Shrew Tenrec, Talazac's Shrew Tenrec, Greater Long-tailed Shrew Tenrec, Shrew-toothed Shrew Tenrec, Cowan's Shrew Tenrec, Drouhard's Shrew Tenrec and other shrew tenrecs (*Microgale* spp.).

RODENTS: **Eastern Red Forest Rat**, Sleek-furred Ground Rat, **White-tailed Tree Rat**, Petter's Tufted-tailed Rat, Webb's Tufted-tailed Rat, Tanala Tufted-tailed Rat, Major's Tufted-tailed Rat and Lesser Tufted-tailed Rat.

BATS: Numerous species recorded, including Madagascar Rousette, **Commerson's Leaf-nosed Bat**, Manavi Long-fingered Bat, Major's Long-fingered Bat, Peters's Sheath-tailed Bat and Madagascar Serotine. Colony at Andasibe Post Office: Malagasy Large White-bellied Free-tailed Bat, Peters's Little Mastiff Bat and Eastern Free-tailed Bat can be seen emerging each evening (Dolch 2021).

Season: Accessible all year; best times for wildlife September–December and April–May.

Photographing reintroduced Diademed Sifakas, Analamazaotra Special Reserve.

Facilities: Several lodges/hotels catering for all levels, from top end to budget. Good-quality guides from Association des Guides d'Andasibe, Association Mitsinjo and Association Tambatra are available. An extensive network of trails runs through Analamazaotra Special Reserve: these can be combined into circuits of varying lengths to suit requirements. Mantadia National Park can be accessed at various points, with trails and circuits of different lengths; the best for wildlife begins at the 15km mark (PK15). The main trail runs along the valley bottom and is relatively easy, but it is often necessary to climb the adjacent steep slopes to track down lemurs. Conditions often wet.

Recommendations: There is realistic potential to see 10+ lemur species (including at sites listed below for night walks). A minimum of three or four days/nights should be devoted to the area. In Analamazaotra and Mitsinjo Forest, with patience really good views of Indri are almost guaranteed. Mantadia National Park warrants at least two morning visits starting at dawn. Enter forest at PK15

and follow trails exploring the valley bottom and adjacent slopes to look for Diademed Sifaka, Indri, Black-and-white Ruffed Lemur, Red-bellied Lemur and Eastern Grey Bamboo Lemur. Mantadia also offers excellent birdwatching with all four rainforest ground rollers (family Brachypteraciidae) and many other endemics possible.

Andasibe: other sites in area

There are a number of NGO and community conservation sites in the Andasibe vicinity. These offer good lemur watching opportunities during the day. Unlike in the national park, night walks are also permitted and a number of key species can be seen.

Mitsinjo Forest Reserve ***

Situated on opposite side of road to Analamazaotra Special Reserve is the Analamazaotra Forest Station administered by NGO Association Mitsinjo (www. associationmitsinjo.wordpress.com) (Dolch 2008). Forest comparable to Analamazaotra Special Reserve with more second growth in places. Good network of trails with some steep slopes. Renowned place to see **Indri**. Night walks possible: an excellent place to look for **Hairy-eared Dwarf Lemur**, along with other nocturnal species, **Furry-eared Dwarf Lemur**, **Goodman's Mouse Lemur**, Weasel Sportive Lemur, Eastern Woolly Lemur, Lowland Streaked Tenrec and Commerson's Leaf-nosed Bat. Can also be excellent for reptiles, especially chameleons and leaf-tailed geckos (*Uroplatus* spp.), and frogs. Association Mitsinjo is also central to a collaborative partnership project with other NGOs and the local community to restore the rainforest between Analamazaotra and Mantadia National Park to the north. Their tree nurseries produce >30,000 saplings annually and to date over one million trees have been planted.

V.O.I.M.M.A. Community Reserve ***

Name is an abbreviation of *vondron'olona ifotony miaro sy mitia ny ala*, meaning 'a local community group for loving and protecting the forest'. A 28ha community forest, adjacent to Analamazaotra Special Reserve, run by local guides. By day good for Brown Lemur and Eastern Grey Bamboo Lemur. Worthwhile at night for Furry-eared Dwarf Lemur, Goodman's Mouse Lemur and sometimes **Hairy-eared Dwarf Lemur**. Also good for frogs, chameleons and invertebrates.

Maromizaha Forest **

Under the stewardship of Malagasy primate study group GERP (www.gerp.mg), an area of *c.*1,600ha of forest south-east of Andasibe. Guides available at park office near km142 (PK142) on main road (*Route Nationale* 2). Forests lie at *c.*900–1,200m and are similar to Mantadia National Park; some difficult terrain with steep slopes. Twelve lemur species recorded: good for Indri, **Diademed Sifaka**, Black-and-white Ruffed Lemur, Brown Lemur, Red-bellied Lemur, Eastern Grey Bamboo Lemur and Eastern Woolly Lemur.

Vohimana Forest **

Located a short distance east of Andasibe at km149 (PK 149) on main road (*Route Nationale* 2) (*c.*13km past turn off for Andasibe). Walk to reserve takes *c.*45 minutes. Community and village partnership with NGO Man and the Environment's Net Positive Impact project: >2,000ha of forest, secondary growth and cultivated areas. Twelve species of lemur recorded but they can be challenging to see: most likely are Indri, Brown Lemur and Red-bellied Lemur, plus at night Furry-eared Dwarf Lemur, **Hairy-eared Dwarf Lemur** and Goodman's Mouse Lemur. Excellent for frogs and reptiles, especially chameleons, including Lance-nosed Chameleon *Calumma gallus*.

Iaroka Forest **

Some 10km south of *Route Nationale* 2 junction for Andasibe, Iaroka covers *c.*3,300ha of mid-elevation rainforest and is managed by local community organisation VOI Firaisankina. Resident lemur species are the same as nearby Maromizaha and include: Indri, Diademed Sifaka, Black-and-white Ruffed Lemur, Brown Lemur, Red-bellied Lemur, Eastern Grey Bamboo Lemur and Eastern Woolly Lemur. Access is via a poor road from RN2 (one hour, 4×4 required), then a further hour on foot to the forest. Trails and circuits 4–10km long. Steep slopes and walking can be challenging. A renowned birdwatching site, most famous for Helmet Vanga *Euryceros prevostii*, Bernier's Vanga *Oriolia bernieri*, Pollen's Vanga *Xenopirostris polleni*, Scaly Ground Roller *Geobiastes squamiger*, Rufous-headed Ground Roller *Atelornis crossleyi* and Brown Emutail *Bradypterus brunneus*.

Hairy-eared Dwarf Lemur, Andasibe.

Anjozorobe-Angavo Forest Corridor ***

In combination with Tsinjoarivo (see below), Anjozorobe-Angavo represents the most westerly remaining portion of eastern rainforest. It spans both eastern and western drainages, contains areas of high-elevation forest on the edge of the Central Highlands and is bounded by the Mangoro River to the east and Onive River to the south. Anjozorobe lies at the north of the corridor and offers an excellent alternative to Andasibe-Mantadia National Park with far fewer visitors. Home to ten lemur species: a good place to see Indri, many of which are very dark, sometimes almost black, and Diademed Sifaka. Night walks are also possible to see a variety of nocturnal species. Good birdwatching and excellent for reptiles and frogs as well.

Location and Access: Some 90km north of Antananarivo on *Route Nationale* 3. Drive takes *c.*3 hours.

Habitat and Terrain: One of the last vestiges of forest in the Central Highlands: a sliver of forest extending for *c.*80km and covering *c.*52,000ha that was protected in 2005. Mainly mid-elevation montane forest, with exposed and higher areas resembling high-elevation montane forest. Canopy rarely exceeds 15–20m, and is often lower. Bamboo and bamboo vines common. Numerous tree-ferns (*Cyathea* spp.) and epiphytes. Rainfall 2,000mm or more per year.

KEY SPECIES

LEMURS: Indri, Diademed Sifaka, Brown Lemur, Red-bellied Lemur, Eastern Grey Bamboo

Greater Bamboo Lemur, Ranomainty Community Forest.

Lemur, Weasel Sportive Lemur, Eastern Woolly Lemur, Furry-eared Dwarf Lemur, Sibree's Dwarf Lemur, Goodman's Mouse Lemur and Aye-aye.

CARNIVORANS: Fosa and Eastern Ring-tailed Vontsira.

TENRECS: Greater Hedgehog Tenrec, **Lowland Streaked Tenrec**, Mole-like Mole Tenrec, Dobson's Shrew Tenrec, Talazac's Shrew Tenrec, Greater Long-tailed Shrew Tenrec, Cowan's Shrew Tenrec, Naked-nosed Shrew Tenrec and several other shrew tenrecs (*Microgale* spp.).

RODENTS: **Eastern Red Forest Rat**, Sleek-furred Ground Rat, **White-tailed Tree Rat**, Tanala Tufted-tailed Rat, Major's Tufted-tailed Rat, Grandidier's Tufted-tailed Rat and **Anjozorobe Naked-tailed Forest Mouse**.

Season: Accessible all year; best times for wildlife September–December and April–May.

Facilities: Two comfortable eco-lodges; one operated by Malagasy NGO FANAMBY (www.association-fanamby.org), who have been instrumental in establishing the protected area. A range of trails and circuits caters for all levels of fitness. Some slopes steep and conditions often wet.

Recommendations: Easily reached and a worthwhile alternative or addition to visiting Andasibe for two or three days/nights. Dark (sometimes black) Indri

and Diademed Sifaka are the highlights, with Eastern Woolly Lemur, Furry-eared Dwarf Lemur and Goodman's Mouse Lemur often seen at night.

Ankeniheny-Zahamena Forest Corridor (CAZ) *

A discontinuous strip of low, mid- and high-elevation forest *c.*180km long and up to 30km wide, covering a total of *c.*550,000ha, and incorporating well-known protected areas such as Andasibe-Mantadia National Park to the south and Zahamena National Park at its northern extremity: collectively one of the largest rainforest areas on the island (Randrianasolo *et al.* 2013). Some western parts are the only known areas where four highly endangered lemurs occur sympatrically: Greater Bamboo Lemur, Indri, Diademed Sifaka and Black-and-white Ruffed Lemur (King *et al.* 2013).

The Ramsar site at Torotorofotsy is close to Andasibe and home to Greater Bamboo Lemur and Aye-aye. It is administered by Association Mitsinjo, who can arrange visits (see above). Community conservation projects at Sakalava and Ranomainty, between Mantadia and Zahamena National Parks operate in partnership with the Aspinall Foundation (www.aspinallfoundation.org). They support some of the largest known subpopulations of Greater Bamboo Lemur. These sites are difficult to reach, but potentially worthwhile for those who are dedicated.

Antanetiambo Nature Reserve **

Created by Marojejy National Park guide, Désiré Rabary, a small (17ha) private reserve located at Matsobe-Sud, 7km north of Andapa. There are fragments of native forest, secondary forest and cultivated areas. Main attraction is **Northern Grey Bamboo Lemur** which can be found feeding on non-native Chinese Dwarf Bamboo *Phyllostachys aurea* (Patel *et al.* 2014). On guided walks, visitors may also see frogs, chameleons and other reptiles, and learn about the cultural values and practices of the Tsimihety people.

Farankaraina Reserve **

Located 9km east of Maroantsetra, Farankaraina is a 1,650ha area of lowland rainforest and second growth bordering the Bay of Antongil; managed by Antongil Conservation, a collaboration between local guides and British and French zoos. Access by pirogue from Maroantsetra (1.5 hours) to village of

Andranofotsy, and walk along beach. Limited facilities: very simple bungalows and self-sufficient camping. Canopy often >25m, with many mature Ramy trees *Canarium madagascariensis* (Burseraceae). Trails and circuits 1–4km long. The main attractions are after dark, with a genuine chance of seeing **Aye-aye**: they feed in the large *Canarium* trees and sometimes it is possible to see three or four in one night, but they are often high in the canopy. Other nocturnal species include **Greater Dwarf Lemur** and mouse lemurs, probably Anjiahely Mouse Lemur. **Commerson's Leaf-nosed Bat** often seen around campsite. By day, **White-fronted Brown Lemur** is readily seen and **Northern Grey Bamboo Lemur** is also possible. Also very good for frogs and reptiles, particularly leaf-tailed geckos (*Uroplatus* spp.) and chameleons, e.g. Panther Chameleon *Furcifer pardalis*.

Lac Alaotra: Park Bandro **

Park Bandro is an 85ha community protected area administered by NGO Madagascar Wildlife Conservation (www.madagascar-wildlife-conservation. org). Located on the south-east shore of Lac Alaotra (Madagascar's largest lake) near the village of Andreba Gare, the marshland is home to the Critically Endangered **Lac Alaotra Reed Lemur** or **Bandro**: the only primate in the world to live permanently in aquatic habitat. Park population *c.*150 individuals (total population around lake *c.*2,000–2,500). Best time is during rainy season (November–April) when water levels are higher allowing canoes to pass through reedbeds. It is necessary to be on the water before sunrise, as the best chance of seeing the lemurs is shortly after dawn. Lac Alaotra is *c.*185km (4.5 hours) north of Moramanga along a poor road (*Route Nationale* 44). Bungalows, basic facilities and guides at Camp Bandro, near Andreba Gare.

Makira Natural Park *

The largest remaining rainforest block on the island, covering >370,000ha of strictly protected forest buffered by a similar area of community-managed forest. Home to more than 60 species of mammals, including 17 species of lemur (Ratelolahy *et al.* 2013). Region has a large rural population of >90,000 people, who are almost exclusively subsistence farmers and also rely on harvesting forest products. Major threats include forest clearance to plant crops (mainly rice), forest fires, selective logging, mining and bushmeat hunting. The latter is particularly

Aye-aye, Farankaraina Reserve.

prevalent in Makira, with several lemurs and carnivorans targeted (Golden 2005, 2009, Golden *et al.* in press). Creation of new protected areas, a collaboration between Madagascar National Parks (MNP) and the Wildlife Conservation Society (WCS) (www.madagascar.wcs.org): projects underway to promote alternative agricultural methods and income-generating activities like growing vanilla and cloves, or farming chickens, ducks or fish (Ratelolahy *et al.* 2013). Also long-term restoration of six forest corridors via tree-planting programme.

Location and Access: Located east and north of Bay of Antongil in north-east. Makira forest block provides a crucial conduit for biodiversity between Masoala to the east, Anjanaharibe-Sud and Marojejy to the north, and Mananara to the south. Access to eastern edge of Makira via boat, upriver from Maroantsetra.

Habitat and Terrain: Lowland and mid-elevation rainforest, with estimates suggesting it may harbour up to 50% of island's floral biodiversity. In lowland areas canopy reaches 30m.

KEY SPECIES

LEMURS: Indri, Silky Sifaka, Northern Black-and-white Ruffed Lemur, Red Ruffed Lemur, White-fronted Brown Lemur, Red-bellied Lemur, Northern Grey Bamboo Lemur, Seal's Sportive Lemur, Eastern Woolly Lemur, **Masoala Fork-marked Lemur**, Greater Dwarf Lemur, Mittermeier's Mouse Lemur, **Anjiahely Mouse Lemur** and Aye-aye.

CARNIVORANS: Spotted Fanaloka, Fosa, Falanouc, Eastern Ring-tailed Vontsira, Broad-striped Vontsira and Brown-tailed Vontsira.

Ring-tailed Vontsira, Marojejy National Park.

TENRECS: Greater Hedgehog Tenrec, Lowland Streaked Tenrec, Web-footed Tenrec, Dryad Shrew Tenrec, Naked-nosed Shrew Tenrec, Mountain Shrew Tenrec and several other shrew tenrecs (*Microgale* spp.).

RODENTS: Eastern Red Forest Rat, White-tailed Tree Rat and several species of tufted-tailed rat (*Eliurus* spp.).

Season: Best time to visit September–December.

Facilities: Infrastructure for ecotourism at Andaparaty in early stages of development. Sahavilory Camp is a community-run site. Operators in Maroantsetra can arrange trips.

Recommendations: Day trips possible. Better to plan an extended visit of at least three nights. Possible to see Indri, White-fronted Brown Lemur and Red-bellied Lemur. Black-and-white Ruffed Lemur often heard but more challenging to see. Birdwatching potentially very good. Frogs and reptiles also excellent.

Manombo Special Reserve *

Once part of a vast rainforest covering the coastal plain, Manombo is now among the last remnants of such forest. The reserve covers *c.*5,000ha, although only *c.*3,000ha is forest. Adjoining Manombo Classified Forest covers *c.*10,000ha, of which *c.*7,000ha is forest. Separated from main eastern rainforest block for at least 60 years. Further deforestation and hunting remain threats. Being coastal, also very susceptible to serious cyclone damage (Ralainasolo *et al.* 2013).

Location and Access: Approximately 30km south of Farafangana along *Route Nationale* 12 on south-

east coast. From village of Manombo routes into forest require 4×4 vehicle or walking. Best areas are 6–7km from village.

Habitat and Terrain: Lowland rainforest and littoral forest covering *c.*3,000ha (although reserve is 5,000ha).

KEY MAMMAL SPECIES

LEMURS: White-collared Brown Lemur, Southern Black-and-white Ruffed Lemur, Southern Grey Bamboo Lemur, **Manombo Woolly Lemur**, **Manombo Sportive Lemur**, Aye-aye, Greater Dwarf Lemur and Jolly's Mouse Lemur.

RODENTS: Lowland Red Forest Rat.

BATS: Madagascar Flying Fox and Commerson's Leaf-nosed Bat.

Season: Access difficult during heavy rain. September–December or April–May are best times.

Facilities: None at reserve. Camping at reserve office on main road, 4km from forest.

Recommendations: Visits arranged by hotels in Farafangana, e.g. Les Cocotiers and Hotel Austral. Permission required from local Madagascar National Parks office, where guides can be arranged. A two-day stay is recommended.

Agnalahaza Forest (formerly Mahabo Forest Reserve) is a little further south (40km from Farafangana) and covers 1,500ha of wetland and littoral forest. It is *c.*7km from main road on a dirt track. Groups of White-collared Brown Lemur can be seen. Permits and guides as for Manombo.

Marojejy National Park **

A UNESCO World Heritage Site, Marojejy is located at the northern extremity of the eastern rainforests. A mixture of verdant forest, craggy cliffs and imposing mountains, it offers one of the best, and most rugged, rainforest experiences in Madagascar. The park covers 55,500ha and harbours incredible biodiversity, in part due to the range of forest types at different elevations, rising beyond the treeline to the summit at 2,132m.

Location and Access: The park lies between the coastal town of Sambava and Andapa, 110km to south-west. Manantenina (at PK66), where the park office is located, is the initial point of access some 65km from Sambava and 45km from Andapa. From here drive to Mandena, where the trail begins: the park boundary is around a 2–3-hour walk.

Habitat and Terrain: Marojejy spans a considerable elevational range, with an unbroken transect from lowland rainforest through mid-elevation and high-elevation montane (cloud) forests to sclerophyllous high-mountain thicket near peaks. Below 750m canopy reaches 30+m, with large buttress trees like *Canarium*; typical mid-elevation montane rainforest between 750m and 1,400m, canopy averaging 20–25m; above 1,400m is high-elevation montane forest, and above 1,800m mainly stunted bush dominated by family Ericaceae. At lower elevations, annual rainfall averages around 3,000mm, rising to >4,000mm at higher altitudes. Slopes often steep and there are many deep valleys with fast-flowing streams.

KEY MAMMAL SPECIES

Lemurs: Silky Sifaka, **White-fronted Brown Lemur**, Red-bellied Lemur, **Northern Grey Bamboo Lemur**, Seal's Sportive Lemur, Eastern Woolly Lemur, Greater Dwarf Lemur, Crossley's Dwarf Lemur, Sibree's Dwarf Lemur, Hairy-eared Dwarf Lemur, Mittermeier's Mouse Lemur and Aye-aye.

Carnivorans: Fanaloka, Fosa, **Eastern Ring-tailed Vontsira** and Broad-striped Vontsira.

Tenrecs: Greater Hedgehog Tenrec, **Lowland Streaked Tenrec**, Dobson's Shrew Tenrec, Talazac's Shrew Tenrec, Montane Shrew Tenrec, Pale Shrew Tenrec, Thomas's Shrew Tenrec, Shrew-toothed Shrew Tenrec, Naked-nosed Shrew Tenrec and several other shrew tenrecs (*Microgale* spp.).

Rodents: Eastern Red Forest Rat, Sleek-furred Ground Rat, White-tailed Tree Rat, Major's Tufted-tailed Rat, Tanala Tufted-tailed Rat, Webb's Tufted-tailed Rat, Grandidier's Tufted-tailed Rat, Lesser Tufted-tailed Rat and Northern Naked-tailed Forest Mouse.

Bats: Among the richest sites for bats in eastern region. Species include: Madagascar Flying Fox, Madagascar Rousette, **Commerson's Leaf-nosed Bat**, Eastern Sucker-footed Bat, Madagascar Free-tailed Bat, Madagascar Sheath-tailed Bat, Madagascar White-bellied Free-tailed Bat and Malagasy Mouse-eared Bat.

Season: Avoid rainy season between late December and March. Months prior to this (September–November) offer best combination of mammal watching (especially for Silky Sifaka) and tolerable weather conditions.

Facilities: Three campsites, each with rustic cabins containing beds and bedding and sheltered eating areas. All lie on main summit trail. Camp Mantella (450m, six cabins) in lowland forest is a 2–3-hour walk from park boundary (4–5 hours from Manantenina), Camp Marojejia (775m, four cabins) is a further 1–2-hour walk and lies at the boundary of lowland and mid-elevation forest, and Camp Simpona (1,250m, two cabins) is 3+ hours beyond Camp Marojejia. Other than the main summit trail, and a trail to *Cascade de Humbert*, trails are few and slopes are steep or very steep. Walking, particularly at higher elevations is tough, and walk to summit is very strenuous (but worth it).

Recommendations: A fantastic wilderness experience that should not be rushed. Guides, porters and minor provisions can be arranged at park office in Manantenina (www.marojejy.com). Supplies should be brought from Andapa or Sambava. Fee structure for guides, porters and camping is administered through park office. Minimum stay of three nights recommended to track down Silky Sifakas, with two nights at Camp Marojejia. The view alone, across to a forest-encrusted craggy outcrop, is worth the effort. Main area to see Silky Sifakas is above Camp Marojejia, where the slopes are very steep and the terrain often treacherous. To reach

View from Camp Marojejia, Ambatotsondrona (leaning rock), Marojejy National Park.

the summit, requires four or five days with two nights at Camp Simpona (where Silky Sifaka also seen occasionally). A complete package can be arranged through Mimi Hotel (www.mimi-hotel.marojejy.com) in Sambava.

Marojejy also offers excellent birdwatching: Scaly Ground Roller and Short-legged Ground Roller (Camps Mantella and Marojejia) and Rufous-headed Ground Roller *Atelornis crossleyi* (Camp Simpona) can be seen, along with Helmet Vanga and Red-breasted Coua *Coua serriana*. There is also a huge diversity of frogs and reptiles, including a number of *Mantella* frogs and several leaf-tailed geckos (*Uroplatus* spp.).

Masoala National Park ***

A UNESCO World Heritage Site, Masoala has it all: glorious rainforest, fabulous wildlife, golden beaches and warm blue seas. The peninsula covers in excess of 400,000ha, with the park occupying *c.*240,000ha primarily on the western side (the eastern side is largely deforested). More remote parts of the park have suffered serious illegal logging, especially for rosewood (*Dalbergia* sp.) and ebony (*Diospyros* sp.), although areas immediately adjacent to the tourism zones appear not to have been significantly affected.

Male Silky Sifaka, Marojejy National Park.

Location and Access: In the far north-east of the island, east of Maroantsetra and forms eastern coastline of Bay of Antongil. All accessible wildlife and tourism areas are on western coastal side of park, reached by boat from Maroantsetra. The journey takes *c.*2–3 hours to Lohatrozona or Tampolo, the best points of access to the forest.

Habitat and Terrain: Largest and best-remaining area of coastal/lowland rainforest on Madagascar, with forest often extending down to shore. In main area (Lohatrozona) canopy is between 25m and 30+m with few emergent trees. Understorey characterised by abundant palms, with many epiphytes and orchids. Slopes often steep. Numerous clear, fast-flowing streams and small rivers. Littoral forest at Tampolo is often characterised by abundance of *Pandanus* spp., with canopy 10–15m. Annual rainfall 5,000–6,000mm with no discernible dry season.

KEY MAMMAL SPECIES

LEMURS: Red Ruffed Lemur, White-fronted Brown Lemur, Northern Grey Bamboo Lemur, **Masoala Sportive Lemur, Masoala Woolly Lemur,** Masoala Fork-marked Lemur, Greater Dwarf Lemur, Hairy-eared Dwarf Lemur, Aye-aye and mouse lemur (*Microcebus* sp.: species yet to be clarified).

CARNIVORANS: Spotted Fanaloka, Fosa, Falanouc, Eastern Ring-tailed Vontsira, Broad-striped Vontsira and **Brown-tailed Vontsira.**

TENRECS: Greater Hedgehog Tenrec, Lowland Streaked Tenrec, Dobson's Shrew Tenrec, Talazac's Shrew Tenrec, Greater Long-tailed Shrew Tenrec and several other shrew tenrecs (*Microgale* spp.).

RODENTS: Eastern Red Forest Rat, **Lowland Red Forest Rat,** Tanala Tufted-tailed Rat, Webb's Tufted-tailed Rat and Lesser Tufted-tailed Rat.

BATS: One of richest rainforest areas for bats, with at least 14 species including **Madagascar Flying Fox,** Madagascar Straw-coloured Fruit Bat, Madagascar Rousette, Eastern Sucker-footed Bat, Peters's Sheath-tailed Bat, Madagascar White-bellied Free-tailed Bat and **Commerson's Leaf-nosed Bat.**

Season: Wet throughout year. Avoid cyclone season (January–March).

Facilities: Hotels in Maroantsetra can arrange boat trips with equipment and guides. Several lodges, of varying standards (good quality to rustic), dotted along coast between Lohatrozona, Ambodiforaha and Tampolo. Main trail network at Lohatrozona is rather limited and paths can be steep. Trails through littoral forest at Tampolo are less demanding.

Recommendations: Takes time and effort: a visit of at least three or four days is recommended. Local guides from Association des Guides Ecotouristiques de Maroantsetra (AGEM) are essential. Red Ruffed Lemur relatively easy to see in accessible forest at Lohatrozona, as is White-fronted Brown Lemur. The latter species is also encountered in littoral forest at Tampolo. Occasionally Eastern Ring-tailed Vontsira and even Brown-tailed Vontsira are also seen behind Chez Arol Ecolodge. Night walks along the coastal path and the trails behind Chez Arol Ecolodge near Ambodiforaha are good for seeing various nocturnal lemurs, especially Masoala Sportive Lemur, Masoala Woolly Lemur, Greater Dwarf Lemur and mouse lemurs. Night walks also good for Greater Hedgehog Tenrec, Lowland Streaked Tenrec and Commerson's Leaf-nosed Bat. On boat crossing to Maroantsetra, worth going via Nosy Ravina (adjacent to Nosy Mangabe) to see colony of Madagascar Flying Foxes. In addition, Humpback Whales *Megaptera novaeangliae* gather offshore in Bay of Antongil between July and September, to give birth to their calves. Masoala is a renowned birdwatching site with high proportion of rainforest endemics, including Scaly Ground Roller, Short-legged Ground Roller, Helmet Vanga and Red-breasted Coua regularly seen at Lohatrozona. Other rarities including Madagascar Serpent Eagle *Eutriorchis astur*, Madagascar Red Owl and Bernier's Vanga. There is also a huge diversity of frogs and reptiles, including leaf-tailed geckos (*Uroplatus* spp.).

Montagne d'Ambre National Park ***

Madagascar's first national park (created in 1958), Montagne d'Ambre, along with adjacent Forêt d'Ambre Special Reserve, forms an isolated enclave of mid-elevation humid forest covering 18,200ha that shows considerable similarities to eastern rainforest regions, with some subtle differences.

Location and Access: Situated towards island's northern tip, c.30km south of Antsiranana (Diego Suarez). Around a one-hour drive on good road to Ambohitra (Joffreville), then a pot-holed dirt road for last 4km to park entrance.

Habitat and Terrain: Name derived from copal, a soft form of amber resin that oozes from trunks of large trees like Ramy *Canarium madagascariensis* (Burseraceae) and Rotra *Eugenia rotra*. Mid-elevation rainforest between c.850m and summit at 1,475m.

Lowland rainforest meets the sea, Masoala National Park.

Canopy height averages 25–30m. Botanically very rich with numerous bird's nest ferns (*Asplenium* sp.), tree-ferns (*Cyathea* sp.), cycads (Cycadaceae), *Pandanus* spp., orchids and lianas. Wide trails and relatively easy walking. Three waterfalls (*Cascade Sacrée*, *Cascade Antankarana* and *Cascade d'Antomboka*) and crater lake (*Lac de la Coupe Vert*) are points of interest, along with several viewpoints over the forest and surrounding area. Often enveloped in cloud (even when lowlands are hot and sunny). Annual rainfall 3,500–4,000mm.

KEY MAMMAL SPECIES

LEMURS: Sanford's Brown Lemur, Crowned Lemur, Northern Fork-marked Lemur, Montagne d'Ambre Dwarf Lemur, Montagne d'Ambre Mouse Lemur, Aye-aye and Ankarana Sportive Lemur.

CARNIVORANS: Falanouc, Spotted Fanaloka, Fosa and Northern Ring-tailed Vontsira.

TENRECS: Tailless Tenrec, Greater Hedgehog Tenrec, Lowland Streaked Tenrec, Dobson's Shrew Tenrec, Talazac's Shrew Tenrec and several other shrew tenrecs (*Microgale* spp.).

RODENTS: Eastern Red Forest Rat, Webb's Tufted-tailed Rat, Major's Tufted-tailed Rat and Lesser Tufted-tailed Rat.

BATS: Commerson's Leaf-nosed Bat and Montagne d'Ambre Long-fingered Bat.

Season: Accessible all year: September–December is best; can be more challenging January–March because of heavy rain.

Spotted Fanaloka, Ranomafana National Park.

Facilities: Several comfortable lodges in Ambohitra (Joffreville), including Domaine de Fontenay, which has its own private area of forest. More adventurous (or budget conscious) will prefer the campsite in the park at *Station des Rousettes*: equipment and provisions must be taken. Numerous well-maintained broad trails (*Sentier Touristique*, *Grande Cascade* trail) link all the main areas of interest. Trails less frequently used (*Voie des Mille Arbes*) are tougher and tend to be overgrown. Some follow sleep slopes, so walking can be difficult and tiring.

Recommendations: Day trips from Antsiranana possible, but better to stay in vicinity or camp for at least one night and preferably two or three. Areas around, and within easy walking distance from, campsite (*Station des Rousettes*) are good for several mammals, e.g. Sanford's Brown Lemur, Crowned Lemur and Northern Ring-tailed Vontsira. Nocturnal walks are excellent, especially for Northern Fork-marked Lemur, Montagne d'Ambre Mouse Lemur and Ankarana Sportive Lemur. Also, the rarely encountered Falanouc is occasionally seen.

Nosy Mangabe Reserve **

An idyllic rainforest-cloaked island covering 520ha. Sheltered landing area in a beautiful sandy cove, with mature rainforest to the shoreline. Forest has largely regenerated after considerable human activity, and selective logging 200–300 years ago, although it is not as diverse as equivalents on mainland. Nonetheless, lemur watching is excellent. Populations of Black-and-white Ruffed Lemurs and Aye-ayes were introduced in the 1960s and have subsequently thrived.

Location and Access: Lies offshore in north-east, 5km off Maroantsetra in Bay of Antongil. Boat journey from Maroantsetra takes *c*.30 minutes.

Habitat and Terrain: Lowland rainforest and second growth. Large buttress-rooted trees reach 35m or more in height. Typical species *Ravensara*, *Canarium*, *Ocotea*, *Ficus* and *Tambourissa*. Tree-ferns, ferns, epiphytes and orchids also common. Slopes rise steeply from sea to summit at 332m.

KEY MAMMAL SPECIES

LEMURS: Aye-aye (introduced), **Northern Black-and-white Ruffed Lemur** (introduced), **White-fronted Brown Lemur**, Greater Dwarf Lemur and **Anjiahely Mouse Lemur**.

TENRECS: Greater Hedgehog Tenrec.

BATS: Madagascar Flying Fox (adjacent Nosy Ravina) and **Commerson's Leaf-nosed Bat**.

Season: Accessible all year, but avoid cyclone season (January–March).

Facilities: Basic campsite with sheltered pitches and covered platforms, shower (fed from waterfall) and flush toilet. All equipment and provisions must be taken from Maroantsetra. The moderate network of footpaths is gradually being improved. Many trails are steep; flat trails with wooden walkways/bridges around campsite and behind beach.

Recommendations: Day trips from Maroantsetra are possible to see diurnal Black-and-white Ruffed Lemur and White-fronted Brown Lemur. However, to fully appreciate this beautiful island a longer camping trip of two days/nights is recommended and necessary for nocturnal walks that offer the slim chance of seeing an Aye-aye. Between July and September also possible to see Humpback Whale *Megaptera novaeangliae* in Bay of Antongil, when they gather to give birth. Also, one of the best places to see large Leaf-tailed Geckos *Uroplatus fimbriatus*, Madagascar Ground Boa and frogs like Green-backed Mantella *Mantella laevigata*.

Ranomafana National Park ***

When created in 1991, Ranomafana became the third national park in Madagascar (after Montagne d'Ambre and Isalo): its inception was in part due to the discovery of Golden Bamboo Lemur in 1986. It is now established as one of the island's premier wildlife tourism locations, as well as a major centre for rainforest research with Malagasy and foreign scientists working in collaboration. Research is based at Centre Valbio (near Talatakely park entrance), where tours and lectures can be arranged with advance notice. The park covers 39,200ha, but tourism access is limited primarily to two zones, Talatakely and Vohiparara.

Location and Access: In the central south-east of the island, *c.*65km north-east of Fianarantsoa and adjacent to village of Ranomafana (which means 'hot water', on account of the nearby thermal springs). From Fianarantsoa access along *Route Nationale* 45 and 25, the latter bisects the park and continues to Mananjary on east coast. The journey from Fianarantsoa to the park takes 1.5–2.5 hours depending on road conditions. Also possible to drive from Antananarivo in a day, but this takes more than ten hours.

Habitat and Terrain: Excellent mid-elevation rainforest and higher-elevation montane forest. Dominated by Namorona River, fed by many streams flowing from hills, and plunges off eastern escarpment near park entrance. Steep slopes covered with a mix of primary and secondary forest, and canopy averages 25–30m. Much of second growth in main tourist zone at Talatakely is dominated by dense areas of introduced Strawberry Guava *Psidium cattleianum*. Stands of Giant Bamboo *Cathariostachys madagascariensis* also prominent. Annual rainfall 2,000–3,000mm.

KEY MAMMAL SPECIES

Lemurs: **Greater Bamboo Lemur, Golden Bamboo Lemur, Eastern Grey Bamboo Lemur, Milne-Edwards's Sifaka, Red-bellied Lemur, Red-fronted Brown Lemur, Southern Black-and-white Ruffed Lemur, Peyrieras's Woolly Lemur, Small-toothed Sportive Lemur**, possibly **Betsileo Sportive Lemur** (Vohiparara), Crossley's Dwarf Lemur, Sibree's Dwarf Lemur, Brown Mouse Lemur and Aye-aye.

Carnivorans: **Spotted Fanaloka**, Falanouc, **Fosa, Eastern Ring-tailed Vontsira** and Broad-striped Vontsira.

Tenrecs: **Greater Hedgehog Tenrec, Lowland Streaked Tenrec**, Web-footed Tenrec, and several other shrew tenrecs (*Nesogale* and *Microgale* spp.)

Rodents: **Eastern Red Forest Rat, Lowland Red Forest Rat**, Sleek-furred Ground Rat, **White-tailed Tree Rat**, Malagasy Mountain-dwelling Mouse, Tanala Tufted-tail Rat, Major's Tufted-tailed Rat and Lesser Tufted-tailed Rat.

Bats: Madagascar Straw-coloured Fruit Bat, Peters's Little Mastiff Bat, Madagascar Free-tailed Bat, Peters's Sheath-tailed Bat, Madagascar White-bellied Free-tailed Bat, Malagasy Mouse-eared Bat and Madagascar Serotine.

Season: Most vibrant during austral summer rainy season from December to March, but access and conditions can be difficult. Otherwise, the periods either side of main rainy season are most rewarding, April–May (when fruiting Strawberry Guava *Psidium cattleianum* attracts lemurs) and September–November.

Facilities: Several lodges, ranging from good quality to rustic, between park entrance and village of Ranomafana (*c.*7km from entrance). At park entrance there is a campsite and very basic accommodation with dormitories and a snack bar. From the park entrance, an extensive system of well-maintained trails cuts through best wildlife areas in main tourism zone (Talatakely), but many are up and down steep and often muddy slopes. The higher-elevation forest at Vohiparara (*c.*5km from park entrance on RN25, towards Fianarantsoa) also has a good trail network.

Recommendations: One of the islands' best mammal sites. Renowned for three species of bamboo lemur (Greater Bamboo Lemur, Golden Bamboo Lemur and Eastern Grey Bamboo Lemur) and three other diurnal species, Milne-Edwards's Sifaka, Red-bellied Lemur and Red-fronted Lemur. During the day, nocturnal Peyrieras's Woolly Lemur is often seen at sleep sites. Between April and June lemurs can be easier to find in Belle Vue area, when Strawberry Guava is plentiful. Higher-elevation

Namorona River, Ranomafana National Park.

Milne-Edwards's Sifaka, Ranomafana National Park.

forests at Vohiparara also well worth exploring for Milne-Edwards's Sifaka, Red-bellied Lemur and sleeping Small-toothed Sportive Lemur (or possibly Betsileo Sportive Lemur). At night Spotted Fanaloka also can be seen. Local guides very knowledgeable and helpful. Minimum of three or four days is recommended, with at least two daytime excursions to Tanatakely and one to Vohiparara.

Also one of the best birdwatching sites, especially to see rarities like Pollen's Vanga *Xenopirostris polleni*, Brown Mesite *Mesitornis unicolor*, Slender-billed Flufftail *Sarothrura watersi*, Yellow-bellied Sunbird Asity *Neodrepanis hypoxantha* and all four species of rainforest ground roller (family Brachypteraciidae). Additionally, an exceptional diversity of reptiles and frogs.

Tsinjoarivo *

Lying at the southern extreme of an isolated and fragmented sliver of highland forest linked to Anjozorobe-Angavo forest corridor to the north (see above). Situated at the eastern edge of the Central Highlands, Tsinjoarivo receives reduced annual rainfall (*c.*2,000mm) compared to rainforest zones at lower elevations further east. At its

western extreme, the high-elevation forests lies between 1,300m and 1,650m and are amongst the last remaining examples of this habitat. Further east, the forests lie between 1,200m and 1,600m, and are coincident with the steep eastern escarpment.

Because of its transitional east/west position, levels of biodiversity are high and unusual. One of the few locations where the Critically Endangered highland **Sibree's Dwarf Lemur** occurs, which hibernates for up to seven months per year. Nine other lemur species recorded, including **Diademed Sifaka**, with some exceptionally dark (almost all-black) individuals, and Aye-aye. Tenrec diversity is the highest of any known site, with 17 species, including 12 sympatric shrew tenrecs (*Nesogale* spp. and *Microgale* spp.) (Irwin *et al.* 2013). Five species of endemic Nesomyine rodents also occur.

NGO Sadabe (the local name for Diademed Sifaka) (www.sadabe.org) is working with local communities to protect areas of forest and develop low-impact tourism centred on village of Mahatsinjo. In 2019 the Tsinjoarivo-Ambalaomby New Protected Area received official temporary protection (with funding from Rainforest Trust). Permanent protected status will follow in due course, with plans to further expand capacity for ecotourism. Tsinjoarivo also offers a rich cultural history, being the site of a royal palace (*rova*) used by three queens during the 18th century (Irwin *et al.* 2013).

Diademed Sifaka, Tsinjoarivo.

Tsitongambarika Protected Area *

Together with adjacent parcel 1 of Andohahela National Park (see below), Tsitongambarika constitutes the southernmost rainforest on the island. Immediately north of Tolagnaro (Fort Dauphin) and stretching for c.70km, the Tsitongambarika massif (Vohimena Mountains) consists of a series of ridges aligned south-west to north-east, with the north-flowing Manampanihy River forming the western boundary. This recently protected area covers c.60,000ha, and includes the last remaining large areas of lowland rainforest in southern Madagascar. At low elevations the canopy reaches 15–25m, at mid-elevations it is 12–20m. Maximum elevation is 1,358m. Many plant species are locally endemic.

Seven species of lemur have been recorded, including Red-collared Brown Lemur *Eulemur collaris*, Southern Grey Bamboo Lemur, Southern Woolly Lemur, Andohahela Sportive Lemur, Aye-aye, a dwarf lemur *Cheirogaleus* sp. and a mouse lemur *Microcebus* sp. Canivorans include Spotted Fanaloka, Falanouc, Fosa, Eastern Ring-tailed Vontsira and Broad-striped Vontsira. Also, an important site for bats, especially Madagascar Flying Fox with four known roosts containing c.2,000 individuals (BirdLife International 2011).

Management of the protected area is a collaboration between local communities and international organisations. Ecotourism is being developed, with a campsite adjacent to the research station. A network of trails has been established. Local trained guides are available.

Northern and Western Deciduous Forest Areas

The deciduous forests of western and northern Madagascar are less diverse than rainforests of the east, but nonetheless contain a wealth of fauna and flora of great importance, with very high levels of endemicity (around 90%). Canopy height and tree density is lower than eastern rainforests, giving the forests a more 'open' feel. This makes wildlife watching easier. Several major protected areas exist in deciduous forest regions; whilst some are remote and remain difficult for visitors to reach, others are among the most popular and rewarding wildlife watching localities on the island.

Of particular note are parks incorporating limestone massifs that have been eroded into spectacular pinnacle formations known as karst (*tsingy*).

Broad-striped Vontsira, Tsitongambarika Forest.

Rivers flowing through these areas have eroded underground passages and caves, and some have subsequently collapsed to form canyons. Within these, deciduous forest flourishes, often supporting isolated faunal communities unlike any others on the island. Mammal watching can be particularly rewarding in such areas: the main examples are at Ankarana Special Reserve, Tsingy de Bemaraha National Park and Tsingy de Namoroka National Park, although the latter site is not readily accessible.

Analamerana Special Reserve/ Andrafiamena-Andavakoera Protected Area **

Oriented south-west to north-east along the Andrafiamena Mountains, this fragmented enclave of forest covers c.110,00ha, with Analamerana Special Reserve extending c.34,700ha and Andrafiamena-Andavakoera covering c.75,000ha.

Location and Access: Analamerana lies south-east of Montagne d'Ambre National Park and east of *Route Nationale* 6, c.85km south of Antsiranana (60km on main road, 25km on very poor dirt road). Andrafiamena-Andavakoera is south-west of Analamerana and east of *Route Nationale* 6. A 4×4 vehicle is recommended to reach both locations.

Habitat and Terrain: A combination of dry deciduous and dry evergreen forest (with elements from both western and eastern regions), between sea level and c.650m, growing predominantly on a limestone massif (Analamerana) and sandstone massif interspersed by limestone (Andrafiamena-

Andavakoera). Three large forests blocks (two in Analamerana, one in Andrafiamena-Andavakoera) interconnected by numerous small fragments and a mosaic of secondary bush and anthropogenic grassland. In main forest blocks canopy reaches 15–20m in valley bottoms, but often 5–10m in drier areas. Pronounced dry season, with annual rainfall averaging 1,250mm, the majority between November and April.

KEY MAMMAL SPECIES

LEMURS: **Perriers's Sifaka**, **Crowned Lemur**, Sanford's Brown Lemur, **Ankarana Sportive Lemur**, Daraina Sportive Lemur, **Ankarana Dwarf Lemur**, **Tavaratra Mouse Lemur** and Aye-aye.

CARNIVORANS: **Northern Ring-tailed Vontsira**, Fosa and Spotted Fanaloka.

TENRECS: Tailless Tenrec, Greater Hedgehog Tenrec and Short-tailed Shrew Tenrec.

RODENTS: Ankarana Tufted-tailed Rat.

Season: Accessible only during drier months from May to November/December. Can be very hot during day, but cooler at night.

Facilities: In Analamerana, self-sufficient camping only: all equipment and provisions must be taken. Excursions with park permits arranged by tour operators in Antsiranana. Andrafiamena-Andavakoera has a good-quality lodge, with guides and various trail circuits to suit all levels.

Perrier's Sifaka, Analamerana Special Reserve.

Recommendations: Andrafiamena-Andavakoera is the best site for seeing Critically Endangered Perriers's Sifaka. Stay at Black Lemur Camp operated by NGO FANAMBY (www.association-fanamby.org) for two or three nights. Also likely are Crowned Lemur, Ankarana Sportive Lemur, Tavarata Mouse Lemur and sometimes Northern Ring-tailed Vontsira. With no facilities, visiting Analamerana Special Reserve requires substantial planning. Birdwatchers will be interested in White-breasted Mesite *Mesitornis variegatus* and Van Dam's Vanga *Xenopirostris damii*.

Anjajavy Private Reserve ***

A luxury private lodge on a remote peninsula with its own reserve covering *c.*1,030ha: the combination of effortless wildlife watching, stunning coastal location, secluded white-sand beaches and high-quality accommodation and hospitality is intoxicating. There is also an underground cave where the subfossil skull of an extinct giant lemur (*Palaeopropithecus kelyus*) can be seen embedded in the rock. May be beyond the budget of many visitors, but well worth a visit if your pockets are deep enough.

Location and Access: Sited on a peninsula north of the Mahajamba River delta and south of Moramba Bay, some *c.*130km from Mahajanga. Access is only via private plane from Antananarivo, arranged by the lodge as part of the package.

Habitat and Terrain: A mixture of dry deciduous forest on sandy soils, coastal baobab forests, limestone outcrops (with caves) and mangroves. Canopy reaches 10–15m, understorey often tangled. Terrain gently undulating or flat. Annual rainfall 1,000–1,200mm.

KEY MAMMAL SPECIES

LEMURS: **Coquerel's Sifaka**, **Brown Lemur**, **Anjiamangirana Sportive Lemur**, **Anbarijeby Mouse Lemur**, Fat-tailed Dwarf Lemur and Aye-aye (translocated).

CARNIVORANS: Fosa.

TENRECS: Greater Hedgehog Tenrec.

RODENTS: Western Tufted-tailed Rat.

BATS: **Madagascar Flying Fox**, Madagascar Straw-coloured Fruit Bat, Madagascar Rousette and **Commerson's Leaf-nosed Bat**.

Season: Accessible year-round, but best times are May to December to avoid the rainy season.

Facilities: The lodge has 24 luxury villas with sea views (www.anjajavy.com). There is an extensive trail network to access various parts of the forest and coast.

Recommendations: Packages are available for two, three or four nights: longer stays recommended. Diurnal lemurs, Coquerel's Sifaka and Brown Lemur, are readily seen in and around the lodge gardens. Concentrate on wildlife watching early morning, late afternoon and after dark, when night walks for nocturnal lemurs can be productive. Birdwatching also very good. One of the best locations to see the Critically Endangered Madagascar Fish Eagle *Haliaeetus vociferoides*, and good for Sickle-billed Vanga *Falculea palliata*, Madagascar Crested Ibis *Lophotibis cristata*, Coquerel's Coua *Coua coquereli* and Grey-headed Lovebird *Agapornis canus*. Interesting reptiles include Madagascar Ground Boa, Madagascar Hog-nosed Snake *Leioheterodon madagascariensis*, twig-mimic snakes (*Langaha alluaudi* and *L. pseudoalluaudi*), and several species of chameleon.

Anjiamangirana Classified Forest *

Located in the north-west, to the south-east of Antsohihy and north of the Sofia River, this Key Biodiversity Area is a fragmented area of deciduous forest where two Critically Endangered species occur, Anjiamangirana Sportive Lemur (originally described from this forest) and Coquerel's Sifaka. Ambarijeby Mouse Lemur also occurs. There are currently no facilities for visiting the area.

Ankarafantsika National Park (Ampijoroa Forestry Station) ***

Arguably one of the two best western dry forest locations (along with Kirindy), Ankarafantsika combines accessible and varied lemur watching, with the opportunity to see other mammals.

Location and Access: Ankarafantsika lies either side of *Route Nationale* 4, *c.*120km south-east of Mahajanga (2–3 hours drive). Also possible to drive directly from Antananarivo on RN4, but this takes 7–8 hours.

Habitat and Terrain: A mixture of dry deciduous forest and karstic limestone plateau covering *c.*136,500ha. Ampijoroa is near centre of reserve and covers *c.*20,000ha. The immediate vicinity is dominated by Lac Ravelobe, on east side of *Route Nationale* 4. On sandy soils around lake grows typical deciduous forest with canopy 15–20m high.

Coquerel's Sifaka, easily seen at both Anjajavy Private Reserve and Ankarafantsika National Park.

Understorey sparse, with virtually no epiphytes. On higher plateau areas canopy averages 5–10m with very little understorey. In open rocky areas succulents like *Pachypodium rosulatum rosulatum* and *Aloe* spp. grow. Terrain gently undulating or flat, with occasional shallow ridges. Annual rainfall 1,000–1,500mm.

KEY MAMMAL SPECIES

LEMURS: **Coquerel's Sifaka**, **Mongoose Lemur**, **Brown Lemur**, **Western Woolly Lemur**, **Milne-Edwards's Sportive Lemur**, **Fat-tailed Dwarf Lemur**, **Grey Mouse Lemur** and **Golden-brown Mouse Lemur**.

CARNIVORANS: Fosa, Falanouc and non-native Small Indian Civet.

TENRECS: Tailless Tenrec, **Greater Hedgehog Tenrec** and Short-tailed Shrew Tenrec.

RODENTS: **Western Tufted-tailed Rat**, **Western Big-footed Mouse** and **Long-tailed Big-footed Mouse**.

BATS: Madagascar Straw-coloured Fruit Bat, Madagascar Rousette, **Commerson's Leaf-nosed Bat**, Western Sheath-tailed Bat, Western Sucker-footed Bat and **Mauritian Tomb Bat**.

Season: Accessible year-round, but roads more difficult in wet season (January–March), especially from Antananarivo. During austral summer can be extremely hot.

Facilities: At the park entrance, a large campsite with water, small toilet block and cooking facilities. Also seven well-appointed bungalows (*Gîte de Ampijoroa*) with en-suite facilities overlooking Lac Ravelobe.

Close to nearby (5km) village of Andranofasika are two modest lodges, each with six bungalows. The village has small shops for basic supplies. Trails on both sides of road are wide and walking generally easy. Very good guides available at park office by campsite.

Recommendations: Lemur watching begins on arrival: both Coquerel's Sifaka and Brown Lemur are regularly seen in trees in and around car park. Look on tree trunks around car park for Mauritian Tomb Bat. Trails on west side of road (*Circuit Coquereli*) are better for lemurs. It's possible to see six lemur species relatively easily by day and night; only Mongoose Lemur and Golden-brown Mouse Lemur are sometimes elusive. Night walks (when possible) also rewarding for Western Big-footed Mouse, Long-tailed Big-footed Mouse and Commerson's Leaf-nosed Bat. At least two or three days recommended.

One of the island's top birdwatching sites, with good chances of seeing rare endemics like Madagascar Fish Eagle, White-breasted Mesite, Schlegel's Asity *Philepitta schlegeli*, Van Dam's Vanga and Coquerel's Coua.

Ankarana Special Reserve ***

Dominated by impressive eroded limestone pinnacle karst formations, known as *tsingy* (derived from Malagasy word *mitsingitsingy* meaning 'to walk on tiptoes'), which form an almost impenetrable fortress in places. Penetrated by canyons and extensive underground river and cave systems. Where large caves have collapsed there are isolated pockets of river-fed forest.

Location and Access: Approximately 110km south of Antsiranana, the reserve lies mostly to the west of *Route Nationale* 6. From park office at east gate in Mahamasina (on RN6) access on foot. Alternatively, by 4×4 from north-west on track south of Anivorano Avaratra, to west gate at Matsaborimanga. Also, possible to approach from Ambilobe in south-west (near Amboandriky), but road is poor and 4×4 essential (not recommended in rainy season).

Habitat and Terrain: Reserve constitutes majority of Ankarana Massif covering 18,225ha. Deciduous forest grows around periphery and penetrates larger canyons. Also isolated pockets of forest growing in

Male Crowned Lemur climbing on tsingy, *Ankarana Special Reserve.*

limestone karst. Canopy reaches 25m, dominant trees include *Cassia*, *Dalbergia*, *Ficus* and baobab *Adansonia madagascariensis*. In canyon bottoms terrain is undulating, but most areas, especially karst formations, are very rugged and potentially hazardous. Annual rainfall averages 1,800mm.

KEY MAMMAL SPECIES

LEMURS: Crowned Lemur, Sanford's Brown Lemur, Northern Grey Bamboo Lemur (?), **Ankarana Sportive Lemur, Ankarana Dwarf Lemur, Northern Fork-marked Lemur, Tavaratra Mouse Lemur** and Aye-aye.

CARNIVORANS: Northern Ring-tailed Vontsira, Fosa and Spotted Fanaloka.

TENRECS: Greater Hedgehog Tenrec and Short-tailed Shrew Tenrec.

RODENTS: Ankarana Tufted-tailed Rat.

BATS: One of richest bat sites on the island (a consequence of the extensive limestone cave systems). Species include: Madagascar Straw-coloured Fruit Bat, **Madagascar Rousette, Commerson's Leaf-nosed Bat**, Grandidier's Trident Bat, Western Sheath-tailed Bat, Madagascar Giant Mastiff Bat, Madagascar Sheath-tailed Bat, Black-and-red Free-tailed Bat, Malagasy Mouse-eared Bat, Madagascar Yellow Bat, Manavi Long-fingered Bat and Glen's Long-fingered Bat.

Season: Only accessible during drier months from May to December. Can be very hot by day, but cooler at night.

Facilities: Lodges of varying standards (from very comfortable to basic) near east gate at Mahamasina. Within the park there are three campsites: all equipment and provisions must be taken. *Camp des Princes* (on east side, access from Mahamasina) gives access on foot to some forest and caves. It is a 2–3-hour walk through reserve to other campsites, *Campement Anilotra* (formerly *Campement des Anglais*) in the *Canyon Grande* and *Campement d'Andrafiabe* (formerly *Campement des Americains*) at foot of west-facing escarpment. *Campement d'Andrafiabe* also has bungalows managed by Océane Adventures. Water is scarce, but available from underground streams near camps. Reasonable trails around *Canyon Grande* but remote areas are much more challenging to explore. Gondwana Park is a private reserve (www.ankarana-parc.com) near Mahamasina (created and owned by Ankarana Lodge) with areas of limestone karst and forest.

Crowned Sifaka, Antrema Reserve.

There are Crowned Lemurs and night walks are excellent for Ankarana Sportive Lemur and Tavaratra Mouse Lemur. The botany is also fascinating, with pachypodiums and locally endemic aloes.

Recommendations: Camping excursions arranged through tour operators in Antsiranana. Minimum stay of two or three nights is recommended. To explore reserve fully it is best to spend nights either at different campsites or accommodation in Mahamasina. Local guides essential. *Campement Anilotra* is best for Crowned Lemur, Sanford's Brown Lemur, Ankarana Sportive Lemur, Northern Ring-tailed Vontsira and occasionally Fanaloka. Areas close to camp at Mahamasina can be good for Fosa. Night walks at Gondwana Park recommended.

Antrema Reserve (Katsepy Forest) *

Lying on opposite (south) side of the Betsiboka River estuary to Mahajanga, and north of the village of Katsepy, Antrema Reserve is a protected biocultural zone, with fragmented forest, lakes, mangroves and sand dunes. Plans are in place to plant forest corridors, creating 'biological bridges' to link the fragments. Six species of lemur have been recorded, including two Critically Endangered species, Crowned Sifaka and Mongoose Lemur, and possibly Antafia Sportive Lemur. The forest below Katsepy lighthouse offers good chances of seeing the lemurs. Recent surveys suggest local populations of Crowned Sifaka are *c.*800 individuals and Mongoose Lemur *c.*100 individuals (Foundation Ensemble 2021).

Kirindy Forest ***

Part of the wider 125,000ha Menabe Protected Area, Kirindy Forest is perhaps the most rewarding site in western Madagascar. The wildlife is varied, accessible and often very visible. Not only is it a premier site for lemurs (eight species, including six nocturnal species) and other mammals (the best place to see Fosa and Narrow-striped Boky), bird and reptile diversity is equally impressive.

Location and Access: Approximately 60km northeast of Morondava and 20km inland from west coast: east of road to Belo-sur-Tsiribihina, north of village of Marofandilia. From Morondava drive takes two or more hours.

Habitat and Terrain: Typical dry deciduous forest growing on sandy soils covering c.12,000ha. Canopy averages 12–15m, but can reach >20m in more humid areas along watercourses. Often a dense understorey and midstorey. Three species of baobab also present, *Adansonia fony*, *A. za* and *A. grandidieri*, the latter being the largest baobab in Madagascar, reaching >30m. Terrain mainly flat, with moderate undulations. Annual rainfall averages 700–800mm, most falling December–March.

KEY MAMMAL SPECIES

Lemurs: Verreaux's Sifaka, **Red-fronted Brown Lemur**, Madame Berthe's Mouse Lemur, **Grey Mouse Lemur**, Red-tailed Sportive Lemur, **Pale Fork-marked Lemur**, Coquerel's Dwarf Lemur and **Fat-tailed Dwarf Lemur**.

Carnivorans: Fosa and **Narrow-striped Boky**.

Male Fosa, Kirindy Forest.

Tenrecs: Tailless Tenrec, **Greater Hedgehog Tenrec**, Lesser Hedgehog Tenrec, **Large-eared Tenrec**, Nasolo's Shrew Tenrec and Grandidier's Shrew Tenrec.

Rodents: **Giant Jumping Rat**, **Western Tufted-tailed Rat** and **Western Big-footed Mouse**.

Bats: One of most diverse sites in western regions. Species include Madagascar Flying Fox, Madagascar Straw-coloured Fruit Bat, **Commerson's Leaf-nosed Bat**, Grandidier's Free-tailed Bat, Madagascar White-bellied Free-tailed Bat and Racey's Pipistrelle.

Season: Marked dry season May–October with virtually no rain. Wettest months in austral summer (December–March). Wildlife watching best in spring and summer (September–April), particularly after rain. Road conditions can be difficult during wetter months and temperatures very high. In winter forests may appear relatively 'lifeless' and nights can be cold.

Facilities: Good-quality lodge accommodation close by on main road (north of Marofandilia), including Camp Amoureux operated by NGO FANAMBY (www.association-fanamby.org) (which also has its own area of forest where lemurs can be seen). In Kirindy Forest, basic lodge with rustic bungalows and block containing six dormitories. Snack bar/restaurant serves cold drinks and simple meals. Paths through forest wide and flat, and walking easy. Trails arranged in grid system, which makes navigation straightforward. Local guides available at park office, next to car park.

Recommendations: Arguably the premier nocturnal site in Madagascar. The only place to see Giant Jumping Rat, plus six nocturnal lemurs. During the day, Verreaux's Sifaka, Red-fronted Brown Lemur and the delightful Narrow-striped Boky readily seen. Kirindy is also best place to see Fosa, which often scavenge around central area, and in October and November can also be seen mating. Day trips and excursions organised by several hotels in Morondava. Minimum two, ideally three/four days/nights recommended. Some 14km north of Morondava, on way to Kirindy, is famous Avenue of Baobabs *Adansonia grandidieri* (*Avenue des Baobabs* or *Baobab Allé*) (a UNESCO World Heritage Site), which is well worth stopping at, especially for sunrise or sunset. Site also run by NGO FANAMBY with tree nursery, snack bar and gift shop.

Kirindy-Mitea National Park (Kirindy Mité) *

Not to be confused with Kirindy Forest further north, Kirindy-Mitea National Park (also a UNESCO Biosphere Reserve and Ramsar wetland site) encompasses 625,000ha and a variety of habitats, including mangroves, sand dunes, lakes and forests. It sits at the boundary of the western and southern floral domains, and as such exhibits elements of both. The western deciduous forest is a 'kingdom of boababs', with three species occurring in significant numbers. Northern limit of Ring-tailed Lemur, although possibly now extirpated.

Location and Access: The park entrance at Manahy is c.80km south of Morondava and 15km south of Belo-sur-Mer (Madagascar National Parks office next to church). The road from Morondava is poor and seasonal (impassible in wet season); the drive takes 4+ hours. Access to Belo-sur-Mer is quicker and easier by boat from Morondava.

Habitat and Terrain: Mainly dry deciduous forest growing on sandy soils in northern areas of the park, with spiny forest elements appearing in the south. Canopy averages 12–15m, but can reach >20m. Three species of baobab, *Adansonia fony*, *A. za* and *A. grandidieri*, with some *A. grandidieri* of prodigious size. Annual rainfall averages 700–800mm, majority falling between December and March.

KEY MAMMAL SPECIES

Lemurs: Verreaux's Sifaka, Red-fronted Brown Lemur, Ring-tailed Lemur (?), Red-tailed Sportive Lemur, Pale Fork-marked Lemur, Coquerel's Dwarf Lemur, Fat-tailed Dwarf Lemur and Grey Mouse Lemur.

Carnivorans: Fosa and Narrow-striped Boky.

Tenrecs: Tailless Tenrec, Greater Hedgehog Tenrec, Lesser Hedgehog Tenrec and Large-eared Tenrec.

Rodents: Western Tufted-tailed Rat and Western Big-footed Mouse.

Season: Marked dry season May–November with virtually no rain. Wettest months in austral summer (December–March). Wildlife watching best in spring and summer (September–April). Road conditions can be difficult during wetter months and temperatures very high.

Facilities: Tourism in its infancy and facilities still being developed. Self-sufficient camping. Main tourism zones in Ambararatra area, where greatest densities of baobabs occur.

Recommendations: Worth visiting to see the extraordinary baobab forests alone. Lemurs in main tourism areas not yet habituated and difficult to see. Ring-tailed Lemur populations have diminished significantly (probably due to hunting). In coastal part of park, dolphins frequently seen between July and September.

Loky-Manambato Protected Area (Daraina) **

Centred on town of Daraina, the Loky-Manambato Protected Area (named after the two rivers that delimit the area) consists of 11 large forest blocks and smaller fragments separated by grassland. The protected area covers c.57,000ha, with forest area no greater than 44,000ha: individual blocks range from 1ha to 14,000ha. Historically the savannas may have undergone periods of expansion and contraction due to climate change (Quéméré et al. 2012, Salmona et al. 2017), but more recently forests have contracted and become more fragmented due to deforestation. The protected area is managed by NGO FANAMBY (www.association-fanamby.org).

Location and Access: Daraina in the far north-east lies on *Route Nationale* 5a, c.55km west of Iharana (Vohemar) or c.115km east of Ambilobe. At time of writing road remains very poor, but is in process of being upgraded. Main access is to Bekaraoka Forest near village of Andranotsimaty (which in Malagasy means, 'place where the water isn't dead'), 5km north-east of Daraina.

Habitat and Terrain: Covering rolling hills, a mosaic of dry deciduous forest, semi-evergreen forest and humid forest, interspersed with degraded bush, grassland, dry scrub and agricultural land. In valley bottoms, canopy may reach c.15m, on slopes and ridgetops it averages c.8–12m. Understorey and intermediate layer sometimes dense. Some forests, particularly those in vicinity of Andranotsimaty, are degraded by artisanal gold and gemstone mining operations. The understorey has been removed and the ground pockmarked with deep excavation pits and shafts.

KEY MAMMAL SPECIES

Lemurs: **Golden-crowned Sifaka**, **Crowned Lemur**, Sanford's Brown Lemur, **Aye-aye**, **Daraina Sportive Lemur**, **Northern Fork-marked Lemur**, Montagne d'Ambre Dwarf Lemur and **Tavaratra Mouse Lemur**.

Photographing Golden-crowned Sifaka, Bekaraoka Forest, Daraina.

Carnivorans: Fosa, Northern Ring-tailed Vontsira, Broad-striped Vontsira and Spotted Fanaloka.

Tenrecs: Greater Hedgehog Tenrec and Pale Shrew Tenrec.

Rodents: Ankarana Tufted-tailed Rat.

Season: Accessible during drier months from May to December. Can be very hot (40°C+) during day. Concentrate efforts during first and last two hours of daylight and after dark.

Facilities: Very basic hotel in Daraina. Best option is good-quality camp on edge of forest operated by NGO FANAMBY (www.association-fanamby. org). Local guides available in Daraina or through FANAMBY.

Recommendations: Forests adjacent to FANAMBY camp excellent for Golden-crowned Sifaka and Crowned Lemur. Habituated groups are regularly seen close by. Also relatively easy to see Dariana Sportive Lemur in daytime sleep sites. Places to see Aye-aye can require longer walks. Use local guides to locate an Aye-aye nest and stake it out prior to dusk to hopefully see one of the world's most bizarre creatures. During night walks also good chance to see Northern Fork-marked Lemur and Tavaratra Mouse Lemur. A minimum stay allowing at least two night walks is recommended for a realistic chance to see Aye-aye.

Mahavavy-Kinkony Wetland Complex (including Tsiombikibo Forest and Anjahamena) **

A protected area covering c.268,000ha, comprising a complex of forests, interconnected lakes, rivers, marshes, mangroves, caves, and anthropogenic grasslands and agricultural areas. The wide variety of ecosystems supports an extraordinary level of biodiversity. It is particularly renowned for its birds (more than 140 species recorded) and also supports a diversity of bats (on account of the extensive caves) and nine species of lemurs.

Location and Access: At the heart of the area, the town of Mitsinjo is c.65km west of Mahajanga and the Betsiboka River estuary, and south of Mahavavy River delta. From Katsepy (opposite Mahajanga on Betsiboka estuary) journey by 4×4 takes 3+ hours. Three nearby forest areas are renowned sites for lemurs; Tsiombikibo Forest to the north-west of the town; vicinity of Lac Kinkony some 15km to south; and Anjamena on the east bank of the Mahavavy River.

Habitat and Terrain: All three areas are typical western dry deciduous forest, similar to those at Ankarafantsika and Kirindy.

KEY MAMMAL SPECIES

Lemurs: Decken's Sifaka (Tsiombikibo and Lac Kinkony only), **Crowned Sifaka** (Anjamena only), **Mongoose Lemur**, Rufous Brown Lemur, Eastern Grey Bamboo Lemur (Tsiombikibo only), **Antafia Sportive Lemur** (Anjamena only), **Tsiombikibo Sportive Lemur** (Tsiombikibo only), Fat-tailed Dwarf Lemur and Grey Mouse Lemur.

Season: Road between Katsepy and Mitsinjo often impassible during wet season. Best time to visit May–December.

Facilities: No facilities at forest locations. Mitsinjo has a modest hotel offering basic accommodation and good food. Camping excursions can be arranged through tour operators in Mahajunga or Antananarivo.

Recommendations: An exciting area best suited to the adventurous mammal watcher. A stay of three or more days visiting at least two sites is recommended. Mahavavy Delta and Lac Kinkony also internationally important birdwatching sites and specialist operators organise tours where several lemurs are also regularly seen. Birdwatching highlights include many western wetland specialties such as Madagascar Fish Eagle, Sakalava Rail *Zapornia olivieri*, Madagascar or Bernier's Teal *Anas bernieri* and Humblot's Heron *Ardea humbloti*.

Mariarano Forest *

Located c.45km north-east of Mahajanga, Mariarano consists of fragmented areas of dry deciduous forest, gallery forest along the Mariarano River and mangroves adjacent to tidal regions of the estuary. Fragmented areas are interspersed with anthropogenic grasslands and agricultural zones. There are currently no tourism facilities.

Mammal communities are similar to those in Ankarafantsika National Park (c.95km due south). Lemur species are: Coquerel's Sifaka, Mongoose Lemur, Brown Lemur, Western Woolly Lemur Milne-Edwards's Sportive Lemur, Pale Fork-marked Lemur, Fat-tailed Dwarf Lemur, Grey Mouse Lemur and Golden-brown Mouse Lemur. Other mammals include: Fosa, Falanouc, non-native Small Indian Civet, Common Tenrec, Greater Hedgehog Tenrec, Western Tufted-tailed Rat, Madagascar Flying Fox, Madagascar Rousette, Madagascar Giant Mastiff Bat, Madagascar White-bellied Free-tailed Bat and Grandidier's Free-tailed Bat (Long 2017).

Sahamalaza-Iles Radama National Park **

A UNESCO Biosphere Reserve, this national park protects both terrestrial and marine ecosystems. The biosphere reserve covers c.152,000ha, of which terrestrial areas of the national park include c.11,000ha of forest. Protected forests are highly fragmented (and show considerable anthropogenic disturbance) and primarily located on the Sahamalaza Peninsula north-west of Befotaka, but also include mangroves east of the peninsula (and west of road between Antsohihy and Ambanja). The marine zones are around the Radama Archipelago, west and north-west of the peninsula.

Two locally endemic lemurs occur on the peninsula, **Blue-eyed Black Lemur** and **Sahamalaza Sportive Lemur**, as well as **Northern Giant Mouse Lemur**. Ankarafa Forest on the west side of the peninsula is the best site to visit. This can be reached by boat from Analalava to the village of Marovato (opposite Nosy Saba). The forest is a two-hour walk inland. Alternatively, Ankarafa can be reached by 4×4 (only in the dry season, May–November). The forest is c.45km west of

Male and female Blue-eyed Black Lemurs, Ankarafa Forest.

Andranosamonta (c.33km south of Maromandia). Ecotourism is in initial stages of development. There is a campsite with covered pitches (www.aeecl.org/index.php/ecotourism). Local guides essential. Visits can be organised from Antsohihy, Analalava, or arranged via the park office in Maromandia.

Ambalavato Reserve, some 15km south of Maromandia offers an alternative site to see Blue-eyed Black Lemur. Night walks are also possible.

Nosy Saba within the marine zone of the park, 6.5km offshore from Marovato on the west side of the Sahamalaza Peninsula, is home to a large colony of Madagascar Flying Foxes. There is a luxury beach lodge on the island.

Tsingy de Bemaraha National Park ***

A UNESCO World Heritage Site covering 158,000ha, the southern portion, some c.66,600ha, being the national park. A 'cathedral' of towering limestone karst and deep canyons, intermingled with forest, Tsingy de Bemaraha is one of the most unusual and spectacular places in Madagascar.

Location and Access: North of Manambolo River in centre-west. Reached by dirt road from Belo-sur-Tsiribihina (and Morondava). Road conditions very poor: a 4×4 journey from Belo-sur-Tsiribihina takes four to five hours, and from Morondava 10–12 hours. Access also possible from Central Highlands down Manambolo River (3–4 days). Entrance to park is close to village of Bekopaka on north bank of Manambolo River, or alternative areas around a one-hour drive north of Bekopaka.

Habitat and Terrain: Majority of limestone massif has been eroded into largest and most impressive pinnacle karst (*tsingy*) formations in Madagascar, some pinnacles reaching >50m in height. In between rock formations grows typical dry deciduous forest dominated by *Dalbergia*, *Commiphora* and *Hildegardia* spp. Numerous succulents include *Kalanchoe* and *Pachypodium* spp. In shady canyons, species like *Pandanus* and various ferns flourish. Annual rainfall *c*.900–1,000mm, predominantly between December and April.

KEY MAMMAL SPECIES

LEMURS: **Decken's Sifaka**, **Red-fronted Brown Lemur**, **Bemaraha Woolly Lemur**, **Bemaraha Sportive Lemur**, Fat-tailed Dwarf Lemur, Grey Mouse Lemur **Peters's Mouse Lemur** and Aye-aye.

CARNIVORANS: **Western Ring-tailed Vontsira** and Fosa.

TENRECS: Greater Hedgehog Tenrec and Grandidier's Shrew Tenrec.

RODENTS: **Western Red Forest Rat**, **Tsingy Tufted-tailed Rat** and **Rock-loving Tufted-tailed Rat**.

BATS: Together with Ankarana, possibly the richest single site for bats on Madagascar. Documenting this diversity is ongoing, but possibly more than 50% of the island's bat species occur, including: Madagascar Straw-coloured Fruit Bat, Madagascar Rousette, **Commerson's Leaf-nosed Bat**, Madagascar Giant Mastiff Bat, Malagasy Mouse-eared Bat, Robust Yellow Bat, Madagascar Yellow Bat, Manavi Long-

fingered Bat, Glen's Long-fingered Bat, Grandidier's Free-tailed Bat and Trouessart's Trident Bat.

Season: Access only feasible during dry season (May–December) when roads are passable. During spring and early summer (October–December) it is very hot (35°C+).

Facilities: At southern end of park (near Bekopaka and park office) a network of trails and boardwalks (with several different circuits) lead to *Petite Tsingy*. Around one hour further north (*c*.17km) is the *Grand Tsingy* with trails incorporating ladders, boardwalks and suspension bridges allowing access to the main areas. Paths form circuits of varying lengths from around two to eight hours of walking. The best wildlife areas lie at southern end of park in forest close to entrance.

Well-maintained campsite near park entrance. All equipment must be brought in. A snack bar serves basic meals and drinks. Also, luxury and good-quality lodges (closed during wet season) in vicinity of Bekopaka at southern end of park. Excursions organised by tour operators in Morondava. Guides are available at park offices near entrance.

Recommendations: Considerable effort is required to reach Tsingy de Bemaraha, so stay for at least three nights. Circuits at southern end of park have best forest for seeing wildlife, especially Decken's Sifaka. Also only site for Bemaraha Woolly Lemur and Bemaraha Sportive Lemur, which may be seen at sleep sites during day. On forest walks keep an eye out for Western Ring-tailed Vontsira and Western

Towering pinnacles of the Grand Tsingy, Tsingy de Bemaraha National Park.

Red Forest Rat. Also worth taking boat trip through gorge of Manambolo River with good birdwatching and botany on walls of gorge, and occasionally lemurs can be seen in trees close to river.

Tsingy Mahaloka *

Located south of Ankarana Special Reserve and north-west of Ambilobe (c.15km west of village of Isesy on *Route Nationale* 6), Tsingy Mahaloka is a smaller version of Ankarana. It is a community managed reserve set up by KOFAMA (*Koperativa Fikambanana Ankarabe Mitsinjo Arivo*) that aims to promote sustainable conservation through ecotourism (Colquhoun 2015). Being more remote, it receives far fewer visitors than Ankarana, but nonetheless has much to offer. The striking limestone massif rises sharply (80–100m) from the coastal plain and is eroded into impressive pinnacles. There are extensive cave systems and some ancient human burial sites. Lemurs include Crowned Lemur, Ankarana Sportive Lemur and possibly Ankarana Dwarf Lemur and Tavaratra Mouse Lemur.

Zombitse-Vohibasia National Park **

Zombitse, along with the adjacent forests of Vohibasia to the north, constitutes the last remnant of transitional forest between the western and southern floristic domains. The forests, which cover c.36,300ha (Zombitse, c.16,845; Isoky Vohimena, c.3,285ha; Vohibasia, 16,170ha), are a stark illustration of the effects of deforestation and extreme fragmentation. They are completely isolated: where once there was forest there is now just barren anthropogenic grasslands that are burned annually, the fires often spreading to the forest edge, further diminishing their extent.

Location and Access: Straddling *Route Nationale* 7 in the south-west, Zombitse is 25km east of Sakaraha and c.140km from Toliara, a drive of c.3 hours. From the east the drive is c.85km from Ranohira (Isalo National Park) and takes 1.5 hours.

Habitat and Terrain: Similar in appearance to western deciduous forests and shares many tree species, along with some species normally associated with areas further south. Canopy averages 15m. Two baobabs, *Adansonia madagascariensis* and *A. za*, are particularly noticeable. Soils sandy and terrain fairly flat. Annual rainfall averages 700mm, with long dry season (May–October) and no permanent watercourses.

KEY MAMMAL SPECIES

LEMURS: **Verreaux's Sifaka**, Red-fronted Brown Lemur, **Zombitse Sportive Lemur**, **Pale Fork-marked Lemur**, **Coquerel's Dwarf Lemur**, Fat-tailed Dwarf Lemur and Grey Mouse Lemur.

CARNIVORANS: Fosa.

TENRECS: Tailless Tenrec, Greater Hedgehog Tenrec, Lesser Hedgehog Tenrec, Large-eared Tenrec and Nosolo's Shrew Tenrec.

RODENTS: Western Tufted-tailed Rat and Western Big-footed Mouse.

Season: Accessible year-round. Best in austral spring and summer, October–April. Forests appear relatively lifeless during winter, although diurnal lemurs and sleeping nocturnal lemurs can still be seen.

Facilities: No camping facilities. Basic lodge 7km to west or modest hotels in Sakaraha. Easily walked trails and circuits originate at main park entrance by main road and lead into forest on both sides of it. Guides available at park entrance.

Recommendations: Most visitors stop briefly en route between Isalo National Park and Toliara. Verreaux's Sifakas are regularly seen (sometimes in trees close to road), and more occasionally Red-fronted Brown Lemurs. Enthusiasts should enter at first light (before it gets too hot) to see the best this intriguing and unusual forest has to offer. Easy to find Zombitse Sportive Lemur in sleeping sites. Sometimes Fosa is seen.

A renowned location for birdwatching with local endemic Appert's Tetraka (formerly greenbul) *Xanthomixis apperti* is a highlight. Also good for Giant Coua *Coua gigas* and several vanga species. Excellent too for reptiles, including local endemic Standing's Day Gecko *Phelsuma standingi* and often exceptionally large Oustalet's Chamelons *Furcifer oustaleti*.

Southern Areas including Xerophytic Spiny Forests

The so-called 'spiny forest' region is perhaps the most bizarre and unusual in Madagascar. These forests occupy the harshest and most arid regions on the island. Consequently, faunal and floral diversity is much reduced compared to other areas, but rates of endemicity remain very high. Vegetation comprises a type of dry deciduous forest and spiny thicket dominated by the Didiereaceae and Euphorbiaceae

Baobabs like Adansonia rubrostipa *are common in some 'spiny forest' localities.*

families, and is commonly referred to as 'spiny forest' or 'thorn thicket'. Mammal diversity is reduced in this region, but that is not to say it is an area the mammal watcher should overlook as some sites offer a realistic chance of seeing rare species, whilst others serve up some of the most intimate experiences Madagascar has to offer, particularly with two species the island is perhaps most famous for, Ring-tailed Lemur and Verreaux's Sifaka.

Anja Community Reserve ***

Located on the south side of *Route Nationale 7*, *c.*13km south of Ambalavao and at the northern extremity of the Andringitra area, Anja is a fragment of forest, just 34ha in extent, interspersed by huge boulders and surrounded by anthropogenic grasslands, rice paddies, other agricultural land and villages. The area is sacred to the local Betsileo people and it is taboo (*fady*) to hunt lemurs. The park is home to a significant population of **Ring-tailed Lemur** (perhaps 225–300 animals, at the highest density recorded for the species), which are

particularly densely furred and beautiful. The forest is dominated by the tree *Melia azedarach*, the leaves and fruits of which comprise the majority of the lemur's diet (Gould & Gabriel 2013). The lemurs also feed outside the forest in adjacent village gardens. The park is operated by Association Anja Miray (AMI), a collaboration between two villages that has developed the site for sustainable community use and ecotourism. There is modest accommodation and a restaurant adjacent to the car park and park office. Local guides are compulsory.

Andohahela National Park **

A complex of three separate forest areas (parcels) either side of the east/west watershed of the Anosyenne Mountains. The forests in each parcel are very different (because of the extreme variation in annual rainfall), so each supports very different faunal and mammal communities.

Location and Access: In the extreme south-east, the Anosyennes Mountains lie at southern tip of the eastern mountain chain. The park is north and west of Tolagnaro (Fort Dauphin). Parcel 3 (see below) is relatively easy to visit, parcels 1 and 2 are more difficult and require a 4×4 to reach.

Habitat and Terrain: Particularly notable as it spans rainforest regions on the Anosyennes Mountains and arid spiny forest in the rain shadow to the west of these mountains. Habitat diversity corresponds to rich mammal diversity.

Parcel 1 covers 63,100ha of lowland, mid- and high-elevation rainforest up to 1,956m, and represents southernmost tip of humid forest zone. Among the most southerly 'tropical forests' in the Old World. Parcel 2 covers 12,420ha, largely of typical spiny and gallery forest. Communities are dominated by Didiereaceae and Euphorbiaceae families. With proximity of humid forest, this area acts as transition zone for many species and is one of the most biologically diverse forests in the south. Parcel 3 covers just 500ha on the Ambolo Massif and is a representative fragment of 'transition forest', largely set aside to protect the locally endemic Triangle Palm *Dypsis decaryi*, but is home also to many other species. For example, troops of Ring-tailed Lemurs can sometimes be seen, even from the road.

KEY MAMMAL SPECIES
* parcel 1 only; ** parcel 2 only

LEMURS: Parcel 1: **Red-collared Brown Lemur, Southern Grey Bamboo Lemur, Southern**

Woolly Lemur, **Andohahela Sportive Lemur**, Aye-aye, a dwarf lemur (*Cheirogaleus* sp.) and a mouse lemur (*Microcebus* sp.). Parcels 2–3: **Verreaux's Sifaka****, **Ring-tailed Lemur**, **White-footed Sportive Lemur**, Grey-brown Mouse Lemur and Grey Mouse Lemur.

CARNIVORANS: Eastern Ring-tailed Vontsira*, Broad-striped Vontsira*, Fosa, Spotted Fanaloka* and Falanouc*. Non-native Small Indian Civet in both rainforest and arid forest parcels 1 and 2.

TENRECS: Tailless Tenrec, Greater Hedgehog Tenrec, **Lesser Hedgehog Tenrec**** and various shrew tenrecs* (*Nesogale* and *Microgale* spp.).

RODENTS: Eastern Red Forest Rat*, Sleek-furred Ground Rat*, White-tailed Tree Rat*, Malagasy Mountain-dwelling Mouse*, Webb's Tufted-tailed Rat*, Tanala Tufted-tailed Rat*, Major's Tufted-tailed Rat*, Lesser Tufted-tailed Rat* and Western Tufted-tailed Rat**.

BATS: Numerous species including Madagascar Flying Fox, Madagascar Rousette, Commerson's Leaf-nosed Bat, Manavi Long-fingered Bat and Malagasy Mouse-eared Bat.

Season: Parcels 2 and 3 accessible most of year. Parcel 1 better visited outside rainy season. The best months are during austral spring and early summer, September–December.

Facilities: There are basic camping facilities in parcel 1 at Malio and a modest network of trails. In parcels 2 and 3, three areas have been developed for tourists, Ihazofotsy in the north-west and Tsimelahy and Mangatsiaka in the south-east. All have basic camping facilities, although it is necessary to be entirely self-sufficient (including water supplies). Mangatsiaka is relatively easy to reach being only 4km off *Route Nationale* 13 (Tolagnaro to Amboasary-Sud road), whilst Tsimelahy is 10km from the main road.

Recommendations: Malio (parcel 1) is off the regular rainforest circuit, but for the committed mammal enthusiast is worth visiting. Lemur watching, especially for Verreaux's Sifaka, is excellent at Ihazofotsy (parcel 2), which is also good for White-footed Sportive Lemur and Southern Grey-brown Mouse Lemur. Day trips to both Ihazofoty and Mangatsiaka are possible. Tsimelahy (parcel 3) is not good for mammal watching, but is nonetheless a beautiful place with a rich diversity of unusual plants and reptiles.

Verreaux's Sifaka basking in spiny forest, Ihazofotsy.

Berenty Private Reserve ***

A private reserve established in 1936 by the owners of the vast surrounding sisal plantations. Although it can feel rather contrived, Berenty offers the guarantee of easy lemur watching. This is *the* place to see Verreaux's Sifakas 'dancing' across open ground, and intimate encounters with troops of Ring-tailed Lemurs are certain.

Location and Access: On banks of Mandrare River, *c.*85km west of Tolagnaro and *c.*10km north of Amboasary-Sud. Journey from Tolagnaro takes around four hours (poor road).

Habitat and Terrain: Surrounded by a sea of sisal, a small (250ha) isolated patch of riverine gallery forest dominated by huge Tamarind trees *Tamarindus indica*, which may exceed 20m. Adjacent to gallery forest are much smaller parcels of spiny forest dominated by *Alluaudia procera* and *A. ascendens*. Terrain flat and walking very easy. North of Berenty are more pristine spiny forest parcels at Anjampolo. Here forest is dominated by various species of Didiereaceae including *Didierea madagascariensis*, *D. trolli*, *Alluaudia procera* and *A. ascendens*.

KEY MAMMAL SPECIES

Lemurs: **Verreaux's Sifaka**, **Ring-tailed Lemur**, brown lemur *Eulemur* sp. (introduced), **Petter's Sportive Lemur** (previously regarded as White-footed Sportive Lemur), **Grey-brown Mouse Lemur** and **Grey Mouse Lemur**.

Carnivorans: Non-native **Small Indian Civet**.

Tenrecs: **Lesser Hedgehog Tenrec** and Tailless Tenrec.

Bats: **Madagascar Flying Fox**.

Season: Accessible year-round. September and October are good for baby lemurs.

Facilities: Comfortable bungalows with a restaurant adjacent to reserve. A network of wide paths and trails covers the gallery forest and spiny forest area. Guides provided.

Recommendations: Day trips from Tolagnaro are possible, but a stay of at least 1–3 days and nights is recommended. In addition to Verreaux's Sifaka and Ring-tailed Lemur, sleeping Petter's Sportive Lemur is also easy to find. It is worth making the effort to visit Anjampolo (short drive north of main reserve), where beautiful Ring-tailed Lemur and Verreaux's Sifaka can be seen in spiny forest. Night walks offer chance to see Grey-brown Mouse Lemur (in spiny forest) and Grey Mouse Lemur (in gallery forest), and sometimes Small Indian Civet can be seen scavenging below Cattle Egret *Bubulcus ibis* colony. There is also a major colony of Madagascar Flying Foxes in the gallery forest. The colony should NOT be approached too closely.

Bezà-Mahafaly Special Reserve *

Offers similar lemur viewing to Berenty, but in a more remote and wild location. Like Berenty, Bezà-Mahafaly is a site of long-term lemur research, so the animals are habituated and easily seen.

Location and Access: South-east of Toliara and south of the Onilahy River. Only accessible in dry season. Roads very poor; 4×4 required. Drive from Toliara on *Route Nationale* 7 to Andranovory (one hour), then south on *Route Nationale* 10 to Betioky (5+ hours). The reserve is a further one to two hours north-east of Betioky on a very poor road.

Habitat and Terrain: Two parcels of distinct habitat. A small patch (100ha) of riverine gallery forest with Tamarind trees *Tamarindus indica* and a larger area (480ha) of spiny forest dominated by *Alluaudia ascendens*. Terrain is flat and walking very easy, although it can be very hot (35°C+).

KEY MAMMAL SPECIES

Lemurs: **Verreaux's Sifaka**, **Ring-tailed Lemur**, **Petter's Sportive Lemur**, Grey-brown Mouse Lemur and Grey Mouse Lemur.

Tenrecs: Tailless Tenrec, **Lesser Hedgehog Tenrec** and **Large-eared Tenrec**.

Bats: **Madagascar Flying Fox**.

Season: Only accessible outside wet season (November–April). In August–October baby lemurs very evident.

Facilities: No facilities. Self-sufficient camping necessary. Trails and circuits in gallery forest and spiny forest. Excursions can be arranged through operators in Toliara.

Spectacular sandstone formations, Isalo National Park.

Recommendations: Requires a full day of travel to get there and back (from Toliara), so recommended to stay at least two nights. Because of the heat, all wildlife watching is at its best very early and later in the day. Because of the research some Verreaux's Sifakas and Ring-tailed Lemurs may be collared.

Isalo National Park **

Isalo receives more visitors than any other park on the island. The extensive and spectacularly eroded sandstone outcrops create beautiful scenery ideally suited to trekking and exploration. In the harsh environment numerous endemic arid-adapted plants have evolved and are easily seen .

Location and Access: In the central south, north of Ranohira on *Route Nationale* 7. From the east, *c.*90km (two hours) from Ihosy, *c.*280km (five hours) from Fianarantsoa. From the west *c.*110km from Sakahara (two hours), *c.*240km (4+ hours) from Toliara. Access to the park requires a vehicle or long walks from Ranohira.

Habitat and Terrain: An eroded sandstone massif, with huge outcrops and deep canyons. Permanent ground water and seasonal watercourses encourage vegetation and patches of forest. Some dry areas dominated by Madagascar Tapia Tree *Uapaca bojeri*. Growing on rocks are Elephant's Foot Plant *Pachypodium rosulatum gracilius* and local endemic *Aloe isaloensis*. Locally endemic palm *Chrysalidocarpus isaloensis* and *Pandanus pulcher* grow in ribbons of forest following watercourses.

KEY MAMMAL SPECIES

Lemurs: Verreaux's Sifaka, **Ring-tailed Lemur** and Red-fronted Brown Lemur.

Tenrecs: Tailless Tenrec and Greater Hedgehog Tenrec.

Rodents: Daniel's Tufted-tailed Rat.

Season: Accessible year-round, but more difficult in rainy season.

Facilities: Wide range of hotels and lodges at south edge of park in and around Ranohira, ranging from budget to luxury. Park office and guides in Ranohira. Treks and circuits to various parts of the park, but drive to appropriate trailheads before starting walks. Lemurs most likely to be seen in and around forest on trail to *Piscene Naturelle* and on *Namaza* and *Canyon des Makis* trails.

Recommendations: Not a park renowned for

Ring-tailed Lemur, Isalo National Park.

mammal/lemur watching, but worth making the effort to see Ring-tailed Lemurs clambering around on rocks in such a spectacular setting. Verreaux's Sifaka and Red-fronted Brown Lemur more likely to be seen in forest areas with water.

Tsimanampetsotse National Park **

A UNESCO Biosphere Reserve, Tsimanampetsotse supports one of the most 'otherworldly' landscapes in Madagascar. It is a fantastical land of jagged rocks, stunted bulbous baobabs, swollen giant *Pachypodium* trees, tangled banyan and an eponymous soda lake with resident breeding Greater Flamingos *Phoenicopterus roseus* and visiting Lesser Flamingos *Phoeniconaias minor*. It is also home to the island's rarest carnivoran and several species of lemur.

Location and Access: In extreme south-west, *c.*80km south of Toliara and 6–8km inland of the Mozambique Channel. Best access via coastal village of Ambola, 6km west of park. Ambola reached by boat from Toliara or Anakao, or from Anakao by sandy road (1.5 hours, 4×4 required). Park office at Efoetse, 2km from Ambola.

Habitat and Terrain: Covering an area of 43,200ha and encompassing the western portion and escarpment of the eroded limestone Mahafaly Plateau, Tsimanampetsotse is one of the driest and hottest (35–40°C) places in Madagascar. At base of the escarpment lies Lac Tsimanampetsotse, a shallow (maximum depth 2m after wet season) soda lake 20km long and 2–3km wide. On the

Ring-tailed Lemurs descending into cave, Tsimanampetsotse National Park.

plateau, vegetation is dominated by Didieraceae (*Didiera madagascariensis* and *Alluaudia procera* with some *A. montagnacii* and *Alluaudiopsis fiherenensis*), Euphorbiaceae, giant *Pachypodium geayi* and the baobab *Adansonia fony*.

KEY MAMMAL SPECIES

LEMURS: **Ring-tailed Lemur**, Verreaux's Sifaka, **Petter's Sportive Lemur**, **Grey-brown Mouse Lemur** and Grey Mouse Lemur.

CARNIVORANS: **Grandidier's Vontsira**.

TENRECS: **Lesser Hedgehog Tenrec**.

BATS: An important area for bats with extensive limestone cave systems. Species include: Straw-coloured Fruit Bat, Commerson's Leaf-nosed Bat, Madagascar Cryptic Leaf-nosed Bat**,** Trouessart's Trident Bat, Mahafaly Long-fingered Bat and Griffiths's Long-fingered Bat.

Season: Very difficult to access by road during wet season (December-April), but Ambola can be reached by boat from Toliara at this time. Cooler in austral winter (June-August). September to November ideal (lemurs with young, greater bird activity).

Facilities: Basic campsite at foot of escarpment near north-east corner of lake. All provisions must be taken. Modest beach hotels in Ambola. Various trails along the foot and top of the escarpment. Some trails lead to subterranean caves and two spectacular viewpoints overlooking the lake.

Recommendations: Worth visiting because it is so unusual. Mammal highlights are locally endemic and rare Grandidier's Vontsira, and Ring-tailed Lemur. Grandidier's Vontsira is nocturnal, so camping required (two nights recommended): easily seen as often visits campsite at night. Groups of Ring-tailed Lemurs easily encountered and often visit nearby caves to drink and sleep among rocks. Petter's Sportive Lemur and Grey-brown Mouse Lemur also often seen after dark. In nearby caves three endemic species of blind cave fish (*Typhleotris madagascariensis*, *T. pauliani* and *T. mararybe*) can be seen.

Tsinjoriake and Amoron'i Onilahy Protected Areas *

Two recently established protected areas immediately south of Toliara in the far south-west. Tsinjoriake borders Saint Augustine's Bay and stretches from *Route Nationale* 7 in the north to the Onilahy River in the south, and includes the Sarodrano Peninsula and Table Mountain. Amoron'i Onilahy is larger and covers the area immediately north and south of the Onilahy River from Tongobory (*c.*65km inland) downstream to Tsinjoriake.

Both areas are community-run projects (with the help of well-known international conservation agencies) focusing on promoting biodiversity, reducing human impacts, and developing ecotourism to generate income. Infrastructure for ecotourism is gradually being developed; currently, there are campsites and some trail networks. Beautiful areas of spiny forest and riverine vegetation along the Onilahy River and other smaller seasonal watercourses. The area is home to an important population of Ring-tailed Lemurs, and nocturnal species like Petter's Sportive Lemur, Grey-brown Mouse Lemur and Grey Mouse Lemur.

The Extinct Mammal Fauna

Over the course of the past several thousand years, there have been waves of extinctions in Madagascar that have signalled the demise of some extraordinary creatures including giant elephant birds (*Aepyornis* spp., *Mullerornis* spp., *Vorombe* spp.), three species of dwarf hippopotamus (*Hippopotamus lemerlei*, *H. laloumena* and *H. guldbergi*), lemurs the size of gorillas and orang-utans (e.g. *Archaeoindris* and *Megaladapis*), large carnivorans (Giant Fosa *Cryptoprocta spelea*), two bizarre 'aardvark-like' mammals (*Plesiorycteropus* spp.) and a huge forest eagle (*Stephanoaetus mahery*). No single factor was responsible and in some cases several may have combined to 'tip the balance' beyond the point of no return.

Early human colonists probably arrived during the Holocene (*c.*2,000–3,000 years ago or perhaps earlier) (Dewer *et al.* 2013, Hansford *et al.* 2018, Douglass *et al.* 2019). Patterns of species abundance differ after human arrival depending on body size and geographic location. Within the initial 500 years after humans arrived, there were major population declines in very large species (>150 kg) (especially in spiny forest regions) and medium to large species (10–150 kg) in the dry deciduous forests and the Central Highlands: these declines were probably caused by over-hunting by humans. Relatively large species continued to survive in spiny forest and succulent forests in the south and south-west, until around 1,000 years ago, when populations again declined sharply, probably caused by ongoing human predation, or a combination of climate change with consequent drought, and human hunting and habitat modification (Crowley 2010). Other contributing factors may have been large-scale fires started by early humans, and disease and pathogens inadvertently introduced by humans and their commensal animals (Burney 1999, Goodman & Jungers 2014). The abundance of endemic species weighing less than 10 kg increased after the decline in large-bodied species (Crowley 2010).

Of course, the origins of mammalian life in Madagascar are rooted much further back in time. However, piecing together a reasonably clear picture of the evolutionary sequence of events has proved extremely challenging, one of the main reasons being the incomplete nature of the fossil record: in Madagascar there are not just gaps in this catalogue, there are chasms. Hence, many considerations of the island's biological heritage, particularly relating to vertebrates, are by necessity highly speculative.

There is a reasonable subfossil record from the post 'Ice Age' Holocene (the last *c.*11,000 years) as far as the Late Pleistocene (*c.*26,000 years ago) (Simons *et al.* 1995a), then a gap of >60 million years to the Late Cretaceous with no fossil evidence (primarily because rocks in Madagascar of this age are almost entirely marine in origin, where no fossilisation of terrestrial mammals occurred).

This gap coincides with the 'Age of Mammals' and the period when it is thought colonisation of the ancestral lineages that ultimately gave rise to contemporary mammalian forms occurred. Successful colonisation of Madagascar by non-volant mammals was exceptionally rare (see Biogeography) with only four such events leading to the evolution of the monophyletic groups represented today (carnivorans, rodents, tenrecs and primates).

The majority of Madagascar's fossil record almost exclusively dates from the Mesozoic (Age of the Dinosaurs), with mammal fossils from the Late Cretaceous, *c.*66 mya (just prior to the Cretaceous-Paleogene, K-Pg boundary), early mammals from the Middle Jurassic (*c.*165 mya) and fossils of mammal-like reptiles (non-mammal cynodonts) from the Early Jurassic (*c.*200 mya), e.g. *Menadon besairiei* from the Isalo region in south-central Madagascar (Flynn *et al.* 2000).

Globally, the earliest mammals evolved in the Late Triassic (*c.*210 mya) and were small, probably had fur, and almost certainly laid eggs like their ancestors, the mammal-like reptiles (non-mammal cynodonts). During the entire Mesozoic, most mammals remained small, rodent to badger-sized, and were probably primarily nocturnal, and their diets herbivorous, omnivorous or dominated by insects and invertebrates. In the Middle Jurassic and Cretaceous (between 174 million and 66 mya) they evolved and diversified into many different groups with a variety of lifestyles and ecologies.

During the Mesozoic, the vast supercontinent Gondwana underwent significant change and gradually began breaking apart under the forces of plate tectonics. Around c.250 mya, the continent was largely intact: Madagascar nestled in a central position, with Africa, Arabia and South America attached to the west and north, and the Indian Subcontinent, Antarctica and Australia attached to the east. By around 165 mya, the major landmasses had begun to separate, with Madagascar and the Indian Subcontinent drifting approximately north and east, reaching Madagascar's current position (relative to the African mainland) around c.90 mya. Shortly afterwards, the Indian landmass broke free and continued to drift north and east before colliding with Eurasia. Prior to fragmentation, the early terrestrial mammals would have been free to roam across the entire continent of Gondwana (other than constraints imposed by topography), hence Middle Jurassic fossils from Madagascar closely resemble species unearthed in Argentina, but as Gondwana fragmented, some may have become more restricted in their range, including mammals 'marooned' on the fragment that became Madagascar.

The Cretaceous-Paleogene (K-Pg) boundary mass extinction 66 mya (caused by a catastrophic asteroid impact in the Yucatán Peninsula, Mexico) changed everything. Not only did this precipitate the disappearance of the dinosaurs (other than those that evolved into birds), but also saw the extinction of the majority of Mesozoic mammal radiations, the only survivors being the ancestors of the placental mammals, monotremes and marsupials we see today. It is likely that all Mesozoic mammalian lineages on Madagascar became extinct during the mass extinction (or subsequently died out leaving no fossil record), hence the island was something of a 'blank canvas' when later evolving mammals arrived, in all likelihood by rafting (see Biogeography).

Middle Jurassic Mammals

Known from a fossil broken jaw with three teeth, *Ambondro mahabo* was a tiny mammal with molars c.1 mm long and was probably similar in size to a small shrew tenrec (*Microgale sp.*), weighing c.5–15g. Its remains date back c.167 mya and were found at a site in the north-west that has also revealed fossils of crocodilians, plesiosaurs and sauropod dinosaurs, but no other mammals (Flynn *et al.* 1999).

Ambondro is the oldest known mammal with a complex tribosphenic dentition (a characteristic of modern mammals), whereby the molar cusps interlock like serrated scissors and there is a basined heel that acts like a mortar and pestle, facilitating the crushing of insects (the basined heel is common to the ancestors of marsupial and placental mammals and is retained in many living opossums). The discovery of *Ambondro* pushed back the origin of such dentition by 25 million years from its previously known first occurrence in the Cretaceous.

Late Cretaceous Mammals

The gondwanatherians are an extinct group of mammal forms that lived in the Southern Hemisphere, including Antarctica, during the Late Cretaceous to Eocene (Goin *et al.* 2020). Some species probably perished at the K-Pg boundary mass extinction 66 mya, but others survived into the Eocene, before dying out c.45 mya. Most are known only from teeth or lower jawbones, except *Vintana* (complete cranium) and *Adalatherium* (complete skeleton). More recently, a species has been described from a more complete skeleton. Gondwanatherians probably fed on hard plant material.

In Madagascar several species of gondwanatherians have been described from fossilised dental and other skeletal material discovered at the Maevarano Formation in the Mahajanga Basin in the north-west, and dating from the Maastrichian (66–72mya).

Lavanify miolaka is known from a complete cheek tooth and second fragmented tooth, and appears to be most closely related to *Bharattherium*, a form discovered in India. *Lavanify* had high-crowned curved teeth (Krause *et al.* 1997).

At the time of its description, *Vintana sertichi* was (and still is) the largest known mammal from the Mesozoic of Gondwana, and the first gondwanatherian species described from near-complete cranial remains (Krause *et al.* 2014). It is thought to have vaguely resembled a large rodent and weighed *c*.9kg. Its cranial anatomy suggests it was large-eyed, agile and herbivorous, with well-developed high-frequency hearing and a keen sense of smell.

The most complete skeleton yet found of any Southern Hemisphere Mesozoic mammal comes from the same Maevarano Formation dated *c*.66 mya. The animal was named *Adalatherium hui*, which is a derivation from Malagasy and Greek meaning, 'crazy beast' (Krause *et al.* 2020). Typical, previously known, mammals from the time were the size of small rodents, whilst the skeleton of *Adalatherium* (thought to be a subadult) suggests it was perhaps 600mm in length.

Reconstructions suggest its appearance was somewhat 'badger-like'; however, many skeletal features suggest the animal was very different from anything else so far discovered. *Adalatherium* had more foramina around its face and snout than any known mammal. Foramina are small holes through which nerves and blood vessels pass, suggesting the snout was very sensitive and covered with whiskers. The skull also has a very large hole on the top of the snout for which there is no parallel for any known mammal, living or extinct.

Additionally, the teeth of *Adalatherium* are very different to any known mammal. Its backbone has more vertebrae than any other Mesozoic mammal and one of its leg bones is strangely curved. Speculation suggests *Adalatherium* might have been a powerful digging animal that was also capable of running, and potentially also possessed other forms of locomotion. Given Madagascar separated from the Indian Subcontinent *c*.88 mya, populations of *Adalatherium* or its ancestors might have been isolated on the island for *c*.20 million years, and hence evolved in ways very different to contemporary gondwanatherians elsewhere.

A further Cretaceous fossil continues to prompt debate and speculation among palaeontologists. The broken lower molar from the same Maastrichian deposits is known as UA8699 (University of Antananarivo specimen 8699), and was initially thought to be of marsupial origin (Krause 2001). However, this view has been questioned, as the characteristics used to identify the specimen as a marsupial are not exclusive to this group and are also found in some placental mammals, most notably Late Cretaceous ungulatomorph zhelestids, a group to which UA8699 may be more closely aligned (Averianov *et al.* 2003).

Quaternary Mammals

While the Paleogene and Neogene have revealed no terrestrial fossils, the more recent Quarternary has yielded a considerable number of subfossils. Their study has provided an extraordinary window into the recent evolutionary history of the island's mammals.

Order Bibymalagasia

Plesiorycteropus madagascariensis Filhol 1895
Plesiorycteropus germainepetterae MacPhee, 1994
On the basis of cranial characteristics, a skull fragment found on the west coast was originally described by Filhol in 1895 and thought to be closely related to contemporary aardvarks (genus *Orycteropus*, family Orycteropodidae, order Tubulidentata); it was described as *P. madagascariensis*. Later examination and comparison with other edentates highlighted various osteological features inconsistent with this alignment, such that it is distinguishable from all other known eutherians, and so warrants separation into a new mammalian order Bibymalagasia, to acknowledge the uniqueness of these creatures (MacPhee 1994). The etymology derives from 'biby' (animal in Malagasy) and 'malagasia' (from Madagascar). A second species, *P. germainepetterae*, was also described (MacPhee 1994).

More recently an alternative interpretation, based on molecular sequencing from bone collagen, concluded *Plesiorycteropus* was more closely aligned with the order Tenrecoidea and might be a highly adapted member of the Tenrecidae (Buckley 2013). However, other experts consider such a shift premature without further evidence (Goodman & Soarimalala in press).

It is clear that the diagnosis and placement of subfossils referable to *Plesiorycteropus* is problematic, as various bone fragments (femur and pelvis) were initially described as belonging to the large rodents *Hypogeomys boulei* (Grandidier 1912) and *Myoryctes rapeto* (Major 1908).

Numerous fragmentary specimens of *Plesiorycteropus* have now been unearthed, but as yet no dental or facial material has been discovered, and it is unknown whether or not this genus possessed dentition. Radiocarbon dating of one subfossil specimen suggests an age of *c*.2,200 years (Crowley 2010).

Two species are currently recognised based on significant scale variation in skull morphology.

P. madagascariensis was described from a partial skull excavated in alluvial sands near Belo-sur-Mer in central-western Madagascar. It is now known from numerous locations on the west coast and inland at Ampasambazimba (MacPhee 1994, Burney *et al.* 1997, 2020). *P. germainepetterae* is also known only from incomplete skull fragments, probably recovered from Ampasambazimba in the Central Highlands (Lamberton 1946). The cranial fragment indicates that this species was probably smaller than its congener. If the provenance of the specimen is Ampasambazimba, then both species of *Plesiorycteropus* were sympatric in the Central Highlands (Goodman & Soarimalala in press).

Order Afrosoricida

Microgale macpheei Goodman, Vasey & Burney, 2007
To date, only a single extinct species from this order has been found, the shrew tenrec *M. macpheei*, from Quarternary deposits excavated in the Andrahomana Cave in the extreme south-east. The species is phylogenetically closely related to the extant Short-tailed Shrew Tenrec (Goodman *et al.* 2007c).

Other species of tenrec have been described from deposits or owl pellets from Andrahomana Cave, but subsequent analysis has concluded that all are referable to extant species. *Cryptogale australis* was described from skulls found in owl pellets (Grandidier 1928), but these were later identified as belonging to Large-eared Tenrec (Heim de Balsac 1972). From the same deposits another presumed extinct species, *Microgale decaryi*, was described (Grandidier 1928): this is now regarded as a synonym of the extant Greater Long-tailed Shrew Tenrec (MacPhee 1987). Similarly, *M. brevipes* was described from cave deposits near Mahajanga in the north-west (Kaudern 1918), but is a synonym of the extant Short-tailed Shrew Tenrec (MacPhee 1987).

Order Chiroptera

Only two extinct species of bat have thus far been described from Madagascar: the remains of both were found in breccia deposits at the Anjohibe Cave in the north-west, and date to <10,000 years BP (Samonds 2007).

Family Hipposideridae
Macronycteris besaoka (Samonds, 2007)
The Anjohibe Leaf-nosed Bat was originally described

as *Hipposideros besaoka* (Samonds 2007), before taxonomic realignment to the genus *Macronycteris* (Foley *et al.* 2017). The species name *besaoka* derives from the Malagasy for 'big chin'. Described from mandibular and dental remains, *M. besaoka* was larger and more robust than the extant Commerson's Leaf-nosed Bat.

Family Rhinonycteridae
Triaenops goodmani Samonds, 2007
Described from mandibular and dental remains, Goodman's Trident Bat *T. goodmani* was larger than any known species of *Triaenops* (Samonds 2007).

Order Primates

Suborder Strepsirrhini
Extinct Lemurs
The impressive adaptive radiation of Malagasy primates includes many species of lemur that are now extinct and known only from subfossil records. To date, the remains of at least 17 such species have been documented. The majority of these have been recovered at Holocene limestone cave, marsh or riverside sites in central, northern, western and south-western regions of the island (Godfrey & Jungers 2003, Goodman & Jungers 2014).

It is interesting to note that the bones of these extinct lemurs have been found alongside not only similarly extinct creatures like pygmy hippopotamuses (*Hippopotamus* spp.) and elephant birds (*Aepyornis* spp., *Mullerornis* spp., *Vorombe* spp.), but also those of extant lemurs, carnivorans, rodents and tenrecs, yet only one of the subfossil species, Giant Aye-aye *Daubentonia robusta*, has a congener that survives to this day, the Aye-aye *D. madagascariensis*. However, the two extinct *Pachylemur* species are very close relatives of the extant genus *Varecia* (ruffed lemurs) and, indeed, have in the past been included within it. The remaining species, representing seven genera, were totally unlike any living lemur. They are arranged in three families with no living members: Palaeopropithecidae, Archaeolemuridae and Megaladapidae.

All extinct lemurs were diurnal (except Giant Aye-aye) and larger (often dramatically so) than the largest extant species, the Indri. They ranged in body mass from *c*.10–180kg and are often collectively referred to as the 'giant' lemurs. Despite their large size, most were primarily arboreal. However, although they obviously share a number of basic and defining Strepsirrhine and primate features, their nearest ecological equivalents (ecotypic analogues) often lie outside the order Primates. For instance, various species have been compared with cave bears, tree sloths, ground sloths, koalas, possums and lorises, and also higher primates like macaques, baboons, orang-utans and gorillas. Most extinct lemur subfossils date from the Holocene (11,650 years or younger), although some date to the late Pleistocene, c.26,000 years BP, whilst the youngest are less than 500 years old (Simons *et al.* 1995a, Burney *et al.* 2004). Indeed, when Etienne de Flacourt travelled to Madagascar in the late 1600s, he documented stories from local people of a giant lemur, the *Tretretretre*, (possibly a *Palaeopropithecus* sp.) living in the humid forested mountains north of Tolagnaro (Godfrey & Jungers in press). Certainly, some of the giant lemurs survived well into the last millennium, but with strong evidence of rapid population declines 1,250–1,050 BP, well after human colonisation of Madagascar (Godfrey *et al.* 2019), which dates back at least 2,000 years and possibly much further towards the early Holocene (Douglass *et al.* 2019).

There is ongoing debate regarding the relative degree to which post-glacial climate change and human-induced factors (over-hunting and forest clearance and burning) triggered the extinction of giant lemurs (and other 'megafauna'). However, we do know they were killed and eaten by various large predators. The majority of subfossil lemurs were forest animals, but because of their large size it is likely they spent some time on the ground, at least to move between trees or forest patches (in the same way large male orang-utans (*Pongo* spp.) do today), Furthermore, one group, the archaeolemurids, were more terrestrial than any extant lemur. On the ground the giant lemurs were undoubtedly more vulnerable and fell victim to crocodiles and the large extinct Giant Fosa: the recovered bones of slow-moving species like *Palaeopropithecus* and *Megaladapis* show obvious traces of predation (Meador *et al.* 2019).

Infraorder Lemuriformes

Family Lemuridae

The prolific extant family Lemuridae also contains a single extinct genus, *Pachylemur*. Some early authorities proposed several different species, based on subfossil material recovered at widely separated sites in central and south-western regions. However, more recent analysis indicates that all the material collected to date is referable to just two species, *P. jullyi* and *P. insignis* (Godfrey & Jungers 2002).

Pachylemur jullyi (G. Grandidier, 1899)
Pachylemur insignis (Filhol, 1895)
Estimated body weight: 11–13kg (Jungers *et al.* 2002, 2008)
There are similarities between *Pachylemur* spp. and the extant genus *Varecia*, most notably in skull structure, as well as striking post-cranial differences between them, which serve to differentiate the genera. *Pachylemur* was perhaps akin to a more robust and deliberate version of *Varecia* (Godfrey & Jungers in press). It was primarily an arboreal quadruped – the major limb bones are robust and the fore and hindlimbs are subequal in length. It probably climbed more slowly and was capable of less bounding and leaping, and more hindlimb suspension than *Varecia* (Jungers *et al.* 2002). The dentition of *Pachylemur* is very similar to *Varecia* suggesting a similar mixed diet based on fruit and foliage, and it was presumably an important seed disperser (Godfrey *et al.* 2004b, Godfrey & Jungers in press).

Pachylemur remains have been found at many subfossil sites distributed widely around Madagascar, including Antsirabe and Ampasambazimba in the Central Highlands, and the Ankarana Massif towards the island's northern tip. *Pachylemur* is thought to have survived in the far south of the island until at least c.1,300–1,400 CE (Burney *et al.* 2004).

Family Archaeolemuridae

Members of this family are characterised by thick nipping incisors (rather than tooth combs) and have a molar morphology that resembles that of Old World monkeys (hence they are sometimes called 'monkey lemurs'). The structure of their skulls is considered primitive, although they possessed relatively large brains. Their build was stocky, with broad terminal phalanges on the digits of both hands and feet, suggesting a more terrestrial lifestyle (perhaps somewhat akin to present-day macaques and baboons). Although they would certainly have climbed trees (Godfrey & Jungers 2002), they are best regarded as semi-terrestrial quadrupeds (Lemelin *et al.* 2008).

Archaeolemur majori Filhol, 1895
Archaeolemur edwardsi (Filhol, 1895)
Estimated body weight: 15–25kg (Jungers *et al.* 2002, 2008)
Both species of *Archaeolemur* are thought to have been semi-terrestrial quadrupeds and generally rather baboon-like in their behaviour. However, they probably spent time in trees to sleep and

Skull and jaw fragments of Palaeopropithecus kelyus, *cave at Anjajavy.*

feed, and were almost certainly less cursorial than contemporary baboons, with shorter limbs and digits, and a wider trunk. The dentition of *Archaeolemur*, which includes enlarged incisors, suggests these species were omnivorous, feeding on fruits, hard-shelled nuts and seeds, and some animal matter (Godfrey *et al* 2004b, Crowley & Godfrey 2013). They may also have foraged in more open savanna as well as forest. Subfossil remains have been recovered at widely dispersed localities in the south, south-west, north-west, north and central regions, where they often appear to be the most abundant species. Probably survived until *c.*1,000–1,200 CE, and possibly later (Burney *et al.* 2004).

Hadropithecus stenognathus Lorenz von Liburnau, 1899
Estimated body weight: *c.*27–35kg (Jungers *et al.* 2002, 2008)
The genus *Hadropithecus* is monotypic, the sole species being *H. stenagnathus*. It is thought to have been more terrestrial than *Archaeolemur*, probably filling a similar niche to that of Gelada *Theropithecus gelada* in central Ethiopia, although it spent some time in trees. Its unusual dentition implies that grasses, small seeds and succulent vegetation formed significant parts of the diet (Godfrey *et al.* 2004b, Godfrey *et al.* 2016). *Hadropithecus* was probably less agile than *Archaeolemur*, possessing no leaping or suspensory capabilities (Walker *et al.* 2008). *Hadropithecus* is a rarely recovered species – current knowledge being based on only partial skeletal remains found at sites in the south-east, south-west

and Central Highlands. Populations in the south-west are thought to have survived until *c.*450–770 CE (Burney *et al.* 2004).

Family Palaeopropithecidae
This family includes species ascribed to four genera, *Palaeopropithecus*, *Archaeoindris*, *Babakotia* and *Mesopropithecus*. All were slow climbers, with adaptations for varying degrees of quadrupedal suspension (especially in the hindlimbs, hands and feet), hence this family is commonly referred to as the 'sloth lemurs'. They had relatively small brains, poor daytime vision and short tails (Godfrey & Jungers in press). Their diets appear to have been dominated by leaves, perhaps with the addition of some fruits and seeds (Godfrey *et al.* 2004b).

Palaeopropithecus ingens G. Grandidier, 1899
Palaeopropithecus maximus Standing, 1903
Palaeopropithecus kelyus Gommery *et al.* 2009
Estimated body weight: 35–50kg (Jungers *et al.* 2002, 2008)
Probably the most specialised members of the family and the most 'sloth-like'. All anatomical features are indicative of a highly developed four-limbed suspensory lifestyle, special adaptations being joints throughout the skeleton constructed for flexibility (not strength and stability), except in the hand and feet digits, where strength is required to hang below branches. The first digit on the hand and foot (pollex and hallux) is reduced or missing, and long, curved phalanges (fingers) have joints that limit movement to one plane, thus maximising grasping power (similar to New World spider monkeys, genus *Ateles*).
    P. ingens and *P. maximus* are considered quite similar, whereas *P. kelyus* is smaller and noticeably distinct (Mittermeier *et al.* 2010). Their diet is thought to have consisted mainly of fruits and seeds (Godfrey *et al.* 2004).
    Examples of this genus have been recovered at subfossil sites throughout the island. Some from Ankililtelo Cave, in the Mikoboka Mountains, north of the Onilahy River in the south-west, have been dated as recently as *c.*510–680 years BP (Simons *et al.* 1995b) and relict populations may have persisted until *c.*1,300–1,600 CE (Burney *et al.* 2004).

Archaeoindris fontoynontii Standing, 1909
Estimated body weight: 160–180kg (Jungers *et al.* 2002, 2008)
This species is perhaps the most poorly known of the subfossil lemurs. *Archaeoindris* has a gorilla-sized skull

and dentition, and long bone dimensions that also suggest a body size approaching that of a male gorilla, making it the largest of all known Lemuriformes. Its very large size suggests a more terrestrial lifestyle, and comparisons have been drawn with South American giant ground sloths. However, evidence from the humerus, femur and hip joint mobility indicate some adaptation, at least in part, to a more arboreal/scansorial existence (Godfrey & Jungers 2002).

Dental wear patterns suggest *Archaeoindris* was folivorous, with fruits and seeds also included in the diet (Godfrey & Jungers 2002, Godfrey *et al.* 2006).

Archaeoindris has been found at just one site, Ampasambazimba, in the Central Highlands. Specimens have been dated c.2,150–8,250 years BP (Godfrey & Jungers 2002, Godfrey *et al.* 2010).

Babakotia radofilai Godfrey *et al.*, 1990
Estimated body weight: *c.*15–20kg (Jungers *et al.* 2002, 2008)
The skull of *Babakotia* is very *Indri*-like, but in contrast to the Indri, the forelimbs, hands and feet are long, whilst the hindfoot is reduced, indicating enhanced grasping capabilities and a style of movement away from leaping towards more suspensory behaviour. The adaptations of the spinal column, pelvis, forelimbs, hindlimbs and digits of *Babakotia* suggest an intermediate position between *Mesopropithecus* and *Palaeopropithecus*, i.e. *Babakotia* was probably more suspensory in its behaviour than *Mesopropithecus*, but less so than *Palaeopropithecus*. Its diet is thought to have consisted mainly of fruits and seeds (Godfrey *et al.* 2004). *Babakotia* was significantly smaller than *Palaeopropithecus* but slightly larger than *Mesopropithecus* (Godfrey & Jungers 2002, Jungers *et al.* 2002, 2008).

Babakotia is known from sites in north and north-west Madagascar, the Ankarana Massif and Anjohibe, with more recent discoveries near Maintirana in the centre-west (Burney *et al.* 2020). Radiocarbon dating of the Ankarana specimens indicates they are c.4,500 years BP (Simons *et al.* 1995a).

Mesopropithecus globiceps (Lamberton, 1936)
Mesopropithecus pithecoides Standing, 1905
Mesopropithecus dolichobrachion Simons *et al.*, 1995
Estimated body weight: 10–14kg (Jungers *et al.* 2002, 2008)
The genus *Mesopropithecus* represents the smallest of the sloth lemurs: they were arboreal quadrupeds, adapted for slow climbing and to a lesser extent suspension. Their skulls closely resemble those of

Skull of Mesopropithecus dolichobrachion *discovered in cave in Ankarana.*

extant sifakas (*Propithecus* spp.). However, the post-cranial skeleton is quite different: in *Mesopropithecus* it is the forelimbs, rather than the hindlimbs (as with sifakas), that are greatly elongated, indicating more quadrupedal behaviour than sifakas, and to a lesser degree suspensory behaviour. *M. pithecoides*, known from the Central Highlands, was mainly folivorous, whilst *M. globiceps* and *M. dolichobrachion* are thought to have been more generalist, with varied diets of leaves, fruits and seeds (Simons *et al.* 1995b, Godfrey *et al.* 2004b). The remains of the three species have been found in deposits mainly in central and southern regions. It is likely *Mesopropithecus* survived until c.250–430 CE (Burney *et al.* 2004).

Family Megaladapidae
Represented by a single genus, *Megaladapis*, and three species. Due to their proportions and posture, they are collectively known as 'koala lemurs'. *Megaladapis* were large-bodied animals, with skulls similar in size to gorillas. However, their skulls were highly elongate (to a degree seen in no other primates), with the face tilted upwards, whilst the body was comparatively small in relation to the head. The connectivity and articulation of the skull to the neck vertebrae suggest the head was able to stretch and rotate considerably, presumably to maximise reach and browsing capabilities from a sitting position. The toothcomb is reduced but was likely used in grooming.

The post-cranial skeleton was also unusual, with extraordinarily long hands and feet, indicating a high degree of adaptation towards strong grasping and an arboreal lifestyle. The arms and legs were relatively short, indicating a significantly reduced leaping capability. Skeletally, they appear closely analogous to marsupial koalas, hence their vernacular name.

Previously, it was thought they were closely related to the extant genus *Lepilemur* (indeed *Lepilemur* was at one time included within the family Megaladapidae), however, more recent DNA studies indicate *Megaladapis* has closer phylogenetic links to Lemuridae (Marciniak *et al.* 2020).

Some of the first *Megaladapis* remains to be found in the Ankarana Massif have been dated to the late Pleistocene, *c.*26,200 years BP, while later discoveries from nearby caves were dated at *c.*12,700 BP (Simons *et al.* 1995a). More recently collected specimens from Ankilitelo Cave, in the south-west, reveal dates *c.*600–680 years BP, and it is thought *Megaladapis* survived in the south-west until at least *c.*1,280–1,420 CE (Burney *et al.* 2004).

Megaladapis madagascariensis Forsyth-Major, 1894
Megaladapis grandidieri Standing, 1903
Estimated body weight: 35–70kg (Jungers *et al.* 2002, 2008)
These large-bodied lemurs had long forelimbs with huge grasping hands and feet. The hallux or great toe was enormous and widely divergent, promoting sufficient grip to support the animals' considerable weight. *Megaladapis* were principally vertical clingers and climbers, and probably quite deliberate in their movements, although it is clear they were also capable of suspensory behaviour, including hindlimb-only suspension while feeding, somewhat similar to extant *Propithecus* and *Varecia* spp.

Although primarily arboreal, these species also descended to ground level where they moved quadrupedally, perhaps in a fashion similar to orang-utans (*Pongo* spp.). The dentition is reminiscent of *Lepilemur* spp., indicating a folivorous diet, and although upper incisors are absent, a leaf-cropping adaptation such as a mobile proboscis has been suggested. Remains of these species have been found at subfossil sites throughout Madagascar.

Megaladapis edwardsi (G. Grandidier, 1899)
Estimated body weight: 75–85 kg (Jungers *et al.* 2002, 2008)
This species was appreciably larger than its congeners. The forelimbs were longer than the hindlimbs, with very large grasping hands and feet. Like *M. madagascariensis* and *M. grandidieri*, this species was primarily a vertical clinger, although

suspensory behaviour was dramatically reduced compared to other *Megaladapis*. It also seems likely that *M. edwardsi* was far more terrestrial, where it moved quadrupedally, again in 'orang-utan' fashion. The diet was folivorous. There were no upper incisors, although a cropping pad was present. Subfossil remains have been found at sites in south-east and south-west Madagascar.

Infraorder Chiromyiformes

Family Daubentoniidae
Daubentonia robusta (Lamberton, 1934)
Estimated body weight: 10–14kg (Jungers *et al.* 2002, 2008)
Daubentonia is the only lemur genus known to contain both an extinct and extant species. The massive and robust limb bones of the Giant Aye-aye *D. robusta* indicate that it was between 2.5 and five times heavier than the extant Aye-aye (Simons 1994). Dental evidence (large isolated incisors) together with the arrangement of the hand bones suggests a lifestyle, mode of locomotion, foraging behaviour and diet (hard-cased seeds and wood-boring insect larvae) very similar to that of its congener (Godfrey *et al.* 2004b, Godfrey & Jungers in press). Remains collected in southern Madagascar suggest this species survived until at least *c.*890–1,030 CE (Burney *et al.* 2004).

Order Carnivora

Family Eupleridae
Cryptoprocta spelea G. Grandidier, 1902
A single extinct carnivoran, the Giant Fosa *C. spelea*, is known from subfossil remains found at a variety of sites around the island. In the past, some authorities have concluded that species differentiation from the extant Fosa is unwarranted (Savage 1978). However, more recent investigation of contemporary skeletal specimens and subfossil remains have demonstrated significant size differences and justified separate species status (Goodman *et al.* 2004). Bone measurements of the Giant Fosa indicate it was noticeably larger (30–40%) than its extant congener, with an estimated body mass of *c.*12–15kg (Goodman & Jungers 2014, Meador *et al.* 2019). Examination of subfossil skeletons of a variety of extinct lemurs has also revealed signs of carnivoran predation, presumably by Giant Fosa, the largest of these being the 'koala lemurs', *Megaladapis* spp., which weighed up to *c.*75–85kg (Meador *et al.* 2019, Goodman & Soarimalala in press).

Remains of Giant Fosa have been found at sites in the far north (e.g. Ankarana), lowlands of the centre-west (e.g. Belo-sur-Mer), a few locations in the Central Highlands, and numerous locations in the extreme south and south-west, indicating the species had a broad distribution (Goodman *et al.* 2004). Specimens range in age from *c.*3,720 years BP to the most recently verifiable radiocarbon date for *C. spelea* of *c.*1,520 years BP excavated at Andolonomby (Crowley 2010, Meador *et al.* 2019).

Furthermore, at a number of sites the subfossil remains of both the extant Fosa and Giant Fosa have been found. Therefore, we can be certain the ranges and habitat preferences of the two species overlapped, although it is not possible to conclusively ascertain whether or not they were contemporaneous (Goodman & Jungers 2014).

The Giant Fosa was probably an opportunistic hunter of medium to large arboreal extinct lemurs (predation records for *Pachylemur*, *Mesopropithecus*, *Archaeolemur*, *Hadropithecus*, *Palaeopropithecus* and *Megaladapis*) (Meador *et al.* 2019), its large skull (length *c.*153mm, *c.*20% longer than 125mm for *C. ferox*) (Goodman *et al.* 2004) and powerful forelimb anatomy alluding to a capability to predate larger prey. Cooperative hunting (as observed in extant Fosa) may have further enhanced its ability to capture large prey species weighing up to approximately six times its own body weight (Meador *et al.* 2019).

It is clear the Giant Fosa survived well beyond the arrival of the first human colonists. We know the two species of Fosa lived sympatrically and perhaps synchronically, i.e. in the same forests at the same time, therefore it seems likely Giant Fosa primarily predated larger (now extinct) lemur species, whilst the Fosa concentrated on relatively small lemurs (as it does today) and other smaller prey. Therefore the extinction of the larger lemurs may have consequentially resulted in the demise of the Giant Fosa, which relied so heavily on them for food (Goodman & Jungers 2014).

Order Artiodactyla

Family Hippopotamidae

At least three species of dwarf hippopotamus occurred in Madagascar until the relatively recent past. The most recent radiocarbon-dated specimen (1,260 BP ± 25) is a bone excavated at Antsirabe (Crowley 2010). Furthermore, fables and stories of these creatures remain part of the cultural and oral fabric in several regions on the island, hence it is thought hippos may have survived on Madagascar until the late 19th century or even more recently (Godfrey 1986, Burney & Ramilisonina 1999).

All three species were similar to extant hippopotamuses on mainland Africa – Common Hippopotamus *Hippopotamus amphibius* and Pygmy Hippopotamus *Choeropsis liberiensis* (formerly genus *Hexaprotodon*).

At one site, Anjohibe Cave near the north-west coast, an excavated area of 2m² revealed the partial skeletons of at least eight individual *Hippopotamus*, including five adults, three immatures of different ages, and one possible neonate or foetus (Burney *et al.* 1997). The bones of several individuals lay upon one another, suggesting the animals died together: the group may have been a herd that somehow got trapped in the cave. These bones showed no signs suggesting humans were responsible for their deaths. However, at several sites, hippopotamus bones have been found showing distinct cut marks, indicating human butchery. These date from *c.*2,000–3,800 years ago (MacPhee & Burney 1991, Gommery *et al.* 2011).

Hippopotamus lemerlei A. Grandidieri, 1868

Estimated measurements: total length: 200cm; height at shoulder: 80cm.

This species is known from numerous sites in the south-west and west, as far north as the region of Mahajanga (Stuenes 1989, Burney *et al.* 1997). *H. lemerlei* probably resembled a small Common Hippopotamus *H. amphibius*. Noticeable variation in body and skull size, and canine length, suggests it was sexually dimorphic (Stuenes 1991). Its lifestyle was probably similar to *H. amphibious*: cranial remains show bone trauma consistent with those seen in *H. amphibious* and associated with males fighting for access to females. Bone remains of both *H. lemerlei* and *H. madagascariensis* have been excavated from the same deposits at many sites (particularly in the Central Highlands) suggesting the species were sympatric (Stuenes 1989).

Hippopotamus laloumena Faure & Guérin, 1990

This species is known only from a lower jaw and limb bones found near Mananjary on the south-east coast (Faure & Guérin 1990). It was the largest of the three Malagasy hippopotamuses, but did not approach the size of the Common Hippopotamus *H. amphibius*, which it otherwise appears to have resembled closely. Previously named *H. amphibius standini* (Monnier & Lamberton 1922).

Hippopotamus madagascariensis Guldberg, 1883
Synonyms:
Choeropsis madagascariensis,
Hexaprotodon madagascariensis
Hippopotamus guldbergi Fovet *et al.* 2011
H. madagascariensis was considerably smaller than *H. lemerlei*, with no appreciable size difference between the sexes. Based on excavations, both species had broad distributions across much of the island, excluding the east and far north, although there are few appropriate subfossil sites known in central-east and north-east regions, so inference that hippos did not occur in these regions cannot be conclusive (Goodman & Soarimalala in press).

Based on bone and cranial characters, *H. madagascariensis* was more terrestrial than *H. lemerlei* (Stuenes 1989, Rakotovao *et al.* 2014) and there was probably habitat and ecological segregation between them. Unlike in *H. lemerlei*, there is little evidence of cranial trauma in *H. madagascariensis*, perhaps suggesting reduced levels of male competition and fighting. Therefore, this species may have lived in pairs and been monogamous, and was perhaps secretive and nocturnal.

Because of its smaller size and presumably more terrestrial habits (perhaps similar to those of the extant Pygmy Hippopotamus *Hexaprotodon liberiensis* in equatorial West Africa), it has been suggested *H. madagascariensis* be placed in the genus *Hexaprotodon* (Harris 1991). However, until molecular studies to assess the phylogenetic relationships of Malagasy hippos are undertaken, most authorities retain this species in *Hippopotamus* (Goodman & Soarimalala in press).

Order Rodentia

Subfamily Nesomyinae
Within Nesomyinae, three now-extinct species have been identified in Holocene deposits.

Hypogeomys australis G. Grandidier, 1903
This species of jumping rat was described from jaw material *c.* 4,400 years old, collected in Andrahomana Cave, some 40km west of Tolagnaro. Subsequent excavations at the same site have revealed more than 40 bones of this animal (Goodman & Rakotondravony 1996) and further material pertaining to *H. australis* has been recovered during more recent work, which has been radiocarbon dated at *c.* 1,550 BP (Burney *et al.* 2008).

H. australis was considerably larger than its extant congener, the Giant Jumping Rat *H. antimena*, with the two species being the largest rodents known to have occurred during the Quaternary in Madagascar, although there is no evidence they were sympatric (Goodman & Soarimalala in press).

Further subfossil material, collected from the Antsirabe region in the Central Highlands, attributable to *H. australis* indicates this species was once widely distributed, with its range covering much of the southern half of the island at least. Nothing is known about the natural history of *H. australis*, although it is presumed to have lived in burrows excavated in areas of sandy soil, much like *H. antimena* does today.

In 1912 Grandidier erroneously described a second species, *H. boulei*, from subfossil bone material, which has been subsequently identified as *Plesiorycteropus* Order Bibymalagasia (MacPhee 1994).

Brachytarsomys mahajambaensis Mein et al., 2010
Described from *c.* 12,000-year-old subfossil remains unearthed north of Mahajunga, including on the Narindra Peninsula. *B. mahajambaensis* was smaller than any extant congener. Today, the genus *Brachytarsomys* is restricted to humid forests, which suggests the area of discovery may once have been much wetter and that western Madagascar has since undergone significant climate change (Mein *et al.* 2010).

Nesomys narindaensis Mein et al., 2010
N. narindaensis was described from *c.* 12,000-year-old deposits from the Narindra Peninsula north of Mahajunga. Other specimens attributable to *N. narindaensis* have been found in Anjohibe Cave (also in the north-west) and in caves on the Ankarana Massif in the far north, suggesting this was a relatively widespread species in north and north-west Madagascar (Goodman & Soarimalala in press).

This species was larger than any extant *Nesomys*. Today, the genus is more normally associated with humid forest regions (although one extant species *N. lambertoni* does occur in the Tsingy de Bamaraha region in the centre-west), and it may be that the areas of discovery were once much wetter than they are today (Mein *et al.* 2010).

Classification of Extinct Subfossil Mammals

Order Bibymalagasia
Plesiorycteropus germainepetterae MacPhee, 1994
Plesiorycteropus madagascariensis Filhol, 1895

Order Afrosoricida
Microgale macpheei Goodman, Vasey & Burney, 2007

Order Chiroptera
Suborder Yinpterochiroptera
    Family Hipposideridae
        Anjohibe Leaf-nosed Bat *Macronycteris besaoka* (Samonds, 2007)
    Family Rhinonycteridae
        Goodman's Trident Bat *Triaenops goodmani* Samonds, 2007
Order Primates
Suborder Strepsirrhini
Infraorder Lemuriformes
    Family Lemuridae
        Pachylemur jullyi (G. Grandidier, 1899)
        Pachylemur insignis (Filhol, 1895)
    Family Archaeolemuridae
        Archaeolemur edwardsi (Filhol, 1895)
        Archaeolemur majori Filhol, 1895
        Hadropithecus stenognathus Lorenz von Liburnau, 1899
    Family Palaeopropithecidae
        Palaeopropithecus ingens G. Grandidier, 1899
        Palaeopropithecus maximus Standing, 1903
        Palaeopropithecus kelyus Gommery, Ramanivosoa, Tombomiadana-Raveloson, Randrianantenaina & Kerloc'h, 2009.
        Archaeoindris fontoynontii Standing, 1909
        Babakotia radofilai Godfrey, Simons, Chatrath & Rakotosamimanana, 1990
        Mesopropithecus dolichobrachion Simons, Godfrey, Jungers, Chatrath & Ravaoarisoa, 1995
        Mesopropithecus globiceps (Lamberton, 1936)
        Mesopropithecus pithecoides Standing, 1905
    Family Megaladapidae
        Megaladapis madagascariensis Forsyth-Major, 1894
        Megaladapis grandidieri Standing, 1903
        Megaladapis edwardsi (G. Grandidier, 1899)

Infraorder Chiromyiformes
    Family Daubentoniidae
        Giant Aye-aye *Daubentonia robusta* (Lamberton, 1934)

Order Carnivora
    Family Eupleridae
        Giant Fosa *Cryptoprocta spelea* G. Grandidier, 1902

Order Artiodactyla
    Family Hippopotamidae
        Hippopotamus lemerlei A. Grandidier, 1868
        Hippopotamus laloumena Faure & Guérin, 1990
        Hippopotamus madagascariensis Guldberg, 1883

Order Rodentia
    Subfamily Nesomyinae
        Brachytarsomys mahajambaensis Mein, Sénégas, Gommery, Ramanivosoa, Randrianantenaina & Kerloc'h, 2010.
        Hypogeomys australis G. Grandidier, 1903
        Nesomys narindaensis Mein, Sénégas, Gommery, Ramanivosoa, Randrianantenaina & Kerloc'h, 2010.

Male Fosa in forest canopy, Andasibe-Mantadia National Park.

Acknowledgements

My first nod, ironically, is to a bundle of protein and DNA, 200 times smaller than the width of a human hair. In January 2020, when Bloomsbury first broached the idea of a new edition of *Mammals of Madagascar*, my diary was overflowing with no time to contemplate another book. Two months later, Covid-19 stopped the world and all my work evaporated. What better time to devote to writing than lockdown? Without Coronavirus this book would not have been possible.

It soon became clear knowledge had moved on significantly, since the previous edition, requiring a completely new volume to be written. The task of distilling and assembling the information has been made easier by the help of numerous friends and eminent colleagues who have pointed me in the direction of published research and reduced the errors I would otherwise have made. Many also commented on early sections and drafts of my manuscript, which has consequently been refined and improved. I could not have written this book without their generosity. I am indebted to (in no particular order):

Ute Radespiel, Rainer Dolch, Steve Goodman, Mia Lana-Lührs, Jörg Ganzhorn, Zach Farris, Melanie Seiler, Tony King, Melanie Dammhahn, Chris Golden, Samuel Merson, Marina Blanco, Lydia Green, Peter Kappeler, Claudia Fitchtel, Brandon Semel, Tim Eppley, Dan Hending, Mitchell Irwin, Brian Gerber, Erin Wampole, Rich Lawler, Michelle Sauther, Karen Samonds, David Krause, P.J. Stephenson, Julia Jones, Andrea Baden, Richard Jenkins, Isabella Mandl, Leslie Wilmet, Gabriele Sgarlata, Daniel Austin, Roger Safford, Rebecca Lewis, Summer Arrigo-Nelson, Johanna Rode-Margono, Patricia Wright, Eric Patel, Lena Reibelt, James Herrera, Chris Birkinshaw, Patrice Antilahimena, Caroline Amoroso, Anne Yoder, Natalie Vasey, Cortni Borgerson, Dominik Schüßler, Charlie Welch, Derick Forbanka, Eileen Wyza, Erin Wampole, Luke Dollar, Kelly Flanigan, Sam Hyde Roberts, Marni La Fleur, Kim Reuter, Karla Biebouw, Blanchard Randrianambinina, Julie Hanta Razafimanahaka, Anna Nekaris, Lucienne Wilme, Emmanuel Do Linh San, Jane Sedgeley-Strachan, Ann Smith, Phil Smith, Anjali Goswami and Cathy Dean.

My gratitude also extends to the numerous photographers (see photo credits) who have allowed me to use important images of rare or rarely photographed species. My grateful thanks to them all.

When my commissioning editor, Alice Ward, first discussed this project, I was daunted. Her support has made the task less daunting and she has patiently cajoled me in the appropriate direction, when necessary. She is one of a team at Bloomsbury involved in bringing this book to fruition. Julian Baker skillfully converted my rather amateurish maps into the precise versions, while Susan McIntyre created the eye-catching design. Thank you to them all.

Helen Gilks (Nature Picture Library) has always been supportive of my book projects, allowing access to my images and those of other photographers. And Laura Barwick's dogged persistence helped track down important images.

On numerous tours over the past 20 years, I have worked closely with my good friend Hery Andrianiantefana and benefited hugely from his help and expertise. It is difficult to imagine such trips without him. His enthusiasm is infectious and I cherish his friendship.

My family deserve special mention. Lynne and Mike, my parents, encouraged me to follow my dreams, even when they seemed devoid of common sense. And my sister Jody has always supported me in her own inimitable way. Dad loved language, but insisted it be used correctly, so would sometimes take a red pen to my drafts (he was a teacher). He died while this book was being written (cancer, not Covid). I'm so sorry he is not around to see it reach his bookshelf.

During two years of pandemic, the thousands of hours spent researching and writing this book has been my salvation and has filled a void, when all other work vanished. But it also became an unhealthy obsession. My partner Nicola Murphy has helped restore balance and made me aware that other aspects of life are also important. As thanks, I allowed her to colour in one of the maps.

Nick Garbutt
Devon, UK 2022

Glossary

Adaptive Radiation A burst of evolution, where an ancestral species undergoes rapid diversification into a multitude of new forms, particularly when environmental changes make new resources available, opening new niches for exploitation (cf. Convergent Evolution).

Aestivate In tropical environments where animals seasonally enter a state of dormancy or torpor in response to periods of water and/or food shortage, usually corresponding with hot dry climatic conditions (often the summer months) (cf. Hibernate).

Afrotherian Belonging to a clade of mammals, Afrotheria, the living members of which are placed in taxonomic groups that are either currently living in Africa or are of African origin: e.g. golden moles, elephant shrews, aardvarks, hyraxes, elephants etc., and several extinct clades.

Agouti Referring to pelage/fur: a grizzled coloration resulting from alternate light and dark barring of each hair ('salt-and-pepper').

Alates Used in entomology and botany to refer to something that has wings or wing-like structures. In entomology, 'alate' usually refers to the winged form of social insects that are normally non-winged, especially ants or termites (although can also be applied to aphids and some thrips).

Allogrooming Strictly means 'grooming another' and usually refers to mutual same-species grooming: a form of care-giving through physical contact, typically where one animal uses its hands, mouth, or other part of its body to touch another individual.

Allopatry (Allopatric) The occurrence of populations of different species (or higher taxonomic units) in different geographical areas, i.e. with non-overlapping ranges or distributions (cf. Sympatry).

Altricial Refers to young animals that are born in a rudimentary state of development and, therefore, require an extended period of nursing by the parents (cf. Precocial).

Anthropogenic Refers to habitats and environments that are primarily a consequence of human interference.

Arboreal Refers to animals that spend the majority of their lives living in trees.

Austral Of the southern hemisphere.

Cathemeral Refers to animals that are active both day and night. The relative proportions of daytime and nighttime activity may vary with the seasons (cf. Diurnal and Nocturnal).

Carnivoran More formally refers to members of the taxonomic order Carnivora (cf. Carnivore).

Carnivore (Carnivorous) More generally refers to a primarily flesh-eating organism, that may not necessarily belong to the order Carnivora (cf. Carnivoran).

Clade A group of organisms believed to comprise all the evolutionary descendants of a common ancestor.

Synonymous with a monophyletic group or a natural group.

Class A taxonomic category that is subordinate to phylum and superior to order, e.g. the mammals are class Mammalia in the phylum Chordata (see Taxonomy).

Cline (Clinal) A graded sequence of differences within a species across its geographical distribution.

Commensal Refers to an animal that lives side-by-side with another species, benefiting from the association while the other species neither benefits nor is harmed. Often applied to species that live in close association with man, e.g. rats and mice.

Congener Individuals from the same or different species that belong to the same genus.

Conspecific Individuals that belong to the same species.

Convergent Evolution The independent acquisition through evolution of similar characteristics and ways of life in unrelated but allopatric taxonomic groupings, as opposed to the possession of similarities by virtue of descent from a common ancestor, e.g. both species of hedgehog tenrec and true hedgehogs (cf. Parallel Evolution).

Crepuscular Refers to animals that are primarily active during the twilight period, around dusk or dawn.

Cursorial Refers to an animal possessing limbs adapted for running.

Dichromatic Where individuals of the same species exhibit two noticeably different colour patterns, e.g. sexually dichromatic refers to differences between adult males and females, while maturationally dichromatic refers to differences between immature and adult individuals (cf. Dimorphic).

Digitigrade Refers to the gait in which only the digits, and not the heel, make contact with the ground (cf. Plantigrade).

Dimorphic Where individuals of the same species exhibit noticeably different sizes, e.g. sexually dimorphic refers to size differences between adult males and females (cf. Dichromatic).

Diurnal Refers to animals that are primarily active during the daytime (cf. Nocturnal and Cathemeral).

Dorsal The upper or top side or surface of an animal (cf. Ventral)

Ecotone A region of transition (often abrupt) between two biological communities or habitats.

Endemic Where a species or other taxonomic grouping is naturally restricted to a particular geographic region; such a taxon is then said to be endemic to that region. In this context, the size of the region will usually depend on the status of the taxon; thus, all other factors being equal, a family will be endemic to a larger area than a genus or a species. For instance, the family Lemuridae is endemic to the island of Madagascar as a whole, while the Ring-tailed Lemur is endemic to the south of the island and the Crowned Lemur is endemic to the far northern regions of the island (cf. Indigenous and Exotic).

Eutherian A clade consisting of all therian mammals, i.e. those possessing a placenta, which equates to all mammals except monotremes (Order Monotremata, comprising the egg-laying mammals, e.g. platypus and echidnas) and marsupials (infraclass Metatheria or Marsupialia comprising mammals characterised by premature birth and continued development of the newborn while attached to nipples on the mother's lower belly, often in a pouch, e.g. opossums, koala, kangaroos and wombats).

Exotic Refers to a taxonomic grouping (usually a species) that has been accidentally or deliberately introduced to a region in which it does not occur naturally. Same as non-native (cf. Indigenous).

Fady A belief in traditional Malagasy culture, where something is considered 'taboo' or not allowed.

Family A taxonomic category that is subordinate to order and superior to subfamily, e.g. the Malagasy Carnivorans – family Eupleridae (*see* Taxonomy).

Fecundity Refers to reproductive output. High fecundity is the ability to produce an abundance of offspring.

Folivore Refers to an animal that feeds primarily on leaves.

Fossorial Refers to an animal with a burrowing lifestyle.

Frugivore Refers to an animal that feeds primarily on fruit.

Genus (plural Genera) A taxonomic category that is subordinate to subfamily and superior to species, e.g. the sifakas – genus *Propithecus* (*see* Taxonomy).

Gestation The period of development within the uterus.

Glabrous Smooth and lacking hairs.

Hallux The 'great toe'. In mammals, on the pentadactyl hindlimb, the digit on the tibial side, which is often shorter than the other digits.

Heterothermic Refers to animals that vary between self-regulating body temperature and allowing their surrounding environment to influence it. In other words, they exhibit characteristics of both poikilothermy and homeothermy.

Hibernate Where an animal becomes dormant, lowering its metabolic rate (slows heart rate and breathing rate) over a period of time. This is usually in response to colder climatic conditions when food and water availability are reduced. It often refers to species that become dormant for prolonged periods (in the winter months), but can refer to species that enter daily periods of torpor. Often individuals survive on stored fat reserves that have been built up during the favourable summer months (cf. Aestivate).

Homeothermic Refers to an animal able to maintain a stable internal body temperature, regardless of external influences. Generally, but not always, internal body temperature is higher than external ambient temperature. Equates to the vernacular 'warm blooded' (cf. Poikilothermic).

Hybrid An offspring produced by parents which are genetically unlike, e.g. belonging to different species.

Hyperphagy Abnormally increased appetite and corresponding increase in consumption of food.

Indigenous Refers to species or other taxonomic groupings that occur naturally in a specified area or region and have, therefore, not been introduced either deliberately or accidentally by man. This term is synonymous with Native (cf. Endemic and Exotic).

Inguinal Refers to the groin region.

Labile Liable to change. Easily altered.

Monogamy A mating system where individuals have only one partner per breeding season. This is often extended to mean the pair mate for life (cf. Polygamy and Polygyny).

Monophyletic In taxonomy, a group of organisms descended from a single common ancestor, and therefore by definition a clade.

Niche The functional position or role of an organism (usually applied to a species) within its community and environment, defined in all aspects of its lifestyle, e.g. food, competitors, predators and other resource requirements. Also referred to as ecological niche.

Nocturnal Refers to animals that are primarily active during the night (cf. Diurnal and Cathemeral).

Omnivore Refers to an animal with a varied diet that feeds on both flesh and vegetation.

Order A taxonomic category that is subordinate to class and superior to family, e.g. the rodents – order Rodentia (*see* Taxonomy).

Parallel Evolution The independent acquisition through evolution of similar characteristics and ways of life in unrelated but sympatric taxonomic groupings. e.g. true sunbirds and sunbird asities in Madagascar (cf. Convergent Evolution).

Pelage The hair covering the body of a mammal. Coat is a synonymous and more popular term.

Penicillate Refers to the tails of mammals that end in a conspicuous brush-like terminal tuft.

Philopatry The tendency of an organism to stay in or habitually return to a particular area.

Plantigrade Refers to the gait in which the soles of the feet, including the heels, make contact with the ground (cf. Digitigrade).

Poikilothermic Refers to an animal whose internal temperature varies considerably. Such animals usually rely on external environmental factors to raise body temperatures. Equates to the vernacular 'cold blooded' (cf. Homeothermic).

Polygamy A mating system where individuals of either sex have more than one partner per breeding season (cf. Monogamy and Polygyny).

Polygyny A mating system where males mate with more than one (and generally several) females during a single breeding season (cf. Monogamy and Polygamy).

Post-cranial Refers to all parts of the skeleton behind/below the cranium (skull).

Precocial Refers to young animals that are born in a relatively advanced state of development and, therefore, require only a brief period of nursing by the parents (cf. Altricial).

Prosimian Literally meaning 'before the monkeys'; used as a collective term referring to relatively primitive primates belonging to the suborder Strepsirhini (formerly Prosimii) which includes the lemurs, galagos, pottos and lorises.

Pygal The region at the base of the back around where the tail joins the body.

Quadrupedal Refers to an animal that walks on all four limbs.

Rhinarium The furless skin surface surrounding the external openings of the nostrils in many mammals. Commonly referred to as the tip of the nose or snout.

Saltatory Refers to animals whose primary mode of movement is leaping.

Scansorial Refers to an animal that is adapted to climbing.

Sclerophyllous Refers to vegetation, typically scrub, but also woodland, in which the leaves of the trees and shrubs are small, hard, thick, leathery with a waxy cuticle and evergreen. These adaptations allow the vegetation to survive a pronounced hot, dry season.

Somatic Relating to the body, especially as distinct from the mind.

sp. (plural spp.) When considering a species, an abbreviation indicating that the genus is known, but not the specific species, e.g. *Eulemur* sp. refers to an unspecified member of the genus *Eulemur*, whereas *Eulemur* spp. refers to two or more members of the genus *Eulemur*.

Species A taxonomic category that is subordinate to genus and superior to subspecies, e.g. Ring-tailed Lemur (*Lemur catta*). This is the fundamental unit of taxonomy and is broadly defined as a population of organisms with identical morphology that are able to interbreed and produce viable offspring, i.e. they have compatible gametes and share a common fertilisation technique (this is the basis of the Biological Species Concept – see Taxonomy and 'The Species Conundrum).

Speciation The process by which new species arise through evolution. It is widely accepted that this occurs when a single species' population becomes divided and then different selection pressures acting on each new population cause them to diverge.

Subfamily A taxonomic category that is subordinate to family and superior to genus, e.g. the furred tenrecs – subfamily Oryzoricinae (see Taxonomy).

Subspecies A taxonomic category that is subordinate to species and denotes a recognisable subpopulation within a single species that typically has a distinct geographical range, e.g. Ring-tailed Vontsira (*Galidia elegans*) is divided into an eastern subspecies, *Galidia elegans elegans* (the nominate form), a northern subspecies, *Galidia elegans dambrensis* and a western subspecies, *Galidia elegans occidentalis*. Generally, subspecies is interchangeable with the term 'race' (see Taxonomy).

Sweepstake Dispersal Coined by G. G. Simpson in 1940, the term describes a possible route of faunal interchange which is unlikely to be used by most animals, but which will, by chance, be used by some. It requires a major barrier that is occasionally crossed. Which groups cross and when they cross are determined at random, e.g. ancestral mammals crossing the Mozambique Channel on mats of floating vegetation debris to reach Madagascar.

Sympatry (Sympatric) The occurrence of populations of different species (or higher taxonomic units) in the same geographical areas, i.e. with overlapping ranges or distributions (cf. Allopatry).

Taxon (plural Taxa) A general term for a taxonomic group whatever its rank, e.g. family, genus, species or subspecies.

Taxonomy The science of classifying organisms, whereby organisms that share common features (anatomical or genetic or both) are grouped together and are, therefore, thought to share a common ancestry. Each individual is thus a member of a series of ever-broader categories (individual – species – genus – family – order – class – phylum – kingdom) and each of these categories can be further divided where convenient and appropriate (e.g. subspecies, subfamily, superfamily or infraorder) – see 'The Species Conundrum.

Terrestrial Refers to animals that spend the majority of their lives living on the ground.

Ventral The lower or bottom side or surface of an animal.

Vibrissae Stiff, coarse hairs, richly supplied with nerves that are especially found around the snout and have a sensory (tactile) function. More commonly referred to as whiskers.

Xerophytic Refers to vegetation, typically a forest, that grows in areas that receive relatively little rainfall. The trees and plants of such forests are generally adapted to protect themselves against browsing animals by having well-developed spines, and reduce water loss through transpiration by having small, leathery leaves with a waxy cuticle.

Zeitgeber Any external or environmental cue that acts to synchronise an organism's biological rhythms; these are usually naturally occurring and correspond to the Earth's 24-hour light/dark and 12-month cycles.

General Bibliography

Austin, D. & Bradt, H. (2020). *Madagascar: the Bradt Travel Guide* (Edition 13). Bradt, Chalfont St Peter.

Benda, P. (2019). Family Rhinonycteridae (trident bats). Pp. 194–209 *in* Wilson, D.E. & Mittermeier, R.A. (eds) *Handbook of Mammals of the World, Vol. 9. Bats*. Lynx Edicions, Barcelona.

Bonaccorso, F.J. (2019). Family Emballonuridae (sheath-tailed bats). Pp. 334–373 in Wilson, D.E. & Mittermeier, R.A. (eds) *Handbook of Mammals of the World, Vol. 9. Bats*. Lynx Edicions, Barcelona.

Garbutt, N. (2007) *Mammals of Madagascar: A Complete Guide*. A. & C. Black, London.

Giannini, N.P. (2019). Family Pteropodidae (Old World fruit bats). Pp. 16–163 in Wilson, D.E. & Mittermeier, R.A. (eds) *Handbook of Mammals of the World, Vol. 9. Bats*. Lynx Edicions, Barcelona.

Golden, C.D. (2009). Bushmeat hunting and use in the Makira Forest, north-eastern Madagascar: a conservation and livelihoods issue. *Oryx* 43: 386.

Goodman, S.M. (2009). Family Eupleridae (Madagascar carnivores). Pp. 330–351 in Wilson, D.E. & Mittermeier, R.A. (eds) *Handbook of Mammals of the World, Vol. 1. Carnivores*. Lynx Edicions, Barcelona, Spain.

Goodman, S.M (2019). Family Myzopodidae (sucker-footed bats). Pp. 388–393 in Wilson, D.E. & Mittermeier, R.A. (eds) *Handbook of Mammals of the World, Vol. 9. Bats*. Lynx Edicions, Barcelona.

Goodman, S.M. (ed.) (in press). *The New Natural History of Madagascar*. Princeton University Press, Princeton, NJ.

Goodman, S.M. & Benstead, J.P. (eds) (2003). *The Natural History of Madagascar*. University of Chicago Press, Chicago.

Goodman, S.M. & Jungers, W.L. (2014). *Extinct Madagascar: Picturing the Island's Past*. University of Chicago Press, Chicago.

Goodman, S.M. & Monadjem, A. (2017). Family Nesomyidae (pouched rats, climbing mice and fat mice). Pp. 156–203 in Wilson, D.E., Lacher, T.E. & Mittermeier, R.A. (eds) *Handbook of Mammals of the World, Vol. 7 Rodents II*. Lynx Edicions, Barcelona.

Ibáñez, C. & Juste, J. (2019). Family Miniopteridae (long-fingered bats). Pp. 674–709 in Wilson, D.E. & Mittermeier, R.A. (eds) *Handbook of Mammals of the World, Vol. 9. Bats*. Lynx Edicions, Barcelona.

Jenkins, P.D. (2018). Family Tenrecidae (tenrecs and shrew tenrecs). Pp. 134–172 in Wilson, D.E. & Mittermeier, R.A. (eds) *Handbook of Mammals of the World, Vol. 8. Insectivores, Sloths and Colugos*. Lynx Edicions, Barcelona.

Krause, D.W. (2010). Washed up in Madagascar. *Nature* 463: 613–614.

Mittermeier, R.A., Louis, E.E., Richardson, M.C., Schwitzer, C., Langrand, O., Rylands, A.B., Hawkins, A.F.A., Rajaobelina, S., Ratsimbazafy, J., Rasoloarison, R., Roos, C., Kappeler, P.M. & Mackinnon, J. (2010) *Lemurs of Madagascar*. Third edition. Conservation International, Washington, D.C.

Mittermeier, R.A., Schwitzer, C., Louis, E.E. & Richardson, M.C. (2013). Family Indriidae (woolly lemurs, sifakas and Indri). Pp. 142–175 in Wilson, D.E. & Mittermeier, R.A. (eds) *Handbook of Mammals of the World, Vol. 3. Primates*. Lynx Edicions, Barcelona.

Mittermeier, R.A., Langrand, O.M., Wilson, D.E., Rylands, A.B., Ratsimbazafy, J., Reuter, K.E., Andriamanana, Louis, E.E., Schwitzer, C. & Sechrest, W. (2021) *Lynx Illustrated Checklist of the Mammals of Madagascar: with the Comoros, the Seychelles, Réunion and Mauritius*. Lynx Edicions, Barcelona.

Monadjem, A. (2019). Family Hipposideridae (Old World leaf-nosed bats). Pp. 210–258 in Wilson, D.E. & Mittermeier, R.A. (eds) *Handbook of Mammals of the World, Vol. 9. Bats*. Lynx Edicions, Barcelona.

Monadjem, A. (2019). Family Nycteridae (slit-faced bats). Pp. 374–386 in Wilson, D.E. & Mittermeier, R.A. (eds) *Handbook of Mammals of the World, Vol. 9. Bats*. Lynx Edicions, Barcelona.

Moratelli, R. & Burgin, C.J. (2019). Family Vespertilionidae (vesper bats). Pp. 716–981 in Wilson, D.E. & Mittermeier, R.A. (eds) *Handbook of Mammals of the World, Vol. 9. Bats*. Lynx Edicions, Barcelona.

Poux, C., Madsen, O., Marquard, E., Vieites, D.R., de Jong, W.W. & Vences, M. (2005). Asynchronous colonization of Madagascar by the four endemic clades of primates, tenrecs, carnivores, and rodents as inferred from nuclear genes. *Syst. Biol.* 54: 719–730.

Reuter, K.E., Gilles, H., Wills, A.R. & Sewall, B.J. (2016). Live capture and ownership of lemurs in Madagascar: extent and conservation implications. *Oryx* 50: 344–354.

Schwitzer, C., Mittermeier, R.A., Louis, E.E. & Richardson, M.C. (2013). Family Cheirogaleidae (mouse, giant mouse, dwarf and fork-marked lemurs). Pp. 28–65 in Wilson, D.E. & Mittermeier, R.A. (eds) *Handbook of Mammals of the World, Vol. 3. Primates*. Lynx Edicions, Barcelona.

Schwitzer, C., Mittermeier, R.A., Louis, E.E. & Richardson, M.C. (2013). Family Lepilemuridae (sportive lemurs). Pp. 66–89 in Wilson, D.E. & Mittermeier, R.A. (eds) *Handbook of Mammals of the World, Vol. 3. Primates*. Lynx Edicions, Barcelona.

Schwitzer, C., Mittermeier, R.A., Louis, E.E. & Richardson, M.C. (2013). Family Lemuridae (bamboo, true and ruffed lemurs). Pp. 90–141 in Wilson, D.E. & Mittermeier, R.A. (eds) *Handbook of Mammals of the World, Vol. 3. Primates*. Lynx Edicions, Barcelona.

Schwitzer, C., Mittermeier, R.A., Louis, E.E. & Richardson, M.C. (2013). Family Daubentoniidae (Aye-aye). Pp. 176–183 in Wilson, D.E. & Mittermeier, R.A. (eds) *Handbook of Mammals of the World, Vol. 3. Primates*. Lynx Edicions, Barcelona.

Taylor, P.J. (2019). Family Molossidae (free-tailed bats). Pp. 598–672 in Wilson, D.E. & Mittermeier, R.A. (eds) *Handbook of Mammals of the World, Vol. 9. Bats*. Lynx Edicions, Barcelona.

A full list of references can be found at researchgate.net/profile/Nick-Garbutt

Photo Credits

Bloomsbury Publishing would like to thank the following for providing photographs and for permission to reproduce copyright material. While every effort has been made to trace and acknowledge all copyright holders, we would like to apologise for any errors or omissions and invite readers to inform us so that corrections can be made in any future editions of the book.

With the exceptions of the photographs listed on the page numbers below, all photographs in this book (including the cover) remain © Nick Garbutt.

Key HA = Hery Andrianiantefana / Aspinall Foundation; LB = Lorraine Beanery / Nature Picture Library; JB = Jon Betz; MB = Dr Marina P. Blanco; AB = Adam Britt; SD = Stephen Dalton / NHPA / Photoshot; ME = Manfred Eberle; ZF & AM = Zach Farris & Asia Murphy; CG = Chris Golden; LG = Dr Lydia K. Greene; AH = Alex Hyde; MI = Mitchell Irwin; LJ = Louise Jasper (louisejasper.zenfolio.com); TK = Tony King / Aspinall Foundation; L = Laingo / Aspinall Foundation; OL = Olivier Langrand / BIOS / Still Pictures; CL = Chien Lee (chienclee.com); AL = Adrià López-Baucells (adriabaucells.com); EL = Ed Louis; MV = Madagasikara Voakajy (Bat Conservation Madagascar); JM = Joan de la Malla / Adrià López-Baucells (joandelamalla.com); SM = Dr Samuel D. Merson; TM = Thorsen Milse (wildlifephotography.de); MN = Martin Nicoll / BIOS / Still Pictures; DN = Dietmar Nill / Nature Picture Library (naturepl.com); GO = Gillian Olivieri; PO = Pete Oxford / Nature Picture Library (naturepl.com);

JP = Julie Pomerantz; BR = Bruno Raveloson; AR = Andry Ravoahangy; IR = Iñaki Relanzón (www.photosfera.com); PS = Peter Schachenmann; DS = Dominik Schüßler; HS = Harald Schütz; MS = Dr Melanie Seiler; RS = Roland Seitre / Nature Picture Library (naturepl.com); BS = Brandon Semel; P. J. S = P. J. Stephenson; PS = Paul D. Stewart / Nature Picture Library (naturepl.com); CT = Colin Taylor; EW & BG = Erin Wampole & Brian Gerber.

Page 47 (bottom) MN; 48 MN; 49 RS; 50 MN; 51 LJ; 52 MN; 53 OL; 56 HS; 57 P. J. S; 59 RS; 60 HS; 61 RS; 62 HS; 63 HS; 64 HS; 65 HS; 66 HS; 67 HS; 68 HS; 73 JM; 74 JM; 79 (bottom) IR; 82 AL; 84 AL; 85 AL; 87 AL; 91 CL; 92 JM; 93 JM; 94 (top) AL, (bottom) JM; 95 MV; 97 MV; 99 AL; 100 AL; 102 MV; 104 (top) MV, (bottom) IR; 105 IR; 106 LJ; 108 AL; 109 LJ; 113 IR; 124 (top) GO, (bottom) LG; 125 LJ; 126 GO; 128 LJ; 132 DS; 133 DS; 134 EL; 137 EL; 138 BR; 141 ME; 142 CL; 147 MB; 154 MB; 157 MB; 163 CL; 171 EL; 172 EL; 173 BR; 179 (right) EL; 184 MS; 185 MS; 191 (top) CL, (bottom) LJ; 203 IR; 205 LJ; 209 IR; 210 IR; 221 TM; 232 AH; 234 CL; 235 CL; 236 IR; 237 IR; 248 MS; 249 MS; 273 CL; 278 BS; 297 L; 298 (top) TK; 318 ZF & AM; 321 (bottom) EW & BG; 326 PO; 327 PO; 331 AB; 332 ZF & AM; 333 (top) BR, (bottom) CL; 340 BR; 341 HS; 342 MN; 344 HS; 345 (top) LJ, (bottom) HS; 346 HS; 349 MN; 350 BR; 351 HS; 354 HS; 355 LB; 357 HS; 361 HS; 364 LJ; 366 SD; 367 SD; 369 SD; 371 AR; 372 EW & BG; 373 (left) JP, (right) SM; 375 DN; 385 JM; 386 CG; 387 BS; 388 JB; 389 JM; 390 IR; 402 HA; 411 BR; 415 TK; 419 MS; 420 CL; 426 LJ; 433 PS.

Nick Garbutt is an award-winning wildlife photographer, author and tour leader, with a particular interest in rainforests and island wildlife. He has been fascinated by Madagascar since childhood and first travelled to the island in 1991 returning almost annually over in the past 30 years. He has visited virtually all the islands' major wildlife locations and seen over 80 species of lemur and a high proportion of the other mammals and fauna in the wild.

Nick has twice been a category winner in the Wildlife Photographer of the Year Competition, and his other books include *Chameleons* (Natural History Museum), *100 Animals to See Before They Die* (Bradt), *Mammals of Madagascar: A Complete Guide* (A&C Black 2007), *Madagascar Wildlife* (Bradt) and *Wild Borneo* (New Holland).

Index